Innovations in Agricultural Water Management

Muhammad Mubeen • Wajid Nasim Jatoi
Muhammad Zaffar Hashmi • Mushtaq Ahmad
Editors

Innovations in Agricultural Water Management

Risks and Solutions

Springer

Editors
Muhammad Mubeen
Department of Biotechnology
COMSATS University Islamabad
Vehari, Pakistan

Muhammad Zaffar Hashmi
Health Services Academy
Islamabad, Pakistan

Wajid Nasim Jatoi
Department of Agronomy
Faculty of Agriculture and Environment
The Islamia University of Bahawalpur
Bahawalpur, Pakistan

Mushtaq Ahmad
Department of Plant Sciences
Quaid-i-Azam University
Islamabad, Pakistan

ISBN 978-3-031-91882-7 ISBN 978-3-031-91883-4 (eBook)
https://doi.org/10.1007/978-3-031-91883-4

This Springer imprint is published by the registered company Springer Nature Switzerland AG
The registered company address is: Gewerbestrasse 11, 6330 Cham, Switzerland

Foreword

Sustainable agricultural water management includes integrated water resources management, the use of water-efficient technologies, and the adoption of water conservation practices. The innovations in agricultural water managements have the key role in achieving short- and long-term water management in agriculture. There are actually two main long-term goals in irrigated agriculture in the era of climate change: (1) to achieve better water use efficiency and (2) to continue the innovations on sustainable basis. So, this book would be a guideline for the policy makers, the researchers as well as the farmers to get thoughtful insight for solving these two problems.

Water harvesting provides as a counterbalance to irrigation requirements, can replenish depleted groundwater sources, and can also enhance resilience to climate change, especially in the face of droughts and high rainfall. The utilization of wastewater for irrigation offers benefits such as conserving freshwater resources, reducing pollution and environmental impact, and providing nutrients for crops.

Data analytics and big data play a crucial role in precision irrigation, with a focus on harnessing big data for water use efficiency, decision support systems, and the application of machine learning algorithms for predictive irrigation management. Unmanned aerial vehicles (UAVs) offer a range of applications that contribute to improving irrigation practices, optimizing water use, and enhancing agricultural water management. Groundwater modeling is discussed as a valuable tool for predicting availability and understanding aquifer dynamics.

Ecohydrology, an interdisciplinary field bridging ecological and hydrological sciences, addresses pressing concerns related to water scarcity, environmental degradation, and ecosystem health in agriculture. It also highlights the importance of ecohydrological tools such as remote sensing, GIS, and hydrological models in decision-making processes.

This book will try to explore various horizons for sustainable water management in agriculture in the scenario of climate change and its drastic effects in terms of water shortage. The book will compile a comprehensive collection on precision agricultural practices with respect to irrigation while keeping in view all the risks involved.

Vehari, Pakistan Muhammad Mubeen
Bahawalpur, Pakistan Wajid Nasim Jatoi
Islamabad, Pakistan Muhammad Zaffar Hashmi
Islamabad, Pakistan Mushtaq Ahmad

Preface

Due to climate change, population growth, and urbanization, the competition for water resources is expected to increase, with a particular impact on agriculture. Sustainable water management in agriculture is essential for ensuring food security and environmental protection. Here comes the role of innovative approaches in agricultural water management.

Precision agriculture is one of such approaches; it is often described as the application of the right amount of input at right time in the right place. Precision irrigation involves soil moisture monitoring through sensor technologies, weather-based irrigation scheduling systems, remote sensing, and GIS applications and the implementation of the Internet of Things (IoT). However, there are a number of limitations and risks involved in adopting precision agriculture including initial investment, technology acquisition, farmers' training and capacity building, technical and infrastructural limitations. Impact-based forecasting is an emerging field and capable of assisting the farming community in smart cultivation practices. Now the international agencies have become aware of the issues and playing supportive roles to manage the farm-level practices for precision irrigation.

Water harvesting, or the deliberate collecting of rainwater and runoff, is another way of water management in agriculture. These strategies include infiltration trenches and basins, micro- and macro-embankments, swales and valley bed system of water harvesting. However, implementing water harvesting strategy has some difficulties, such as labour, building and maintenance expenses, land degradation, equitable access to runoff and related resources. These challenges highlight the needs for alternatives and a variety of solutions, including financial incentives to promote the most appropriate water collecting techniques customized to local socioeconomic and environmental conditions. Another relevant concept is terrace farming; conservation tillage, grassed water ways, and contouring can be adopted to improve water management on sloppy lands.

The utilization of wastewater for irrigation has become important due to the increasing demand for water in agriculture and the generation of wastewater. However, it also presents challenges regarding water quality, health risks, and potential soil and crop contamination. To overcome these challenges, modern irrigation practices use various treatment methods to improve wastewater quality before irrigation. These methods include physical, chemical, and biological processes to remove contaminants and pathogens, ensuring the safety of crops and consumers. Digitalization, precision farming, and artificial intelligence can work together to improve efficiency, reduce risks, and maximize resource use in wastewater reuse practices.

The invention and cultivation of cultivars that use less water and the application of rhizobacteria that promote drought-tolerant plant growth is possible through adopting biotechnological approaches. Gene identification and amplification through polymerase chain reaction (PCR), gene cloning and gene transformation (recombinant DNA technology), and marker-assisted selection are used to develop genetically modified plant cultivars that require less water for growth than the conventionally developed cultivars.

The remote sensing (RS) explores the topics such as detecting water stress in crops, estimating crop water requirements through evapotranspiration modelling, assessing water availability and mapping, optimizing irrigation scheduling, and monitoring and mitigating water-related issues like floods, salinity, and waterlogging.

The book would be a comprehensive sourcebook for graduate students and researchers working in the field of Agronomy, Irrigation Science, On-farm water management, Precision agriculture, Soil Science, Big data analysis, Biotechnology, and other Agricultural, Environmental, and Social Sciences. Practitioners, policy makers, and different organizations relevant to these fields would also be the key beneficiaries.

Vehari, Pakistan Muhammad Mubeen

Acknowledgements

This book is the outcome of the dedication and untiring efforts of Editors Dr. Muhammad Mubeen, Dr. Wajid Nasim Jatoi, Dr. Muhammad Zaffar Hashmi, and Dr. Mushtaq Ahmad. We would like to sincerely thank our valuable reviewers as well. Many renowned researchers and academicians provided excellent guidance and support for preparation of this international book.

The first editor (Dr. Muhammad Mubeen) is highly indebted to many colleagues from COMSATS University Islamabad, Vehari Campus including the departments of Environmental Sciences and Biotechnology and administration of the university who helped a lot in achieving this milestone. Dedication and appreciation go to my family including my parents (late), my wife, my children Haroon, Memoona, Hassaan, Hafsa, Haleema, Hasnain, my granddaughter Javeria, my elder brothers Muhammad Yousaf, Muhammad Saleem Ghafoor, my elder sisters and their families whose moral support has been a source of inspiration for me throughout.

We are highly indebted to our respected teachers whose training and guidance helped us to reach the horizons of knowledge in general and in this specialized field of study in particular.

COMSATS University Islamabad
Vehari, Pakistan

Muhammad Mubeen

Contents

Editors and Contributors

About the Editors

Muhammad Mubeen is working as Tenured Associate Professor at COMSATS University Islamabad, Vehari Campus (Pakistan). His areas of interest include Agronomy, Climate change and Agro-meteorology, Agricultural Biotechnology, Crop Growth Modelling, Remote sensing, GIS, and Water management. He has expertise in cropping system model "Decision Support System for Agro-technology Transfer (DSSAT)" and GIS softwares (ArcGIS and ERDAS). He has over 100 publications in national and international peer-reviewed journals and books over his areas of expertise. He is reviewer of many national and international well-reputed journals. He has edited two international books *Building Climate Resilience in Agriculture-Theory, Practice and Future Perspective* and *Climate Change Impacts on Agriculture: Concepts, Issues and Policies for Developing Countries*.

Wajid Nasim Jatoi is Professor, Department of Agronomy, Faculty of Agriculture and Environment, the Islamia University of Bahawalpur (IUB), Pakistan. He also served COMSATS University Islamabad, Vehari Campus (Pakistan), and Quaid-i-Azam University Islamabad (Pakistan). He completed his PhD training from the College of Agriculture and Environmental Sciences, the University of Georgia, USA; he completed his Post Doctorate one from France and one from Australia. He is working as Focal Person in Consortium on Climate Change, Sustainability, and Conservation and as Director at the International Center for Climate Change, Food Security, and Sustainability, IUB. He has completed many projects in collaboration with Columbia University, USA; IFPRI, USA; CIHEAM-IAMM, France; etc. He published more than 100 peer-reviewed research publications. He organized and attended more than 100 national and international conferences, seminars, and workshops. He published 02 books *Building Climate Resilience in Agriculture* and *Climate Change Impacts on Agriculture* in Springer Nature as first editor.

Muhammad Zaffar Hashmi is an Associate Professor of Environmental Science at Health Services Academy Islamabad, Pakistan. Dr. Hashmi works on multiple water, soil, air, and health-related issues and has expertise in developing environmentally sustainable and feasible technologies at the Health Services Academy Islamabad, Pakistan. Dr. Hashmi works on multiple water, soil, air and health-related issues and has expertise in developing environmentally sustainable and feasible technologies.

Mushtaq Ahmad is renowned world Top Agrobiologist and Director of Sino-Pak Biodiversity Research Centre for Agro-Biological Resources, Department of Plant Sciences, Quaid-i-Azam University, Islamabad, Pakistan. He is Honorary Distinguished professor in China and Uzbekistan. He is author of 997 articles (citations ±21,405, H-index 72, i10-index 430) including 60 international books and book chapters. He has successfully supervised/produced 47 PhDs and Post Doctorates, 180 MPhil/MSc/BS research scholars. He is recipient of the highest Civil Award of Pakistan (Tamgha-e-Imtiaz), High Foreign talent award by China, International Scientist Award by Uzbekistan, Fellow of TWAS by UNESCO, Fellow of PIFI-Chinese Academy of Sciences, and Fellow of Pakistan Academy of Sciences (PAS). He is the Member and Awardee of international bodies, i.e. Chairman Silkroad Biohealth Agricultural Industrial Alliance, China, IUCN, IBPES, Association of federation of biotechnologists Asia Japan, Flora of Pan Himalaya-China, PAS, TWAS, PIFI-CAS, NAS, TUBITAK, MEVLANA, HEC, PBS, BSP.

Contributors

Farrukh Tanzeel Abbas Department of Agronomy, College of Agriculture, University of Sargodha, Sargodha, Pakistan

Touqeer Abbas College of Horticulture and Agronomy, University of Nebraska, Lincoln, NE, USA

Aneela Afzal National Center of Industrial Biotechnology, PMAS-Arid Agricultural University, Rawalpindi, Pakistan

Department of Agricultural Extension, PMAS-Arid Agriculture University, Rawalpindi, Pakistan

Ashfaq Ahmad Ghazi University, Dera Ghazi Khan, Pakistan

Shakeel Ahmad Institute of Agronomy, FAS&T, Bahauddin Zakariya University, Multan, Pakistan

Tanveer Ahmad Department of Agricultural Engineering, Khwaja Fareed University of Engineering and Information Technology, Rahim Yar Khan, Pakistan

Shoaib Akhtar Department of Agricultural Business and Marketing, Bahauddin Zakariya University, Multan, Pakistan

Aqdus Ali Department of Environmental Sciences, COMSATS University Islamabad, Vehari Campus, Vehari, Pakistan

Ehsan Ali Department of Forestry and Range Management, Bahauddin Zakariya University, , Multan, Pakistan

Mazhar Ali Cholistan Institute of Business Administration, CUVAS Bahawalpur, Bahawalpur, Pakistan

Department of Biotechnology, COMSATS University Islamabad, Vehari Campus, Vehari, Pakistan

Shaukat Ali Centre for Global Child Health, The Hospital for Sick Children, Toronto, ON, Canada

Sikandar Ali Department of Irrigation and Drainage, University of Agriculture Faisalabad, Faisalabad, Pakistan

Asad Amin Queensland Alliance for Agriculture and Food Innovation, St. Lucia, QLD, Australia

Muhammad Amjad Department of Environmental Sciences, COMSATS University Islamabad, Vehari, Pakistan

Muhammad Naveed Anjum Department of Land and Water Conservation Engineering, Faculty of Agricultural Engineering and Technology, PMAS Arid Agriculture University, Rawalpindi, Pakistan

Muhammad Anwar-ul-Haq Institute of Soil and Environmental Sciences, University of Agriculture, Faisalabad, Pakistan

Rao Husnain Arshad Department of Irrigation and Drainage, University of Agriculture Faisalabad, Faisalabad, Pakistan

Sundas Ashraf Department of Agronomy, College of Agriculture, University of Sargodha, Sargodha, Pakistan

Atique-ur-Rehman Department of Agronomy, Bahauddin Zakariya University, Multan, Pakistan

Muhammad Awais University of Agriculture, Faisalabad Sub-Campus, Depalpur, Pakistan

Pir Mahir Ali Shah Arid Agriculture University, Rawalpindi, Pakistan

Usman Khalid Awan Center for Development Research (ZEF), University of Bonn, Bonn, Germany

Muhammad Ayaz Pakistan Meteorological Department, National Agromet Centre, Islamabad, Pakistan

Muhammad Farooq Azhar Department of Forestry and Range Management, Bahauddin Zakariya University, Multan, Pakistan

Abida Aziz Department of Botany, The Women University, Multan, Pakistan

Tariq Aziz Ausenco, Burlington, ON, Canada

Ecohydrology Research Group, Water Institute and Department of Earth and Environmental Sciences, University of Waterloo, Waterloo, ON, Canada

Zaheer Ahmed Babar Pakistan Meteorological Department, National Weather Forecasting Centre, Islamabad, Pakistan

Hafiz Faiq Bakhat Department of Environmental Sciences, COMSATS University Islamabad, Vehari, Pakistan

Tanzeela Bashir Department of Biotechnology, COMSATS University Islamabad, Vehari Campus, Vehari, Pakistan

Mumtaz Cheema School of Science and the Environment, Memorial University of Newfoundland, St. John's, NL, Canada

Muhammad Azhar Ehsan International Research Institute for Climate and Society (IRI), Columbia Climate School, Palisades, NY, USA

Ayman El Sabagh Department of Agronomy, Faculty of Agriculture, Kafrelsheikh University, Kafrelsheikh, Egypt

Department of Field Crops, Faculty of Agriculture, Siirt University, Siirt, Turkey

Sania Fareed Department of Environmental Sciences, COMSATS University Islamabad, Vehari Campus, Vehari, Pakistan

Hafiz Umar Farid Department of Agricultural Engineering, Bahauddin Zakariya University, Multan, Pakistan

Abu Bakr Umer Farooq Department of Biotechnology, COMSATS University Islamabad, Vehari Campus, Vehari, Pakistan

Maham Fatima Department of Environmental and Polymer Engineering, Inha University, Incheon, South Korea

Raqash Fatima Department of Environmental Sciences, COMSATS University Islamabad, Vehari, Pakistan

Muhammad Habib ur Rahman Crop Science Group, Institute of Crop Science and Resource Conservation (INRES), University of Bonn, Bonn, Germany

Department of Agronomy, MNS-University of Agriculture, Multan, Multan, Pakistan

Zulqarnain Haider Department of Agricultural Economics, College of Agriculture, University of Sargodha, Sargodha, Pakistan

Hafiz Mohkum Hammad Department of Agronomy, Muhammad Nawaz Sharif University of Agriculture, Multan, Pakistan

Department of Agricultural and Biological Engineering, University of Florida, Gainesville, FL, USA

Asma Jawad Hashmi Pakistan Meteorological Department, National Agromet Centre, Islamabad, Pakistan

Nazim Hussain Faculty of Agricultural Sciences and Technology, Bahauddin Zakariya University, Multan, Pakistan

Sajjad Hussain Department of Environmental Sciences, COMSATS University Islamabad, Vehari Campus, Vehari, Pakistan

Irfan Iftikhar University of Agriculture, Faisalabad Sub-Campus, Depalpur, Pakistan

Ministry of Commerce, Climate Division, Islamabad, Pakistan

Muhammad Imran Department of Environmental Sciences, COMSATS University Islamabad, Vehari, Pakistan

Department of Sanitary Engineering, Faculty of Civil and Environmental Engineering, Gdansk University of Technology, Gdańsk, Poland

Muhammad Azhar Inam Department of Agricultural Engineering, Bahauddin Zakariya University, Multan, Pakistan

Wajid Nasim Jatoi Department of Agronomy, Faculty of Agriculture and Environment, The Islamia University of Bahawalpur (IUB), Bahawalpur, Punjab, Pakistan

Sidra Javed National Center of Industrial Biotechnology, PMAS-Arid Agricultural University, Rawalpindi, Pakistan

Hafiz Muhammad Rashad Javeed Department of Environmental Sciences, COMSATS University Islamabad, Vehari, Pakistan

Rana Muhammad Asif Kanwar Department of Agricultural Engineering, Bahauddin Zakariya University, Multan, Pakistan

Shankar Karuppannan Department of Applied Geology, College of Applied Natural Science, Adama Science and Technology University, Adama, Ethiopia

Department of Research Analytics, Saveetha Institute of Medical and Technical Sciences (SIMATS), Saveetha University, Chennai, Tamil Nadu, India

Dildar Hussain Kazmi Pakistan Meteorological Department, National Agromet Centre, Islamabad, Pakistan

Sana Khalid Department of Environmental Sciences, COMSATS University Islamabad, Vehari Campus, Vehari, Pakistan

Bilal Ahmad Khan Department of Soil and Crop Sciences, Texas A&M University, College Station, TX, USA

Urooj Khan Department of Biotechnology, COMSATS University Islamabad, Vehari Campus, Vehari, Pakistan

Zahid Mahmood Khan Department of Agricultural Engineering, Bahauddin Zakariya University, Multan, Pakistan

Isaya Kisekka Department of Land, Air and Water Resources, University of California, Davis, CA, USA

Samina Kousar Department of Environmental Sciences, COMSATS University Islamabad, Vehari, Pakistan

Shehryar Liaqat Department of Environmental Sciences, COMSATS University Islamabad, Vehari Campus, Vehari, Pakistan

Tariq Mahmood Pakistan Meteorological Department, National Agromet Centre, Islamabad, Pakistan

Qurat-ul-ain Maken Department of Environmental Sciences, COMSATS University Islamabad, Vehari Campus, Vehari, Pakistan

Nasir Masood Memorial University of Newfoundland, St. John's, NL, Canada

Shehzad Mehmood Department of Biotechnology, COMSATS University Islamabad, Vehari Campus, Vehari, Pakistan

Iqra Mubeen Department of Environmental Sciences, COMSATS University Islamabad, Vehari, Punjab, Pakistan

Department of Renewable Resources, University of Alberta, Edmonton, AB, Canada

Muhammad Mubeen Department of Biotechnology, COMSATS University Islamabad, Vehari Campus, Vehari, Pakistan

Department of Environmental Sciences, COMSATS University Islamabad, Vehari, Pakistan

Tayyaba Mumtaz Department of Environmental Sciences, COMSATS University Islamabad, Vehari Campus, Vehari, Pakistan

Behzad Murtaza Department of Environmental Sciences, COMSATS University Islamabad, Vehari, Pakistan

Ghulam Murtaza Institute of Soil and Environmental Sciences, University of Agriculture Faisalabad, Faisalabad, Pakistan

Muhammad Usama Mustafa University of Agriculture, Faisalabad Sub-Campus, Depalpur, Pakistan

Muhammad Ather Nadeem Department of Agronomy, College of Agriculture, University of Sargodha, Sargodha, Pakistan

Muhammad Nadeem Institute of Agro-Industry and Environment, The Islamia University of Bahawalpur, Bahawalpur, Pakistan

Muhammad Asad ur Rehman Naseer Department of Agricultural Business and Marketing, Bahauddin Zakariya University, Multan, Pakistan

Wajid Nasim Department of Agronomy, Faculty of Agriculture and Environment (FA&E), Islamia University of Bahawalpur (IUB), Bahawalpur, Pakistan

Parveen University of Agriculture, Faisalabad Sub-Campus, Depalpur, Pakistan

Sobia Parveen Department of Biotechnology, COMSATS University Islamabad, Vehari Campus, Vehari, Pakistan

Saeed Ahmad Qaisrani Department of Environmental Sciences, COMSATS University Islamabad, Vehari Campus, Vehari, Pakistan

Department of Biotechnology, COMSATS University Islamabad, Vehari Campus, Vehari, Pakistan

Plant Breeding Institute (PBI), The University of Sydney, Sydney, NSW, Australia

Marryam Qamar Department of Environmental Sciences, COMSATS University Islamabad, Vehari Campus, Vehari, Pakistan

Rafi Qamar Department of Agronomy, College of Agriculture, University of Sargodha, Sargodha, Pakistan

Faiz Rabbani Department of Environmental Sciences, COMSATS University Islamabad, Vehari Campus, Vehari, Pakistan

Beenish Nazakat Raja Pakistan Meteorological Department, National Agromet Centre, Islamabad, Pakistan

Fahd Rasul Agro-Climatology Lab, Department of Agronomy, University of Agriculture, Faisalabad, Pakistan

Muhammad Razaq Department of Entomology, Faculty of Agricultural Sciences and Technology, Bahauddin Zakariya University, Multan, Pakistan

Abdul Rehman Department of Botany, University of Agriculture Faisalabad, Faisalabad, Pakistan

Muhammad Ahmad Rustam Department of Agricultural Engineering, Khwaja Fareed University of Engineering and Information Technology, Rahim Yar Khan, Pakistan

Farhan Saeed Department of Environmental Sciences, COMSATS University Islamabad, Vehari, Pakistan

Alishba Sajjad Department of Environmental Sciences, COMSATS University Islamabad, Vehari Campus, Vehari, Pakistan

Mazhar Saleem Department of Biotechnology, COMSATS University Islamabad, Vehari Campus, Vehari, Pakistan

Usman Saleem Deptartment of Plant Breeding and Genetics, University of Sargodha, Sargodha, Pakistan

Aneeza Sardar Department of Environmental Sciences, COMSATS University Islamabad, Vehari, Pakistan

Asif Sardar National Center of Industrial Biotechnology, PMAS-Arid Agricultural University, Rawalpindi, Pakistan

Qamar Sarfaraz Department of Soil Science, Faculty of Agriculture, Lasbela University of Agriculture, Water, and Marine Sciences, Uthal, Balochistan, Pakistan

Khadija Shabbir Department of Environmental Sciences, COMSATS University Islamabad, Vehari Campus, Vehari, Pakistan

Ghulam Mustafa Shah College of Engineering (Key Laboratory for Clean Renewable Energy Utilization Technology, Ministry of Agriculture), China Agricultural University, Beijing, People's Republic of China

Department of Environmental Sciences, COMSATS University Islamabad, Vehari, Pakistan

Muhammad Adnan Shahid College of Life Sciences and Agriculture, University of New Hampshire, Durham, NH, USA

Khurram Shahzad Department of Crop and Soil Sciences, Washington State University, Pullman, WA, USA

Department of Soil Science, Faculty of Agriculture, Lasbela University of Agriculture, Water, and Marine Sciences, Uthal, Balochistan, Pakistan

Aamir Shakoor Department of Agricultural Engineering, Bahauddin Zakariya University, Multan, Pakistan

Muhammad Shoaib Department of Agricultural Engineering, Bahauddin Zakariya University, Multan, Pakistan

Muhammad Suffian Department of Biochemistry, The Islamia University of Bahawalpur, Bahawalpur, Pakistan

Henry Y. Sintim Department of Crop & Soil Sciences, University of Georgia, Tifton, GA, USA

Muhammad Tahir Department of Biotechnology, COMSATS University Islamabad, Vehari Campus, Vehari, Pakistan

Maham Tariq Department of Environmental Sciences, COMSATS University Islamabad, Vehari Campus, Vehari, Pakistan

Samra Tariq Department of Environmental Sciences, COMSATS University Islamabad, Vehari Campus, Vehari, Pakistan

Urban Unit, Lahore, Pakistan

Salma Tasneem Department of Environmental Sciences, COMSATS University Islamabad, Vehari, Pakistan

Sana Ullah Department of Agronomy, Faculty of Agriculture, University College Dera Murad Jamali, Lasbela University of Agriculture, Water, and Marine Sciences, Uthal, Balochistan, Pakistan

Ahmad Abdul Wahab Institute of Agronomy, FAS&T, Bahauddin Zakariya University, Multan, Pakistan

Syed Aftab Wajid Agro-Climatology Lab, University of Agriculture, Faisalabad, Pakistan

Muhammad Mohsin Waqas Department of Agricultural Engineering, Khwaja Fareed University of Engineering and Information Technology, Rahim Yar Khan, Pakistan

Muhammad Yaseen Department of Agricultural Extension and Rural Studies, College of Agriculture, University of Sargodha, Sargodha, Pakistan

Nadeem Zubair Department of Agricultural Engineering, Bahauddin Zakariya University, Multan, Pakistan

Abbreviations

AI	Artificial intelligence
ANFIS	Adaptive neuro-fuzzy inference system
AOPs	Advanced oxidation processes
APX	Ascorbate peroxidase
ARIES	Artificial intelligence for environment and sustainability
BDA	Big data analytics
BNR	Biological nutrient removal
BOD	Biochemical oxygen demand
CLARANS	Clustering large applications based upon RANdomized search
COD	Chemical oxygen demand
CRISPR	Clustered regularly interspaced short palindromic repeats
EBPR	Enhanced biological phosphorus removal
ELM	Extreme learning machine
EPS	Exopolysaccharides
ET	Evapotranspiration
EVI	Enhanced vegetation index
FAO	Food and Agriculture Organization
GGCRI	German watch global climate risk index
GIA	Groundwater irrigation advisory
GIS	Geographic information system
GIS	Geographic information systems
GPx	Glutathione peroxidase
GR	Glutathione reductase
HEIS	High efficiency irrigation systems
InVEST	Integrated valuation of ecosystem services and tradeoffs
IoT	Internet of Things
ISO	Organization for Standardization
IWMI	International Water Management Institute
IWRM	Integrated water resource management
LiDAR	Light detection and ranging
LIMS	Laboratory Information Management Systems

LUCI	Land utilisation and capability indicator
MAR	Managed aquifer recharge
MBRs	Membrane bioreactors
MCU	Microcontroller unit
NDVI	Normalized difference vegetation index
OFWM	On farm water management
PCD	Programmed cell death
PES	Payments for ecosystem services
PGPR	Plant growth-promoting rhizobacteria
PI	Precision irrigation
PRIAT	Punjab resilient and inclusive agriculture transformation
RBCs	Rotating biological contactors
RBD	River Basin District
ROS	Reactive oxygen species
RS	Remote sensing
SBRs	Sequential batch reactors
SCADA	Supervisory control and data acquisition
SEBAL	Surface energy balance algorithm for land
SMAP	Soil moisture active passive
S-SEBI	Simplified surface energy balance index
UAVs	Unmanned aerial vehicles
VRI	Variable rate irrigation
WBICs	Weather-based irrigation controllers
WEAP	Water Evaluation and Planning
WFD	European Union's Water Framework Directive
WHO	World Health Organization
WLE	Water, land, and ecosystems
WQI	Water Quality Index
WSN	Wireless Sensor Networks
WUE	Water use efficiency

Prospects of Agricultural Water Management

Muhammad Mubeen, Salma Tasneem, Mazhar Ali, Wajid Nasim Jatoi, Hafiz Mohkum Hammad, and Shaukat Ali

Abstract Agricultural water management systems are critical to achieving food production sustainability. Although a number of agricultural water management practices are in operation since long; these practices may not be able to meet the current challenges of water scarcity under the era of climate change. Here comes the role of innovative approaches in modern agriculture. However, these innovations carry inherent risks that must be identified and mitigated in order for them to be implemented successfully. This abstract delves into major innovations such as drip irrigation, precision irrigation, sprinkler and furrow irrigation systems, water distribution networks, conservation agriculture, desalination, and mulching. The risks of each innovation are identified, such as technological malfunctions, financial constraints, environmental impacts, and social disparities. Furthermore, potential solutions are proposed, such as regular maintenance and calibration, the use of renewable energy sources, farmer education and training, equitable policy frameworks, and

M. Mubeen (✉) · M. Ali
Department of Biotechnology, COMSATS University Islamabad, Vehari Campus, Vehari, Pakistan
e-mail: muhammadmubeen@cuivehari.edu.pk; mazharali@cuivehari.edu.pk

S. Tasneem
Department of Environmental Sciences, COMSATS University Islamabad, Vehari, Pakistan

W. N. Jatoi
Department of Agronomy, Faculty of Agriculture and Environment, The Islamia University of Bahawalpur (IUB), Bahawalpur, Punjab, Pakistan
e-mail: wajid.nasim@iub.edu.pk

H. M. Hammad
Department of Agronomy, Muhammad Nawaz Sharif University of Agriculture, Multan, Pakistan

Department of Agricultural and Biological Engineering, University of Florida, Gainesville, FL, USA

S. Ali
Centre for Global Child Health, The Hospital for Sick Children, Toronto, ON, Canada

M. Mubeen et al. (eds.), *Innovations in Agricultural Water Management*,
https://doi.org/10.1007/978-3-031-91883-4_1

stakeholder involvement. Addressing these risks through comprehensive strategies is critical to realizing the full potential of agricultural water management innovations and ensuring the long-term viability of water resources in agriculture.

Keywords Agricultural prospects · Water management · Water scarcity · Irrigation techniques · Water risks · Solutions · Future prospective

1 Introduction

Water is an essential natural resource for the survival of all ecosystems on our planet. The amount of liquid freshwater in the world's total water is equivalent to a spoon of water in approximately 1 L of water. The oceans hold almost 97% of the world's water. More than 98% of liquid freshwater exists as groundwater, with less than 2% in more visible forms such as groundwater-fed streams and lakes (Qadir et al., 2024; Hussain et al., 2024a). In contrast, 8–15% of freshwater resources will be taken out of agriculture in order to meet the growing demand for industrial and residential use. Up to 70% of freshwater used worldwide is consumed by agriculture, making it clear how crucial wise water management is to maintaining global food and water security (Hussain et al., 2024b; Srivastav et al., 2021). Field application techniques and irrigation systems play a major role in crop cultivation. In an effort to prevent productivity loss brought on by water stress (under-irrigation), farmers over-irrigate, which wastes water and energy, in addition to creating productivity issues.

Furthermore, the crop uses only 55% of the available water due to the extremely low irrigation efficiency (Hussain et al., 2020, 2024c). Over 40% of the world's population lives in nearly 80 countries where water supply is already insufficient due to unequal distribution of water resources and population densities.

In turn, precision irrigation discourages both under- and over-irrigation, and encourages more effective and efficient use of water. For precision irrigation in agriculture to increase crop yield, reduce costs, and support environmental sustainability, smart water management is essential.

In order to achieve sustainable water management in agriculture, water availability and demand must be balanced in terms of quantity and quality, space and time, cost, and environmental impact. Technological challenges, social behavior in rural communities, economic limitations, institutional and legal frameworks, and agricultural practices all have an impact on its implementation (Hussain et al., 2023a; Hammad et al., 2023; Rolandi et al., 2021).

Both human and natural systems can use existing knowledge and technology to adapt to change to varied degrees. They are called self-adapting adaptations. To adjust for variations in precipitation amount and timing, farmers, for instance, can alter the crop mix and planting dates over time. Additional funding and institutional adjustments are needed for other adaptations (Novoa et al., 2019). Using our example again, when changes get too big for autonomous adaptation, farmers might need completely new crop varieties, irrigation systems, and facilities for processing and education. Therefore, there are many options for agricultural water management in the modern day.

2 Agricultural Water Management

The goal of agricultural water management systems is to effectively manage water resources in agricultural settings in order to minimize water waste and environmental impact, while simultaneously ensuring optimal crop growth and productivity. These systems usually combine technologies, practices, and infrastructure to collect, store, distribute, and use water in agricultural operations as efficiently as possible (Fig. 1).

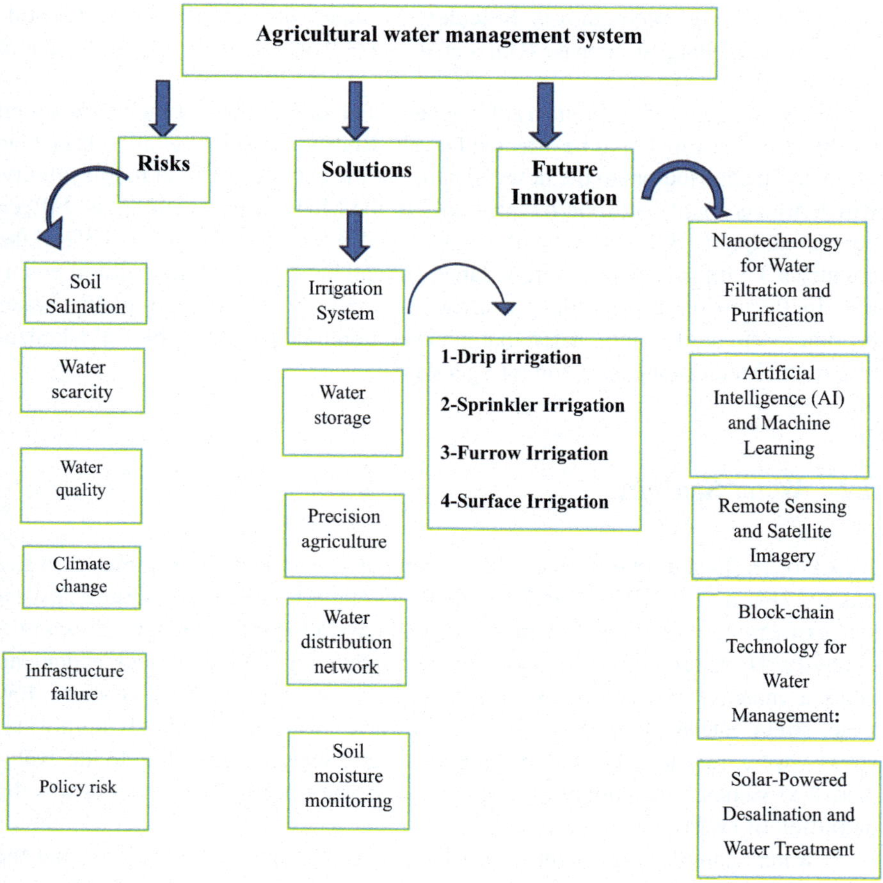

Fig. 1 Perspective of agriculture water management

Numerous risks affect agricultural water management systems and can affect crop yield, water availability, and environmental quality. Some common risks include:

3 Common Risks to Agriculture Water Management System

3.1 Soil Salination

The presence of water-soluble salts such as Na^+, K^+, Cl^-, and SO_4^{2-} is referred to as soil salinity. Worldwide crop production is severely impacted by salinity, which causes 20% of cultivated land and 33% of irrigated land to become degraded. This process may be made worse by climate change and excessive groundwater use, particularly in areas close to the ocean (Stavi et al., 2021). One of the main reasons cultivated soil productivity is declining is soil salinization. The area of salinized soils is increasing, although it is difficult to estimate with precision; irrigated soils are showing the largest increase (Haris et al., 2023; Hussain et al., 2023b; Kumar & Sharma, 2020).

Twenty percent (45 million ha) of irrigated land, or one-third of the world's food production, is estimated to be affected by salt (Rodríguez-Berbel et al., 2022). One of the main contributors to the desertification in the European Union is soil salinity, which affects about 1 million ha, primarily in the Mediterranean countries. Fifteen percent of Spain's 3.5 million ha of irrigated land are in grave danger, and 3% of the country's 3.5 million ha of severely damaged land have reduced agricultural potential. In the upcoming decades, as irrigated areas grow and high-quality water becomes scarcer, it is anticipated that land degradation brought on by soil alkalization in the Mediterranean region will get worse.

3.2 Water Scarcity

A lack of freshwater that is renewable in comparison to demand is known as water scarcity (Farid et al., 2023; Damkjær, 2020). Drought-related water scarcity, excessive extraction, or conflicting demands can all have a big effect on agricultural productivity. There are only so many water resources in a given area, and as demand and use increase, resource stress increases. Five categories—no water stress, low water stress, intermediate water stress, high water stress, and extremely high water stress—were traditionally used to categorize water scarcity, according to the evaluated WSI values. The amount of usable water in the primary water supply was the definition of IWMI water scarcity.

In water scarcity assessments, the Blue Water Scarcity (BWS) indices and the International Water Management Institute (IWMI) indicator are frequently used. First, per capita water was used to calculate the water scarcity level (Huang et al.,

2022). Drip irrigation techniques, both surface and subsurface, can lessen water scarcity, particularly in regions that are vulnerable to drought. In an effort to combat water scarcity, the Pakistani government recently (2018) implemented a water policy (Haseeb et al., 2023; Dawar et al., 2021; Ali et al., 2020, 2021; Bukhari et al., 2021; Rehman et al., 2021; Rasool et al., 2020, 2021; Shah et al., 2021; Mehmood et al., 2021; Hussain et al., 2020; Muzammil et al., 2020).

The goal of this policy is to use the idea of more crop per drop as the foundation for future strategy and action plans. This concept has been supported by the proposal of various management techniques. A few examples are the use of conservation tillage to increase crop water efficiency, the cultivation of drought-resistant crops, the substitution of less water-demanding crops for more water-intensive ones, the preservation of soil moisture through mulching, and deficit irrigation to increase water productivity.

3.3 Water Quality

Groundwater and surface water systems are essential to more than five billion people globally for a range of uses, including manufacturing, housing, agriculture, and drinking water (Akhtar et al., 2021). Degradation of water resources is a well-researched phenomenon that can be brought on by human activity (urban waste and agricultural practices), natural processes (climate change, water-rock interactions, and geological factors), and the presence of important chemical compounds after the industrial revolution. In spite of this, managing groundwater and surface water resources is still difficult, and pertinent data is often ambiguous.

The main contributors to surface water pollution are industrial discharge, agriculture, and urbanization, especially in rivers and streams (Alsuhaibani et al., 2023; Anh et al., 2023; Yang et al., 2023). Moreover, heated water released from power plants can cause temperature fluctuations, and physical environmental factors can contaminate aquatic environments. Moreover, if hydro-heated water is rapidly diluted by mixing with surface water, it may not be an issue at any time of the year if it is contaminated with certain substances.

Moreover, pollutants from agricultural practices include metals, pesticides, pathogens, nutrients, and salts, all of which have an impact on surface water. Hazardous materials that are discharged into river water as a consequence of urban activity can also be found in untreated and partially treated sewage, construction waste, and solid/liquid waste components (Muhammad et al., 2023; Chukwuemeka et al., 2021). Flooding, excessive irrigation, and precipitation combine chemical residues with river water, which eventually finds its way into the food chain. These substances contaminate a wide variety of fruits and vegetables, and are fatal to living things. Insecticides, herbicides, fungicides, nematicides, and rodenticides are among the chemicals of concern when it comes to the deterioration of groundwater quality because of agriculture, along with nitrate (Yadav et al., 2022).

The loss of phosphorus from agricultural fields has been recognized as a significant factor in the eutrophication of water bodies in recent decades, especially when phosphorus fertilizer applications surpass crop requirements. In addition to causing soil compaction, which lowers cereal production, too much P in the soil can affect the quality of surface water and groundwater through runoff and infiltration, respectively (Hussain et al., 2023c; Liu et al., 2021).

3.4 Climate Change

Extreme weather events, temperature changes, and changes in precipitation patterns can all affect water availability and exacerbate water stress in agriculture. Significant climate risk exists for rainfed agriculture in South Asia, Sub-Saharan Africa, and other nations (Aryal et al., 2021). In northern China, annual precipitation is usually sparse and concentrated in a few months. The North China plain experiences approximately 600 mm of annual rainfall, whereas the Loess Plateau experiences 440 mm and the northwest of China experiences less than 300 mm. Roughly 70% of the precipitation falls in June, July, August, and September each year is caused (Zhu & Xu, 2021; Hammad et al., 2010a, 2010b; Nasim et al., 2011; Nasim & Bano, 2012; Nasim et al., 2011; Wajid et al., 2010; Usman et al., 2010; Ahmad et al., 2009; Usman et al., 2009; Wajid et al., 2007).

Climate change affects human health and ecosystem survival in addition to the effectiveness and availability of water resources. Every phenomenon—natural, technological, and human—is interrelated. Every fraction in the intricate system has a unique, irreplaceable component.

A 3–4 °C temperature increase, according to Šulyová et al. (2021), could cause a 90% decline in population, limited adaptation because of water scarcity, and system collapse.

The fundamental problems are:

- *Variations in rainfall frequency and intensity will impact coastal cities like Singapore, leading to hot and muggy weather. Rising sea levels and flooding could affect other cities, like Amsterdam.*
- *The 200-year-old worldwide trend of rising migration brought on by a lack of resources and social injustices will not abate.*
- *The growing population cannot be supported by limited resources.*

3.5 Infrastructure Failure

Networks known as infrastructure make it possible to move people, products, and ideas across geographic boundaries. The water infrastructure includes water treatment facilities, levees, canals, pipelines, pumps, and dams.

According to Moravej et al., 2021, this infrastructure also comprises the links between the devices that offer access to water alternatives. For instance, the replacement of Pakistan's extensive water infrastructure is expected to cost USD60 billion.

Due to age and neglect, a large portion of the water infrastructure has deteriorated, leading to significant system losses and subpar water transportation to the canal commands' tail-reaches.

Many parts of the massive hydraulic system need to be renewed or replaced because they are getting close to the end of their design life (Qureshi, 2022). An enormous backlog of incomplete maintenance and repairs exists. Restoration of badly maintained systems has taken precedence over the development of new infrastructure in recent "investments" related to irrigation and water supply. The well-being of the country and sustainable irrigated agriculture depends on this infrastructure. These services are only possible, though, if the structures are kept up properly and replaced when necessary. As a result, significant funding for the system's restoration is desperately needed.

3.6 Policy Risks

The financial and human resources needed to manage groundwater are scarce in many developing nations. Aquifers that are located across provincial administrative boundaries have not received the same level of attention as surface water resources.

Solutions for institutional groundwater management have also shown to be far more complicated than first thought (Qureshi, 2020).

Especially in rural areas where over 70% of the population resides, governments are often under pressure to produce and secure food in order to feed the populace and reduce poverty.

Since significant investments in irrigation systems and surface water have dropped dramatically, the development of groundwater resources has made it possible for irrigated agriculture to continue growing. Since a lot of farming families now mainly depend on groundwater for their operations, the federal government was hesitant to put the rules into effect.

The upland Baluchistan region of Pakistan is losing groundwater at a rapid rate, endangering agricultural production that depends on it. It is possible to hold the government and water users responsible for the looming catastrophe. The government's groundwater development policy, which led to an overestimation of groundwater reserves, and the inefficient use of water by users are the main causes of declining water tables.

Throughout the past 40 years, the impact of groundwater policies and other factors on the decline in the water table has been examined using the ARIMA model, a time series data analysis technique. The results showed that groundwater declines have been made worse by government policies supporting tubewell development and power subsidies. The decline in the water table was significantly impacted by the tubewell area-to-total area ratio.

Most likely, a highly interactive and fruitful process involving pertinent stakeholders produced the eagerly awaited policy (Mumtaz et al., 2021; Ahmad et al., 2009; Usman et al., 2009; Wajid et al., 2007).

4 Solutions

Long-term food production depends on better agricultural water management, especially in light of climate change and growing water scarcity. The following are a few answers to the issues surrounding agricultural water management:

4.1 Irrigation Systems

These systems use a range of techniques, such as flood irrigation, furrow irrigation, sprinkler irrigation, and drip irrigation, to supply water to crops. Choose between these methods based on crop type, climate, soil type, and water availability; each has advantages.

4.1.1 Drip Irrigation

This can be accomplished either directly onto the root zone through a subsurface drip irrigation system or on the soil's surface through a network of emitters, pipes, and valve (Cahn & Hutmacher, 2024).

Although it is more expensive than conventional irrigation systems, it delivers a precise amount of water at a time (Jarwar et al., 2019).

In recent decades, there has been a notable increase in the global use of drip irrigation and sprinkles irrigation systems. These systems' main benefit is their capacity to increase crop yields while lowering the costs of cultivation, fertilizer, and water (Usman et al., 2010; Seleiman et al., 2020).

Several factors affect the pattern of soil moisture distribution surrounding a water emitter, such as the following:

- The total amount of water applied
- The emitter's flow rate
- The initial boundary conditions and the source configuration (surface, subsurface, point, or line). In addition, the physical characteristics of the soil and its spatial distribution are important.
- Controlling irrigation and keeping an eye on plant root activity.

If the drip irrigation system is appropriately designed, with sufficient emitter spacing, flow rate, and installation depth, it can achieve high water efficiency. Subsurface drip irrigation systems are very effective and can be used as workable alternatives for water management (Hammad et al., 2010a, 2010b; Valentín et al., 2020; Nasim et al., 2011).

These systems do have certain drawbacks, though, most notably the variation in flow rate brought on by the properties of the soil. Particularly in Brazil's irrigated coffee crops, subsurface drip irrigation systems with a thin soil layer have been used; nevertheless, most related research has concentrated on flow rate variation at relatively deep soil levels.

4.1.2 Sprinkler Irrigation

Proper irrigation system design and management are essential for irrigation agriculture in windy and semi-arid regions. Water is sprayed into the air by sprinkler irrigation, which disperses it over the irrigated area. While a great deal of research has been done to better understand the factors influencing sprinkler irrigation systems' uniformity and efficiency, most of those studies have concentrated on medium-sized sprinklers. Large sprinklers that operate at high flow rates (more than 3 L/s) and large layout spacing (more than 24 m sprinkler spacing) have not, however, been the subject of any study (Zhu et al., 2024; Mubeen et al., 2021; Rasool et al., 2020; Hussain et al., 2020).

When it comes to irrigation systems in semi-arid climates, semi-portable large sprinklers are a common choice. This system irrigates 85% of the sprinkler irrigation surface (660,000 ha) and 48.7% of the total pressurized irrigation surface in Iran. Compared to conventional surface irrigation, sprinkler irrigation has a number of advantages. By evenly dispersing water in the form of rain across the land surface when needed, in the necessary quantity, and according to the same pattern, it promotes natural rainfall. Water is applied at a rate that matches the rate at which the soil infiltrates in order to prevent surface runoff associated with irrigation. Sprinkler irrigation systems are suitable for sandy or shallow soils, uneven water distribution, undulating terrain, and limited water availability (Kumar et al., 2023a).

4.1.3 Furrow Irrigation

Furrow irrigation is a method of watering crops where trenches or furrows are dug between rows of plants. Water is then allowed to flow through these furrows, typically by gravity, directly to the root zones of the plants. This method is one of the oldest and simplest forms of irrigation, and is commonly used in agriculture, particularly for row crops like corn, cotton, and vegetables (Walia et al., 2024). The key components of furrow irrigation include the furrows themselves, which are usually created using plows or other implements, and a method of delivering water to the furrows, such as open ditches or pipes. Farmers typically manage the flow of water by controlling gates or valves, allowing them to regulate the amount of water applied to each furrow based on factors like soil type, crop needs, and weather conditions (Gupta et al., 2022). While furrow irrigation is relatively easy and inexpensive to implement, it can have drawbacks such as uneven water distribution, soil erosion in the furrows, and water loss due to runoff. However, with proper management practices like land leveling, soil amendments, and conservation measures, these issues can be minimized, making furrow irrigation a viable option for many farmers around the world.

4.1.4 Surface Irrigation

Unlike pressurized systems, surface irrigation systems (SI), also called gravity irrigation, rely only on gravity to transport, distribute, and apply water to the land surface. They developed over millennia, passing through the pre-Columbian American and Chinese civilizations, as well as the ancient civilizations of the Harappans in the Indus and the Egyptians in the Nile (Gupta et al., 2022).

Surface irrigation techniques such as borders, furrows, and basins are used to irrigate most of this land. The most popular method of applying water to agricultural land is surface irrigation, which uses overland flow to distribute water throughout the field (Ghimire et al., 2021). Poor design and management are often the root cause of inefficient irrigation, which leads to water waste, logging, salinization, and contamination of surface and ground water resources. When designed and operated correctly, surface irrigation systems can attain application efficiencies that are comparable to those of pressurized irrigation systems. Mathematical models for surface irrigation can aid in the design and administration of systems.

Numerous computer programs based on surface irrigation have already been created to mimic the complexity of design. Each of the three surface irrigation systems (furrow, border, and basin) was the focus of development for the software on the above list. But for designing the three different types of surface irrigation systems (furrow, border, and basin), SIRMOD and SRFR/WinSRFR are the most complete programs that have been created to date (Deekshithulu et al., 2020).

4.2 Water Storage

Water is stored for later use in times of scarcity in reservoirs, tanks, ponds, and other structures during periods of excess supply (such as rainy seasons). Variability in the water supply is a long-standing issue. Large-scale water infrastructure projects (big dams, reservoirs, and canals) were started during the twentieth century as the western United States' population and agricultural growth accelerated. These infrastructure projects provided drinking water, hydroelectric power, flood protection, and the ability for water managers to meet historical agricultural irrigation demand by reducing the variability associated with seasonal water supplies.

Reducing water waste is the goal of the Internet of Things (IoT)-based dam water management system (IoT-DWM). The suggested IoT-DWM is broken up into multiple sections, including ones for dam control, field sensing, and IoT networks (Krishnan et al., 2022). After obtaining actual data from the region, the dam controller makes an estimate of the required water volume. The crop grown in that area will determine how much water is needed. The controller takes into account various factors such as wind speed, temperature, humidity, and crop types in the area when estimating the amount of water required. The outcomes of the simulation demonstrate that the suggested IoT-DWM enhances performance, conserves a substantial amount of water, and helps reduce water scarcity.

4.3 Precision Agriculture

Leading the way in reshaping agriculture to tackle the broad problems facing the sector is precision agriculture. Precision agriculture is referred to as the "use of technologies that integrate sensors, information systems", enhanced machinery, and informed management to improve production by accounting for dynamics within sustainable agricultural systems. Smart irrigation, a component of precision agriculture, allows farmers to conserve precious resources without depriving plants of moisture. Applying water to a field at the appropriate time, volume, and location is known as smart irrigation (Abbas et al., 2022; Mubeen et al., 2022; Akram et al., 2022; Rasul et al., 2022; Nawaz et al., 2022; Kazmi et al., 2022; Ahmed et al., 2022; Bakhsh et al., 2022; Mehmood et al., 2022; Bwambale et al., 2022).

Even though it has not quite finished integrating all the technologies needed to make it function flawlessly in practice, the Internet of Things (IoT) becomes the clear choice for smart water management applications. Before precision irrigation becomes commonplace, there are still certain challenges to be addressed with IoT. First off, there is still some manual labor involved in the creation of software for IoT-based smart applications like irrigation systems for farms. Second, to deploy pilot applications for smart water management, advanced IoT software platforms are still insufficient for automating portions of the process and integrating multiple technologies like IoT, big data analytics, cloud computing, and fog computing. Third, the creation of suitable standards and information models is required for the integration of various sophisticated sensors.

4.4 Water Distribution Network

Water can be used for both social and economic reasons thanks to water distribution systems, which link water in lakes, rivers, and storage areas to plumbing in houses, irrigation pipes in farmers' fields, or factory outlets (Siddha & Sahu, 2022). The performance of collective irrigation networks' water delivery services can be enhanced by a variety of strategies, from more costly structural works to better water allocation (Jovanovic et al., 2020). A nationally representative overview of regional water management organizations that directly supply farms with irrigation water or have an impact on the use of groundwater on farms can be found in the USDA's 2019 Survey of Irrigation Organizations (SIO).

The irrigation water transportation infrastructure is managed by irrigation districts, ditch companies, acequias (community irrigation ditches), and other comparable organizations (Hrozencik et al., 2021). We provide water delivery services. Groundwater districts and delivery organizations are examples of groundwater organizations. Rebuilding catchment structures (structural method), enhancing conventional water distribution techniques (nonstructural method), and automating the surface water distribution system (combined method) are some of the ways to modernize the surface water distribution system and lessen water withdrawal from aquifers (Tork, 2021). MATLAB software was used to simulate surface water distribution, and the MODFLOW code was used to look into the effects on the aquifer.

4.5 Soil Moisture Monitoring

Soil moisture levels are measured using sensors and monitoring systems, which allows farmers to better match crop water requirements to irrigation schedules and quantities. Soil moisture determination is divided into four major sections, starting with soil moisture measurement metrics and laboratory-based testing and moving on to in-situ, remote, and proximal sensing techniques.

Soil moisture sensing methodologies are divided into five subsections:

- *Soil moisture metrics and laboratory method*
- In-situ *soil moisture sensing methods*
- *Remote sensing methods for measuring soil moisture*
- *Proximal in-field soil moisture monitoring, and other soil moisture sensing methods*

A limited number of "trusted" technologies currently dominate the field of soil moisture monitoring in agriculture: time domain reflectometry (TDR), gypsum block sensors, frequency domain reflectometry (FDR) or capacitance, and, in certain industries, the neutron moisture meter (NMM) and amplitude domain reflectometry (ADR) (Weinkove et al., 2020).

5 Future Prospects

These solutions and taking a holistic approach to agricultural water management can help mitigate water-related challenges. In the future, several innovations are expected to further revolutionize agricultural water management systems, addressing emerging challenges and opportunities. Here are some possible future innovations.

Water disinfection processes such as sedimentation, filtration, and chemical or biological degradation are insufficient to eliminate emerging contaminants or certain types of bacteria. Nanotechnology could be used to address the challenges posed by various contaminants and microorganisms (Iravani, 2023).

5.1 *Nanotechnology for Water Filtration and Purification*

In the twenty-first century, nanoparticles, which are one billionth (10–9) of a meter in size, will play a big role in the development of contemporary technologies. Research teams are interested in nanotechnology because the surface characteristics of these materials are crucial for the design of technologies in many different domains. Hospitals, industries, farms, and domestic activities are the main sources of wastewater pollution, which is primarily caused by nonmetabolized and metabolized particles (Jedla et al., 2022).

Effluent, surface water, groundwater, and other natural materials can be cleaned of pathogens, metal oxides, toxic metal ions, and inorganic and organic solutes using novel nanoparticles developed by (Kumar et al., 2023b). These methods include ultrafiltration membranes, carbon nanotubes, reverse osmosis, nanofiltration, and nanofiber filters. In order to destroy and adsorb different wastewater pollutants, the most recent method is nanotechnology, more especially green synthesis nanoparticles. XRD, SEM, and EDAX were used to characterize Green Synthesis nano Zero Valent Iron (GT-nZVI), which was extracted from soft black tea (Saminathan et al., 2022).

5.2 *Artificial Intelligence (AI) and Machine Learning*

It has been shown that machine learning and artificial intelligence techniques can optimize, model, and automate crucial water and wastewater treatment applications, natural system management and monitoring, and water-based agriculture techniques like hydroponics and aquaponics. Applications of artificial intelligence and machine learning (AI/ML) should enhance water-based applications while lowering capital expenses (Lowe et al., 2022). There are numerous AI-based techniques for treating wastewater.

It was discovered that Federated Learning (FL) and Artificial Neural Network (ANN) were the two AI techniques most successfully applied in wastewater treatment.

Using Genetic Algorithms (GA), the main parameters to be measured in wastewater were discussed in a similar survey. These parameters included pH levels, biological oxygen demand (BOD), chemical oxygen demand (COD), nitrogen, turbidity, and sulfur. When used in conjunction with other AI techniques, ANN can determine the trace of heavy metals and other effluents with an accuracy of 85–90%. A different study estimated BOD and COD at the Tabriz wastewater treatment plant (WWTP) using the adaptive neuro-fuzzy inference system (ANFIS), feed forward neural network (FFNN), and support vector regression (SVR). The autoregressive integrated moving average (ARIMA) was used by the author to differentiate between the predictive powers of linear and nonlinear models in order to forecast effluents.

5.3 *Remote Sensing and Satellite Imagery*

Thus, these sensors can offer vital data to assist with water management and track the development of hazards and their effects (Markogianni, 2023). Additionally, there is a long history of using remote sensing for agricultural monitoring and management to retrieve variations in vegetation state and, consequently, plant productivity and health.

The broad coverage (global level) of satellite data makes risk assessment possible when considering regional food production, storage, and trade, as well as water security. With their large spatial coverage and high temporal resolution (subdaily for geostationary and equatorial orbiting satellites), these satellite remote sensing products—some of which are still in their early stages and have significant limitations, challenges, and caveats for use in WRM—allow them to provide near-global information in near real time. Many studies also use images from other satellites in addition to Landsat images (Wulder et al., 2022).

5.4 *Blockchain Technology for Water Management*

The decentralized nature of urban water supply plants and water sources results in ineffective information sharing and distribution. Adequate data security is a problem with traditional centralized water information systems. Information can be twisted or faked, and network congestion makes it difficult to obtain it quickly (Bhushan et al., 2021). Blockchain technology's dependable network that unifies processes and data allows it to address these problems, especially increasing data reliability and transmission efficiency. First, the distributed storage network of the blockchain prevents participants from making biased decisions based on their own data and enables each link to query water data quickly, guaranteeing a scientific response in the event of an accident. Second, data is protected and tampering is difficult thanks to encryption technology.

Additionally, blockchain technology can be economical while fostering trust and safe information exchange. And lastly, the creative decentralized information management model raises the efficiency of managing water resources. The blockchain network serves as a means of safeguarding information and promoting confidence among neighbors (Suhail et al., 2021).

5.5 *Solar-Powered Desalination and Water Treatment*

Desalination of brackish and seawater is a commonly acknowledged substitute method for providing freshwater to numerous water-stressed regions. Large-scale fossil fuel-fueled desalination plants have been operational for many years in water-stressed nations, especially those in the oil-rich Middle East. Desalination plants can be directly or indirectly powered by solar energy (He et al., 2022).

Solar-powered desalination offers a sustainable way to meet water demands in areas with abundant solar radiation. The type and availability of energy needed for each desalination process determines whether or not it is compatible with solar technology. Desalination and solar energy are becoming more and more popular as a way to increase energy efficiency due to the rapid advancements in solar energy technologies, both photovoltaic and solar thermal.

6 Conclusion

Innovative agricultural water management systems provide promising solutions to the challenges of water scarcity, inefficiency, and sustainability in food production. However, successful implementation of these innovations necessitates a proactive strategy for mitigating associated risks. During this investigation, we discovered a variety of risks, including technological failures, financial constraints, environmental consequences, and social disparities. Stakeholders must give priority to a number of important solutions in order to get past these challenges and guarantee the successful adoption of agricultural water management innovations. Regular maintenance, calibration, and proper management of these systems are critical for avoiding technological failures and maximizing performance. Integrating renewable energy sources can help reduce the operational costs and environmental impact of innovations like desalination and smart irrigation systems. Furthermore, farmer education and training programs are critical for facilitating the adoption of new practices and technologies, allowing farmers to make more informed decisions about water management on their land. Equitable policy frameworks and transparent governance structures are required to address social disparities and ensure equitable access to water resources, especially in water banking and trading systems.

By implementing these solutions in a comprehensive and collaborative manner, we can fully realize the potential of agricultural water management innovations and achieve sustainable water use in agriculture. This requires ongoing commitment from governments, businesses, researchers, farmers, and civil society to prioritize water sustainability as a shared goal and collaborate to achieve it. Through collaborative efforts, we can create resilient agricultural water management systems that meet the needs of current and future generations while protecting our precious water resources and food security.

References

Abbas, M., et al. (2022). Infirmity to climate change and regional impacts. In W. N. Jatoi, M. Mubeen, A. Ahmad, M. A. Cheema, Z. Lin, & M. Z. Hashmi (Eds.), *Building climate resilience in agriculture*. Springer.

Ahmad, A., et al. (2009). Seasonal growth, radiation interception, its conversion efficiency and aerial dry biomass production of *Oryza sativa* L. under diverse agro-environments in Pakistan. *Pakistan Journal of Botany, 41*(3), 1241–1257.

Ahmed, M., et al. (2022). Nutrient dynamics and the role of modeling. In W. N. Jatoi, M. Mubeen, A. Ahmad, M. A. Cheema, Z. Lin, & M. Z. Hashmi (Eds.), *Building climate resilience in agriculture*. Springer.

Akhtar, N., Syakir Ishak, M. I., Bhawani, S. A., & Umar, K. (2021). Various natural and anthropogenic factors responsible for water quality degradation: A review. *Water, 13*(19), 2660.

Akram, R., et al. (2022). Research on climate change issues. In W. N. Jatoi, M. Mubeen, A. Ahmad, M. A. Cheema, Z. Lin, & M. Z. Hashmi (Eds.), *Building climate resilience in agriculture*. Springer.

Ali, S., et al. (2020). Assessment of arsenic exposure by drinking well water and associated carcinogenic risk in peri-urban areas of Vehari, Pakistan. *Environmental Geochemistry and Health, 42*, 121–133.

Ali, S., et al. (2021). Future climatic changes, extreme events, related uncertainties, and policy recommendations in the Hindu Kush sub-regions of Pakistan. *Theoretical and Applied Climatology, 143*, 193–209.

Alsuhaibani, A. M., et al. (2023). Green buildings model: Impact of rigid polyurethane foam on indoor environment and sustainable development in energy sector. *Heliyon, 9*, 14451.

Anh, N. T., Nhan, N. T., Schmalz, B., & Le Luu, T. (2023). Influences of key factors on river water quality in urban and rural areas: A review. *Case Studies in Chemical and Environmental Engineering, 8*, 100424.

Aryal, J. P., Rahut, D. B., & Marenya, P. (2021). Climate risks, adaptation and vulnerability in sub-Saharan Africa and South Asia. In *Climate vulnerability and resilience in the global south: Human adaptations for sustainable futures* (pp. 1–20). Springer.

Bakhsh, K., et al. (2022). Effects of climate change on the socioeconomic conditions of farmers: A case study. In W. N. Jatoi, M. Mubeen, A. Ahmad, M. A. Cheema, Z. Lin, & M. Z. Hashmi (Eds.), *Building climate resilience in agriculture*. Springer.

Bhushan, B., Sahoo, C., Sinha, P., & Khamparia, A. (2021). Unification of Blockchain and Internet of Things (BIoT): Requirements, working model, challenges and future directions. *Wireless Networks, 27*, 55–90.

Bukhari, M. A., et al. (2021). Silicon mitigates drought stress in wheat (Triticum aestivum L.) through improving photosynthetic pigments, biochemical and yield characters. *Silicon, 13*, 4757–4772.

Bwambale, E., Abagale, F. K., & Anornu, G. K. (2022). Smart irrigation monitoring and control strategies for improving water use efficiency in precision agriculture: A review. *Agricultural Water Management, 260*, 107324.

Cahn, M., & Hutmacher, R. (2024). Subsurface drip irrigation. In *Microirrigation for crop production* (pp. 257–301). Elsevier.

Chukwuemeka, I., Njoku, V., Arinze, C., Chizoruo, I., & Blessing, E. (2021). A review: Effects of air, water and land dumpsite on human health and analytical methods for determination of pollutants. *AMEC J, 4*, 80–106.

Damkjær, S. S. (2020). The characterisation of water scarcity: Developing a storage-based indicator framework for the great Ruaha River Catchment, Tanzania UCL ().

Dawar, M., et al. (2021). The effect of biochar and nitrogen inhibitor on ammonia and nitrous oxide emissions and wheat productivity. *Journal of Plant Growth Regulation, 40*, 2465–2475.

Deekshithulu, N., Tejaswini, V., Surekha, D., & Shanti, Y. P. (2020). Development of different irrigation systems calculator using VB6. *Indian Journal Of Ecology, 47*(4), 902–909.

Farid, H. M., et al. (2023). Groundwater quality risk assessment using hydro-chemical and geospatial analysis. *Environment, Development and Sustainability, 25*, 8343–8365.

Ghimire, C. P., Appels, W. M., Grundy, L., Ritchie, W., Bradley, S., & Snow, V. (2021). Towards predicting the initiation of overland flow from relatively flat agricultural fields using surface water coverage. *Journal of Hydrology, 596*, 126125.

Gupta, A., Singh, R. K., Kumar, M., Sawant, C. P., & Gaikwad, B. B. (2022). On-farm irrigation water management in India: Challenges and research gaps. *Irrigation and Drainage, 71*(1), 3–22.

Hammad, H. M., et al. (2010a). Optimizing water and nitrogen requirement in maize under semi arid conditions of Pakistan. *Pakistan Journal of Botany, 43*(6), 2919–2923.

Hammad, H. M., et al. (2010b). Organic farming in wheat crop under arid conditions of Pakistan. *Pakistan Journal of Agricultural Sciences, 48*(2), 97–102.

Hammad, H. M., et al. (2023). Impacts of COVID-19 pandemic on environment, society, and food security. *Environmental Science and Pollution Research, 30*, 99261–99272.

Haris, A., et al. (2023). Investigating on-site energy consumption patterns using vertical electrical sounding (VES) and GIS techniques. *Sustainability, 2023*(15), 16782. https://doi.org/10.3390/su152416782

Haseeb, M., et al. (2023). Quantifying irrigation water demand and supply gap using remote sensing and GIS in Multan, Pakistan. *Environmental Monitoring and Assessment, 195*, 990.

He, Q., Zheng, H., Ma, X., Wang, L., Kong, H., & Zhu, Z. (2022). Artificial intelligence application in a renewable energy-driven desalination system: A critical review. *Energy and AI, 7*, 100123.

Hrozencik, R. A., Wallander, S., & Aillery, M. (2021). *Irrigation organizations: Water storage and delivery infrastructure*. EB-32. U.S. Department of Agriculture, Economic Research Service.

Huang, X., Fang, H., Wu, M., & Cao, X. (2022). Assessment of the regional agricultural water-land nexus in China: A green-blue water perspective. *Science of the Total Environment, 804*, 150192.

Hussain, S., et al. (2020). Study of land cover/land use changes using RS and GIS: A case study of Multan district, Pakistan. *Environmental Monitoring & Assessment, 192*(2), 1–15.

Hussain, S., et al. (2023a). Assessment of land use/land cover changes and its effect on land surface temperature using remote sensing techniques in southern Punjab, Pakistan. *Environmental Science and Pollution Research, 30*, 99202–99218.

Hussain, S., et al. (2023b). Relation of land surface temperature with different vegetation indices using multi-temporal remote sensing data in Sahiwal region, Pakistan. *Geoscience Letters, 10*, 33.

Hussain, S., et al. (2023c). Investigation of irrigation water requirement and evapotranspiration for water resource management in Southern Punjab, Pakistan. *Sustainability, 15*(15), 1768.

Hussain, S., et al. (2024a). Assessing the impact of land use land cover changes on soil moisture and vegetation cover in southern Punjab, Pakistan using multi-temporal satellite data. *Geology, Ecology, and Landscapes*. https://doi.org/10.1080/24749508.2024.2338970.

Hussain, S., et al. (2024b). Agricultural land suitability analysis of southern Punjab, Pakistan using analytical hierarchy process (AHP) and multi-criteria decision analysis (MCDA) techniques. *Cogent Food & Agriculture, 10*(1), 2294540.

Hussain, S., et al. (2024c). Assessment of future prediction of urban growth and climate change in district Multan, Pakistan using CA-Markov method. *Urban Climate, 53*, 101766.

Iravani, S. (2023). Silica-based nanosystems against antibiotic-resistant bacteria and pathogenic viruses. *Critical Reviews in Microbiology, 49*(5), 598–610.

Jarwar, A. H., Wang, X., Long Wang, L. Z., Zhaoyang, Q., Mangi, N., Pengjia, B., Jinjin, W., Ma, Q., & Shuli, F. (2019). Performance and evaluation of drip irrigation system, and its future advantages. *Journal of Biology, Agriculture and Healthcare, 9*(9).

Jedla, M. R., Koneru, B., Franco, A., Rangappa, D., & Banerjee, P. (2022). Recent developments in nanomaterials based adsorbents for water purification techniques. *Biointerface Research in Applied Chemistry, 12*, 5821–5835.

Jovanovic, N., Pereira, L., Paredes, P., Pôças, I., Cantore, V., & Todorovic, M. (2020). A review of strategies, methods and technologies to reduce non-beneficial consumptive water use on farms considering the FAO56 methods. *Agricultural Water Management, 239*, 106267.

Kazmi, D. H., et al. (2022). Agro-meteorological aspect of climate change. In W. N. Jatoi, M. Mubeen, A. Ahmad, M. A. Cheema, Z. Lin, & M. Z. Hashmi (Eds.), *Building climate resilience in agriculture*. Springer.

Krishnan, S. R., Nallakaruppan, M., Chengoden, R., Koppu, S., Iyapparaja, M., Sadhasivam, J., & Sethuraman, S. (2022). Smart water resource management using artificial intelligence—A review. *Sustainability, 14*(20), 13384.

Kumar, P., & Sharma, P. K. (2020). Soil salinity and food security in India. *Frontiers in Sustainable Food Systems, 4*, 533781.

Kumar, M., Sahoo, P., Kushwaha, D., Gudi, S., Singh, G., Mahore, A., Nalawade, R., Patel, A., Pradhan, N., & Rahimi, M. (2023a). Tackling the constraints of cumin cultivation and management practices. *Annals of Agricultural & Crop Sciences, 8*(3), 1134.

Kumar, S., Yadav, S., Kataria, N., Chauhan, A. K., Joshi, S., Gupta, R., Kumar, P., Chong, J. W. R., Khoo, K. S., & Show, P. L. (2023b). Recent advancement in nanotechnology for the treatment of pharmaceutical wastewater: Sources, toxicity, and remediation technology. *Current Pollution Reports, 9*(2), 110–142.

Liu, L., Zheng, X., Wei, X., Kai, Z., & Xu, Y. (2021). Excessive application of chemical fertilizer and organophosphorus pesticides induced total phosphorus loss from planting causing surface water eutrophication. *Scientific Reports, 11*(1), 23015.

Lowe, M., Qin, R., & Mao, X. (2022). A review on machine learning, artificial intelligence, and smart technology in water treatment and monitoring. *Water, 14*(9), 1384.

Markogianni, V. V. (2023). *Water quality assessment in Greek lakes by using remote sensing and statistical modelling*.

Mehmood, A., et al. (2021). S exploring the potential of moringa leaf extract as bio stimulant for improving yield and quality of black cumin oil. *Scientific Reports, 11*, 24217.

Mehmood, M. Z., et al. (2022). Internet of things (IoT) and sensors technologies in smart agriculture: Applications, opportunities, and current trends. In W. N. Jatoi, M. Mubeen, A. Ahmad, M. A. Cheema, Z. Lin, & M. Z. Hashmi (Eds.), *Building climate resilience in agriculture*. Springer.

Moravej, M., Renouf, M. A., Lam, K. L., Kenway, S. J., & Urich, C. (2021). Site-scale urban water mass balance assessment (SUWMBA) to quantify water performance of urban design-technology-environment configurations. *Water Research, 188*, 116477.

Mubeen, M., et al. (2021). Effect of PGPR and drought on spring maize (Zia mays L.). *Pakistan Journal of Botany, 53*(2), 731–739.

Mubeen, M., et al. (2022). Climate change-induced irrigation water problems and resolution strategies: A case study. In W. N. Jatoi, M. Mubeen, A. Ahmad, M. A. Cheema, Z. Lin, & M. Z. Hashmi (Eds.), *Building climate resilience in agriculture*. Springer.

Muhammad, H., et al. (2023). A paradigm shift towards beneficial microbes enhancing the efficiency of organic and inorganic nitrogen sources for a sustainable environment. *Land, 12*, 680.

Mumtaz, F., Arshad, A., Mirchi, A., Tariq, A., Dilawar, A., Hussain, S., Shi, S., Noor, R., Noor, R., & Daccache, A. (2021). Impacts of reduced deposition of atmospheric nitrogen on coastal marine eco-system during substantial shift in human activities in the twenty-first century. *Geomatics, Natural Hazards and Risk, 12*(1), 2023–2047.

Muzammil, M., Zahid, A., & Breuer, L. (2020). Water resources management strategies for irrigated agriculture in the Indus Basin of Pakistan. *Water, 12*(5), 1429.

Nasim, W., & Bano, A. (2012). Impact of nitrogen and plant growth promoting Rhizobacteria on yield and yield components of sunflower in glasshouse environment. *Journal of Crop Science and Biotechnology, 15*(4), 319–324.

Nasim, W., et al. (2011). Nitrogen effects on growth and development of sunflower hybrids under agro-climatic conditions of Multan, Pakistan. *Journal of Botany, 43*(4), 2083–2092.

Nawaz, H., et al. (2022). Water management in era of climate change. In W. N. Jatoi, M. Mubeen, A. Ahmad, M. A. Cheema, Z. Lin, & M. Z. Hashmi (Eds.), *Building climate resilience in agriculture*. Springer.

Novoa, V., Ahumada-Rudolph, R., Rojas, O., Sáez, K., De La Barrera, F., & Arumí, J. L. (2019). Understanding agricultural water footprint variability to improve water management in Chile. *Science of the Total Environment, 670*, 188–199.

Qadir, M., Jones, E., & Drechsel, P. (2024). *Domestic wastewater generation, treatment, and agricultural reuse.*

Qureshi, A. S. (2020). Groundwater governance in Pakistan: From colossal development to neglected management. *Water, 12*(11), 3017.

Qureshi, A. U. (2022). *Life cycle evaluation of rehabilitation of residential structures, subjected to earthquake*. Concepts Books Publication.

Rasool, A., et al. (2020). Microbial diversity response in thallium polluted riverbank soils of the Lanmuchang. *Ecotoxicology and Environmental Safety, 187*, 109854.

Rasool, A., et al. (2021). Microbial diversity and community structure in alpine stream soil. *Geomicrobiology Journal, 38*(3), 210–219.

Rasul, F., et al. (2022). Role of modeling in assessing climate change. In W. N. Jatoi, M. Mubeen, A. Ahmad, M. A. Cheema, Z. Lin, & M. Z. Hashmi (Eds.), *Building climate resilience in agriculture*. Springer.

Rehman, M. H., et al. (2021). Influence of semi-arid environment on radiation use efficiency and other growth attributes of lentil crop. *Environmental Science and Pollution Research, 28*, 13697–13711.

Rodríguez-Berbel, N., Soria, R., Ortega, R., Lucas-Borja, M. E., & Miralles, I. (2022). Agricultural land degradation in Spain. In *Impact of agriculture on soil degradation II: A European perspective* (pp. 263–297). Springer.

Rolandi, S., Brunori, G., Bacco, M., & Scotti, I. (2021). The digitalization of agriculture and rural areas: Towards a taxonomy of the impacts. *Sustainability, 13*(9), 5172.

Saminathan, A., Zajac, M., Anees, P., & Krishnan, Y. (2022). Organelle-level precision with next-generation targeting technologies. *Nature Reviews Materials, 7*(5), 355–371.

Seleiman, M. F., Almutairi, K. F., Alotaibi, M., Shami, A., Alhammad, B. A., & Battaglia, M. L. (2020). Nano-fertilization as an emerging fertilization technique: Why can modern agriculture benefit from its use? *Plants, 10*(1), 2.

Shah, M. Z., et al. (2021). Synthesis of silver nanoparticles using Plantago lanceolata extract and assessing their antibacterial and antioxidant activities. *Scientific Reports, 11*, 20754.

Siddha, S., & Sahu, P. (2022). Traditional and existing methods of urban water supply and their loopholes. In *Current directions in water scarcity research* (Vol. 6, pp. 245–271). Elsevier.

Srivastav, A. L., Dhyani, R., Ranjan, M., Madhav, S., & Sillanpää, M. (2021). Climate-resilient strategies for sustainable management of water resources and agriculture. *Environmental Science and Pollution Research, 28*(31), 41576–41595.

Stavi, I., Paschalidou, A., Kyriazopoulos, A. P., Halbac-Cotoara-Zamfir, R., Siad, S. M., Suska-Malawska, M., Savic, D., Roque de Pinho, J., Thalheimer, L., & Williams, D. S. (2021). Multidimensional food security nexus in drylands under the slow onset effects of climate change. *Land, 10*(12), 1350.

Suhail, S., Hussain, R., Jurdak, R., & Hong, C. S. (2021). Trustworthy digital twins in the industrial internet of things with blockchain. *IEEE Internet Computing, 26*(3), 58–67.

Šulyová, D., Vodák, J., & Kubina, M. (2021). Effective management of scarce water resources: From antiquity to today and into the future. *Water, 13*(19), 2734.

Tork, D. G. (2021). *A perennial flax (Linum spp.) breeding program using ideotype models to select for oilseed, garden, and cut flower cultivars*. University of Minnesota.

Usman, M., et al. (2009). Development and application of crop water stress index for scheduling irrigation in cotton (*Gossypium hersutumn* L.) under semiarid environment. *Journal of Food Agriculture and Environment, 07*(3&4), 386–391.

Usman, M., et al. (2010). Lower and upper baselines for crop water stress index and yield of *Gossypium hirsutum* L. under variable irrigation regimes in irrigated semiarid environment. *Pakistan Journal of Botany, 42*(4), 2541–2550.

Valentín, F., Nortes, P., Domínguez, A., Sánchez, J., Intrigliolo, D., Alarcón, J., & López-Urrea, R. (2020). Comparing evapotranspiration and yield performance of maize under sprinkler, superficial and subsurface drip irrigation in a semi-arid environment. *Irrigation Science, 38*, 105–115.

Wajid, A., et al. (2007). Maize response to yield under semi arid conditions of Pakistan. *Pakistan Journal of Agricultural Sciences, 44*(2), 217–220.

Wajid, A., et al. (2010). Quantification of growth, yield and radiation use efficiency of promising cotton cultivars at varying nitrogen levels. *Pakistan Journal of Botany, 42*(3), 1703–1711.

Walia, S. S., Kaur, K., & Kaur, T. (2024). Soil and water conservation techniques in rainfed areas. In *Rainfed agriculture and watershed management* (pp. 115–124). Springer.

Weinkove, R., McQuilten, Z. K., Adler, J., Agar, M. R., Blyth, E., Cheng, A. C., Conyers, R., Haeusler, G. M., Hardie, C., & Jackson, C. (2020). Managing haematology and oncology patients during the COVID-19 pandemic: Interim consensus guidance. *Medical Journal of Australia, 212*(10), 481–489.

Wulder, M. A., Roy, D. P., Radeloff, V. C., Loveland, T. R., Anderson, M. C., Johnson, D. M., Healey, S., Zhu, Z., Scambos, T. A., & Pahlevan, N. (2022). Fifty years of Landsat science and impacts. *Remote Sensing of Environment, 280*, 113195.

Yadav, M., Singh, N. K., Singh, N. K., Jagwani, T., & Yadav, S. (2022). Pesticides and fertilisers contamination of groundwater: Health effects, treatment approaches, and legal aspects. In *Contaminants of emerging concerns and reigning removal technologies* (pp. 349–376). CRC Press.

Yang, X., et al. (2023). 2023 quantitative evaluation of soil water and wind erosion rates in Pakistan. *Remote Sensing, 15*, 2404.

Zhu, T., & Xu, X. (2021). *Watershed erosion processes*. Springer.

Zhu, Z., Zhu, D., & Ge, M. (2024). The generation processes, spatiotemporal distribution heterogeneity, and interpretation of the spatial variation mechanism of the net sprinkler water under maize canopy. *Irrigation Science, 42*(2), 305–326.

Environmental Aspects of Water Management in Agriculture

Muhammad Imran, Behzad Murtaza, Muhammad Amjad, Aneeza Sardar, and Samina Kousar

Abstract Agriculture is a major user of water and accounts for about 70% of global freshwater withdrawals. This substantial dependence of agriculture on water has led to various environmental problems including water scarcity, soil erosion, and water pollution. The present chapter reviews the environmental aspects of water management in agriculture by investigating the various strategies to use water. Sustainable water management in agriculture is crucial to food security and environmental protection. Numerous approaches for sustainable management of water in agriculture including water conservation, water efficiency, and water reuse have been focused. It is concluded with the future challenges and opportunities for sustainable water management in agriculture. These challenges include the increasing water demand for food production, the effects of climate change on water availability, and the need to protect water quality. Recommendations have been suggested for achieving sustainable water management in agriculture including integrated water resources management, water-efficient technologies, and the adoption of water conservation practices.

Keywords Water · Agriculture · Environment · Challenges · Management · Sustainable · Conservation

---

M. Imran (✉)
Department of Environmental Sciences, COMSATS University Islamabad, Vehari, Pakistan

Department of Sanitary Engineering, Faculty of Civil and Environmental Engineering, Gdansk University of Technology, Gdańsk, Poland
e-mail: imranrb@cuivehari.edu.pk

B. Murtaza · M. Amjad · A. Sardar · S. Kousar
Department of Environmental Sciences, COMSATS University Islamabad, Vehari, Pakistan

M. Mubeen et al. (eds.), *Innovations in Agricultural Water Management*,
https://doi.org/10.1007/978-3-031-91883-4_2

1 Introduction

1.1 Background

Agriculture is the largest consumer of freshwater worldwide; it accounts for about 70% of all withdrawals (Naresh et al., 2021; Wu et al., 2022). This is because agriculture requires a large quantity of water to grow crops and to raise livestock (Steensland & Zeigler, 2021; Wisser et al., 2024). Water use in agriculture varies widely from region to region depending on climate, crops grown, and irrigation method. In arid and semi-arid regions, irrigation is essential for growing crops, while irrigation may be used to supplement rainfall during dry periods or to grow water-intensive crops in humid regions (Muratoglu et al., 2022; Wisser et al., 2024). Water use in agriculture has increased significantly in recent decades due to population growth at alarming rate and increasing food demand (Hemathilake & Gunathilake, 2022), increased production of water-intensive crops, for example, rice, corn, and soybeans (Kulkarni et al., 2023), and expansion of agricultural land into arid and semi-arid regions (Muratoglu et al., 2022; Hammad et al., 2018; Tariq et al., 2023). This increase in water use has put pressure on freshwater resources, particularly in water-stressed regions where competition for water is high among different sectors (du Plessis, 2023). Water consumption intensification in agriculture can also have adverse consequences, for example, loss of biodiversity, soil flooding and salinization (Ramirez-Contreras et al., 2022), inequality between users, increased vulnerability to climate change and deterioration of ecosystem and water resources (Kulkarni et al., 2023; Niazi et al., 2024; Pinho et al., 2022).

Climate change is also threatening the availability of agricultural systems and water resources. Climate change is expected to bring about significant changes in soil moisture, evapotranspiration patterns, surface runoff, and precipitation cycles (Liu et al., 2021). These changes coupled with an increased occurrence of extreme weather events, lead to a substantial threat to agricultural production and water availability.

1.2 Environmental Impacts of Agricultural Water Use

Modern agriculture faces various unprecedented challenges to provide nutritious and wholesome food for a rapidly increasing global population considering the impacts of urbanization and rising incomes (Ramirez-Contreras et al., 2022). As a critical source of employment, agriculture provides employment to billions of people worldwide. Besides, agriculture plays a pivotal role in the maintenance of biodiversity and use of natural resources on a sustainable basis specifically under a changing climate. To keep pace with the growing demand for food across the globe, agricultural production and trade must continue to expand (Muratoglu et al., 2022). Global agricultural trade has witnessed significant growth in recent decades at a 6%

growth rate between 2000 and 2016 (Hassan et al., 2020; Wang et al., 2019; Findell et al., 2019; Shah et al., 2019; Shahzad et al., 2018; Rahman et al., 2019; Ahmad et al., 2019; Sabagh et al., 2019; Danish et al., 2019; Iqbal et al., 2019). In 2018, the EU, the USA, Brazil, China, Canada, Indonesia, Thailand, India, Australia, and Mexico collectively accounted for 72% of total global exports. China (9%), Brazil (6%), and Mexico (6%) were prominent in agricultural product exports among the top exporters in 2018 (Hassan et al., 2020). Expansion of international trade has induced several environmental issues on a global scale (Balogh & Jámbor, 2020). International trade can intensify environmental externalities manifesting in various forms such as production growth-driven externalities, transboundary pollution, resource trade-based externalities, transportation-induced externalities, and production relocation to evade ecological regulations (Fig. 1).

Water usage in agriculture is a major concern due to its multifaceted environmental impacts (Zhang et al., 2022). On the positive side, agriculture is very important in maintaining the water cycle and provides valuable services, for example, water filtration and carbon sequestration. However, agricultural water use can also lead to several detrimental environmental impacts (Yang et al., 2023; Zhang et al., 2022). Extensive land conversion for agricultural purposes has led to the degradation of terrestrial ecosystems. Agricultural water use can contribute to a range of problems (Hora, 2022; Yang et al., 2023).

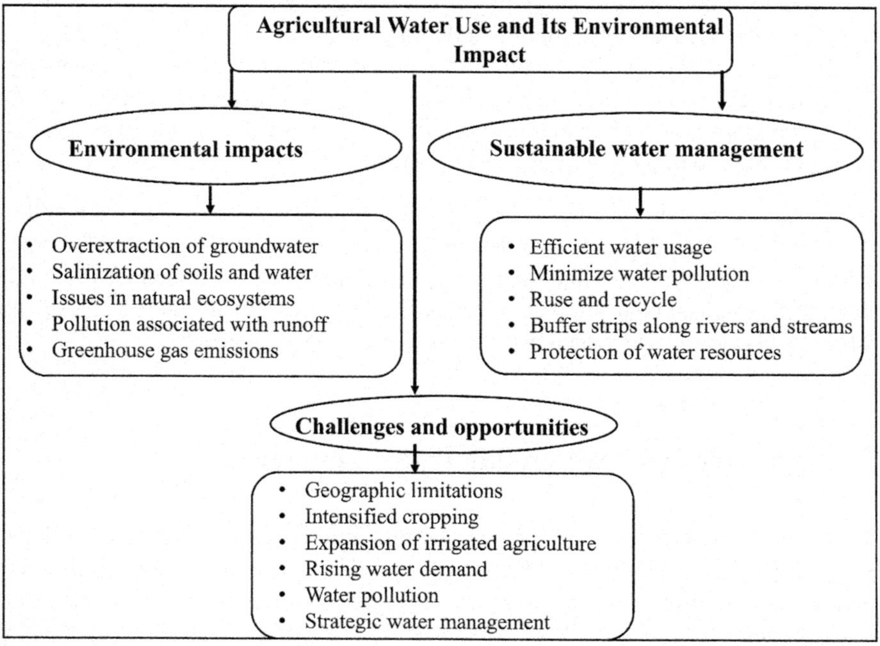

Fig. 1 Agricultural water usage, its environmental impacts, challenges, and opportunities

(a) Agriculture's reliance on groundwater for irrigation is not without consequences. Overextraction of groundwater can deplete aquifers, reducing water availability for drinking water and industry. This unsustainable practice can have far-reaching implications, threatening water security for both human consumption and industrial processes.
(b) Irrigation, while essential for crop production, can lead to salinization of soils and water, rendering them unsuitable for agriculture. This phenomenon occurs when irrigation water evaporates, leaving behind concentrated salts in the soil. Over time, these salts accumulate to toxic levels, rendering the land unproductive. Salinization is particularly problematic in coastal regions, resulting in saltwater intrusion.
(c) Agricultural runoff poses a significant threat to water quality as a form of nonpoint pollution source. As rainwater flows across cultivated land, it picks up pollutants such as fertilizers, pesticides, and manure, carrying them into nearby waterways and groundwater sources. This influx of pollutants can have devastating consequences for aquatic ecosystems. The excessive growth of algae that reduces oxygen levels in water, and the introduction of harmful pathogens that can contaminate drinking water supplies.
(d) The relentless expansion of agriculture often comes at the expense of natural ecosystems, particularly wetlands and forests. These ecosystems are crucial for maintaining biodiversity, providing habitats for a vast array of plant and animal species. They also perform valuable services to the ecosystem, for example, water purification and flood control. Agricultural activities can fragment habitats, disrupting wildlife migration patterns and reducing overall biodiversity.
(e) Agriculture contributes significantly to greenhouse gas emissions. Methane exhibits a warming effect about 25 times higher than that of carbon dioxide and is primarily produced from livestock and manure management. Nitrous oxide has a warming impact about 300 times more than that of CO_2 and is mainly associated with the use of nitrogen fertilizers.

Addressing the environmental impacts of agricultural water use requires a collaborative effort from farmers, governments, and consumers (Haque et al., 2023). Farmers can adopt new technologies and practices, while governments can support them through financial incentives and extension programs. Consumers can also play a role by choosing food grown using sustainable water practices.

2 Approaches to Sustainable Water Management

To tackle the challenges of water scarcity, numerous studies have considered sustainability science, a discipline that provides a roadmap toward a sustainable society and seeks to comprehend the complex interactions between natural, human, and social systems (Imran et al., 2023; Mishra et al., 2021; Mishra, 2023). In the context of agriculture, sustainability focuses on developing environment-friendly practices that do not harm the ecosystem (Zhang et al., 2022). Sustainable water management in agriculture results in using water efficiently, minimizing water losses, and safeguarding water quality to ensure the fulfillment of food production and environmental needs (Kulkarni et al., 2023). The goals of sustainable water resources management encompass the continuity of the agricultural system from a physical and biological perspective, economic efficiency in resource utilization, and social participation in decision-making processes (Ghodsvali, 2022; Nungula et al., 2023).

Following approaches may be adopted for sustainable water management in agriculture (Fig. 1):

(a) Improve water use efficiency
(b) Reduce water pollution
(c) Reuse and recycle water
(d) Protect groundwater resources
(e) Develop buffer strips by plantation along rivers and streams

2.1 *Water Efficiency*

Water efficiency is the cornerstone of water management in agriculture, encompassing a range of practices and techniques aimed at maximizing crop productivity while minimizing water consumption. By adopting water-efficient strategies, farmers can ensure the long-term sustainability of their operations while contributing to the preservation of precious water resources.

2.1.1 Adopting Efficient Irrigation Technologies

Efficient irrigation technologies, for example, drip irrigation, sprinkler irrigation, and subsurface irrigation, deliver water directly to the roots of plants, significantly reducing evaporation and runoff (Chathuranika et al., 2022; Ghodsvali, 2022). These targeted irrigation methods optimize water use and enhance crop yields, making them crucial tools for sustainable agriculture.

2.1.2 Enhancing Irrigation Management Practices

Precision irrigation scheduling is a key factor in water conservation that considers the practice of irrigating crops only when they require water. Soil moisture sensors can provide real-time data and enable the farmers to make informed irrigation decisions and avoid overwatering (Greeshma et al., 2023; Laha et al., 2023). Additionally, regular maintenance and repair of irrigation systems can minimize water losses due to leaks and ensure optimal water distribution.

2.1.3 Cultivating Water-Efficient Crops

Crop selection plays a vital role in water efficiency. Farmers can optimize water usage by choosing crops that are well-adapted to their local climate and soil conditions. Drought-tolerant varieties, specifically bred to withstand dry periods, offer a promising solution for water-scarce regions (Yahaya & Shimelis, 2022). By incorporating water-efficient crop choices into their agricultural practices, farmers can significantly reduce their water footprint.

2.1.4 Using Recycled Water and Wastewater for Irrigation

Recycled water and wastewater, treated to meet specific quality standards, can serve as alternative sources for irrigation, reducing the demand for freshwater resources (Ghodsvali, 2022; Rasool et al., 2016; Alghabari et al., 2016; Fahad et al., 2016a, 2016b, 2016c, 2016d; Shahid et al., 2014; Masood et al., 2013; Nasim et al., 2012). This approach not only promotes water conservation but also contributes to environmental sustainability by diverting wastewater from landfills and waterways.

2.1.5 Restoring and Protecting Wetlands and Natural Ecosystems

Wetlands and other natural ecosystems play a critical role in water management acting as natural filters, purifying water, and replenishing groundwater reserves. Restoring and protecting these ecosystems is essential for maintaining healthy watersheds and ensuring long-term water availability (Mishra, 2023; Sharma & Singh, 2021).

By embracing these water-efficient practices and technologies, farmers can transform their operations into models of sustainable agriculture, ensuring food security for future generations while safeguarding the planet's precious water resources (Chathuranika et al., 2022; Laha et al., 2023; Mishra, 2023).

2.2 *Water Conservation*

Conserving water is essential for achieving sustainable water management in agriculture. Several approaches can be implemented to achieve water conservation in agriculture.

2.2.1 Improving Irrigation Efficiency

Various measures may be adopted to enhance irrigation efficiency (Bwambale et al., 2022):

(a) Adopting more efficient irrigation technologies: Drip irrigation and sprinkler irrigation are highly efficient methods in providing water directly to plant roots by decreasing water consumption by 50% and 25%, respectively, compared to traditional flood irrigation methods.
(b) Improving irrigation management practices: Implementing precise irrigation scheduling and using soil moisture sensors to determine irrigation timing can minimize water waste.
(c) Growing water-efficient crops: Choosing crops adapted to local climate and soil conditions, along with cultivating drought-tolerant varieties, can significantly reduce water usage.

2.2.2 Reducing Evaporation

Evaporation is a major source of water loss, which can be minimized (Liao et al., 2021; Thevs & Aliev, 2023; Thevs et al., 2021):

- Using mulches: Organic mulches like straw or wood chips help reduce evaporation and suppress weeds.
- Planting windbreaks: Windbreaks, such as rows of trees or tall shrubs, can reduce wind erosion and evaporation (Image of windbreaks in agriculture).
- Shading irrigation canals: Covering irrigation canals with shade cloth or vegetation can minimize evaporation losses.

2.2.3 Collecting and Storing Rainwater

Rainwater harvesting is an eco-friendly method that involves capturing rainwater from rooftops, greenhouses, and other surfaces, then storing it in tanks, cisterns, or similar containers for use in irrigation or watering livestock.

2.2.4 Reusing and Recycling Water

Water reuse is the process of using wastewater or other nontraditional water sources for irrigation and other purposes. It is a promising strategy to sustainable water management in agriculture, as it can help reduce the demand for freshwater and protect water resources (Quon & Jiang, 2023). This alternative practice is adopted in different countries and regions to overcome the water shortage for growing urban populations (Salehi, 2022). It helps in safeguarding the water resources (surface and underground) caused by climate variability and climate change (Quon & Jiang, 2023; Salehi, 2022). Water resource availability is further threatened by wastewater-sourced pollution and the associated aquifer contamination that arises when inadequately treated wastewater reaches surface water channels (Imran et al., 2023).

The major benefit associated with the use of wastewater is the reduction of pressure on the sources of freshwater as this wastewater is an alternative source of irrigation (Quon & Jiang, 2023). On the other hand, agricultural production increases in areas with water shortage with the reuse of wastewater (Salehi, 2022). Extraction from the groundwater resources avoided by wastewater reuse is one of the other advantages. The cost of groundwater pumping can account for up to 65% of irrigation expenses, making it an energy-intensive practice (Siyal et al., 2021). However, reuse of wastewater for irrigation offers several economic and environmental benefits. The nutrients naturally existing in wastewater can reduce fertilizer costs and help create an environment-friendly nutrient cycle (Ye et al., 2020). Wastewater can serve as a valuable source of both macronutrients (N, P, and K) and micronutrients (Ca, Mg, B, Mg, Fe, Mn, or Zn) (Saliu & Oladoja, 2021).

3 Challenges of Water Reuse in Agriculture

There are a few challenges that need to be addressed to expand the reuse of water in agriculture (Shoushtarian et al., 2020).

- Public acceptance: There is some public concern about the safety of using recycled water for irrigation, particularly for food crops. It is important to educate the public about the safety of recycled water and the benefits of water reuse.
- Cost: The cost of treating wastewater and other nontraditional water sources to a high level of quality for irrigation can be high. This can make water reuse less cost-effective than using traditional freshwater sources in some cases.
- Regulation: There is a need for clear and consistent regulations to govern the use of recycled water for irrigation. This will help ensure the safety of recycled water and protect the environment.

4 Challenges and Opportunities

Water is a critical resource for agriculture, which accounts for 65–75% of the total freshwater consumption. However, the increasing shortage of water is becoming a significant issue in numerous areas due to uneven distribution of water resources, population growth, and unsustainable water management practices. Several factors contribute to water scarcity in agriculture (Fig. 1) (Niazi et al., 2024; Wisser et al., 2024):

1. Geographic limitations: In some areas, rainfall is scarce, groundwater is inaccessible, and water treatment options are economically constrained.

2. Intensified cropping: Increased cropping intensities on existing farmland consume more water per unit area, potentially degrading land, and water resources.
3. Expansion of irrigated agriculture: Bringing new lands under cultivation requires additional water, which can deteriorate water quality if management practices are inadequate.
4. Rising water demand: Industrial and domestic water consumption is increasing due to population growth and rising living standards.
5. Water pollution: Contamination of surface and groundwater resources further limits the availability of clean water for agriculture.
6. Strategic water management is essential to promoting sustainable farming methods and address the food needs of an expanding global population.

5 Effects of Climate Change on Water Resources

Human activities have induced a significant rise in greenhouse gases, including carbon dioxide, methane, and nitrous oxide (Jones et al., 2023). Fossil fuel combustion and land-use changes are the primary drivers of carbon dioxide emissions, while agricultural activities are the main contributors to increased methane and nitrous oxide levels (Sonwani & Saxena, 2022). These alterations in greenhouse gas and aerosol concentrations, along with changes in solar radiation and inland surface properties, disrupt the Earth's climate and energy balance. The combined effects of these changes are expected to induce climate change and global warming.

Global warming is now widely accepted as a reality, although there is considerable uncertainty about the magnitude and rate of temperature rise (Huggel et al., 2022; Shahzad et al., 2022; Din et al., 2022; Balal et al., 2022; Abbas et al., 2020; Rehman et al., 2020; Hussain et al., 2020; Sabagh et al., 2020a, 2020b, 2020c; Ahmed et al., 2020). Climate fluctuations driven by greenhouse gas emissions have been the focus of extensive research in recent years (Jones et al., 2023; Sonwani & Saxena, 2022). This research has used various sophisticated climate models to predict future climate changes under different emission and concentration scenarios for greenhouse gases. The encountered emission of greenhouse gases at or above current rates will inevitably lead to further warming and induce a multitude of transformations within the Earth's climate system.

The diverse emission scenarios reveal a global average surface air temperature increase of 1.8–4 °C by the end of the twenty-first century as compared with the baseline period of 1980–1999 (IPCC, 2007). Current climate models anticipate that precipitation patterns will shift under a future warmer climate, with increased precipitation in tropical regions and decreased precipitation in the subtropics. The average amounts of water vapor, evaporation, and precipitation across the globe are expected to increase (Findell et al., 2019).

6 Future Recommendations

6.1 Determining Crop Water Requirements

The foundation of sustainable agricultural practices lies in accurately determining irrigation water requirements for crops under specific climatic conditions. This critical information can be acquired through various techniques, including (Gebremedhin et al., 2023):

1. Soil monitoring: This method involves measuring soil moisture content using tensiometers, gypsum blocks, or other sensors to assess crop water needs.
2. Lysimeters: These are buried tanks that allow for the direct measurement of water movement through the soil profile, providing precise estimates of evapotranspiration rates.
3. Eddy covariance: This technique uses sophisticated instruments to measure water vapor and heat fluxes, enabling the calculation of evapotranspiration rates over large areas.
4. Bowen ratio: This method relies on the principle of energy balance to estimate evapotranspiration rates by measuring the ratio of sensible heat to latent heat.
5. Surface renewal methods: These techniques involve measuring the rate of evaporation from a water surface, which can be used to infer evapotranspiration rates from adjacent crops.

While these monitoring methods may necessitate sophisticated sensors and expert assistance, the application of soil water balance models offers a cost-effective alternative for on-farm and regional water management systems. These models, when coupled with appropriate climate data and crop characteristics, can effectively estimate crop water requirements and the depth of water storage needed to meet agricultural demands.

6.2 Enhancing Irrigation Efficiency

Once crop water requirements are determined, optimizing irrigation practices becomes a crucial approach to conserving water in agricultural settings. The term "efficiency" typically refers to the effectiveness of a system's performance. In the context of agriculture, water use efficiency (WUE) specifically relates crop growth with the amount of water used. However, it is essential to differentiate between plant physiology-based water use efficiency and the agronomic practices to apply irrigation water efficiently to maximize crop yield. Water productivity is expressed as the ratio of crop output per unit of water (rainfall and/or irrigation) consumed to produce a crop (Fig. 2) (Liu et al., 2021).

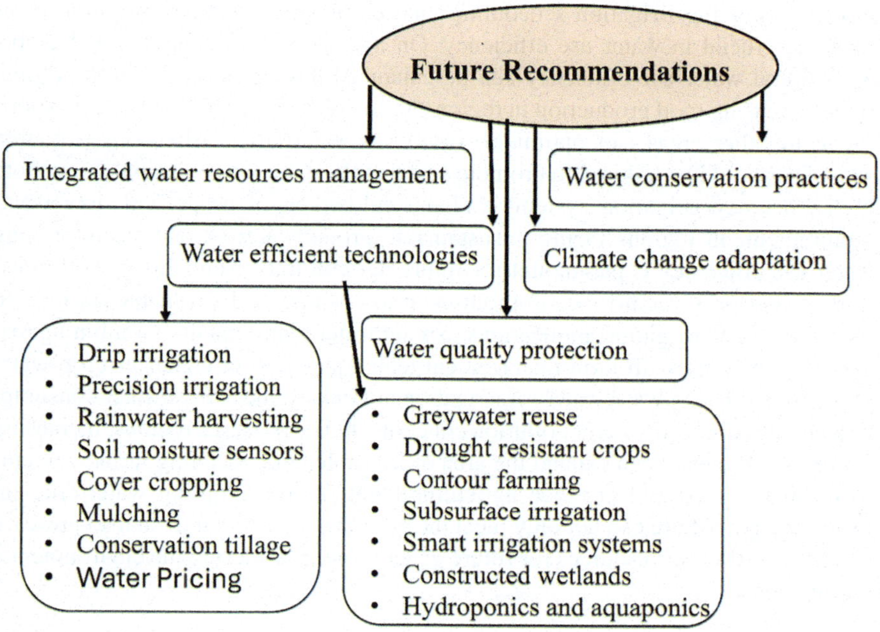

Fig. 2 Possible future recommendations for sustainable water management in agriculture

6.3 *Strategies for Improving Water Productivity*

Increasing water productivity is a potential solution to meet the rising food demands associated with population growth. Numerous strategies are currently being implemented to improve water productivity, including (Lopez et al., 2022):

- Proper irrigation scheduling: It aims to determine the optimal timing and quantity of water to meet crop water requirements.
- Crop management: Selecting appropriate crop varieties and implementing crop management practices that align with soil and climate conditions can significantly improve water productivity.
- Shifting planting dates: Adjusting planting dates due to climate change can be useful. Shortening the growing season and planting earlier have been shown to enhance crop yields.

7 Conclusion

Agriculture uses more water than any other sector. It can significantly reduce its water footprint by implementing various strategies to improve crop water productivity. On-farm water management practices, for example, supplementing rainfed crops, optimizing irrigation scheduling, and employing effective irrigation practices are crucial in water use efficiency. On a broader scale, national and global agricultural water use efficiency can be enhanced through strategic crop production. Increasing food production in regions with high water productivity and exporting it to other areas can optimize overall water resource utilization. Irrigation subsidies should be tailored to promote sustainable water usage, incentivizing both high-efficiency irrigation systems and the cultivation of crops with low water requirement. In regions facing unsustainable groundwater extraction, improving water use efficiency is paramount for future agricultural sustainability. Soil moisture sensors deliver real-time information on soil wetness, which enables farmers to precisely match irrigation applications to plant water requirements. Combining irrigation scheduling with farm management techniques further enhances crop water productivity. Using wastewater for irrigation not only reduces freshwater consumption but also alleviates water pollution concerns. In water-scarce regions, increasing water use efficiency can expand the area of irrigable land, enabling farmers to cultivate more crops and enhance agricultural output. By adopting water-efficient practices, agriculture can not only meet the growing food demand but also preserve precious freshwater resources for future generations considering the environmental implications.

References

Abbas, G., et al. (2020). Applications of crop modeling in cotton production. In S. Ahmad, & M. Hasanuzzaman (Eds.), *Cotton production and uses*.

Ahmad, A., Shahid, M., et al. (2019). Residues of endosulfan in cotton growing area of Vehari, Pakistan: An assessment of knowledge and awareness of pesticide use and health risks. *Environmental Science and Pollution Research, 26*, 20079–20091.

Ahmed, M., et al. (2020). Abiotic stress tolerance in wheat and the role of silicon: An experimental evidence. In M. Hasanuzzaman (Ed.), *Agronomic crops*. Springer.

Alghabari, F., et al. (2016). Gibberellin-sensitive Rht alleles confer tolerance to heat and drought stresses in wheat at booting stage. *Journal of Cereal Science, 70*, 72–78.

Balal, R. M., et al. (2022). Morphological, physiological, and biochemical modulations in crops under salt stress. In W. N. Jatoi, M. Mubeen, A. Ahmad, M. A. Cheema, Z. Lin, & M. Z. Hashmi (Eds.), *Building climate resilience in agriculture*. Springer.

Balogh, J. M., & Jámbor, A. (2020). The environmental impacts of agricultural trade: A systematic literature review. *Sustainability, 12*, 1152.

Bwambale, E., Abagale, F. K., & Anornu, G. K. (2022). Smart irrigation monitoring and control strategies for improving water use efficiency in precision agriculture: A review. *Agricultural Water Management, 260*, 107324.

Chathuranika, I., Khaniya, B., Neupane, K., Rustamjonovich, K. M., & Rathnayake, U. (2022). Implementation of water-saving agro-technologies and irrigation methods in agriculture of Uzbekistan on a large scale as an urgent issue. *Sustainable Water Resources Management, 8*, 155.

Danish, S., et al. (2019). Allevation of Cr toxicity in maize by Fe fortification and Cr tolerant ACC deaminase producing PGPR. *Ecotoxicology & Environmental Safety, 185*, 109706.

Din, M. S. U., et al. (2022). World nations priorities on climate change and food security. In W. N. Jatoi, M. Mubeen, A. Ahmad, M. A. Cheema, Z. Lin, & M. Z. Hashmi (Eds.), *Building climate resilience in agriculture*. Springer.

du Plessis, A. (2023). *Water resources from a global perspective, South Africa's water predicament: Freshwater's unceasing decline* (pp. 1–25). Springer.

Fahad, S., et al. (2016a). Exogenously applied plant growth regulators affect heat-stressed Rice pollens. *Journal of Agronomy and Crop Science, 202*, 139–150.

Fahad, S., et al. (2016b). A combined application of biochar and phosphorus alleviates heat-induced adversities on physiological, agronomical and quality attributes of rice. *Plant Physiology and Biochemistry, 103*, 191–198.

Fahad, S., et al. (2016c). Responses of rapid Viscoanalyzer profile and other rice grain qualities to exogenously applied plant growth regulators under high day and high night temperatures. *PLoS One, 11*(7), e0159590.

Fahad, S., et al. (2016d). Exogenously applied plant growth regulators enhance the morpho-physiological growth and yield of rice under high temperature. *Frontier in Plant Sciences, 07*, 1250.

Findell, K. L., Keys, P. W., Van Der Ent, R. J., Lintner, B. R., Berg, A., & Krasting, J. P. (2019). Rising temperatures increase importance of oceanic evaporation as a source for continental precipitation. *Journal of Climate, 32*, 7713–7726.

Gebremedhin, T., Haile, G. G., Gebremicael, T., Libsekal, H., & Reda, K. W. (2023). Balancing crop water requirements through supplemental irrigation under rainfed agriculture in a semi-arid environment. *Heliyon, 9*, e18727.

Ghodsvali, M. (2022). *A transdisciplinary decision-making approach to food-water-energy nexus: A guide towards sustainable development.*

Greeshma, M., Yadav, A., Aryaan, A. S., Deshpande, P. S., & Konguvel, E. (2023). Revolutionizing farming with IoT: Smart irrigation system for sustainable agriculture. In *2023 4th international conference on electronics and sustainable communication systems (ICESC)* (pp. 420–425). IEEE.

Hammad, H. M., Abbas, F., Saeed, S., Fahad, S., Cerdà, A., Farhad, W., Bernardo, C. C., Nasim, W., Mubeen, M., & Bakhat, H. F. (2018). Offsetting land degradation through nitrogen and water management during maize cultivation under arid conditions. *Land Degradation & Development, 29*(5), 1366–1375.

Haque, F., Fan, C., & Lee, Y.-Y. (2023). From waste to value: Addressing the relevance of waste recovery to agricultural sector in line with circular economy. *Journal of Cleaner Production, 20*, 137873.

Hassan, M. S., Bukhari, S., & Arshed, N. (2020). Competitiveness, governance and globalization: What matters for poverty alleviation? *Environment, Development and Sustainability, 22*, 3491–3518.

Hemathilake, D., & Gunathilake, D. (2022). *Agricultural productivity and food supply to meet increased demands, future foods* (pp. 539–553). Elsevier.

Hora, T. (2022). *Addressing groundwater over-extraction in India: Assessments, monitoring methods and interventions.*

Huggel, C., Bouwer, L. M., Juhola, S., Mechler, R., Muccione, V., Orlove, B., & Wallimann-Helmer, I. (2022). The existential risk space of climate change. *Climatic Change, 174*(1), 8.

Hussain, S., et al. (2020). Irrigation scheduling for cotton cultivation. In S. Ahmad, & M. Hasanuzzaman (Eds.), *Cotton production and uses.*

Imran, M., Ali, L., Ali, L., Wakeel, M., Siddique, M. H., Khan, Z. U. H., Murtaza, B., Iqbal, J., Al-Kahtani, A. A., & Shahid, M. (2023). Remediation potential of biochar/copper oxide nanoparticles composite for lead-and cadmium-contaminated wastewater. *Environmental Earth Sciences, 82*, 574.

Intergovernmental Panel on Climate Change (IPCC). (2007). Climate change 2007: The physical science basis. In S. Solomon, D. Qin, M. Manning, Z. Chen, M. Marquis, K. B. Averyt, M. Tignor, & H. L. Miller (Eds.), *Contribution of working group I to the fourth assessment report of the intergovernmental panel on climate change.* Cambridge University Press.

Iqbal, J., et al. (2019). Purple nutsedge (Cyperus rotundus) control through interference by summer crops. *International Journal of Agriculture Biology, 21*, 1083–1088.

Jones, M. W., Peters, G. P., Gasser, T., Andrew, R. M., Schwingshackl, C., Gütschow, J., Houghton, R. A., Friedlingstein, P., Pongratz, J., & Le Quéré, C. (2023). National contributions to climate change due to historical emissions of carbon dioxide, methane, and nitrous oxide since 1850. *Scientific Data, 10*, 155.

Kulkarni, S., Kumar, V., Bhanage, V., & Gedam, S. (2023). Exploring the association between changing crop types and water scarcity: A case study over west-Central India. *Climate, 11*, 93.

Laha, S. R., Pattanayak, B. K., Pattnaik, S., Mishra, D., Nayak, D. S. K., & Dash, B. B. (2023). An IOT-based soil moisture management system for precision agriculture: Real-time monitoring and automated irrigation control. In *2023 4th international conference on smart electronics and communication (ICOSEC)* (pp. 451–455). IEEE.

Liao, Y., Cao, H.-X., Liu, X., Li, H.-T., Hu, Q.-Y., & Xue, W.-K. (2021). By increasing infiltration and reducing evaporation, mulching can improve the soil water environment and apple yield of orchards in semiarid areas. *Agricultural Water Management, 253*, 106936.

Liu, Y., Zhuo, L., Varis, O., Fang, K., Liu, G., & Wu, P. (2021). Enhancing water and land efficiency in agricultural production and trade between Central Asia and China. *Science of the Total Environment, 780*, 146584.

Lopez, J. R., Winter, J. M., Elliott, J., Ruane, A. C., Porter, C., Hoogenboom, G., Anderson, M., & Hain, C. (2022). Sustainable use of groundwater may dramatically reduce irrigated production of maize, soybean, and wheat. Earth's. *Futures, 10*, e2021EF002018.

Masood, N., et al. (2013). Whitefly (Aleurolobus barodensis mask) Spatio-temporal trends in semi-arid agro-ecological zones of Pakistan. *Pakistan Journal of Agricultural Sciences, 50*(1), 69–74.

Mishra, R. K. (2023). Fresh water availability and its global challenge. *British Journal of Multidisciplinary and Advanced Studies, 4*, 1–78.

Mishra, B. K., Kumar, P., Saraswat, C., Chakraborty, S., & Gautam, A. (2021). Water security in a changing environment: Concept, challenges and solutions. *Water, 13*, 490.

Muratoglu, A., Iraz, E., & Ercin, E. (2022). Water resources management of large hydrological basins in semi-arid regions: Spatial and temporal variability of water footprint of the upper Euphrates River basin. *Science of the Total Environment, 846*, 157396.

Naresh, R. K., Bhatt, R., Chandra, M. S., Laing, A. M., Gaber, A., Sayed, S., & Hossain, A. (2021). Soil organic carbon and system environmental footprint in sugarcane-based cropping systems are improved by precision land leveling. *Agronomy, 11*, 1964.

Nasim, W., et al. (2012). Effect of nitrogen on growth and yield of sunflower under semiarid conditions of Pakistan. *Pakistan Journal of Botany, 44*(2), 639–648.

Niazi, H., Wild, T. B., Turner, S. W., Graham, N. T., Hejazi, M., Msangi, S., Kim, S., Lamontagne, J. R., & Zhao, M. (2024). Global peak water limit of future groundwater withdrawals. *Nature Sustainability, 7*, 1–10.

Nungula, E. Z., Mugwe, J., Nasar, J., Massawe, B. H., Karuma, A. N., Maitra, S., Seleiman, M. F., Dindaroglu, T., Khan, N., & Gitari, H. I. (2023). Land degradation unmasked as the key constraint in sunflower (Helianthus annus) production: Role of GIS in revitalizing this vital sector. *Cogent Food & Agriculture, 9*, 2267863.

Pinho, P. F., Canova, M. T., Toledo, P. M., Gonzalez, A., Lapola, D. M., Ometto, J. P., & Smith, M. S. (2022). Climate change affects us in the tropics: Local perspectives on ecosystem services and Well-being sensitivity in Southeast Brazil. *Regional Environmental Change, 22*, 89.

Quon, H., & Jiang, S. (2023). Decision making for implementing non-traditional water sources: A review of challenges and potential solutions. *NPJ Clean Water, 6*, 56.

Rahman, H. M., et al. (2019). Application of CSM-CROPGRO-cotton model for cultivars and optimum planting dates. *Field Crops Research, 238*, 139–152.

Ramirez-Contreras, N. E., Fontanilla-Díaz, C. A., Pardo, L. E., Delgado, T., Munar-Florez, D., Wicke, B., Ruíz-Delgado, J., van der Hilst, F., Garcia-Nuñez, J. A., & Mosquera-Montoya, M. (2022). Integral analysis of environmental and economic performance of combined agricultural intensification & bioenergy production in the Orinoquia region. *Journal of Environmental Management, 303*, 114137.

Rasool, A., et al. (2016). Arsanic and heavy metal contaminations in the tube well water of Punjab, Pakistan and risk assessment: A case study. *Ecological Enginee., 95*, 90–100.
Sabagh, A. E. L., et al. (2019). Drought and salinity stresses in barley: Consequences and mitigation strategies. *Australian Journal of Crop Science, 13*(6), 810–820.
Sabagh, A., et al. (2020a). Drought and heat stress in cotton (Gossypium hirsutum L.): Consequences and their possible mitigation strategies. In M. Hasanuzzaman (Ed.), *Agronomic crops*.
Sabagh, A., et al. (2020b). Nitrogen fixation of legumes under the family Fabaceae: Adverse effect of abiotic stresses and mitigation strategies. In M. Hasanuzzaman, S. Araújo, & S. Gill (Eds.), *The plant family Fabaceae*. Springer.
Sabagh, A., et al. (2020c). *Elevated CO2 concentration improves heat-tolerant ability in crops*.
Salehi, M. (2022). Global water shortage and potable water safety; today's concern and tomorrow's crisis. *Environment International, 158*, 106936.
Saliu, T., & Oladoja, N. (2021). Nutrient recovery from wastewater and reuse in agriculture: A review. *Environmental Chemistry Letters, 19*, 2299–2316.
Shah, G. M., et al. (2019). Composting of mu municipal solid waste by different methods improved the growth of vegetables and reduced the health risks of cadmium and lead. *Environmental Science and Pollution Research, 26*(06), 5463–5474.
Shahid, M., et al. (2014). EDTA-enhanced phytoremediation of heavy 1 metals: A review. *Soil and Sediments Contamination, 23*, 389–416.
Shahzad, K., et al. (2018). Tillage and biochar effects on wheat productivity under arid conditions. *Crop Science, 59*, 1–9.
Shahzad, K., et al. (2022). Importance of carbon sequestration in the context of climate change. In W. N. Jatoi, M. Mubeen, A. Ahmad, M. A. Cheema, Z. Lin, & M. Z. Hashmi (Eds.), *Building Climate Resilience in Agriculture*. Springer.
Sharma, S., & Singh, P. (2021). *Wetlands conservation: Current challenges and future strategies*. Wiley.
Shoushtarian, F., & Negahban-Azar, M. (2020). Worldwide regulations and guidelines for agricultural water reuse: A critical review. *Water, 12*(4), 971.
Siyal, A. W., Gerbens-Leenes, P. W., & Nonhebel, S. (2021). Energy and carbon footprints for irrigation water in the lower Indus basin in Pakistan, comparing water supply by gravity fed canal networks and groundwater pumping. *Journal of Cleaner Production, 286*, 125489.
Sonwani, S., & Saxena, P. (2022). *Introduction to greenhouse gases: Sources, sinks and mitigation* (pp. 1–7). Springer.
Steensland, A., & Zeigler, M. (2021). *Productivity in agriculture for a sustainable future, the innovation revolution in agriculture*. Springer.
Tariq, A., & Qin, S. (2023). Spatio-temporal variation in surface water in Punjab, Pakistan from 1985 to 2020 using machine-learning methods with time-series remote sensing data and driving factors. *Agricultural Water Management, 280*, 108228.
Thevs, N., & Aliev, K. (2023). Agro-economy of tree wind break systems in Kyrgyzstan, Central Asia. *Agroforestry Systems, 97*, 319–334.
Thevs, N., Aliev, K., & Lleshi, R. (2021). Water productivity of tree wind break agroforestry systems in irrigated agriculture–an example from Ferghana Valley, Kyrgyzstan. *Trees, Forests and People, 4*, 100085.
Wang, D., Fahad, S., et al. (2019). Morphological acclimation to agronomic manipulation in leaf dispersion and orientation to promote "Ideotype" breeding: Evidence from 3D visual modeling of "super" rice. *Plant Physiology and Biochemistry, 135*, 499–510.
Wisser, D., Grogan, D. S., Lanzoni, L., Tempio, G., Cinardi, G., Prusevich, A., & Glidden, S. (2024). Water use in livestock Agri-food systems and its contribution to local water scarcity: A spatially distributed global analysis. *Water, 16*, 1681.
Wu, B., Tian, F., Zhang, M., Piao, S., Zeng, H., Zhu, W., Liu, J., Elnashar, A., & Lu, Y. (2022). Quantifying global agricultural water appropriation with data derived from earth observations. *Journal of Cleaner Production, 358*, 131891.

Yahaya, M. A., & Shimelis, H. (2022). Drought stress in sorghum: Mitigation strategies, breeding methods and technologies—A review. *Journal of Agronomy and Crop Science, 208*, 127–142.

Yang, M., Chen, L., Wang, J., Msigwa, G., Osman, A. I., Fawzy, S., Rooney, D. W., & Yap, P.-S. (2023). Circular economy strategies for combating climate change and other environmental issues. *Environmental Chemistry Letters, 21*, 55–80.

Ye, Y., Ngo, H. H., Guo, W., Chang, S. W., Nguyen, D. D., Zhang, X., Zhang, J., & Liang, S. (2020). Nutrient recovery from wastewater: From technology to economy. *Bioresource Technology Reports, 11*, 100425.

Zhang, H., Zhang, J., & Song, J. (2022). Analysis of the threshold effect of agricultural industrial agglomeration and industrial structure upgrading on sustainable agricultural development in China. *Journal of Cleaner Production, 341*, 130818.

Modern Irrigation Methods

Hafiz Umar Farid, Nadeem Zubair, Muhammad Shoaib, Muhammad Azhar Inam, Zahid Mahmood Khan, Rana Muhammad Asif Kanwar, Muhammad Naveed Anjum, Muhammad Mubeen, Aamir Shakoor, and Khurram Shahzad

Abstract Water scarcity problems around the world are becoming more and more serious over time. Therefore, there is a need for farmers to focus on reducing water consumption, and increasing crop yield and profitability. Modern irrigation methods can significantly improve water use efficiency, crop productivity, and sustainability. In this chapter, various modern irrigation methods and their impacts on water use efficiency, crop production, and net profit for the farmers were investigated. Drip irrigation, sprinkler irrigation, canter pivot irrigation, bubble irrigation, precision irrigation, mobile irrigation, adjustable deficit irrigation, hydroponics, surface and subsurface irrigation, rainfall harvesting, vertical farming, and intelligent controlled irrigation are some of the modern irrigation methods discussed. These techniques use advanced sensors, real-time data, and automation to optimize water delivery, reduce waste, and tailor irrigation to specific crop needs. In addition, innovative water management practices play a key role in promoting resource-efficient agriculture. As the world faces water scarcity and climate challenges, adopting modern irrigation methods is crucial for sustainable food production and environmental protection.

Keywords Water scarcity · Water losses · Modern irrigation systems · Sustainable agriculture

---

H. U. Farid (✉) · N. Zubair · M. Shoaib · M. A. Inam · Z. M. Khan · R. M. A. Kanwar
A. Shakoor
Department of Agricultural Engineering, Bahauddin Zakariya University, Multan, Pakistan
e-mail: hufarid@bzu.edu.pk

M. N. Anjum
Department of Land and Water Conservation Engineering, Faculty of Agricultural Engineering and Technology, PMAS Arid Agriculture University, Rawalpindi, Pakistan

M. Mubeen
Department of Environmental Sciences, COMSATS University Islamabad, Vehari, Pakistan

K. Shahzad
Department of Crop and Soil Sciences, Washington State University, Pullman, WA, USA

M. Mubeen et al. (eds.), *Innovations in Agricultural Water Management*,
https://doi.org/10.1007/978-3-031-91883-4_3

1 Introduction

Pakistan, once considered a water-surplus country, is now facing water scarcity issues. Insufficient irrigation applications and irregular rainfall fail to meet the increasing water demand. Approximately 60% of annual rainfall occurs between June and September (Ma et al., 2023). Pakistan's surface water primarily comprises flows from the Indus River and its tributaries, which provide around 97–186 million-acre feet (MAF) of water annually (Kalair et al., 2019). The Indus River alone contributes 65% of total river flows, with the Jhelum and Chenab rivers contributing 17 and 19%, respectively. Peak months for water availability coincide with the summer crop season from June to August. Presently, 84% of total surface water is used during the Kharif (summer) season and 16% during the Rabi (winter) season (Akhtar & Athar, 2019). Pakistan consists of alluvial plains, which possess extensive groundwater resources. About 50 MAF groundwater is currently being exploited by over 1.2 million private and public tubewells (Sharif et al., 2016; Qureshi, 2020; Shahzad et al., 2020). It has been reported that some aquifers are already being overexploited. On the other hand, irrigation systems, comprising canal networks, distributors, watercourses, and irrigated fields, suffer significant water losses due to leakage and evaporation. Studies estimate conveyance losses in channels and watercourses at approximately 25 and 30%, respectively, with demand losses in fields ranging from 25 to 40%. Outdated irrigation practices contribute to these losses, resulting in an overall irrigation efficiency of around 30%. Different strategies have been proposed including launching mass awareness campaigns, activating water user organizations, adjusting cropping patterns, implementing conjunctive water management approaches, regulating groundwater abstraction, enhancing flood and drought forecasting mechanisms, and promoting highly efficient irrigation systems. Improving irrigation efficiency and conserving water necessitate the development of pressurized irrigation systems and better water management practices. Despite challenges, agricultural advancements in the past century have significantly boosted food security, quality food production, and socioeconomic development, although with ecological trade-offs.

2 Global Distribution of Water Resources

As always, amid the additional resources, there remains only one place for unplanned materials within the irrigation infrastructure. A company and the corresponding type have been allocated to that specific position. Furthermore, there is an increasing emotional need for clean salts. The country can achieve irrigation coverage of up to 1.4 billion kilometers, with approximately 35 million kilometers being attributed to clean water (about 2.5% of the total contribution). Conversely, the relationship between research and practical use has resulted in overexploitation of water resources. Excessive pumping and human activities are affecting water quality at its

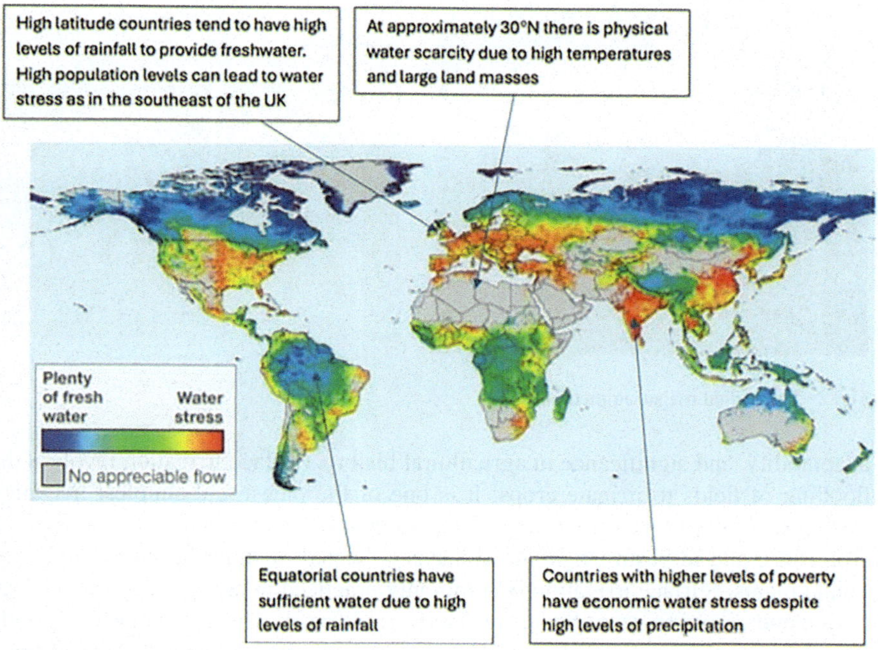

Fig. 1 Global freshwater availability

source. Globally, the total investment in water resources is estimated at $8 billion, with a projected growth of $2 billion by 2025. However, the economic outlook for 2025 remains uncertain. Water tests conducted in 1995 indicate a 22% increase in demand. This highlights the escalating need for irrigation, with 17% of water allocated for domestic use, 20% for agricultural purposes, and 70% for urban consumption. Meanwhile, the pressure on water resources persists due to environmental warming (Fig. 1). Climate change projections suggest a 20% reduction in snow and glacier meltwater from the Himalayas by 2030, potentially impacting water availability in Asia. Currently, agriculture accounts for approximately 40% of global food production. While irrigation can boost crop yields by 100–400%, the disparity between water availability and demand may lead to a release of around 10% of the water.

3 Traditional Irrigation Methods

The traditional irrigation method involves manually bringing water from a water source to the agriculture field. The most common traditional irrigation methods are surface irrigation methods such as border irrigation, furrow irrigation, and basin irrigation (Fig. 2). These methods have served communities around the world for centuries by providing water to crops. Each method has its own diversity,

Fig. 2 Traditional irrigation methods

adaptability, and significance in agricultural history. Surface irrigation involves the flooding of fields to irrigate crops. It is one of the oldest and simplest irrigation methods. Similarly, in furrow irrigation, water is directed along furrows between crop rows, and in basin irrigation, water is collected in small basins around individual plants. Surface irrigation is commonly practiced in areas with relatively flat terrain and access to water bodies or rivers. Traditional irrigation methods embody the wisdom of generations, harmonizing with local ecosystems and cultural practices. While modern technologies offer efficiency and scalability, traditional methods continue to play a vital role, especially in resource-constrained environments and in preserving biodiversity and cultural heritage. Their integration with modern practices can lead to sustainable water management and resilient agricultural systems.

4 Need for Modern Irrigation Methods

Agriculture plays an important role in world food security, livelihood, and economic development. While the sustainability of agricultural systems is threatened by various factors such as water, energy, and other crop inputs, inefficient management of water resources is considered a primary challenge, particularly in water-scarce regions of the world (Nikolaou et al., 2020). The use of traditional irrigation methods, while historically significant, often suffers from inefficiencies, water loss, and environmental degradation. In contrast, modern irrigation methods (drip and sprinkler irrigation systems) offer innovative and sustainable solutions to optimize water use efficiency, minimize water loss, enhance crop yield, and mitigate environment impacts (Yang et al., 2023; Hammad et al., 2018; Awais et al., 2018; Rasool et al., 2018; Fahad et al., 2018; Afzal et al., 2018; Amin et al., 2018; Khan et al., 2018; Hashmi et al., 2018; Adnan et al., 2018; Awais et al., 2017). The drip and sprinkler irrigation systems significantly improve water use efficiency compared to traditional irrigation methods. Modern irrigation methods deliver water directly to the root zone of the plant, which can minimize the water loss from runoff and evaporation. These irrigation methods can reduce water consumption by 50% or more. The adoption of these methods in water-stressed regions of the world is critical for sustaining agricultural production and safeguarding the interests of future generations. These irrigation methods also help promote soil health, fertility, and structure, supporting sustainable crop production and ecosystem resilience.

5 Benefits of Modern Irrigation Systems

These irrigation methods provide adaptive strategies to cope with the changing climatic conditions (drought and rainfall patterns). These methods offer flexibility in water management practices. These can also allow farmers to adjust the irrigation schedule and water application rates according to the weather forecast and crop water requirement. It has been reported that an effective irrigation plan is necessary to increase the cultivated area to meet the food demand from the growing population (Khor & Feike, 2017; Abbas et al., 2017; Fahad et al., 2017; Saud & e al., 2017; Zia et al., 2017; Amin et al., 2017a, 2017b; Shakeel et al., 2017; Jabran et al., 2017; Mirza et al., 2017; Hammad et al., 2017; Moursy et al., 2023). According to an estimate, in conventional irrigation methods, 50–70% water is wasted during conveyance and field application phases. These irrigation losses can be minimized, and water productivity can be increased using modern irrigation practices such as drip, sprinkler, and precision irrigation systems (Fig. 3). Some other studies have also showed that farmers can achieve significant returns on investment by increasing irrigation efficiency from 60 to 90% by adopting modern irrigation technologies (Table 1). In some countries like Spain, irrigation efficiency has increased from 65

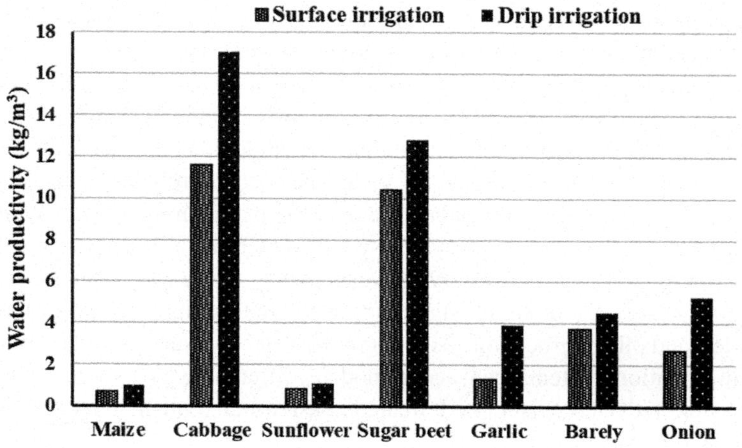

Fig. 3 Water productivity of different crops under various irrigation systems. (Source: Moursy et al., 2023)

Table 1 Comparison of water and energy efficiencies for various irrigation methods

Sr. no.	Irrigation method	Water efficiency	Energy efficiency
1	Surface irrigation	50–60	Low
2	Level basin	60–80	Low
3	Sub-irrigation	50–75	Low to medium
4	Overhead irrigation	60–80	Medium
5	Sprinkler irrigation	60–85	Medium
6	Drip irrigation	80–90	Medium to high

Source: Munir et al. (2018)

Table 2 Cost–benefit analysis for various crops under different irrigation methods

	Surface irrigation system				Drip irrigation system			
Crops	Total cost ($/ha)	Unit price ($/kg)	Total income ($/ha)	Net profit ($/ha)	Total cost ($/ha)	Unit price ($/kg)	Total income ($/ha)	Net profit ($/ha)
Maize	1013.00	0.32	1200.44	0187.42	0869.23	0.32	1163.75	0294.52
Cabbage	1414.90	0.03	2114.64	0699.75	1237.99	0.03	2141.81	0903.81
Sunflower	0935.72	0.38	1295.07	0390.46	0832.82	0.39	1223.28	0390.46
Sugar beet	0522.86	0.04	2085.78	1562.92	0468.39	0.05	2010.27	1541.88
Garlic	0564.91	0.28	1518.43	0953.52	0499.25	0.35	2919.32	2420.38
Baley	0431.22	0.26	2755.60	2324.38	0418.26	0.27	2193.33	1775.07
Onion	1964.56	0.16	1902.64	−61.92	1512.86	0.18	1884.44	0371.58

Source: Moursy et al. (2023)

to 87% following the use of modern irrigation systems (Hussain et al., 2022a, 2022b, 2022c; Khaliq, 2022; Hammad et al., 2022; Tahir et al., 2022; Sabagh et al., 2022; Berbel et al., 2019). The modern irrigation systems also strengthen the adaptive capacity of agricultural systems and long-term sustainability by enhancing the resilience and reducing vulnerability to climate-related risk.

In addition to other benefits, these modern irrigation techniques offer economic advantages for farmers, businesses, and communities (Table 2). While initial investment may be higher than that for traditional irrigation methods, the long-term benefits in terms of water savings, crop yield, and net return outweigh the upfront expenses. Moreover, by enhancing market competitiveness and diversification opportunities, modern irrigation systems contribute to rural livelihoods, economic growth, and poverty alleviation. By delivering water to the root zone and minimizing surface runoff, the modern irrigation systems reduce the leaching of fertilizers, pesticides, and other agrochemicals into groundwater and surface water bodies. The modern irrigation systems can accommodate different water sources, including fresh water, recycled water, and treated wastewater, enhancing resilience and resource sustainability.

6 Transition from Traditional to Modern Irrigation Methods for Improved Efficiency

The transition from traditional to modern irrigation methods indicated a pivotal shift in agricultural systems for improving water efficiency, productivity, and sustainability (Table 3). The drivers, challenges, and implications of this transition highlight the opportunities for enhancing agricultural resilience. The potential of traditional and modern irrigation systems in agriculture has been investigated to identify the adoption constraints for both traditional and modern irrigation systems. It was observed that education, training, construction of strong irrigation networks, age, and years of experience are the key driving factors for the adoption of modern irrigation systems (Al-Subaiee et al., 2013). By synthesizing existing knowledge and identifying gaps in research and practice, this chapter provides insights into strategies and pathways for facilitating the widespread adoption of modern irrigation methods and achieving transformative outcomes in agriculture. Implementing

Table 3 Parameters for suitability and selection of different irrigation systems

Sr. No.	Parameters	Flood irrigation	Drip irrigation	Sprinkler irrigation	
				Hand-move	Fixed
1	Soil type	Heavy clay to medium	Light sandy to heavy clay	Light sandy to medium	Light sandy to medium
2	Suitable location	Valley to Delta	Desert, valley, and delta land	Dessert land	Dessert land
3	Land leveling	Required	Not required	Not required	Not required
4	System costs	Low	Medium above medium	Medium	Medium above medium
5	Suitable crop	Most crops	Trees, vegetables, and row crops	Except long crops and trees	Except trees
6	Labor and efforts	High	Low	Medium	Low
7	High wind	Not affected	Not affected	Affected with spacing	Not affected
8	Irrigation water salinity	Up to 1100 ppm	Up to 1750 ppm	Up to 800 ppm	Up to 800 ppm
9	Fertilizer injection	Not use	Preferable	Possible	Possible
10	Required pressure	–	1–2 bar	3–4 bar	3–4 bar
10	Irrigation efficiency	Low (<50%)	High (>85%)	Medium (60–65%)	Above medium (>75%)
11	Maintenance	Low	Medium	Below medium	Above medium

Source: Samir M Ismail

innovative irrigation methods and adopting sustainable agriculture practices are necessary for ensuring the long-term viability of the agriculture industry and safeguarding our natural resources (Abubakar & Attanda, 2013).

7 Modern Irrigation Systems

The country is currently grappling with a water shortage crisis. In response to this pressing issue, the President of the Islamic Republic of Pakistan announced the implementation of a drip/sprinkler and rain gun system in 2004. In accordance with the directives, a project has been developed to conserve water and boost productivity across the country through the deployment of high-efficiency irrigation systems (HEIS) in both rainfed and canal command areas (Fig. 4). Under the HEIS project, micro irrigation systems were installed. The micro irrigation systems can be further classified into drip irrigation systems (including lines/tapes, bubblers, and family drip systems) and sprinkler irrigation systems, encompassing center pivot, impact sprinkler, and rain gun technologies (Fig. 4). The adoption of HEIS is more when water prices are higher than the use of water-saving technologies (Mattoussi & Mattoussi, 2022).

Fig. 4 Modern irrigation technologies

8 Drip Irrigation

Drip irrigation, also known as trickle irrigation, entails the slow dripping of water onto the soil at very low rates (2–20 L/h) through a network of small-diameter plastic pipes equipped with an outlet called emitter or dripper. Drip irrigation has remarkable potential to conserve water by 70–95% and enhance irrigation efficiency by as much as 95%. Farmers are becoming increasingly inclined to adopt drip irrigation technology due to the notable increase in crop production. It has been found that plant parameters such as plant height, root depth, fruit length, and fruit diameters are also affected by the application of water through different irrigation methods (Tables 4 and 5). The adoption of the drip irrigation system has the potential to boost farmers' net returns by USD 4000 to 9000 per hectare, particularly when cultivating high-value crops. The drip irrigation system saved 50% of irrigation cost. In comparison with the furrow irrigation system, the drip irrigation increased farmers' net returns by 27–54% (Aziz et al., 2021).

Drip irrigation is highly suitable for row crops (such as vegetables and soft fruits), tree crops, and vine crops, where individual plants can be equipped with one or more emitters. Typically, only high-value crops are considered due to the significant capital investment required for drip system installation. The adaptability of drip irrigation extends to farms with varying slopes, where crops are often planted along contour lines to minimize changes in emitter discharge resulting from changes in land elevation. One of the primary challenges associated with drip irrigation is

Table 4 Plant attributes affected by different irrigation methods

	Drip irrigation				Furrow irrigation			
Crops	Plant height (m)	Root depth (m) Leaf Area	Fruit length (mm)	Fruit diameter (mm)	Plant height (m)	Root depth (m) leaf area	Fruit length (mm)	Fruit diameter (mm)
Potato	0.60	0.33	88.00	55.00	0.52	0.36	75.00	48.00
Onion	0.51	0.27	65.00	70.90	0.44	0.31	59.20	62.40
Chilies	0.92	0.39	55.50	–	0.85	0.42	50.40	44.40
Olive	02.10	15.00	14.90	02.03	01.90	12.00	12.80	01.75
Peach	03.50	130.00	06.31	116.00	03.00	100.00	05.94	105.50
Citrus	02.00	95.00	77.00	120.00	01.85	75.00	07.00	115.00

Source: Aziz et al. (2021)

Table 5 Average yield (kg/ha) by irrigation systems from 2009 to 2015

Irrigation system	Barley	Forage corn	Wheat	Tomato
Surface	2992.86	55884.84	4469.33	64773.93
Drip	3591.44	64084.80	6480.53	77728.81
Sprinkler				
Permanent	3160.46	57884.63	5452.59	–
Semi-permanent	3160.46	57084.63	5452.59	–

Source: Jahangirpour and Zibaei (2022)

emitter blockage, which can occur due to sediment accumulation or the presence of algae, fertilizer deposits, and dissolved chemicals in the water. Proper filtration is essential in preventing blockages, especially when using poor-quality water sources like saline water. Despite these challenges, the precise delivery of water to individual plants makes drip irrigation highly efficient in water usage, making it particularly suitable for regions facing water scarcity (Jahangirpour & Zibaei, 2022; Hussain et al., 2021a, 2021b; Mehmood et al., 2021; Sabagh et al., 2021; Javeed et al., 2021a, 2021b; Ali et al., 2021; Mubeen et al., 2021). The system operates effectively at low pressures ranging from 10 to 20 psi, although higher pressure is required to compensate for losses through filters, valves, pressure regulators, and piping, typically requiring a minimum of 40 psi from the pump. Drip irrigation can be adapted to various environments and situations, reducing the risk of diseases since it avoids wetting the plant foliage. Drip irrigation, when coupled with appropriate fertilization, not only conserves groundwater but also addresses issues associated with traditional methods such as soil salinity. Fertilization contributes to enhanced water efficiency, improved soil health, high-quality crop yields, and optimal fertilizer use.

9 Drip Irrigation for Arid Soils

The conventional approach to salt management, characterized by a “loose” methodology, is no longer a viable alternative to subsurface drip irrigation (SDI). This is because SDI lacks the ability to alter the depth of the water belt and leach salts during the crop growing season. While SDI effectively delivers water to the plant’s root zone, it can lead to salt accumulation in the soil in the absence of rainfall or sufficient leaching. Encouraging the adoption of integrated drainage systems (IDS) and proper plant selection can contribute significantly to the long-term sustainability of SDI irrigation systems. The accumulation of salts at crop root levels poses a serious threat to plant productivity. Addressing soil salinity requires not only adequate water provision but also the replenishment of essential nutrients lost due to salt accumulation (e.g., through filtration requirements). Effective long-term agricultural planning involves the judicious use of salts and appropriate fertilizer management practices. Some irrigation methods, such as those using saline water, effectively manage salt levels in the soil.

10 Surface Drip Irrigation

Recreational water offers numerous solutions, notably in repurposing “domestic water” for freshwater purification, a cost-effective alternative to developing new water sources. Furthermore, it mitigates waste disposal issues and reduces the demand for fertilizers. Sustainable Development and Soil Reduction (SDI) practices are ideally suited for managing clean water, leading to increased water utilization and crop growth compared to traditional methods. Nonetheless, ongoing research is imperative to ensure sustainable water usage in agricultural production. Water management serves as the cornerstone of success in Israel's dry, arid agricultural landscape. Pioneering technologies, such as soil perforation introduced in the 1960s, have significantly enhanced farmers' productivity while reducing water consumption. This technology enhances water efficiency by minimizing runoff and other constraints, delivering nutrients to plants while maintaining a dry surface. The SDI system's drip kit is engineered to streamline water management, weed control, and optimize fertilization, making it challenging for machinery to reach the soil surface and reducing workers' direct contact with water. Another driving factor behind SDI adoption is the substantial labor savings associated with installing and monitoring conventional irrigation systems. The reuse of treated (unpolluted) water is common in developing countries across Asia and Africa, while water reclamation initiatives are prevalent in developed nations such as the United States, Australia, and those in the Middle East. SDI serves as a potent tool for addressing health, odor, and pollution concerns while also enhancing water management, particularly for plant growth and land cover. It offers a unique opportunity to regulate root and soil conditions, thereby exerting better control over biodiversity, salinity, oxygen levels, and temperature. Municipal wastewater, particularly from suburban areas, has become a crucial water source for Israeli farmers. Agricultural investments were projected to increase to 37% by 2010 and 46% by 2020. Forest irrigation in Israel has seen significant growth, expanding from 5100 ha in 1975 to 36,300 ha.

11 Sprinkler Irrigation

Sprinkler irrigation emulates natural rainfall by distributing water through a network of pipes, typically powered by pumping. Water is then dispersed into the air through sprinklers, breaking into small droplets that descend to the ground. The design of the pump supply system, sprinklers, and operational parameters must ensure uniform water application (Daccache et al. 2010). This method is well-suited for a variety of crops, including row, field, and tree crops, with the flexibility to spray water over or under the crop canopy (Table 6, Fig. 5). However, caution is advised when using large sprinklers for delicate crops like lettuce, as the forceful water droplets may cause damage. Sprinkler irrigation is adaptable to various slopes, whether uniform or undulating, with lateral pipes ideally laid along the land contour to minimize pressure fluctuations and ensure uniform irrigation. While sandy soils with high infiltration rates are ideal, sprinklers can be adjusted for most soil types. The application rate from sprinklers should always be lower than the soil's basic infiltration rate to prevent surface ponding and runoff. Sprinklers may not be suitable for soils prone to crusting, but if used, lighter fine sprays are recommended over larger droplets to mitigate crust formation. A clean water supply,

Table 6 Summary statistics of simulated net return by irrigation systems (1000 Rials/ha)

	Barley		Forage corn		Wheat		Tomato	
Irrigation system	Mean	Std.	Mean	Std.	Mean	Std.	Mean	Std.
Surface	917	592	2580	999	1021	773	30,531	11,441
Drip	1027	710	3319	1241	1687	1255	39,050	15,113
Sprinkler								
Permanent	848	640	2393	956	1142	914	–	–
Semi-permanent	859	600	2515	990	1198	940	–	–

Source: Jahangirpour and Zibaei (2022)

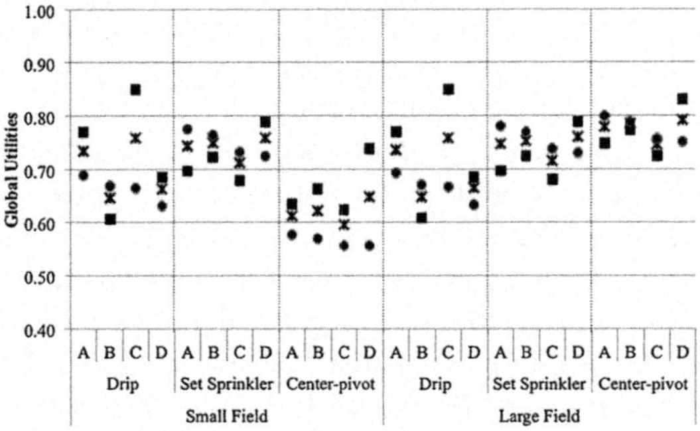

Fig. 5 Global utilities relative to the prioritization scheme adopted. (Source: Rodrigues et al. (2013)). Or Balance between both; ¤ Farm Economic Return; × When considering various deficit irrigation treatments, A through D

devoid of suspended sediments, is crucial to preventing nozzle blockages and avoiding crop damage caused by sediment coating.

12 Rain Gun Sprinkler

Rain guns can be employed for various crops, including sugarcane, coffee, tea, cardamom, onion, cabbage, potato, cauliflower, carrot, beetroot, tapioca, and groundnut. However, the suitability of the rain gun for specific crops depends on numerous factors. Rain gun systems use large-volume sprinklers that operate at relatively high pressures (3–6 bars), either mounted on a riser stand or towed through the field. Three types of rain gun systems are available (Fig. 6). The minimum requirement for these systems is a 5 HP motor. This requirement is contingent upon the Kg/Sq. cm (KSC) pressure generated by the pump. Our standard minimum requirement is a 2 KSC pressure, which cannot be achieved with a motor/pump less than 5 HP. Therefore, our requirement ranges from >5H.P. to 20 H.P. Optimal results are typically obtained with KSC pressure exceeding 2.75.

Fig. 6 Rain gun sprinkler irrigation system

Fig. 7 Bubbler of mini sprinkler system

13 Bubbler or Mini Sprinkler

Mature trees necessitate bubbler or mini-sprinkler systems capable of delivering water to the entire root system (Fig. 7). Conversely, for young trees, point-source drippers can suffice, with the flexibility to add more drippers as the trees mature and require greater water supply. In densely planted orchards and vineyards, drip lines may also serve as a viable option. Hussien et al. (2013) conducted experiments to study the water use pattern, growth parameters, fruit quality, and mineral contents of Washington Navel orange trees using two irrigation methods. The obtained results pointed out that consumptive use for surface irrigation was higher than that of bubbler irrigation methods. As a result of surface irrigation, higher soil moisture was maintained, and water consumption increased (Table 7). The bubbler irrigation methods were found more effective in root development compared with the surface ones. Root development was reflected in the improved yield and quality of orange production.

Table 7 Applied water for orange using two irrigation methods

Season	Bubbler irrigation				Surface irrigation			
	2011		2012		2011		2012	
Months	m^3/ha/day	m^3/ha/month	m^3/ha/day	m^3/ha/month	m^3/ha/day	m^3/ha/month	m^3/ha/day	m^3/ha/month
January	13.33	400	13.10	405	0	0	0	0
February	16.42	493	15.48	448	19.54	586	16.43	493
March	18.33	550	18.33	564	33.10	993	32.38	971
April	27.14	814	26.67	800	42.14	1264	42.14	1264
May	36.43	1093	35.48	1100	63.81	1914	61.43	1843
June	43.10	1293	37.14	1117	109.52	3286	108.81	3264
July	43.10	1293	37.14	1148	109.76	3293	109.76	3293
August	41.90	1257	37.14	1150	102.14	3064	94.52	2836
September	32.38	971	29.29	881	57.62	1729	57.86	1736
October	28.81	864	24.52	760	31.43	943	30.71	921
November	21.19	636	18.57	560	14.29	429	15.95	479
December	14.04	421	12.38	386	0	0	0	**0**
Seasonal (m^3/ha)		10,086		9317		17,500		17,100

Source: Hussien et al. (2013)

Fig. 8 Center pivot system

14 Center Pivots

In extensive agricultural areas, center pivot systems are the preferred choice (Fig. 8). This technology is robust, well-established, and cost-effective per unit of irrigated land. However, its implementation necessitates a flat terrain, a dependable and abundant water source (ideally a well, as on-site storage would be prohibitively costly), and a reliable energy supply. Center pivots should only be operated by trained professionals. Additionally, the upfront investment for medium and large center pivot systems often surpasses the maximum grant amount provided under this program.

15 Regulated Deficit Irrigation

Declining water resources and escalating water demand raised concerns about its availability for irrigation. Increasing food demand from growing populations necessitates boosting crop production and conserving water resources. In this regard, deficit irrigation emerges as a crucial strategy to enhance water use efficiency (WUE). Deficit irrigation serves as a water-saving technique during the maize growing season. Implementing regulated deficit irrigation (RDI) can lead to significant water savings with minimal impact on crop yield. RDI is used to conserve water by subjecting crops to moisture stress, allowing for varying levels of management allowed depletion (MAD). Suitable MAD levels for specific crops were identified, which have negligible effects on yields. In 2003, field experiments were conducted on maize fodder production at a research site of the Department of Irrigation and Drainage, University of Agriculture, Faisalabad, Pakistan to find the effects of different water stress treatments. An irrigation schedule was prepared based on the specified MAD levels. Water depletion in the upper 200 mm of soil was used as a criterion for initiating irrigation. Field experimental data were analyzed using the GLEAM model. The significant treatment effect was observed at 30% MAD level, which resulted in significantly higher yields compared to all other treatments. The highest water use efficiency (WUE) was observed at 60% MAD level. Management scenario simulations using the validated GLEAMS model indicated that maintaining soil moisture within the range of 30 to 90% of available water led to improved WUE. Future field experiments should be designed to identify the most effective management practices for MAD and potential soil moisture deficit, in order to maximize water use efficiency, optimize yield with minimal water usage and contribute to water conservation for future generations.

16 Precision Irrigation

Precision irrigation is an advanced water-saving approach that uses technology- and data-driven strategies to optimize water use in agriculture. This is possible through the integration of advances technologies such as global positioning system, sensors, machine learning, and data analytics, which allow farmers to adjust water use based on spatial and temporal variabilities (Fig. 9). Precision irrigation provides accurate information about crop water requirement, soil moisture content, and weather conditions, so that precise amount of water can be delivered where and when it is needed. This can reduce water wastage and energy cost. It was found that precision irrigation has energy-saving potential of 4–8% and average potential for increase in net returns equal to 16.4%. Precision irrigation not only enhances net profit, but also contributes toward the conservation of water resources and protection of the environment. Precision irrigation has gained significant attention from the farmers and stakeholders in water-scarce regions such as the Americas, Australia, parts of Africa, and the Middle East. However, it will still take a long time to adopt precision irrigation as a mainstream solution for sustainable agriculture globally. The high initial cost and complexity in design pose a barrier for many small farmers. Continuous research is critical for making precision irrigation a viable and affordable solution for farmers. Research on precision irrigation may also unlock its full potential in sustainable agriculture to ensure food security in future (Liu et al., 2023; Violino et al., 2023).

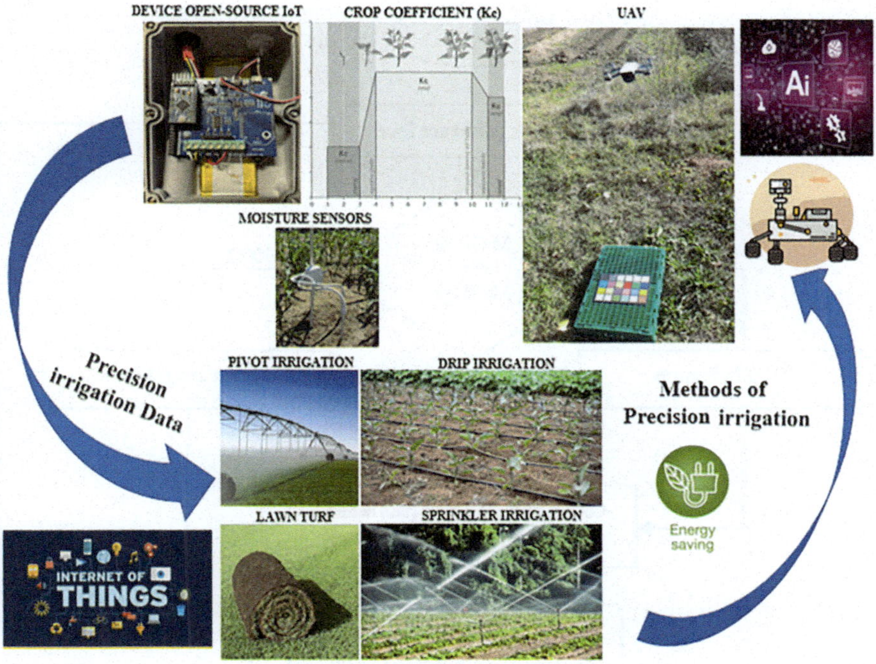

Fig. 9 Precision irrigation cycle. (**Source:** Violino et al., 2023)

17 Irrigation Technologies and Automation

Irrigation is considered necessary for plant growth and health. However, it was accomplished manually. This can increase the quantity of water used and cost of irrigation. Automating irrigation could bring many benefits to the farmers (Yang et al., 2007). Automation involves controlling the irrigation system through means other than human operation. For example, the automatic drip and sprinkler irrigation systems include various components that differ from traditional system. These components include electronic controllers, automatic sand media filters, automatic fertigation injectors, and automatic valve opening devices. The automatic irrigation system can enhance the efficiency of water use in irrigation and decrease water utilization. Ramachandran et al. (2018) combined the Internet of Things (IoT), cloud computing, and optimization tools (Fig. 10). The results demonstrate that the proposed methodology can reduces water utilization and increase data availability. Phasinam et al. (2022) described that an irrigation system that was designed based on cloud computing architecture and the Internet of Things (IoT) can save significant amounts of fresh water because of smart irrigation. Champness et al. (2023) evaluated the performance and opportunity cost of a sensor-based

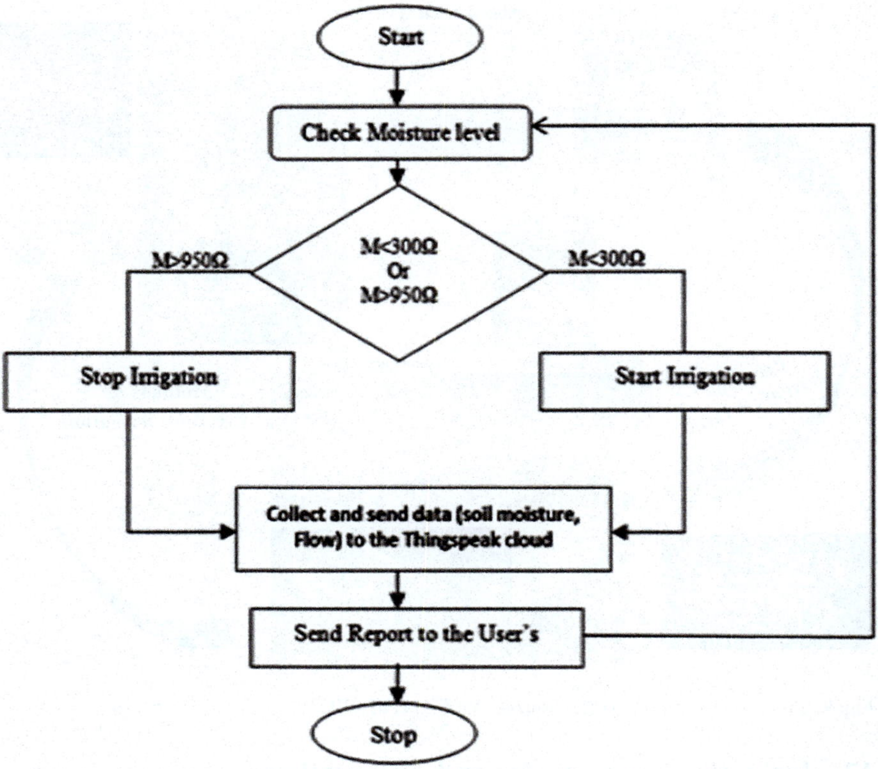

Fig. 10 Soil moisture-based irrigation system automation. (Source: Ramachandran et al. (2018)

smart irrigation system for water saving in rice cultivation in temperate Australia. It was found that the automated surface irrigation system has potential to save significant labor cost (82–88%) in traditional flooded rice. Al-Ghobari et al. (2017) compared automated irrigation systems with conventional irrigation systems to investigate the electronic controller in irrigation systems. For wheat, water use efficiency and irrigation water use efficiency were 1.64 and 1.37 kg m^{-3}, respectively, under the automated irrigation system. Similarly, the automated irrigation system provided a higher crop yield with 26% less water utilization compared with conventional irrigation systems. These results demonstrated that automated irrigation system is a valuable tool for irrigation scheduling and conserving water (Dursun & Ozden 2011; Lamaddalena et al., 2012; Abioye et al., 2020; Al-Ghobari et al., 2017; Champness et al., 2023). For instance, it optimizes water use and provides a critical water application rate and fertility rate to increase production efficiency, reduce manpower involvement, and reduce crop diseases (Nandanwar et al., 2020).

18 Mobile Irrigation Apps

To optimize irrigation practices, mobile irrigation apps have been considered as a valuable tool to manage precious water resources and improve water use efficiency, as well as to enhance crop yield. For informed decisions and effective water resource management, these apps contain various features such as real-time weather and climate records, soil moisture content status, irrigation application scheduling, and automated control of irrigation (Bartlett et al., 2015; Migliaccio et al., 2015; Ale et al., 2023). Migliaccio et al. (2015) developed mobile irrigation apps for providing real-time irrigation scheduling of various crops such as cotton, citrus, peanut, strawberry, vegetable, and avocado. These mobile irrigation apps provid irrigation schedule based on real-time weather datacollected from sources such as the Florida Automated Weather Networks and Georgia Environmental Monitoring Networking. The mobile irrigation apps include features such as splitting irrigation events, water conservation options, spreadsheets for data management, output emails, and notifications. Additionally, the irrigation scheduling tool Bluleaf® has been developed and evaluated in comparison with conventional irrigation methods. The Bluleaf® mobile irrigation app uses weather, soil, crop growth, and irrigation system data. It helps the farmers in decision-making on the timing and quantity of irrigation. Irrigation scheduling done using the Bluleaf® mobile irrigation app enhanced water saving by 25.7% with respect to conventional irrigation scheduling (Abi Saab et al., 2019). The use of smartphone technology enables farmers to access relevant information anytime and anywhere. These techniques support timely irrigation based on crop's water requirements and environmental conditions. However, challenges such as data accuracy, connectivity issues, and user adoption mustbe addressed to maximize the effectiveness of mobile irrigation apps. Despite these challenges, mobile irrigation apps have emerged as promising tools for promoting sustainable agricultural.

19 Vertical Farming and Hydroponics

Vertical farming and hydroponic systems have gained significant attention as innovative approaches to agricultural production, offering potential solutions to food security challenges, resource constraints, and urbanization pressures. Vertical farming involves the cultivation of crops in vertically stacked layers or structures, using artificial lighting, climate control, and hydroponic or aeroponic techniques to optimize growing conditions. Hydroponic systems, on the other hand, rely on nutrient-rich water solutions to nourish plants' roots, eliminating the need for soil (Fig. 11). By maximizing space utilization, minimizing water usage, and reducing environmental impacts, vertical farming and hydroponics offer opportunities for year-round production, local food supply chains, and reduced carbon footprints. Efficient and cost effective vertical farming and hydroponic systems take into account factors

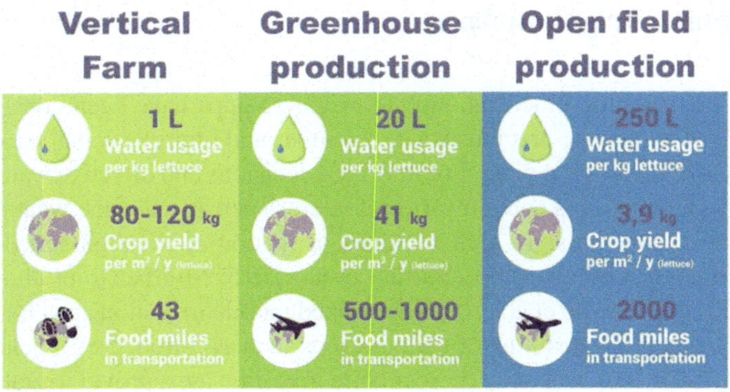

Fig. 11 Comparison of different farming practices. (Source: Jenks (2024), https://green.org/2024/01/30/vertical-farming-and-the-evolution-of-sustainable-farm-to-table)

such as initial investment costs, operational expenses, resource efficiency, and crop yields. Although vertical farming and hydroponics require substantial upfront investments in infrastructure, technology, and energy, they offer potential long-term benefits, including higher crop yields, faster growth rates, and reduced land use. Moreover, vertical farming and hydroponic systems can be tailored to specific crop varieties, consumer preferences, and market demands, enhancing product quality, consistency, and freshness. Challenges associated with vertical farming and hydroponics include energy consumption, technological complexity, and scalability issues. Additionally, the economic viability of these systems depends on factors such as market demand, pricing dynamics, regulatory frameworks, and access to financial resources. Despite these challenges, advancements in automation, precision agriculture, and renewable energy technologies are driving down costs and improving the efficiency and sustainability of vertical farming and hydroponic operations. In conclusion, the efficient and cost-effectiveness of vertical farming and hydroponic systems highlight their potential to revolutionize agricultural production and help tackle global food security challenges. By leveraging advanced technologies, data-driven insights, and innovative business models, vertical farming and hydroponics can enhance resource efficiency, reduce environmental impacts, and create new opportunities for sustainable and resilient food systems in a rapidly urbanizing world.

20 Sensor-Based Decision-Making

Sensor-based decision-making is a rapidly evolving field that leverages the capabilities of sensors and data analytics to inform real-time decision-making processes across various domains. Sensors are devices that detect and measure changes in physical parameters such as temperature, humidity, pressure, and motion. When integrated into systems and processes, sensors continuously collect and transmit data, enabling stakeholders to monitor conditions, detect anomalies, and optimize operations (Loebbecke & Boboschko, 2020; Colaço et al., 2021; Berge et al., 2023; Haseeb et al., 2023). The use of automatic control and remote operation in irrigation systems is enabled by prototypes that integrated sensors, node controllers, and mobile phones. Sensors are placed in the soil to measure moisture levels, and the collected data is transmitted to the control node for processing. The control node compares it with a predefined threshold value. If the soil moisture falls below the required level, the irrigation system is activated, and a notification is sent to the registered mobile phone. The small-scale irrigation system, managed by the control unit, represents an advance mechanism that precisely monitors and regulates all irrigation activities. This system promotes energy efficiency optimizes water usage, and enhances overall profitability.

21 Conclusions

Agriculture plays an important role in global food security, livelihood, and economic development. However, the sustainability of agricultural systems is increasingly threatened by factors such as water scarcity, energy demands, and the rising costs of crop inputs. Efficient management of water resources is considered a primary challenge, particularly in water-scarce regions of the world. Therefore, there is a need for farmers to focus on reducing water consumption, and increasing crop yield and profitability. Modern irrigation methods can significantly improve water use efficiency, crop productivity, and sustainability. The various modern irrigation methods and their impacts on water use efficiency, crop production, and net profit for the farmers were investigated. Drip irrigation, sprinkler irrigation, canter pivot irrigation, bubbler irrigation, precision irrigation, mobile irrigation apps, adjustable deficit irrigation, hydroponics, surface and subsurface irrigation, and vertical farming and intelligent controlled irrigation are some of the modern irrigation methods discussed. These techniques use advanced sensors, real-time data, and automation to optimize water delivery, reduce waste, and tailor irrigation to specific crop needs. In addition, innovative water management practices play a key role in promoting resource-efficient agriculture.

References

Abbas, G., et al. (2017). Quantification the impacts of climate change and crop management on phenology of maize-based cropping system in Punjab, Pakistan. *Agricultural Forest Meteorology, 247*, 42–55.

Abi Saab, M. T., Jomaa, I., Skaf, S., Fahed, S., & Todorovic, M. (2019). Assessment of a smartphone application for real-time irrigation scheduling in Mediterranean environments. *Water, 11*(2), 252.

Abioye, E. A., Abidin, M. S. Z., Mahmud, M. S. A., Buyamin, S., Ishak, M. H. I., Abd Rahman, M. K. I., Otuoze, A. O., Onotu, P., & Ramli, M. S. A. (2020). A review on monitoring and advanced control strategies for precision irrigation. *Computers and Electronics in Agriculture, 173*, 105441.

Abubakar, M.S. and Attanda, M.L., (2013). The concept of sustainable agriculture: Challenges and prospects. In IOP conference series: Materials science and engineering (Vol. 53, No. 1, 012001). IOP Publishing.

Adnan, M., et al. (2018). Phosphate-solubilizing bacteria nullify the antagonistic effect of soil calcification on bioavailability of phosphorus. *Scientific Reports, 2018*, 07–22653.

Afzal, M., et al. (2018). Current status and future possibilities of molecular genetics techniques in brassica napis. *Biotechnology Letters, 40*(3), 479–492.

Akhtar, I. U. H., & Athar, H. (2019). Contribution of changing precipitation and climatic oscillations in explaining variability of water extents of large reservoirs in Pakistan. *Scientific Reports, 9*(1), 19022.

Ale, S., Su, Q., Singh, J., Himanshu, S., Fan, Y., Stoker, B., Gonzalez, E., Sapkota, B. R., Adams, C., Biggers, K., & Kimura, E. (2023). Development and evaluation of a decision support mobile application for cotton irrigation management. Smart Agricultural. *Technology., 5*, 100270.

Al-Ghobari, H. M., Mohammad, F. S., El Marazky, M. S., & Dewidar, A. Z. (2017). Automated irrigation systems for wheat and tomato crops in arid regions. *Water SA, 43*(2), 354–364.

Ali, A., et al. (2021). Temporal analysis for anomaly detection in precipitation pattern over the selected area in Indus Basin of Pakistan. *Pure and Applied Geophysics., 178*, 651–669.

Al-Subaiee, F. S., Al-Ghobari, H. M., Baig, M. B., Ei-Hag, E. A., & Abu-Riziga, M. T. (2013). *Studies on adoption of irrigation methods by the date palm farmers in Al-Qassim area-Kingdom of Saudi Arabia.*

Amin, A., et al. (2017a). Comparaison of future and base precipitation anomalies by SimCLIM statistical projection through ensemble approach in Pakistan. *Atmospheric Research, 194*, 214–225.

Amin, A., et al. (2017b). Optimizing the phosphorus use in cotton by using CSM-CROPGRO-cotton model for semi-arid climate of Vehari-Punjab, Pakistan. *Environmental Science and Pollution Research, 24*(6), 5811–5823.

Amin, A., et al. (2018). Simulated CSM-CROPGRO-cotton yield under projected future climate by SimCLIM for southern Punjab, Pakistan. *Agricultural Systems, 167*, 213–222.

Awais, M., et al. (2017). Nitrogen and plant population change radiation capture and utilization capacity of sunflower arid environment. *Environmental Science and Pollution Research, 24*(21), 17511–17525.

Awais, M., et al. (2018). Potential impacts of climate change climate change impact assessment and adaptation strategies for sunflower in Pakistan. *Environmental Science and Pollution Research, 25*(14), 13719–13730.

Aziz, M., Rizvi, S. A., Iqbal, M. A., Syed, S., Ashraf, M., Anwer, S., Usman, M., Tahir, N., Khan, A., Asghar, S., & Akhtar, J. (2021). A sustainable irrigation system for small landholdings of rainfed Punjab, Pakistan. *Sustainability, 13*(20), 11178.

Bartlett, A. C., Andales, A. A., Arabi, M., & Bauder, T. A. (2015). A smartphone app to extend use of a cloud-based irrigation scheduling tool. *Computers and Electronics in Agriculture, 111*, 127–130.

Berbel, J., Expósito, A., Gutiérrez-Martín, C., & Mateos, L. (2019). Effects of the irrigation modernization in Spain 2002–2015. *Water Resources Management, 33*, 1835–1849.

Berge, T. W., Urdal, F., Torp, T., & Andreasen, C. (2023). A sensor-based decision model for precision weed harrowing. *Agronomy, 14*(1), 88.

Champness, M., Vial, L., Ballester, C., & Hornbuckle, J. (2023). Evaluating the performance and opportunity cost of a smart-sensed automated irrigation system for water-saving rice cultivation in temperate Australia. *Agriculture, 13*(4), p.903.

Colaço, A. F., Richetti, J., Bramley, R. G. V., & Lawes, R. A. (2021). How will the next-generation of sensor-based decision systems look in the context of intelligent agriculture? A case-study. *Field Crops Research, 270*, 108205.

Daccache, A., Lamaddalena, N., & Fratino, U. (2010). On-demand pressurized water distribution system impacts on sprinkler network design and performance. *Irrigation Science, 28*(4), 331–339.

Dursun, M., & Ozden, S. (2011). A wireless application of drip irrigation automation supported by soil moisture sensors. *Scientific Research and Essays, 6*(7), 1573–1582.

Fahad, S., et al. (2017). Crop production under drought and heat stress: Plant responses and management options. *Frontier in Plant Sciences., 08*(1147), 01–16.

Fahad, S., et al. (2018). Consequences of high temperature under changing climate optima for rice pollen characteristics concept and perspectives. *Archives of Agronomy and Soil Science, 64*(11), 1473–1488.

Hammad, H. M., et al. (2017). Maize plant nitrogen uptake dynamics at limited irrigation water and nitrogen. *Environmental Science & Pollution Research, 24*(3), 2549–2557.

Hammad, H. M., et al. (2018). Offsetting land degradation through nitrogen and water management during maize cultivation under arid conditions. *Land Degradation and Development, 29*(5), 1366–1375.

Hammad, H. M., et al. (2022). Assessing the impact of varying temperatures on growth of maize and nitrogen use efficiency under semi-arid environment. *Frontiers in Plant Science., 13*, 885479.

Haseeb, K., Siraj, M., Alzahrani, F. A., & Jeon, G. (2023). Sensor-based optimization multi-decision model for sustainable smart cities. *Sustainable Energy Technologies and Assessments, 60*, 103452.

Hashmi, M. Z., et al. (2018). PCB118-induced cell proliferation mediated by oxidative stress and MAPK signaling pathway in HELF cells. *Dose-Response, 16*.

Hussain, S., et al. (2021a). Satellite-based evaluation of temporal change in cultivated land in southern Punjab (Multan region) through dynamics of vegetation and land surface temperature. *Open Geosciences, 13*, 1561–1577.

Hussain, S., et al., et al. (2021b). Using space–time scan statistic for studying the effects of COVID-19. In Punjab, Pakistan: A guideline for policy measures in regional agriculture. *Environmental Science and Pollution Research., 30*, 42495–42508.

Hussain, S., et al. (2022a). Land use and land cover (LULC) change analysis using TM, ETM+ and OLI Landsat images in district of Okara, Punjab, Pakistan. *Physics and Chemistry of the Earth, 126*, 103117.

Hussain, S., et al. (2022b). Monitoring the dynamic changes in vegetation cover using Spatio-temporal remote sensing data from 1984 to 2020. *Atmosphere, 13*, 1609.

Hussain, S., et al. (2022c). Spatiotemporal variation in land use land cover in the response to local climate change using multispectral remote sensing data. *Land, 11*, 595.

Hussien, S. M., Ismail, E. S. A., Ismail, M. N., & Eid, T. A. (2013). Effect of applying surface and bubbler irrigation systems on fruitful Washington navel orange trees production. *Egyptian Journal of Agricultural Research, 91*(4), 1565–1580.

Jabran, K., et al. (2017). Water-saving technologies affect the grain characteristics and recovery of fine-grain rice cultivars. *Environmental Science and Pollution Research, 24*, 12971–12981.

Jahangirpour, D., & Zibaei, M. (2022). Farmers' decision to adoption of modern irrigation systems under risk condition: Application of stochastic efficiency with respect to a function approach. *Frontiers in Water, 4*, 931694.

Javeed, H. M. R., et al. (2021a). Lipoic acid combined with melatonin mitigates oxidative stress and promotes root formation and growth in salt-stressed canola seedlings (Brassica napus L.). *Molecules, 26*, 3147.

Javeed, H. M. R., et al. (2021b). Potential utilization of diluted seawater for the cultivation of certain summer vegetable crops: Physiological and nutritional implications. *Agronomy, 2021*(11), 1826.

Jenks. (2024). *Vertical farming and the evolution of sustainable farm-to-table*. Retieved from https://green.org/2024/01/30/vertical-farming-and-the-evolution-of-sustainable-farm-to-table.

Kalair, A. R., Abas, N., Hasan, Q. U., Kalair, E., Kalair, A., & Khan, N. (2019). Water, energy and food nexus of Indus water treaty: Water governance. *Water-Energy Nexus, 2*(1), 10–24.

Khaliq, M. A. (2022). Assessment of heavy metal accumulation and health risks in okra (Abelmoschus Esculentus L.) and spinach (Spinacia Oleracea L.) Fertigated with wastewater. *International Journal of Food Contamination, 9*, 11.

Khan, N., et al. (2018). Interaction between PGPR and PGR for water conservation and plant growth under drought condition. *Biologia, 73*(11), 1083–1098.

Khor, L. Y., & Feike, T. (2017). Economic sustainability of irrigation practices in arid cotton production. *Water Resources and Economics, 20*, 40–52. https://doi.org/10.1016/j.wre.2017.10.004

Lamaddalena, N., Khadra, R., & Tlili, Y. (2012). Reliability-based pipe size computation of on-demand irrigation systems. *Water Resources Management, 26*(2), 307–328.

Liu, H., Peng, D., Yang, H. J., Mu, Y., & Zhu, Y. (2023). Exploring the evolution of sustainable fisheries development: Focusing on ecological, environmental and management issues. *Ecological Informatics, 75*, 102004.

Loebbecke, C., & Boboschko, I. (2020). Reflecting upon sensor-based data collection to improve decision making. *Journal of Decision Systems, 29*(sup1), 18–31.

Ma, Y., Hu, X., Chen, Y., Hu, Z., Feng, T., & Feng, G. (2023). Different characteristics and drivers of the extraordinary Pakistan rainfall in July and august 2022. *Remote Sensing, 15*(9), 2311.

Mattoussi, W., & Mattoussi, F. (2022). Adoption of modern irrigation technologies in the presence of water theft and corruption: Evidence from public irrigated areas in Medjez-el-Bab. *Water Policy, 24*(3), 534–568.

Mehmood, A., et al. (2021). Exploring the potential of moringa leaf extract as bio stimulant for improving yield and quality of black cumin oil. *Scientific Reports, 11*, 24217.

Migliaccio, K. W., Morgan, K. T., Vellidis, G., Zotarelli, L., Fraisse, C., Rowland, D. L., Andreis, J. H., Crane, J. H., & Zurweller, B. A. (2015). Smartphone apps for irrigation scheduling. In *2015 ASABE/IA irrigation symposium: Emerging technologies for sustainable irrigation—A tribute to the career of Terry Howell, Sr. conference proceedings* (pp. 1–16). American Society of Agricultural and Biological Engineers.

Mirza, N., et al. (2017). Constitutional tolerance and chlorophyll fluorescence of Boehmeria nivea L in response to the antimony (Sb) and arsenic (As) co-contamination. *Toxicological and Environmental Chemistry, 99*(2), 265–272.

Moursy, M. A. M., ElFetyany, M., Meleha, A. M. I., & El-Bialy, M. A. (2023). Productivity and profitability of modern irrigation methods through the application of on-farm drip irrigation on some crops in the northern Nile Delta of Egypt. *Alexandria Engineering Journal, 62*, 349–356.

Mubeen, M., et al. (2021). Effect of plant growth promoting bacteria and drought on spring maize. *Pakistan Journal of Botany, 53*(2), 731–739.

Munir, M. S., Bajwa, I. S., Naeem, M. A., & Ramzan, B. (2018). Design and implementation of an IoT system for smart energy consumption and smart irrigation in tunnel farming. *Energies, 11*(12), 3427.

Nandanwar, H., Chauhan, A., Pahl, D., & Meena, H. (2020). A survey of application of ML and data mining techniques for smart irrigation system. In *2020 second international conference on inventive research in computing applications (ICIRCA)* (pp. 205–212). IEEE.

Nikolaou, G., Neocleous, D., Christou, A., Kitta, E., & Katsoulas, N. (2020). Implementing sustainable irrigation in water-scarce regions under the impact of climate change. *Agronomy, 10*(8), 1120.

Phasinam, K., Kassanuk, T., Shinde, P. P., Thakar, C. M., Sharma, D. K., Mohiddin, M. K., & Rahmani, A. W. (2022). Application of IoT and cloud computing in automation of agriculture irrigation. *Journal of Food Quality, 2022*(1), 8285969.

Qureshi, A. S. (2020). Groundwater governance in Pakistan: From colossal development to neglected management. *Water, 12*(11), 3017.

Ramachandran, V., Ramalakshmi, R., & Srinivasan, S. (2018). An automated irrigation system for smart agriculture using the internet of things. In *2018 15th international conference on control, automation, robotics and vision (ICARCV)* (pp. 210–215). IEEE.

Rasool, A., et al. (2018). A review of global outlook on fluoride contamination in groundwater with prominence on the Pakistan current situation. *Environmental Geochemistry and Health, 40*(4), 1265–1281.

Rodrigues, G. C., Paredes, P., Gonçalves, J. M., Alves, I., & Pereira, L. S. (2013). Comparing sprinkler and drip irrigation systems for full and deficit irrigated maize using multicriteria analysis and simulation modelling: Ranking for water saving vs. farm economic returns. *Agricultural Water Management, 126*, 85–96.

Sabagh, A. E. L., et al. (2021). Potential role of plant growth regulators in administering crucial processes against abiotic stresses. *Frontiers in Agronomy, 3*, 648694.

Sabagh, A. E. L., et al. (2022). Phytohormones as growth regulators during abiotic stress tolerance in plants. *Frontiers in Agronomy, 4*, 765068.

Saud, S., & e al. (2017). Effects of nitrogen supply on water stress and recovery mechanisms in Kentucky bluegrass plants. *Frontier in Plant Sciences, 8*(983), 01–18.

Shahzad, H., Farid, H. U., Khan, Z. M., Anjum, M. N., Ahmad, I., Chen, X., Sakindar, P., Mubeen, M., Ahmad, M., & Gulakhmadov, A. (2020). An integrated use of gis, geostatistical and map overlay techniques for spatio-temporal variability analysis of groundwater quality and level in the Punjab province of Pakistan, South Asia. *Water, 12*(12), 3555.

Shakeel, M., et al. (2017). Environment polluting conventional chemical control compared to an environmentally friendly IPM approach for control of diamondback moth, Plutella xylostella (L), in China: A review. *Environmental Science and Pollution Research, 24*, 14537–14550.

Sharif, M., Jabbar, A., Niazi, M. A., & Mahr, A. B. (2016). Managing water availability and requirements in Pakistan: Challenges and way forward. *Journal of Agricultural Research, 54*, 117–131.

Tahir, M., et al. (2022). Efficacy of organic-based carrier material for plant beneficial rhizobacteria application in okra under normal and salt-affected soil conditions. *Journal of Applied Microbiology, 133*(2), 943–959.

Violino, S., Figorilli, S., Ferrigno, M., Manganiello, V., Pallottino, F., Costa, C., & Menesatti, P. (2023). A data-driven bibliometric review on precision irrigation. *Smart Agricultural Technology, 5*, 100320.

Yang, L. R., O'Connor, J. T., & Chen, J. H. (2007). Assessment of automation and integration technology's impacts on project stakeholder success. *Automation in Construction, 16*(6), 725–733.

Yang, P., Wu, L., Cheng, M., Fan, J., Li, S., Wang, H., & Qian, L. (2023). Review on drip irrigation: Impact on crop yield, quality, and water productivity in China. *Water, 15*(9), 1733.

Zia, Z., et al. (2017). Effect of water management and silicon on germination, growth, phosphorus and arsenic uptake in rice. *Ecotoxicology and Environmental Safety, 144*, 11–18.

Precision Agriculture: Navigating Water Scarcity with Data-Driven Solutions

Muhammad Mohsin Waqas, Muhammad Ahmad Rustam, Isaya Kisekka, Sikandar Ali, and Wajid Nasim Jatoi

Abstract This book chapter explores the pivotal role of precision agriculture in addressing the challenges of water scarcity in modern agricultural practices. The chapter will describe the significance of precision agriculture in the context of contemporary farming and its vital role in managing water resources efficiently in water-scarce environments. The objectives of the chapter are outlined to guide the readers through the key topics covered. The fundamentals of precision agriculture are discussed, encompassing its definition, evolution, and integration of precision irrigation into farming systems. Various irrigation techniques, from traditional methods to modern technologies like drip irrigation and sprinkler systems, are examined, highlighting their advantages in enhancing water distribution efficiency. The chapter will cover precision agriculture technologies for water management, including soil moisture monitoring through sensor technologies, weather-based irrigation scheduling systems, remote sensing, and Geographic Information Systems (GIS) applications. Further, the implementation of the Internet of Things (IoT) in water management for precision irrigation will be discussed. Data analytics and big data play a crucial role in precision irrigation, hence the chapter focuses on harnessing big data for water use efficiency, decision support systems, and the application of machine learning algorithms for predictive irrigation management. The chapter

M. M. Waqas (✉) · M. A. Rustam
Department of Agricultural Engineering, Khwaja Fareed University of Engineering and Information Technology, Rahim Yar Khan, Pakistan
e-mail: mohsin.waqas@kfueit.edu.pk

I. Kisekka
Department of Land, Air and Water Resources, University of California, Davis, CA, USA
e-mail: ikisekka@ucdavis.edu

S. Ali
Department of Irrigation and Drainage, University of Agriculture, Faisalabad, Pakistan

W. N. Jatoi
Department of Agronomy, Faculty of Agriculture and Environment, The Islamia University of Bahawalpur (IUB), Bahawalpur, Punjab, Pakistan

M. Mubeen et al. (eds.), *Innovations in Agricultural Water Management*,
https://doi.org/10.1007/978-3-031-91883-4_4

further explores precision irrigation's impact on crop water requirements, emphasizing tailoring irrigation practices to crop growth stages and managing water stress through precision irrigation. Challenges and barriers in adopting precision irrigation are addressed, such as initial investment, technology acquisition, farmer training, capacity building, and infrastructural limitations. Case studies and success stories of precision irrigation implementations in different crops are presented, highlighting real-world impacts on water conservation and crop yields. The chapter will be concluded by outlining future prospects and opportunities in precision agriculture and irrigation and exploring the policy implications and government support required to promote sustainable agricultural water management.

Keywords Agriculture · Irrigation · Management · Remote sensing · Sustainability

1 Introduction

Smart farming or precision agriculture, a new and revolutionary agricultural concept rooted in data analysis along with tailored implementation of strategies using technology, is always there to revolutionize the traditional form of cultivation. It helps in optimizing farming procedure, increasing productivity, and promoting sustainable resources management. Through the use of tools such as Global Positioning System (GPS), remote sensing, Internet of Things (IoT) sensors, and data analytics, precision agriculture enables farmers to achieve a level of accuracy and efficiency in their decision-making processes that is unrivalled. Uniform strategy has been used in all aspects of traditional agriculture, but this mode is slowly changing to precision agricultural pattern that accounts for the inherent variability on which fields are based. This replaces the practice of applying treatments uniformly across a large field; for example, it will enable farmers to apply herbicides more precisely. This will then increase profits and reduce soil destruction.

GPS technology is crucial for accurate field mapping and autonomous machinery guidance, which are required to monitor crop health on a large scale. Satellite remote sensing, in addition to airborne unmanned aerial vehicles (UAVs) can serve shift-wise continuous monitoring for detection and timely warning on various crops staring biotic/abiotic stress/diseases/invasion of pests. This information allows farmers to intervene quickly when they need most and prevent significant crop losses. To this end, the authors leverage IoT and state-of-the-art sensor technologies for gathering comprehensive data regarding various environmental features such as soil moisture, temperature, or humidity. Thereafter, this data is put through next-level scrutiny using advanced data analytics and Machine Learning (ML) algorithms that then are able to detect comprehensible patterns, trends, and insights as well. These findings form important edicts for the farmers, which allow them to further improve their agriculture practices.

However, while the implementation of precision agriculture may be challenging in terms of initial investment costs as well as technical and data management capabilities, available tools encourage proceeding to it even if barriers exist. With the need for sustainability as the global population grows, precision agriculture paints a bright light among future farming prospects. Impacted by further technological advancement and availability of new age tools, precision agriculture is set to transform agri-landscape by promising food security and environment sustainability along with prosperity for farmers and communities across the world (Fig. 1) (Cisternas et al., 2020; Sabagh et al., 2020; Akram et al., 2019a, 2019b; Fahad et al., 2019; Amanet et al., 2019; Sami Ul Din et al., 2019; Ali et al., 2019; Tariq et al., 2019; Zahoor et al., 2019).

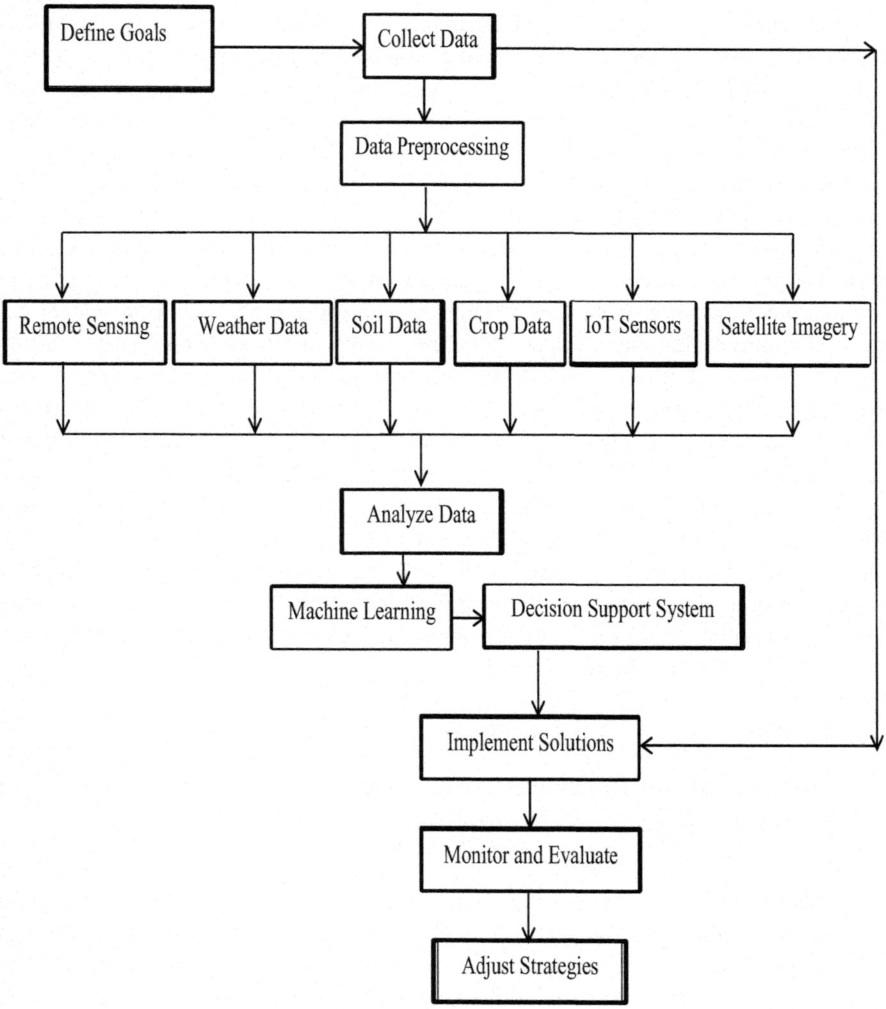

Fig. 1 Precision agriculture framework

2 The Role of Precision Agriculture in Modern Agriculture

Agrochemicals, fertilizers, and irrigation are applied to create a precision agriculture system. By taking orders for live data from sensors and satellite imagery, we have the necessary information to detect soil gradients or differences in plant health (yield) across fields with high accuracy at a very precise resolution, variability, moisture content, etc. (Vecchio et al., 2020). With this information on hand, input applications can be tailored to individual zones, keeping waste and environmental impact as low as possible to yield a maximum crop. This not only reduces input costs but also helps to minimize unintended negative consequences from the overuse of chemicals, ensuring the quality of water and soil. Several data are shaping the future of precision agriculture. Data analytics in precision agriculture gives farmers an important edge over those who have to rely only on intuition. By analyzing trends and patterns in crop performance, weather conditions, and a host of other variables, farmers can plant the right fields at the correct times forecasting yields to make better management decisions as needed. The data-centric perspective broadens and endows precision agriculture to improve the overall efficiency and profitability of agricultural operations, further strengthening its pivotal role in contemporary farming practices (Vecchio et al., 2020).

Additionally, precision agriculture contributes to greater crop supervision and farming practices. Farmers have been able to watch crop health and growth more continuously because drones are capable of capturing high-resolution imagery at near-synoptic-scale temporal and spatial resolution. It allows for an early warning system of pests, diseases, and nutrient deficiencies in time to act. Farmers can prevent issue propagation by pinpointing the problematic areas and a passive intervention through, for example, localized treatments to maintain overall crop potency. Leveraging GPS-guided machinery, IoT sensors, and data analytics for precision agriculture leads to smarter application of inputs based on need rather than traditional methods as well as real-time decision-making, which enhances resource efficiency and sustainability. Using technology and data-driven insights, it enables extensive crop monitoring, yield mapping, and water conservation, encouraging economic viability and driving this into environmental sustainability, which is a game changer for smallholder farmers.

3 Importance of Precision Irrigation in Water-Scarce Environments

The precision irrigation in agricultural practices under climate change are crucial to address the challenges of water scarcity (Liang et al., 2020). One of the first things around precision irrigation that comes into our mind is water optimization. The adoption of modern technologies like soil moisture sensor devices, weather data services, and automatic irrigation systems allows farmers to accurately decide on the when-and-how-much water is needed for their crops (Canaj et al., 2022). This aimed approach saves water from over-irrigation and avoids moisture stress, preserving every drop of water for the crop.

4 Evolution of Precision Agriculture

It can be defined as an advanced farming technique that uses technology, data analysis, and other precision tools to ensure the right amount of crop at the right place on the farm resulting in optimal production with minimal wastage, which makes farmers' lives a bit easier. Precision agriculture should increase greenhouse gas (GHG)-related efficiency and food productivity while minimizing resource use, which is just tip of the solution. The main objective is to improve crop yields (Barnes et al., 2019). It is a cropping system that uses inputs (water, fertilizers, and pesticides) with precise application for maximum food production efficiency. Precision agriculture allows farmers to use data analytics, automatically identify zones within a field, and customize farm operations to deliver more sustainable farming with higher profit.

Precision agriculture is the evolution of traditional farming practices, making it smarter, more productive, and more sustainable. Precision crop farming began with computer science and technology that started in the 1980s. Its initial applications were limited to basic tools like GPS, which allowed for precise field and equipment mapping. As a result, it enabled farmers to avoid overlaping of the inputs and optimize the inputs based on the right decision with accurate spatial variability.

The sensor technology and data collection methods became more advanced in the 1990s and early 2000s for precision agriculture. Thanks to advanced technologies such as yield monitors, soil moisture sensors, and aerial imagery, these all became cheaper and more available. These allowed farmers to take immediate soil readings, make plant health assessments, and assess localized weather patterns so that they could be more nimble with data decision-making. Using Geographic Information Systems (GIS) and other data analysis tools, farmers could map their fields in great detail to perform variable rate applications of fertilizer, pesticides, or water based on the needs for particular areas within each field.

The history of precision agriculture started with the introduction of different technologies and concepts that held the building blocks for modern precision farming practices in the twentieth century. Precision agriculture is made up of five main components:

- Global Positioning System (GPS): GPS technology allows farmers to measure acreage and identify lot boundaries. Vehicle GPS is enabling precision in your fields, even on a small scale: from creating detailed field maps and planning operations to tracking equipment movements for more informed decision-making.

- Remote Sensing: Precision agriculture utilizes a number of remote sensing technologies, like drones, satellites, and aerial imagery, to monitor crop health status, detect stress factors, along with field variability.
- Variable Rate Technology (VRT): This technology allows farmers to apply inputs like fertilizers, pesticides, and irrigation at different rates across a field based on real-time data about soils and other crop factors so that resources are not wasted or the environment is not contaminated due to excess use of chemicals. This makes it custom resource implementation to fit correctly where needed and reduce waste, or optimize the usage of resources.
- Automated Machinery: Precision agriculture in return uses automated machinery such as self-operated tractors, and robotic harvesters. Such machines can run in predetermined paths with the help of GPS and reduce manpower as well as increase the efficiency of field duties.
- IoT Sensors: It gathers the real-time data of soil moisture, humidity, temperature, and other surroundings by deploying sensors based on the Internet of Things. The information collected by these sensors is used to make predictive maintenance schedules on irrigation scheduling and nutrient dosing decisions.

5 Integration of Precision Irrigation in Farming Systems

In precision agriculture, one of its components is precision irrigation in which water can be more accurately applied through an irrigation system to the crops according to the information obtained from real-time data and field variability. Precision irrigation in farming systems brings multiple benefits and transforms the farmers, environment, as well agricultural sustainability (Kamienski et al., 2019). The technology etched into this kind of irrigation system is designed to customize water and nutrient application for each plant while integrating the use of soil moisture sensors, weather reports, as well as advanced algorithms (Kamienski et al., 2019; Akram et al., 2017a, 2017b, 2018a, 2018b; Hasnain et al., 2017; Hashmi et al., 2017; McDermid et al., 2015; Ahmad et al., 2015).

6 Advancements in Irrigation Techniques

6.1 Traditional Irrigation Methods: Limitations and Challenges

Older surface irrigation methods have been practiced for thousands of years on large fields that are relatively flat. This type of extensive application to crops can still be an important part of crop cultivation in Europe and where there is more water than other modern techniques from (e.g., Calif.). These systems use gravity to move

Fig. 2 Flood irrigated farmer field

water across the landscape and have adapted in different ways as they head into their second century on working lands. Although traditional methods of surface irrigation are easy to perform and also more cost effective, when it comes to water use they can be less efficient as compared with the contemporary precision method.

6.2 Flood Irrigation

Flood irrigation is a simple and old form of surface irrigation, depicted in Fig. 2. The water is piped on the field and permitted to flow freely across the entire surface of crop in this method. This method is suitable for level or nearly level soils and is often used in paddy rice lands with some types of vegetable crops (Li, 2020).

The significant drawbacks include substantial water wastage through evaporation and deep percolation, uneven water distribution, and soil erosion. It is labor-intensive, inefficient in water use, can damage crops, and lacks adaptability to changing conditions.

6.3 Furrow Irrigation

Furrow small channels run between crop rows, and water is applied directly into furrows (Fig. 3). Furrow irrigation water is supplied to the root zones of crops by flow from furrows. This method is often used for row crops such as corn, cotton, and

Fig. 3 Furrow irrigation at KFUEIT research farm

potatoes. Although furrow irrigation can use less water than flood irrigation, the introduction of extreme slopes or soil variability may see issues around uneven spread of water (Ippolito et al., 2019).

Significant water losses through runoff and deep percolation, soil erosion, and compaction, offer limited suitability for certain crops and lacks the precision of modern irrigation methods, leading to inefficient water use and limited control over water application.

6.4 Basin Irrigation

In basin irrigation, the small flat areas in the field where water is kept and it enters into soil by means of seepage are called basins as represented in Fig. 4. Basin irrigation is a type of surface irrigation that can be simple and well suited for areas with soils having low infiltration rates (Kassem & Hamadamin, 2023). On the other hand, it does have a tendency to use up water at an exorbitant rate. If not managed correctly, H_2O may accumulate and will be lost because of evaporation mapping network.

The problems associated with basin irrigation are waterlogging and soil salinity, very high evaporation losses, as well as its unsuitability to implement on steep field. Due to irregular water control, seafood, and vegetable obstruction in the sluice, it is

Fig. 4 Basin irrigation in farmer field

usually confined to certain sectors of crop irrigation causing increased management complexity and lower efficiency compared with some modern watering methods.

6.5 Border Irrigation

Border irrigation is also a free flooding system where long narrow strips or borders are made traversal across the field with raised earth bunds on the sides (Singh et al., 2018) as shown in Fig. 5. This is how it should be; at one of its ends the water comes out and goes over the entire surface wetting a strip where we will have crops. This technique is suitable for fields with gentle slopes and it is largely practiced in the case of fodder crops, vegetables, or small fruit orchards. Border irrigation systems can have lower or non-uniform water application along the border and losses of water at the edges of fields.

However, border irrigation can lead to water losses at the field edge, is limited by slope, and is vulnerable to both evaporation and soil erosion. In addition, plants may obstruct its passage; then water becomes unevenly distributed or even wasted altogether.

6.6 Drip Irrigation: Precision Water Delivery at the Root Zone

Drip irrigation employs drip emitters, tubing, filters, and weather-sensitive controllers to bring water, drop by drop, into individual plants or precisely defined areas. In this way, loss due to evaporation is minimized. Farmers can control the flow of water exactly, which prevents excessive watering but also underwatering plants,

Fig. 5 Border irrigation at KFUEIT research farm

Fig. 6 Drip irrigation fields installed by Rabail Technologies

which are harmful for crop health and yield levels (Çetin & Kara, 2019). Here are the main benefits, and the system is displayed in Fig. 6.

1. Drip irrigation ensures water efficiency, further reducing losses from evaporation and runoff. It directs water directly to the root zone, decreasing surface evaporation and thereby allowing for a more efficient absorption by plants, increasing crop yield per volume of irrigation.
2. Drip irrigation leads to improved crop health as it provides the right amount of water and nutrients at all times for each plant, which allows farmers to benefit from increased yields along with better-quality produce.

3. Drip irrigation doses water directly to the crop root zone and leaves between inter-row spaces where weeds often exploit for shelter dry, thus discouraging weed growth.
4. Nutrients are placed in the plant root zone, minimizing fertilizer washout and nutrient losses.
5. Compared to other high-efficiency irrigation methods, drip irrigation requires lower water pressure, which saves energy for pumping.
6. Hilly topography can also be addressed by implementing drip irrigation.
7. It is applicable to a wide range of crops like fruits, vegetables, field crops, orchards, and nurseries.
8. Drip irrigation is a techniques of saving water, reducing soil erosion and runoff, and decreasing the leaching out of nutrients in soils into local bodies of water.
9. These are easy to set up, and can be controlled automatically using timers or controllers, which offer far more control over irrigation time.
10. The installation costs of drip irrigation systems may be higher than some other methods at the outset, but substantial cost savings are earned through prudent water use and lower labor needs with increased yield that a properly implemented rain hose provides in the long term.

Although efficient, drip irrigation has knack for being high-priced to set up and needs constant maintenance to avoid clogging that can be due to the presence of minerals from harsh water. You will need a good design and layout system in place, especially on sloping land, as well as less is doable for large farms unless you have the knowledge or training.

6.7 Sprinkler Systems: Enhancing Water Distribution Efficiency

Sprinkler irrigation works by pressurizing water and distributing through a network of pipes, risers, and sprinklers, as shown in Fig. 7. Sprinkler heads are placed at a specific distance throughout the ground so as to ensure complete coverage. As the system operates, it rotates or emits water in a specific manner and provides rainfall-like coverage to evenly wet crops and soil. Lower distribution accuracy decreases overspray and runoff, making water reaching plant root zones more effective (Maroufpoor et al., 2019). Spray unit of sprinklers allows to control water application rates, which can perform localized irrigation with better uniform growth in each area (Maroufpoor et al., 2019).

1. Improved water application rates and uniformity through a combination of system design and nozzles are key advantages that result in reduced runoff as well as evaporation.
2. The water is applied slowly by sprinkling, as opposed to a flood or furrow method of irrigation that forces it on quickly causing washout.

Fig. 7 Sprinkler irrigation installed by Rabail Technologies

3. This method of irrigation is suitable for field crops, vegetables, and fruits as well as lawns and gardens having ornamental plants.
4. It can be used in areas that are prone to frost events, providing crops with protection against the cold.
5. Sprinkler irrigation systems are relatively easy to control by regular automatic controllers as well as modernized microprocessor-controlled solenoids.
6. Sprinkler irrigation provides an even distribution of water over the field so that each plant receives uniform moisture, which leads to development.
7. In a hilly area, sprinkler irrigation can be used by mounting the heads of adjustable sprinklers on permanently placed stands or rollers.
8. The sprinkler irrigation application of water is controlled and uniform, consequently improving crop productivity significantly along with yield. As the system irrigates consistently, it makes sure that the crops have enough moisture to grow and develop.
9. Sprinkler irrigation systems when automated do not require a lot of man-power, in turn reducing the human power over the process of irrigation.

6.8 Smart Irrigation Controllers: Automating Precision Irrigation

The dearth of drinkable fresh water has become a pressing concern, and so are the numbers staggering when we turn our eyes toward Mediterranean countries or down south in Asian countries such as Pakistan and India (Khanna & Kaur, 2019). Usage of IoT in agriculture is essential to improve productivity and preserve water for the future (Khanna & Kaur, 2019). Components of IoT-based agriculture system are shown in Fig. 8.

Fig. 8 Components of IoT-based agriculture system (AGEN Project)

To address these difficulties, accurate flow and soil moisture data using ultrasonic sensors can help optimize irrigation techniques (Shafi et al., 2019). In addition, effective agricultural resources development and management will open jobs in rural areas that can reduce trend toward rural to urban migration, which is also an alarming sign. Investment by Pakistan on indigenously developed IoT-based sensor technology for balancing sustainability and encouraging transition from conventional agricultural methods to precision agriculture is crucial. Working principle for the project is described by the flow chart depicted in Fig. 9.

7 Precision Agriculture Technologies for Water Management

7.1 Sensor Technologies for Soil Moisture Monitoring

Soil moisture, irrigation, and water usage: Sensor technology, in turn, can give land managers an up to date picture of how much moisture is actually present in the soil profile when compared with what might be happening on a broader scale. Various sensor technologies are used for soil moisture monitoring, each of which has its strengths and limitations. Below, we get into some of the top recommended sensor technologies for soil moisture monitoring.

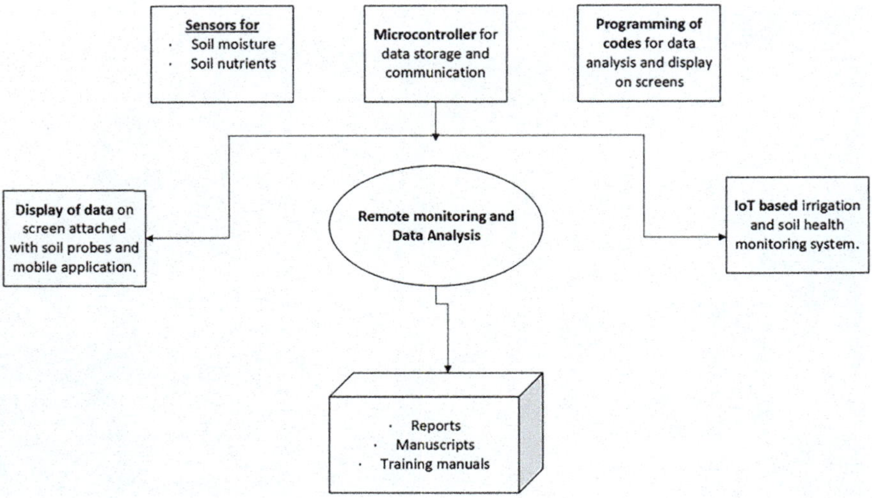

Fig. 9 Flow chart describes the working principle for the project

Pictures from the Internet

Time Domain Reflectometry (TDR) Time domain reflectometry is one of the most widely used and accurate soil moisture measurement techniques. TDR sensors work based on the principle that the propagation time of an electromagnetic pulse is affected by the dielectric constant of the soil, which is related to its moisture content	
Frequency Domain Reflectometry (FDR) Frequency domain reflectometry is another electromagnetic-based soil moisture measurement method. FDR sensors determine soil moisture by analyzing the frequency shift of an electromagnetic wave as it passes through the soil	
Capacitance Sensors Capacitance sensors measure soil moisture by monitoring changes in electrical capacitance between two electrodes inserted into the soil. As soil moisture increases, the dielectric constantly changes, resulting in a measurable change in capacitance	

Resistive Sensors Resistive sensors, also known as gypsum blocks or granular matrix sensors, measure soil moisture based on the electrical resistance between two electrodes embedded in a porous medium. As soil moisture changes, the resistance of the porous medium changes, allowing the sensor to estimate soil moisture levels	
Dielectric Sensors Dielectric sensors use the interaction between an electromagnetic field and the dielectric properties of the soil to estimate soil moisture levels. These sensors can be either capacitance-based or impedance-based	

7.2 *Remote Sensing and GIS Applications in Precision Irrigation*

In agriculture, precision irrigation is a field that has undergone incredible transformation with the advent of remote sensing and Geographic Information Systems (GIS). Precision irrigation is the need of the time as the fresh water supply is reducing while the demand of irrigation to feed the swiftly incrseasing pouplation is increasing. There is need to educate the farmers for effective decisions making for improving crop productivity, sustainable exploitation of water resources, and reduction in environmental consequences (Sishodia et al., 2020). There are certain parameters related to crops (such as vegetation vigor, moisture content, and the level of stress) that produce invaluable crop parameter data that can be determined by using remote sensing. With medium to high-quality imagery such as satellite or aerial, you can capture extensive agricultural areas in a scalable way that can provide key intelligence establishing the crops condition. This information is then easily accommodated into GIS to allow farmers to create spatial cartographic illustrations of variations in crop health and requirements for irrigation over their farm fields (Sishodia et al., 2020). The uses of remote sensing in agriculture are as follows:

1. **Crop Health**: Remote sensing technologies like multispectral/hyperspectral sensors onboard satellite imagery or aerial drones are used to capture reflected light from crops for analysis of crop health. Capable of detecting subtle changes in plant reflectance that serve as an indication of the chlorophyll content and

stress levels, farmers can, therefore, identify poor growing or stressed areas and apply localized irrigation practices while also scheduling for supplementary water applications elsewhere.

2. **Water Stress Detection**: Using remote sensing, farmers can locate water-stressed areas in the field by observing vegetation indices such as Normalized Difference Vegetation Index (NDVI) or Crop Water Stress Index (CWSI). The indices define the level of water stress in plants, helping farmers target irrigation efforts and manage scarce water resources more effectively on a field-by-field basis.
3. **Soil Moisture Monitoring**: Radar and passive microwave sensors will let us estimate soil water content in large agricultural areas. That information is key for planning precision irrigation, which allows farmers to determine when and where they should apply water and in what amounts so that they can make the best use of available land-water resources.
4. **Irrigation Scheduling**: This involves the use of remote sensing data, manipulated in conjunction with GIS maps and weather forecasts to create detailed irrigation-scheduling models for farmers. Based on the crop water demands, actual soil moisture status, and weather forecasts, water availability models generate schedules for irrigation to minimize wastage of water and maximize agricultural productivity.
5. **Variable Rate Irrigation (VRI)**: GIS helps in the execution of variable rate irrigation (VRI) where irrigation is practiced at different rates across the field due to spatial variation in soil properties and crop water requirements. VRI prescriptions are maps created based on data gathered from remote sensing and other sources, which is then used to guide irrigation equipment in dispensing custom water applications (from very little or none at all) on specific areas of the field just when it is needed.
6. **Water Resource Mapping and Management**: Water resource mapping is essential to precision irrigation water management, for which GIS appends all unique qualities. Farmers can use it to map out water sources, create plans for the deliverance of the resource, as well reinforce their planned-out irrigation. Remote sensing merged with GIS guarantees the real-time monitoring of water bodies, reservoir levels, and quality to practice sustainability.
7. **Remote Sensing and GIS for Drought Monitoring and Early Warning Systems**: This constant surveillance of the health and moisture levels in vegetation also allows systems to predict a drought early so that farmers can take action before the crop fails due to lack of water.
8. **Crop Yield Prediction**: Remote sensing data with GIS and Machine Learning methods can be used to predict crop yields so that based upon the yield prediction farmers could make decisions on harvest time schedules or prepare in advance for supply scaling, etc. Most of the agricultural management practices like when to irrigate, how often we need irrigation, and harvesting based on

crop growth functions can be derived by looking at historical data as well as real-time in some cases.

9. **Eligibility and insurance**: Remote sensing, using GIS-based tools for risk assessment, is used by the banks to check crop health conditions as well as yield loss possibilities due to different factors like water stress, pest infestation, or diseases. This data helps farmers get better insurance operating across a region.
10. **Environmental Impact Assessment**: Remote sensing techniques help to locate locations with the potential for over-irrigation or waterlogging and also evaluate irrigation consequences on aquatic resources, directing toward sustainable irrigation implementations.

7.3 Internet of Things (IoT) in Water Management

Over the years, IoT has undoubtedly been a great technological enabler across industries, and water management is in no way an exception. The platform is efficiently using the Internet of Things (IoT), sensors, and data analytics for real-time water resource monitoring, control, and conservation. The use of IoT-enabled water management adds spontaneous analytics and transparency to a more sustainable approach in the deployment, usage, or distribution of high-quality water (Singh & Ahmed, 2020). As you know, the most popular use of IoT in water management is smart metering. Manual readings are often required with traditional water meters, and they do not provide real-time data. On the hand, IoT-based smart meters can continuously monitor your water consumption and automatically send readings to utility companies. This allows for precise billing, leak localization, and prediction of water demands, which all help to minimize waste since it aids in the conservation (Sardar et al., 2020; Hammad et al., 2020; Zamin et al., 2020; Shahzad et al., 2020; Awais et al., 2020; Abbas et al., 2020; Mubeen et al., 2020; Gonzalez-Vidal et al., 2019).

One critical use of IoT in managing water resources includes the adoption of intelligent and smart water networks. This involves networks of sensors and actuators embedded throughout the water infrastructure (e.g., pipelines, reservoirs, treatment plants). These sensors measure metrics like pressure, flow rates, and water quality, while those actuator valves and pumps can be controlled from a far. This data will be used to monitor water distribution, determine leak locations, and enable a rapid response in order to reduce losses of drinking water while guaranteeing proper supply.

IoT water management is the combination of smart sensors, devices, and communication technologies with data related to the use of water. The use of IoT solutions in water management improves the efficiency, conservation, and sustainability related to protected drinking supply chains. So, these may be the uses of IoT in water management.

7.4 Data Analytics and Big Data in Precision Irrigation

Precision irrigation is a data science practice in agriculture that uses big data to optimize the watering of crops, improve crop yields, and increase the overall sustainability of agricultural systems (Luqman, 2023). Precision irrigators access data on the water status of soils, likely weather for their region, and crop growth patterns produced by powerful analytics that process huge volumes of collated information from a multiplicity of sources including soil moisture sensors. This data is very precious for farmers to know about the soil, crop health, and water requirements, by analyzing this data (Luqman, 2023). They can use water-use data to determine how they might modify their irrigation schedule and what kind of precision irrigation could be useful—by applying more efficient systems such as drip or using a soil moisture-based system. By continuously analyzing data and receiving feedback, precision irrigation solutions can adjust and adapt, making it possible for these systems to ensure the best use of water throughout its various states while reducing tremendous amounts of wasted resources. Farmers can use big data to accurately plan the irrigation, choose a suitable crop, and even decide what land management practices would increase their yield per unit area with reduced resource requirements (Kamienski et al., 2019).

7.5 Machine Learning Algorithms for Predictive Irrigation Management

One of the most significant contributions has come from Machine Learning algorithms in predictive irrigation management in precision agriculture. The aforementioned special issue contains many relevant papers dealing with different types of algorithms used to analyze historical or real-time data from weather stations, soil moisture sensors, crop health monitoring, and other sources for predicting the requirement of water by crop per unit of time as well as to make more effective irrigation scheduling (Sharma et al., 2021). Machine Learning thus saves water for farmers, enables analysis-based decision-making, and consequently increases crop yields. Machine Learning algorithms have become powerful tools for predictive irrigation management in agriculture (Baburao et al., 2023). This innovative technology uses historical and real-time data to determine how water resources can be best allocated such that crop yield is maximized, while minimizing the overall amount of fresh or irrigation water used. While it is too powerful to cover all Machine Learning algorithms in under one post, some key ML algorithms have been essential for most of the coding challenges so far (Hussain et al., 2020a, 2020b; Hashmi et al., 2020; Zaman et al., 2020; Farid et al., 2019; Danish et al., 2019; Hammad et al., 2019; Imran et al., 2019).

Ensemble learning, of which Random Forest is the most predominant algorithm in predictive irrigation management, specializes in large datasets and was better able to model the intricate inter-factor relationships between weather patterns, soil moisture levels, and crop types. RL_Random Forest can make the optimum irrigation scheduling and it produces precise predictions based on these variables so the crops receive water at each interval of time.

Support vector machines (SVMs) are another powerful ML algorithm employed in predictive irrigation management. SVMs are particularly adept at classification tasks, which makes them useful for distinguishing between different irrigation zones within a field based on diverse data inputs. By categorizing areas with similar water requirements, farmers can tailor their irrigation strategies to minimize water wastage and maximize crop productivity. *Artificial neural networks* (ANNs) are deep learning models capable of capturing intricate patterns in data. In predictive irrigation management, ANNs can process vast amounts of information, including historical crop performance, soil characteristics, and weather data. They can then generate precise irrigation recommendations, optimizing water distribution across the farm and adapting to changing environmental conditions.

8 Precision Irrigation and Crop Water Requirements

Precision irrigation is a modern watering technique that targets to maximize the water use efficiency of an agroecosystem and ensures applying the correct amount against crops right when they needed, where they required. The data-driven decision-making and real-time monitoring allow the implementation of smart irrigation technologies that guarantee efficient use of water, an economic benefit for the farmers who have limited resources to buy fuel or electricity pumps, as well as a way to reduce waste and improve productivity ensuring sustainable management using less computational technology investment that allows our system to be accessible in remote locations. One of the major components in precision irrigation is knowledge about crop water requirements, which helps farmers to know when and how much amount should be provided at what intervals. The following are detailed notes on precision irrigation and crop water requirement.

8.1 Crop Water Requirements and Evapotranspiration

Crop water requirements include the sum of necessities crops need for growth and development, including soil surface evaporation from earth to atmosphere. It is the release of vapor from leaves as they absorb water at roots, that is, transpiration. The crop water needs are varying according to the crops and growth stage, climate condition, soil types, and local environmental factors.

Crop Water Requirement: It speaks of the whole requirement of water for a crop that can support it in reproduction, development, and physiological functions. This includes the need of water for transpiration, which is a process that involves plant roots absorbing water and then sweating it as vapor into air through their leaves, as well as soil evaporation from surface. The crop water requirement is specific to the type of crop, stage of growth, climate, soil characteristics, and management procedures employed. Crop water requirements are commonly defined in units of depth (such as millimeters or inches) that refer to the volume of water necessary not only to restore soil moisture deficit but also used by plants and that evaporate into atmosphere.

Evapotranspiration (ET): ET is affected by weather conditions such as temperature, humidity, wind speed, and solar radiation; the type of crop being grown; its canopy size; and soil moisture. Correct estimation of evapotranspiration is the address to efficient irrigation programming and for proper planning in water resources, which undoubtedly evolves conservation agriculture.

8.2 Water Stress Management Through Precision Irrigation

With the expanding global demand for food on the one hand and dwindling water resources on the other, paddycrop irrigation is a major area of concern in agriculture. But, one of the key strategies to do this successfully and sustainably is precision irrigation. The practice in modern irrigation methods includes the exact amount of water application to crops that are peculiarly required by them, which should be not too little, not much, and with minimal wastage (Borah & Borah, 2023).

Data-driven technology represents the bedrock of precision irrigation, as it allows you to use devices such as sensors, weather forecasts, and soil moisture monitoring. Such technologies offer instantaneous data concerning soil conditions and weather, which allows growers to gauge when—or not—water should be applied. This avoids a lot of water wastage and there is no need for over-irrigation if irrigation scheduling can be tailored to meet specific field or crop requirements. In addition, Geographic Information Systems (GIS) and precision irrigation go hand in hand. GIS tools enable farmers to isolate plane fields of topography, soil types, crop type variety. Such maps are needed to perform irrigation zoning, which helps in proper allocation of water resources by the farmers (Rowshon & Amin, 2010). You can deliver water exactly where you need it, hitting different zones with varying degrees of thirst.

9 Future Prospects and Opportunities

Technological development in precision agriculture and irrigation brings about a paradigm shift from tradition farming practices to modern mechanized agriculture (Kata & Aparna, 2022). These changing times have been instrumental for the evolved form of agriculture that is steered by technologies such as Internet of Things (IoT), remote sensing, data analytics, and machine learning These trends are changing the face of agriculture by using resources more efficiently, decreasing environmental damage, and boosting crop yields.

A big trend is the rising use of IoT devices and sensors in agriculture. Some of these devices gather in-situ data including soil moisture, weather patterns, and crop health along with the machinery performance (Fatima et al., 2021), after which this data is processed and evaluated for making decisions. For example, at the University of California, Davis, researchers incorporated sensor networks that measure soil moisture and wirelessly transmit data to a farmer's smartphone. This is helpful in proper irrigation scheduling and not wasting the water, ultimately leading to better water use efficiency in ETc.

Policymakers need to ensure an enabling environment for precision agriculture by facilitating a legal and supportive framework. Moreover, strict data ownership frameworks as well as practices related to privacy and sharing can help give farmers the confidence they need in investing further while eliminating fears of misuse on their own agricultural operation. In addition, governments can support the adoption of these technologies by providing tax relief and offering subsidies or grants that will lower production costs for farmers. Take for instance the European Commission's (EU's) Common Agricultural Policy, which earmarks funds to support agriculture modernization and innovation called precision farming practices. Government support also concerns the implementation of essential infrastructure, notably in remote and underserved areas. Real-time data transport, remote to central sensing, and cloud data analytics of precision agriculture demand a robust Internet connection with near 100% availability and more or less uninterrupted power supply. Governments may choose to make large investments in rural infrastructure development that would afford all farmers an equal opportunity for precision agriculture technologies.

10 Conclusion

Precision irrigation practices, supported by technologies like IoT (Internet of Things), automation, and remote sensing, provide real-time data to help farmers in making precision decisions. This level of precision saves water wastage and reduces the chances of any unforeseen shortages for future use leading to increased using efficiency at global scales. In water scarce environments, precision irrigation is a vital tool to improving crop water productivity. This way the farmers can apply their irrigation practices considering crop growth stages and water need and a variety of factors in this region, to maximize yields without wasting this resource. On top of that, the capacity of precision agriculture to identify soil variability and undertake variable rate irrigation enables creating tailored water management practices by area across a field. In this way, water is brought to bear exactly where it is necessary and scarce, thus conserving available supplies of water for use on the land.

References

Abbas, Y., et al. (2020). Valorization of solid waste biomass by inoculation for the enhanced yield of biogas. *Clean Technologies and Environmental Policy, 22*, 513–522.

Ahmad, A., et al. (2015). *Impact of climate change on the rice–wheat cropping system of Pakistan* (pp. 191–221). World Scientific Publishing Co Pt. Ltd.

Akram, R., et al. (2017a). Role of antibiotics in climate change. In M. Hashmi, V. Strezov, & A. Varma (Eds.), *Antibiotics and antibiotics resistance genes in soils. Soil biology* (Vol. 51). Springer.

Akram, R., et al. (2017b). Fate of antibiotics in soil. In M. Hashmi, V. Strezov, & A. Varma (Eds.), *Antibiotics and antibiotics resistance genes in soils. Soil biology* (Vol. 51). Springer.

Akram, R., et al. (2018a). Fate of organic and inorganic pollutants in Paddy soils. In M. Hashmi & A. Varma (Eds.), *Environmental pollution of Paddy soils. Soil biology* (Vol. 53). Springer.

Akram, R., et al. (2018b). Paddy land pollutants and their role in climate change. In M. Hashmi & A. Varma (Eds.), *Environmental pollution of Paddy soils. Soil biology* (Vol. 53). Springer.

Akram, R., et al. (2019a). Plant growth and morphological changes in rice under abiotic stress. In H. Mirza, M. Fujita, & J. K. Biswas (Eds.), *Advances in Rice research for abiotic stress tolerance*.

Akram, R., et al. (2019b). Global trends of E-waste pollution and its impact on environment. In Zaffar Hashmi, M., & Varma, A. (Eds.), *Electronic waste pollution*.

Ali, M., et al. (2019). Role of ICT in crop management. In Mirza, H., (Ed.), *Agronomic crops*.

Amanet, K., Mubeen, M., et al. (2019). *Cotton production in Africa* (pp. 359–369). Wiley.

Awais, M., et al. (2020). Adaptation strategies for sunflower production under climate change for arid environment by using CROPGRO and APSIM-sunflower models. *Pakistan Journal of Agricultural Sciences., 57*(6), 1529–1538.

Baburao, P. S., Kulkarni, R. B., Kharade, P. A., & Patil, S. S. (2023). *Review of machine learning model applications in precision agriculture (issue June)*. Atlantis Press International BV. https://doi.org/10.2991/978-94-6463-136-4_81

Barnes, A. P., Soto, I., Eory, V., Beck, B., Balafoutis, A., Sánchez, B., Vangeyte, J., Fountas, S., van der Wal, T., & Gómez-Barbero, M. (2019). Exploring the adoption of precision agricultural technologies: A cross regional study of EU farmers. *Land Use Policy, 80*, 163–174. https://doi.org/10.1016/j.landusepol.2018.10.004

Borah, I. D., & Borah, S. R. (2023). Precision irrigation systems: Enhancing water efficiency and sustainability in agriculture. *Vigyan Varta, 4*(7). Retrieved from www.vigyanvarta.comwww.vigyanvarta.com.

Canaj, K., Parente, A., D'imperio, M., Boari, F., Buono, V., Toriello, M., Mehmeti, A., & Montesano, F. F. (2022). Can precise irrigation support the sustainability of protected cultivation? A life-cycle assessment and life-cycle cost analysis. *Water (Switzerland), 14*(1). https://doi.org/10.3390/w14010006

Çetin, O., & Kara, A. (2019). Assesment of water productivity using different drip irrigation systems for cotton. *Agricultural Water Management, 223*(June), 105693. https://doi.org/10.1016/j.agwat.2019.105693

Cisternas, I., Velásquez, I., Caro, A., & Rodríguez, A. (2020). Systematic literature review of implementations of precision agriculture. *Computers and Electronics in Agriculture, 176*, 105626. https://doi.org/10.1016/j.compag.2020.105626

Danish, S., et al. (2019). Alleviation of chromium toxicity in maize by Fe fortification and chromium tolerant ACC deaminase producing plant growth promoting rhizobacteria. *Ecotoxicology and Environmental Safety, 185*, 109706.

Fahad, S., et al. (2019). Abiotic stress and rice grain quality. In H. Mirza, M. Fujita, & J. K. Biswas (Eds.), *Advances in rice research for abiotic stress tolerance*.

Farid, H. U., et al. (2019). Assessing seasonal and long-term changes in groundwater quality due to over-abstraction using geostatistical techniques. *Environmental Earth Sciences, 78*, 386.

Fatima, N., Memon, K. F., Ahmed, J., Khand, Z. H., Gul, S., Kumari, M., & Mujtaba, G. (2021). Precision agriculture using Internet of thing with artificial intelligence: A systematic literature review. *University of Sindh Journal of Information and Communication Technology, 5*(2), 101–110.

Gonzalez-Vidal, A., Cuenca-Jara, J., & Skarmeta, A. F. (2019). IoT for water management: Towards intelligent anomaly detection. In *IEEE 5th world forum on internet of things, WF-IoT 2019—Conference proceedings* (pp. 858–863). https://doi.org/10.1109/WF-IoT.2019.8767190

Hammad, H. M., et al. (2019). Environmental factors affecting the frequency of road traffic accidents: A case study of sub-urban area of Pakistan. *Environmental Science and Pollution Research, 26*(12), 11674–11685.

Hammad, H. M., et al. (2020). Comparative effects of organic and inorganic fertilizers on soil organic carbon and wheat productivity under arid region. *Communications in Soil Science and Plant Analysis, 51*(10), 1406–1422.

Hashmi, M. Z., et al. (2017). Management and regulation of antibiotics and antibiotics resistance genes in soils. In M. Hashmi, V. Strezov, & A. Varma (Eds.), *Antibiotics and antibiotics resistance genes in soils. Soil biology* (Vol. 51). Springer.

Hashmi, M. Z., et al. (2020). Arsenic distribution and metabolism genes abundance in Paddy soils from Punjab and Sindh provinces, Pakistan. *Arabian Journal of Geosciences, 13*, 526.

Hasnain, A., et al. (2017). Antibiotics resistance genes. In M. Hashmi, V. Strezov, & A. Varma (Eds.), *Antibiotics and antibiotics resistance genes in soils. Soil biology* (Vol. 51). Springer.

Hussain, S., et al. (2020a). Study of land use/land cover changes using RS and GIS: A case study of Multan district, Pakistan. *Environmental Monitoring and Assessment, 192*, 2.

Hussain, S., et al. (2020b). Using GIS tools to detect the land use/land cover changes during forty years in Lodhran District of Pakistan. *Environmental Science and Pollution Research, 27*, 39676–39692.

Imran, M., et al. (2019). Kinetic and equilibrium studies for cadmium biosorption from contaminated water using Cassia fistula biomass. *International journal of Environmental Science and Technology, 16*(7), 3099–3108.

Ippolito, J. A., Bjorneberg, D. L., Blecker, S. W., & Massey, M. S. (2019). Mechanisms responsible for soil phosphorus availability differences between sprinkler and furrow irrigation. *Journal of Environmental Quality, 48*(5), 1370–1379. https://doi.org/10.2134/jeq2019.01.0016

Kamienski, C., Soininen, J. P., Taumberger, M., Dantas, R., Toscano, A., Cinotti, T. S., Maia, R. F., & Neto, A. T. (2019). Smart water management platform: IoT-based precision irrigation for agriculture. *Sensors (Switzerland), 19*(2). https://doi.org/10.3390/s19020276

Kassem, A. A., & Hamadamin, K. K. (2023). Design of Basin Irrigation System using multilayer perceptron and radial basic function methods. *Tikrit Journal of Engineering Sciences, 30*(2), 60–66. https://doi.org/10.25130/tjes.30.2.7

Kata, L., & Aparna, K. (2022). Technology trends in modern agriculture. March. https://www.researchgate.net/publication/359414587

Khanna, A., & Kaur, S. (2019). Evolution of internet of things (IoT) and its significant impact in the field of precision agriculture. *Computers and Electronics in Agriculture, 157*, 218–231. https://doi.org/10.1016/j.compag.2018.12.039

Li, D. (2020). Quantifying water use and groundwater recharge under flood irrigation in an arid oasis of northwestern China. *Agricultural Water Management, 240*, 106326. https://doi.org/10.1016/j.agwat.2020.106326

Liang, Z., Liu, X., Xiong, J., & Xiao, J. (2020). Water allocation and integrative management of precision irrigation: A systematic review. *Water (Switzerland), 12*(11), 1–23. https://doi.org/10.3390/w12113135

Luqman, S. (2023). *Big data processing in the cloud: Scalable and real-time data*. Department of Computer Science, University of Panjab Lahore.

Maroufpoor, S., Maroufpoor, E., & Khaledi, M. (2019). Effect of farmers' management on movable sprinkler solid-set systems. *Agricultural Water Management, 223*(June), 105691. https://doi.org/10.1016/j.agwat.2019.105691

McDermid, S. P., et al. (2015). *The AgMIP coordinated climate-crop modeling project [C3MP]: Methods and protocols* (pp. 191–221). World Scientific Publishing Co Pt. Ltd.

Mubeen, M., et al. (2020). Evaluating the climate change impact on crop water requirement of cotton-wheat in semi-arid conditions using DSSAT model. *Journal of Water and Climate Change, 11*(4), 1661–1675.

Rowshon, M. K., & Amin, M. S. M. (2010). GIS-based irrigation water management for precision farming of rice. *International Journal of Agricultural and Biological Engineering, 3*(3), 27–35. https://doi.org/10.3965/j.issn.1934-6344.2010.01.027-035

Sabagh, A., et al. (2020). Prospective role of plant growth regulators for tolerance to abiotic stresses. In Aftab, T., & Hakeem, K. R. (Eds.), *Plant growth regulators*.

Sami Ul Din, M., et al. (2019). Agronomic cropping systems in relation to climatic variability. In Mirza, H., (Ed.), *Agronomic crops*.

Sardar, A., et al. (2020). Risk assessment of heavy metal(loid)s via Spinacia oleracea ingestion after sewage water irrigation practices in Vehari District. *Environmental Science and Pollution Research, 27*, 39841–39851.

Shafi, U., Mumtaz, R., García-Nieto, J., Hassan, S. A., Zaidi, S. A. R., & Iqbal, N. (2019). Precision agriculture techniques and practices: From considerations to applications. *Sensors (Switzerland), 19*(17), 1–25. https://doi.org/10.3390/s19173796

Shahzad, H., et al. (2020). An integrated use of GIS, geostatistical and map overlay techniques for Spatio-temporal variability analysis of groundwater quality and level in the Punjab Province of Pakistan, South Asia. *Water, 12*, 3555.

Sharma, A., Jain, A., Gupta, P., & Chowdary, V. (2021). Machine learning applications for precision agriculture: A comprehensive review. *IEEE Access, 9*, 4843–4873. https://doi.org/10.1109/ACCESS.2020.3048415

Singh, M., & Ahmed, S. (2020). IoT based smart water management systems: A systematic review. *Materials Today: Proceedings, 46*, 5211–5218. https://doi.org/10.1016/j.matpr.2020.08.588

Singh, A., Jhorar, R. K., Kumar, S., & Kumar, N. (2018). Performance evaluation of border irrigation method for cotton field. *Journal of Applied and Natural Science, 10*(1), 222–229. https://doi.org/10.31018/jans.v10i1.1608

Sishodia, R. P., Ray, R. L., & Singh, S. K. (2020). Applications of remote sensing in precision agriculture: A review. *Remote Sensing, 12*(19), 1–31. https://doi.org/10.3390/rs12193136

Tariq, M., et al. (2019). Fundamentals of crop rotation in agronomic management. In Mirza, H., (Ed.), *Agronomic crops*.

Vecchio, Y., De Rosa, M., Adinolfi, F., Bartoli, L., & Masi, M. (2020). Adoption of precision farming tools: A context-related analysis. *Land Use Policy, 94*, 104481. https://doi.org/10.1016/j.landusepol.2020.104481

Zahoor, S. A., et al. (2019). Improving water use efficiency in agronomic crop production. In M. Hasanuzzaman (Ed.), *Agronomic crops*. Springer.

Zaman, G., et al. (2020). *Utilization of compost from bio-municipal solid waste improves saline-sodic soils and crop productivity in rice-wheat* (Vol. 28, p. 16). Compost Science & Utilization.

Zamin, M., et al. (2020). Developing the first halophytic turfgrasses for the urban landscape from native Arabian desert grass. *Environmental Science and Pollution Research., 27*, 39702–39716.

Innovative Soil Health Practices for Better Water Management

Behzad Murtaza, Raqash Fatima, Aneeza Sardar, Samina Kousar, Muhammad Imran, Hafiz Faiq Bakhat, Muhammad Amjad, Abdul Rehman, Ghulam Murtaza, and Mumtaz Cheema

Abstract Land and soil "play a key role in the climate system" as an essential carbon sink; however, land and soil resources are under increasing pressure from deforestation, urbanization, industrial development, agricultural expansion, and unsustainable farming practices that are undermining their ability to sustain food production, maintain freshwater and forest resources, and regulate the climate and air quality. Hence, there is a dire need to improve and maintain soil productivity in the twenty-first century with increasing agricultural needs and demands. The current book chapter elaborates farms' novel approaches to improve soil health and quality, and water management. These approaches enhance water use and nurture soil productivity. For instance, planting cover crops provides soil cover, which reduces water loss and erosion, while "no-till farming" preserves soil structure and its capacity to sustain microflora and fauna. Mulching prevents rapid water evaporation. Modifying crop choices and incorporating agro-forestry conserve water distribution. Additionally, amending the soil with calcitrant carbon materials such as biochar enhances water conservation and nutrient availability, and improves the soil's physiochemical and biological properties and greenhouse gas mitigation.

B. Murtaza (✉) · R. Fatima · A. Sardar · S. Kousar · H. F. Bakhat · M. Amjad
Department of Environmental Sciences, COMSATS University Islamabad, Vehari, Pakistan
e-mail: behzadmurtaza@cuivehari.edu.pk

M. Imran
Department of Environmental Sciences, COMSATS University Islamabad, Vehari, Pakistan

Department of Sanitary Engineering, Faculty of Civil and Environmental Engineering, Gdansk University of Technology, Gdańsk, Poland

A. Rehman
Department of Botany, University of Agriculture Faisalabad, Faisalabad, Pakistan

G. Murtaza
Institute of Soil and Environmental Sciences, University of Agriculture Faisalabad, Faisalabad, Pakistan

M. Cheema
School of Science and the Environment, Memorial University of Newfoundland, St. John's, NL, Canada

M. Mubeen et al. (eds.), *Innovations in Agricultural Water Management*,
https://doi.org/10.1007/978-3-031-91883-4_5

Intelligent methods such as “precision irrigation” smart water supply systems, and tools to gauge soil moisture aid decision-making. These collective advances holistically encourage soil and water conservation, strengthen soil flexibility, and support agricultural sustainability to enhance food security to feed the world’s burgeoning population.

Keywords Soil health · Innovative technology · Water management · Sustainability · Precision agriculture

1 Introduction

The ability of soil to perform its role in an ecosystem to maintain the quality of the air and water, increase the productivity of plants, and improve the health of animals is referred to as soil health (Tahat et al., 2020). The primary and fundamental source of nutrients, water, and other four-dimensional components that give plants support comes from the soil (Wang et al., 2024). The biological health of soil is the capacity of the soil to suppress pathogens and promote beneficial microbial and plant health (Brackin et al., 2017). Soil is a complex habitat for different kinds of microorganisms, that is, bacteria, algae, and fungi (Bardgett & Van Der Putten, 2014). The availability and balanced concentration of plant nutrients are known as the chemical health of the soil (Spanner & Napolitano, 2015). It is reported that about 95% of food production is through soil. The Food and Agriculture Organization (FAO) report indicates that excessive use of land for agricultural yield purposes to cope with human food demand causes the degradation of soil and threatens the food production systems and their security (ITPS, 2015; Labrière et al., 2015). Heavy metal accumulation is also one of the major causes of soil degradation (Sardar et al., 2020).

Different agroecological methods, such as soil conservation and organic farming, are beneficial for the improvement of soil and ecosystem, which ensure food productivity (Fig. 1). Soil properties can be enhanced by improving its physical health, including the maintenance of soil ability, and air and water requirements for plants (Spanner & Napolitano, 2015). Organic soil amendment practices, including the application of animal manure compost, biochar, cover crop, green manure, and crop residue (straws), are used to improve soil health (Fig. 2) (Ansari et al., 2017). There are a number of innovative practices that have gained economic importance for the improvement of crop yield by supplying lower water quantities and also play a crucial role in suppressing various environmental burdens, including abstraction of water, use of energy, and reduced pollutant concentration (Molden, 2013). Farmers can adjust cropping patterns to minimize water demand and usage, adopt additional technologies, improve their knowledge of soil and water management, make better use of already-installed technological systems, use fewer agrochemical inputs, etc. Water-efficient techniques have the potential to improve irrigated agriculture, environmental sustainability, and economic viability without necessarily using luxurious/excessive water. Although experts have created a variety of

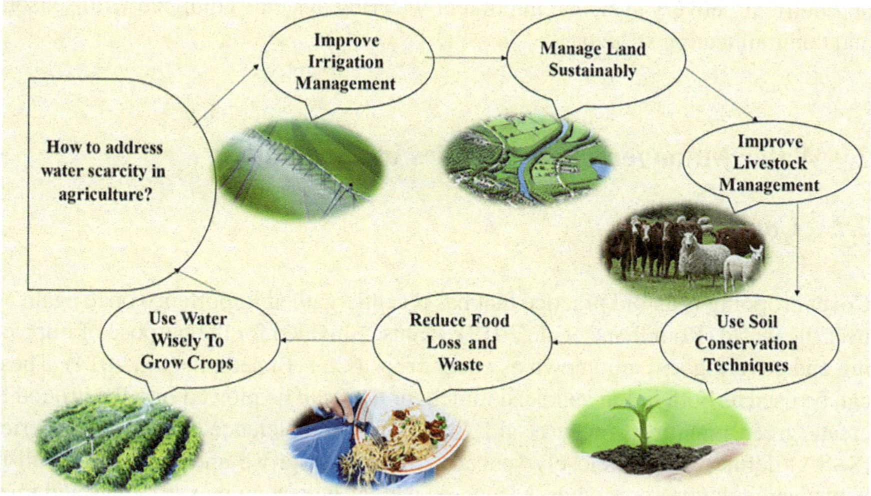

Fig. 1 Agroecological methods for improving soil health

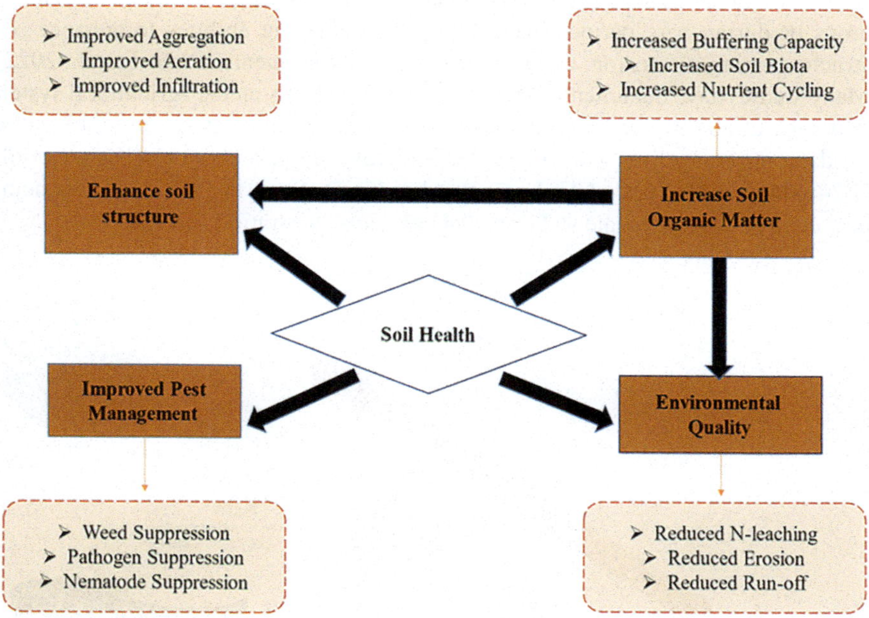

Fig. 2 Maintaining soil health through sustainable agricultural practices

water-efficiency models to guide such practices, farmers hardly ever use them (Levidow et al., 2014; Sabagh et al., 2019; Farid et al., 2019a, 2019b; Tariq et al., 2008; Awais et al., 2018; Amin et al., 2018a, 2018b, 2018c; Hammad et al., 2018). Irrigation-related technological advancements will include high-efficiency water and nutrient delivery systems, intelligent watering systems equipped with sensors, and communication systems.

2 Water Management Strategies in Agriculture

2.1 Cover Crops

Cover cropping is an old practice that has recently gained importance once again all over the world (Romdhane et al., 2019). Plants cultivated for the purpose of improving and covering soil are known as cover crops (CC) (Benedict et al., 2014). These can be used as both living and dead mulch, or they can be plowed into the ground to create green manure (Finney et al., 2017). The Soil Science Society of America (SSSA) defines CCs as closely spaced crops that improve and protect the soil in between regular crop-growing seasons, as well as between vineyard vines and trees in orchards. Crop cover, a keystone of viable soil health practices, offers a convincing data-driven sign of their effectiveness in improving water management (Vanlauwe et al., 2023). These deliberately planted crops, grown between primary crops or during bare periods, have established amazing abilities to enhance soil structure, mitigate erosion, and optimize water requirements (Prabha et al., 2022). Many studies have documented the diverse CCs' impacts on the agricultural system (Coombs et al., 2017). CCs inhibit weed growth by either releasing allelopathic exudates or competing with weeds for nutrients, light, and water (Kumar et al., 2024; Matos et al., 2019). CCs have other beneficial effects in addition to enhancing pest control, soil aggregate stability, and carbon sequestration (Fig. 3).

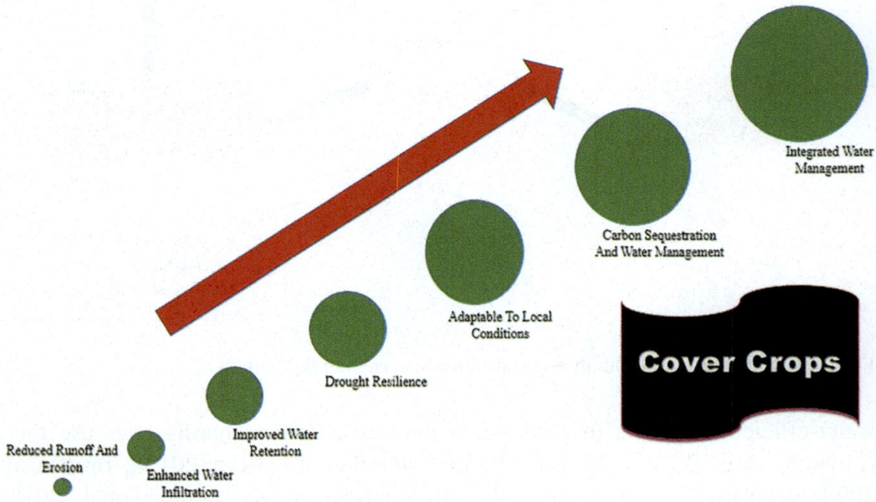

Fig. 3 Maintaining soil health and functionality

2.1.1 Reduced Runoff and Erosion

One of the main strategies for reducing soil erosion is to cover the soil with vegetation (Gobin et al., 2004). Among the various measures that are used to reduce soil erosion in different types of farmland, the application of cover crops is a significant and environmentally friendly way to cover the soil surface with crop residue (López-Vicente et al., 2020). Winter CCs reduce rainfall impact on soil surface, enhance soil structure, and optimize conservation tillage systems by slowing down aggregate breakdown and enhancing soil structure (Price & Norsworthy, 2013). Studies have shown that CCs can slow down the soil erosion by up to 90% compared to bare fallow fields (Kelly et al., 2021). Research conducted in the Midwest United States found that CCs reduced water runoff by as much as 56%, effectively impeding nutrient and sediment runoff into water bodies (Hussain et al., 2023a, 2024a, 2024b, 2024c; Naz et al., 2023; Muhammad et al., 2023; Hansen et al., 2021). Using CCs lessens sediment loss and dissolved nutrient erosion in water; however, CC species and biomass production are necessary for this phenomenon to occur. Thus, using CCs can offer a workable solution for lowering soil erosion, boosting agricultural productivity, and improving soil health. CCs help to prevent soil erosion, improve soil conditions, protect seedlings, and water use for next crops. They are most effective in humid and sub-humid regions due to limited precipitation.

2.1.2 Improved Water Retention

Cover crops increase soil organic matter (SOM), enhancing water holding capacity of the soil. Soil with 3.5% organic matter can hold up to 27,000 gallons/acre (Mauget et al., 2021). This stored water becomes accessible to crops during dry spells, decreasing irrigation needs. Literature shows that CCs, particularly legumes, can reduce nitrate leaching by up to 85% (Hussain et al., 2022a, 2022b, 2022c, 2023b; Khurshid et al., 2023; Quintarelli et al., 2022). This reduction in nutrient loss contributes to improved water quality by minimizing the pollution of groundwater and surface water. Soil amendment with CCs such as rye resulted in preserving the moisture content and improving ground water storage. Raindrop kinetic energy is decreased by the CCs, which lowers the sealing of the soil's surface and is ultimately responsible for preserving the moisture content from irrigation and rainfall (Nyawade et al., 2020).

2.1.3 Drought Resilience and Carbon Sequestration

Many regions of the world are prone to an elevated risk of extended drought spells due to fluctuations in the climate. CCs in arid areas enhance soil structure and organic matter, making them drought tolerant. Fields with CCs can experience a 25–50% increase in water-holding capacity during scarcity conditions (Guo, 2021).

The meta-analysis of several field studies reveals that the presence of CCs enhances soil carbon sequestration (McDaniel et al., 2014; Schipanski et al., 2014). Although there is considerable variation between sites, it appears that CCs have a greater impact on soil C when there is less tillage (Lal, 2015). This develops soil structure, aggregation, and water-holding capacity, ultimately aiding in better water management (Ghimire et al., 2023). Carbon sequestration can be influenced by CCs by reducing soil erosion, which is determined by the balance between accelerated decomposition during transport, delayed sediment decomposition, and variations in new carbon addition and stabilization rates.

2.2 *No-Till Farming*

The no-till farming practice aims to protect soil surfaces from rainfall, maintaining soil structure, enhance biological processes, and safely dispose surface runoff using crop management strategies (Fig. 4).

No-till farming uses unprepared soil for plantation, opening narrow slots for seed coverage and fertilizer placement, ensuring proper soil preparation for cultivation purposes (Derpsch, 2001). No-till farming is a revolutionary method that improves soil health and water management by minimizing traditional plowing and tilling, thereby reducing erosion and enhancing water retention (Ning et al., 2022).

No-till farming preserves soil structure, allowing for water movement and root penetration, reducing erosion risk (Abiye, 2022). Traditional farming disrupts soil structure, leading to increased surface runoff and erosion. No-till farming preserves soil structure by allowing water movement and root penetration, reducing erosion

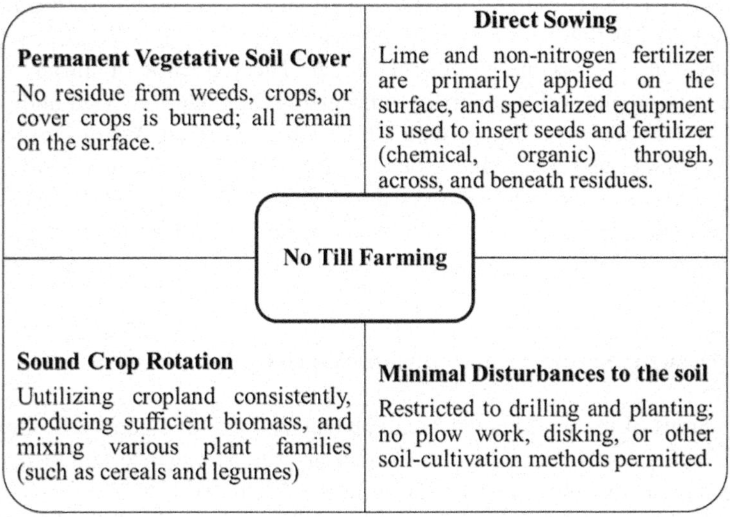

Fig. 4 No-till farming and crop rotation for soil health

risk. No-till farming enhances water use by minimizing soil evaporation, thereby enhancing crop productivity in regions with limited water resources and conserving soil moisture. No-till farming, which incorporates crop residues into the soil, can reduce up to 80% of CO_2 emissions by reducing soil organic matter oxidation (Pieri et al., 2002).

2.2.1 Mitigating Runoff

Conventional farming practices often lead to significant runoff of water and nutrients, causing pollution in nearby water bodies and aquatic ecosystems. No-till farming addresses this issue by reducing runoff through improved soil structure and reduced disruption. The crop residues that remain on the soil's surface function as a natural barrier, reducing water flow and allowing for slow infiltration (Junhao et al., 2022). This reduces the transport of nutrients, pesticides, and sediment into water bodies, enhancing water quality and supporting healthier aquatic ecosystems.

2.2.2 Promoting Carbon Sequestration and Climate Resilience

No-till farming enhances climate elasticity, water management, and carbon absorption, aligning with sustainability goals by reducing greenhouse gas (GHG) emissions and improving soil fertility and structure (Rehberger et al., 2023). One of the no-till farming system's most significant direct benefits is the reduced CO_2 release, which eventually boosts carbon sequestration both above and below the ground surface (Jayaraman et al., 2024).

2.2.3 Challenges and Considerations

No-till farming is a sustainable approach to balance agricultural productivity with water management. However, successful implementation requires careful planning, weed control, crop residue management, and equipment selection. It also requires adjustments in crop rotation and pest management (Kalantari et al., 2023). As agriculture navigates the complexities of a changing world, no-till farming emerges as a beacon of hope, fostering resilient ecosystems that thrive in tandem with enhanced water management and overall soil health.

2.3 *Mulching*

Mulching is a proven agricultural practice that promotes soil health and water management, ultimately enhancing plant growth (Kader et al., 2017). It entails covering the soil's surface with natural or artificial materials to form a protective layer that

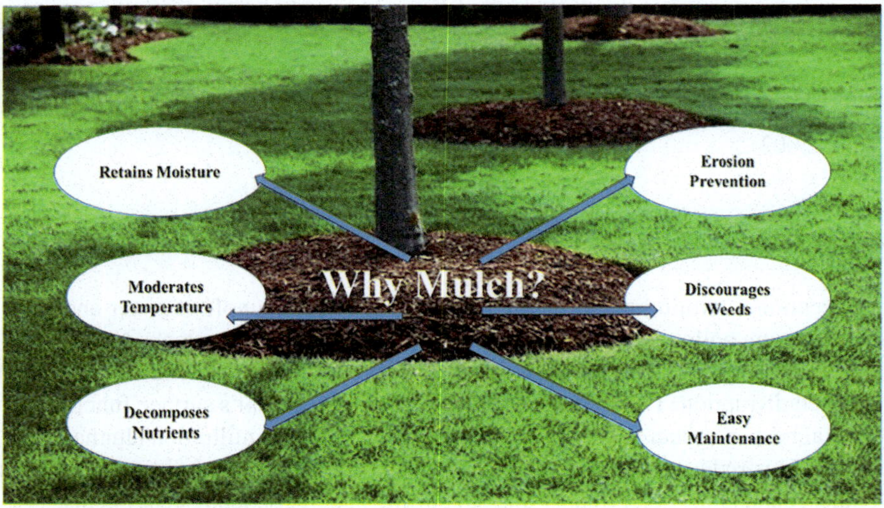

Fig. 5 How mulch can improve soil health

affects water infiltration, retention, and soil quality overall (Jordán et al., 2010). Various materials are used for mulching, including vegetation leftovers, biological geotextiles, gravel, and crushed stone (Prosdocimi et al., 2016). There are several benefits associated with mulching (Fig. 5). Among all these benefits, the most attractive is a decline in soil-water losses, weed suppression, and thermal regulation. Mulching is a versatile practice that can be used in various agricultural systems, including gardens, orchards, row crops, and container plants, to deepen soil, prevent weed growth, and conserve moisture, making it a water-wise approach (Kumar et al., 2020). Mulch is a sustainable, effective soil health practice that improves water.

2.3.1 Types of Mulch

2.3.1.1 Organic Mulch

Organic mulch is made up of natural materials or industrial waste such as sawdust, bark, wood chips, pine needles, dry grasses, paddy straw, dry leaves, and grass clippings. But organic mulch attracts insects, and the cutworms that eat it (Shojaei et al., 2019).

2.3.1.2 Inorganic Mulch

For perennial crops, gravel, pebbles, and crushed stones are used as mulch (Zahed et al., 2022). Weed control is effectively achieved by a 3–4 cm small rock layer. During the summer, however, they can reflect solar radiation and make the soil extremely hot.

2.3.1.3 Plastic Mulch

For mulching, both transparent and black films are typically used. Progress in plastic chemistry has led to the creation of films with optical characteristics that are optimal for a particular crop in a particular area. There are two types of plastic mulch: photo-degradable plastic mulch (under the sun, this type of plastic mulch film breaks down more quickly) and biodegradable plastic mulch.

2.3.2 Reduced Evaporation and Water Loss

Mulch serves as a shield against the Sun's rays, reducing direct exposure and slowing water vapor escape, making it crucial in semi-arid or arid areas. This minimizes evaporation, promotes plant development, and preserves soil moisture levels (Gebretsadikan et al., 2023). Plastic film reduced bell pepper irrigation water requirements by 14–29%, and improved tomato yield, and water use efficiency (Asif et al., 2024).

2.3.3 Erosion Prevention and Soil Health

Mulching considerably mitigates soil erosion by acting as a protective fence against the erosive forces of wind and water (Joshi et al., 2021). Mulching prevents soil particles from being carried away by rainfall or runoff, preserving topsoil and preventing land deprivation. It decomposes organic matter, enhancing topsoil's water-holding capacity and nutrient content. Applying crop residues improves soil quality, including organic carbon, structural stability, and mineralization-induced N levels. It also retains moisture for shallow-rooted crops (Pedda Ghouse Peera et al., 2020).

2.4 Crop Rotation

Crop rotation enhances soil health for sustainable water management (Shah et al., 2021). Crop rotation is a traditional agricultural practice that involves alternating crops over successive growing seasons to improve soil structure, pest and disease management, nutrient cycling, and water retention, ensuring sustainable crop production and efficient water use (Sun et al., 2018). Crop rotation effectively reduces pest and disease infections, maximizes nutrient use, and increases crop yields by efficiently using soil water and minimizing soil nutrient loss (Dias et al., 2015; Sainju et al., 2021; Irfan et al., 2022; Ullah et al., 2022; Naz et al., 2022; Hussain et al. 2022; Waleed et al., 2022). Water productivity should be considered when evaluating crop rotation benefits, as yield and profitability alone are insufficient metrics. Other factors like climatic conditions, input/output, and price fluctuations also impact yields and profits (Schlegel et al., 2016, 2019). Crop yields and profits are influenced by various factors such as crop rotation strategies, climatic conditions, total input and output, and price fluctuations (Schlegel et al., 2019).

2.5 *Agroforestry*

Agroforestry is an integrative approach to soil health and water management. It is a workable land-use system that blends vegetation such as trees and shrubs with livestock or agricultural crops. It stands out for being a comprehensive strategy for promoting soil health and improving water management (Kisaka et al., 2023).

2.5.1 Enhancing Soil Structure and Organic Matter

Agroforestry systems contribute to soil health by introducing a different range of plants, each with diverse root structures that interact with the soil in unique ways. Trees and shrubs, with their deep-reaching root systems, break up condensed soil layers, allowing for improved water infiltration (Hartmann & Six, 2023). The organic matter shed by leaves, branches, and fallen fruit boosts soil structure, promoting soil aggregation and preventing erosion. This encourages the development of stable soil aggregates, which create pathways for water association, reducing surface runoff and preventing soil degradation.

2.5.2 Nutrient Cycling and Water Quality Improvement

The capacity of trees and shrubs to improve soil productivity and protection forms a significant scientific basis for agroforestry systems (Rhodes, 2024). Agroforestry systems use diverse plant species to promote soil microbial community, nutrient cycling, organic matter decomposition, and nitrogen capture, reducing nutrient runoff and water pollution (Li et al., 2023). By minimizing nutrient loss, agroforestry indirectly supports water quality improvement, ensuring that water resources remain clean and uncontaminated.

2.5.3 Climate Resilience and Adaptation

Agroforestry enhances climate resilience by acting as windbreaks, reducing soil evaporation, and conserving soil humidity. Carbon sequestration helps stabilize global temperatures and mitigate climate-change effects by reducing extreme weather events (Leal Filho et al., 2023; Hussain et al., 2022a, 2022b, 2022c; Khaliq et al., 2022; Mubeen et al., 2021; Ahmad et al., 2021). Agroforestry, a holistic approach to soil health and water management, integrates trees and shrubs into agricultural landscapes, improving water infiltration, reducing erosion, and addressing climate adaptation and water resource conservation challenges (Khangura et al., 2023).

2.6 *Precision Irrigation, Benefits, and Challenges*

Precision irrigation manages water supply to the crop's root zone at specific intervals, maximizing efficiency in evapotranspiration and metabolism (Anjum et al., 2023). The primary goal of precision irrigation techniques is to enhance crop yield and productivity by efficiently applying the right amount of water and using resources (Brahmanand & Singh, 2022). Automated sensors measure soil moisture, crop water requirements, weather data, soil components, crop models, and growth stages for water precision irrigation systems, ensuring efficient and accurate water supply (Abioye et al., 2020; Fernández García et al., 2020).

A variety of irrigation systems are used for precision irrigation, including drip irrigation, sprinkle irrigation, micro-sprinkle irrigation (Jarwar et al., 2019). There are numerous benefits of precision irrigation such as increased water efficiency, improved crop yield, improved environmental sustainability, and reduced labor/energy costs. Despite the number of benefits, there are various challenges associated with precision irrigation, including the cost of installation and maintenance, the need for accurate data, and the need for specialized knowledge.

2.7 *Soil Microbial Enhancement*

Soil microbial enhancement is a pioneering approach in soil health practices that positively impacts water management. There is decline in evaporation rate from a soil when microbial activity increases (Choudhury et al., 2018). Microbial activity has a major impact on the hydraulic properties and soil structure (Colica et al., 2014; Helliwell et al., 2014). This technique improves soil health by promoting an active microbial community, enhancing water movement, porosity, and infiltration, thereby enhancing resilience and productivity in agricultural systems (Manda et al., 2023).

2.7.1 Effects of Microbes on Soil Water Management

2.7.1.1 Decomposing Organic Matter

Microorganisms break down organic matter (OM), providing nutrients, promoting healthy growth, and improving soil structure. They (cyanobacteria, rhizobia, Frankia, Azospirillum, Azotobacter, Klebsiella, etc.) fix N, release essential nutrients, and enhance their bioavailability and soil stability (Manda et al., 2023).

2.7.1.2 Association of Microbes with Plant Roots

Microbes can form an association with plant roots, providing nutrients and water in exchange for carbon. Plants use water more effectively and efficiently with this microbial association. Carbon released by the roots as root exudates is a significant microbial activity stimulant. Due to the surfactant characteristics, root exudates exert instantaneous effects on soil water release properties (Gabriele et al., 2021).

2.7.1.3 Effect on Soil Permeability

The ability of the soil to permit the passage of water into and through the soil profile is known as soil infiltration. Poor soil management hinders water penetration, causing it to travel downslope as runoff or accumulate on the surface, where it evaporates. Plant production declines as a result, storing less water in the soil for plant growth. This lowers the amount of OM in the soil and weakens the soil structure, which can further lower the infiltration rate (Haghnazari et al., 2015). Soil microbes produce organic acids that can modify soil permeability, allowing more water to seep into it. Enhancing soil water management through OM enrichment and cover crop cultivation can nourish soil microorganisms and improve soil structure, making it a sustainable and environment-friendly approach.

2.8 Soil Salinity Management for Water Uptake

Soluble salt content in the soil is known as salinity, and it is one of the harsh environmental problems that negatively affects crop productivity (Zaman et al., 2020). Salinity in the soil reduces crop yield by roughly 50% (Murtaza et al., 2017). Salinity decreases soil productivity and fertility, leading to vegetation destruction and soil microbial activity inhibition. Various factors including natural processes, such as weathering of parent material, evaporation, rain fall pattern, geomorphic pattern, or human activities, such as irrigation with saline water, causes the formation of saline soil (Yuvaraj et al., 2021). Salinization can be natural or anthropogenic, with primary processes causing weathering, climatic factors, and hydrological issues. Secondary processes, such as untreated industrial waste and improper irrigation, can lead to nutrient imbalances and plant growth issues, affecting cell turgidity and concentration gradient (Fageria et al., 2011). To enhance water uptake and lessen the detrimental effects of salinity on plant growth, several methods for managing soil salinity can be employed. These methods include scrapping, subsoiling, deep ploughing, leveling, sand mixing, leaching, soil amendments, crop selection, irrigation scheduling, and drainage.

2.9 *Soil Organic Matter Management*

Soil water management relies on proper handling of OM, with compost being a key element for sustainable agriculture. Organic fertilizers preserve soil fertility, especially in tropical regions with delicate soils. Land management practices affect SOM content, affecting chemical, biological, and physical characteristics (Zaman et al., 2020). The roles that organic matter plays in soil and biological functioning are intrinsically linked.

Soil organic matter (SOM) includes organic materials containing O, N, P, S, H, and O, as well as humus, dissolved and particulate OM, inert or highly carbonized OM, and proteins, lipids, and carbohydrates (Murindangabo et al., 2023). Organic matter, including plant and animal wastes, is classified as aboveground, while belowground organic matter includes partially decomposed remains, living soil fauna, and microflora.

Compost is an organic material that decomposes under controlled conditions, with restrictions on its use. It may not be as nutrient-rich as animal excreta, but is beneficial for building soil organic matter (SOM) and reducing water pollution. Nyamangara et al. (2003) stated that the secret to sustainable agriculture is managing the OM in the soil by using composted organic waste. Soil fertility and nutrient loss are crucial, and compost, a nutrient-rich organic fertilizer, is an excellent choice for maintaining these properties.

2.9.1 Effects of Organic Matter on Soil Properties

The soil's capacity to hold and release the water and oxygen required for plant growth is a crucial indicator of its physical fertility. Wang et al. (2022) discovered that the water content increased as the SOC content did, and claimed that 1.5% more moisture by volume of FC can be added with a 1% increase in SOM.

2.9.1.1 Cation Exchange Capacity (CEC)

High CEC is beneficial for soils as it helps retain plant nutrient cations, and an increase in CEC is linked to increased soil organic material.

2.9.1.2 Buffer Capacity and Soil pH

Since soil buffering ensures a reasonable level of pH stability, it is regarded as a crucial component of soil health. The main source of buffering at intermediate pH levels (5–7.5) is the functional groups of SOM, which act as H^+ and OH^- sinks. The importance of SOM in preserving pH levels in the face of acidifying agents was reported by Cayley et al. (2002).

2.9.1.3 Adsorption and Complexation

An essential mechanism for preventing SOM from breaking down is the adsorption of SOM on clay particles. Significant effects on soil fertility result from SOM's complexation of inorganic materials, which may boost P availability by obstructing possible Fe, Al, and Ca adsorption sites. Jiang et al. (2023) noted that higher SOM concentrations lower the concentrations of heavy metals like cupric ions. The formation of thermodynamically stable Ca-phosphate has been demonstrated to be inhibited by humic acids.

2.9.2 Biological Function of SOM

The SOM plays a crucial role in providing nutrients, storing energy, and promoting sustainability, impacting soil redox reactions and structure. Murindangabo et al. (2023) stated that one of SOM's most essential functions is to supply metabolic energy, which powers soil biological processes.

2.10 Conservation Tillage

Conservation tillage reduces soil erosion and runoff by leaving crop residue on fields before and after planting. It covers at least 30% of the soil surface, with some techniques leaving 70% residues. This technique is particularly effective on erosion-prone cropland.

2.10.1 Agricultural Methods of Soil Conservation

The four types of conservation tillage are mulch-till, ridge-till, strip-till, and no-till.

No-Till and Strip-Till: These techniques involve planting crops straight onto the residues of previous crop that has either been tilled sparingly (no-till) or not at all (strip-till), leaving the remaining field tilled.

Ridge-Till: The process entails sowing row crops 4–6 inches high on permanent ridges. The residue from the previous crop is removed from ridge tops and deposited into neighboring furrows, requiring specialized machinery to maintain ridges for new crop planting.

Mulch-Till: It is similar to any other reduced tillage method where crop residues cover at least one-third of the soil's surface.

Zero Tillage: Drilling a hole that can only contain the seed is all that is required.

2.10.2 Benefits of Conservation Tillage Practice

Crop residues depending on the conservation tillage technique shield soil particles from wind and rain until new plants form a protective canopy over the soil, which can reduce soil erosion by as much as 60–90%. Introducing organic matter breaks down crop residue, and increases soil and water content. Moreover, conservation tillage (i) reduces water evaporation at the soil's surface, saving water, (ii) reduces energy consumption by causing tractor trips across the field, and (iii) provides cover and food for wildlife.

2.11 Biochar Incorporation

It involves using carbonized materials or their extracts, which are more stable carbon compounds, in place of organic amendments (Naeem et al., 2018). A similar case is presumed regarding the build-up of the terra preta soil in the Brazilian Amazon area where carbonized materials from incomplete combustion of organic materials maintain high levels of SOM. Up to 12% of annual carbon emissions can be offset with slash-and-char substitution (Lehmann et al., 2006). Applying compost and biochar simultaneously increased crop yield, water retention capacity, soil fertility, and the benefits of carbon sequestration, according to recent studies (Agegnehu et al., 2016).

2.11.1 Biochar-Compost and Soil Health

Soil health refers to the soil's capacity to perform various agronomic and environmental tasks, including withstand biotic and abiotic stresses, increase biomass and productivity, and respond to inputs and management (Naeem et al., 2018). A comprehensive analysis of biochar application is crucial to understanding its impact on soil functionality. Soil quality refers to the soil's capacity to fulfill its three primary functions: regulation of the environment, provision of aesthetic and cultural values, and economic productivity. Soil resilience and quality have a major influence on productivity and environmental quality, and tillage practices and crop residue management have a direct impact on soil quality (Powlson et al., 2011).

2.11.2 Effect on Soil Physical Properties

Restoring the physical and chemical properties of degraded or nutrient-depleted soils can be facilitated by adding biochar as a soil amendment. The ability of biochar to retain soil moisture is determined by both its surface functionality and porosity (Suliman et al., 2017). Biochar improves soil porosity because of its highly porous internal structure. A higher porosity of the soil improves water penetration by increasing the soil's surface area. According to earlier studies, adding biochar to unproductive soils increases their ability to hold water, lowers their bulk density, and increases their total pore volume (Fischer et al., 2019). According to Oguntunde et al. (2008), the charcoal site experienced 9% lower soil density compared to nearby fields, resulting in increased total porosity and saturated hydraulic conductivity, darkened biochar color, and decreased surface albedo. The charcoal site experienced 9% lower soil density compared to nearby fields, resulting in increased total porosity and saturated hydraulic conductivity, darkened biochar color, and decreased surface albedo.

2.11.3 Effect on Soil Chemical Properties

Biochar improves soil chemical properties, affecting pH, organic carbon, exchangeable cations, and N fertilizer use efficiency, while tensile strength declines with higher biochar rates (Bera et al., 2016). Ferrosol and calcarosol exhibit significant changes in pH, CEC, exchangeable Ca, total C, and exchangeable K upon the addition of paper-mill biochar at 10 t ha^{-1} (Van Zwieten et al., 2010). Biochar improves plant nutrient uptake, affecting soil pH, while plant growth, and fertilizer application, and crop harvesting contributes to soil acidification, influenced by ion source, distribution, and differential uptake (Fageria & Baligar, 2008). Agricultural lime is often used to boost soil pH in acidic soils, but high-pH biochar has been found to increase soil pH by about one-third, reduce aluminum toxicity, and increase calcium levels in red ferritic soils.

2.11.4 Effect on Stability and Nutrient Retention

Research is exploring the use of biochar for improving soil fertility, health, and agronomic benefits, while sequestering carbon. Studies show that biochar retains cations better and remains in soil longer than other organic matter (LONE et al., 2015). The duration of biochar's stay in soil is a topic of debate, impacting its value in carbon trading. Biochar can decompose slowly, and their maximum adsorption capacity is unknown. Fresh biochar retains cations less effectively than aged biochar. Biochar and other aromatic black carbons are more persistent than other organic carbons, even in humid tropical soils like the Amazon.

Charred biomass, along with its recalcitrant aromatic ring structures, also contains oxidized and aliphatic carbon structures that degrade more readily. Charring conditions, plant cell structure, and formation methods can impact biochar particles' carbon content. Some biochar may mineralize quickly, causing inaccurate stability projections. Surface oxidation may start quickly, but remains restricted to outer regions even after centuries in soils (Zimmerman, 2010). Thus, short-term experimental results suggest that estimating the fresh biochar degradation rate may overestimate long-term degradation, as aged char releases 40–55 times more condensed structures than fresh char, indicating low initial soluble fraction and increasing over time in soil (Benbi & Brar, 2021).

2.11.5 Effect on Soil Biological Properties

Application of biochar to soils affects soil biota differently than when fresh organic matter is added, potentially impacting the variety, quantity, and activity of soil biotic communities (Lehmann et al., 2011). Biochar, a less biologically available carbon source, alters biological functionality by affecting substrate availability, enzyme activity, and microorganism homes. Studies in India, Thailand, and the UK show soil properties and crop type influence biochar's impact on crop output. The application of biochar for phosphate mobilization in phosphate-rich soils was unsuccessful, but it significantly increased crop yield in phosphate-deficient soils (Deb et al., 2016). Biochar, due to its large surface area and porosity, supports microbes like bacteria, arbuscular mycorrhizal fungi, and actinomycetes. It boosts microbial abundance but reduces arbuscular mycorrhizal fungal (AMF) abundance in roots, suggesting soil P availability may also decrease. Peanut shell biochar increased P by 101%, decreased extra-radical hyphal lengths by 95%, and reduced AMF root colonization by 74%. Mango wood biochar increased P by 163 and 208%, but mycorrhizal fungus turned parasitic. Soil solution buffering may maintain pH levels (Nasim et al., 2018; Awais et al., 2017; Amin et al., 2017a, 2017b; Farid et al., 2017; Wang et al., 2016).

3 Conclusion and Future Prospect

This overview delves into a range of agricultural water management techniques, offering solutions from nature's bounty (cover crops, mulching, biochar amendment, crop residue, crop rotation) to innovative technology (precision irrigation). Each approach tackles water scarcity with unique strengths, such as covering crops, which protect soil from erosion, boost water infiltration, and store moisture for thirsty times. No-till farming improves water retention and carbon storage but may require extra weed control. Mulching shields soil from evaporation, nourishes microbes, and keeps weeds at bay. Crop rotation: Diverse root systems build healthy soil, improve water use efficiency, and combat pests and diseases. These are the tech-savvy solutions; for example, precision irrigation delivers water exactly where and when needed, maximizing efficiency and saving precious resources. However, tech costs and expertise requirements need consideration, and soil microbial boost: encouraging beneficial microbes improves water management, aiding in decomposition, root interaction, and soil structure. Additional techniques are soil salinity management: tackles salty soils and improves water uptake for healthy crops; and soil organic water management: enriching soil with organic matter increases its ability to hold and utilize water. Conservation tillage: minimizing disturbance preserves soil moisture and health, and biochar incorporation: charcoal-like material enhances water retention and nutrient availability. The techniques provide a useful toolbox, and their future lies in convergence, adaptation, and innovation.

Acknowledgement The author (Muhammad Imran) is thankful to Polish National Agency for Academic Exchange (NAWA), Poland.

References

Abioye, E. A., Abidin, M. S. Z., Mahmud, M. S. A., Buyamin, S., Ishak, M. H. I., Abd Rahman, M. K. I., Otuoze, A. O., Onotu, P., & Ramli, M. S. A. (2020). A review on monitoring and advanced control strategies for precision irrigation. *Computers and Electronics in Agriculture, 173*, 105441.

Abiye, W. (2022). Soil and water conservation nexus agricultural productivity in Ethiopia. *Advances in Agriculture, 2022*, 1.

Agegnehu, G., Nelson, P. N., & Bird, M. I. (2016). The effects of biochar, compost and their mixture and nitrogen fertilizer on yield and nitrogen use efficiency of barley grown on a nitisol in the highlands of Ethiopia. *Science of the Total Environment, 569*, 869–879.

Ahmad, M., et al. (2021). Hydrogen sulfide: A novel gaseous molecule for plant adaptation to stress. *Journal of Plant Growth Regulation, 40*, 2485–2501.

Amin, A., et al. (2017a). Optimizing the phosphorus use in cotton by using CSM-CROPGRO-cotton model for semi-arid climate of Vehari-Punjab, Pakistan. *Environmental Science and Pollution Research, 24*, 5811–5823.

Amin, A., et al. (2017b). Comparison of future and base precipitation anomalies by SimCLIM statistical projection through ensemble approach in Pakistan. *Atmospheric Research, 294*, 214–225.

Amin, A., et al. (2018a). Regional climate assessment of precipitation and temperature in southern Punjab (Pakistan) using SimCLIM climate model for different temporal scales. *Theoretical and Applied Climatology, 131*(1–2), 121–131.

Amin, A., et al. (2018b). Evaluation and analysis of temperature for historical (1996-2015) and projected (2030-2060) climates in Pakistan using SimCLIM climate model: Ensemble application. *Atmospheric Research, 213*, 422–436.

Amin, A., et al. (2018c). Simulated CSM-CROPGRO-cotton yield under projected future climate by SimCLIM for southern Punjab, Pakistan. *Agricultural Systems, 167*, 213–222.

Anjum, M. N., Cheema, M. J. M., Hussain, F., & Wu, R.-S. (2023). Precision irrigation: Challenges and opportunities. *Precision Agriculture*, 85–101.

Ansari, R. A., Mahmood, I., Rizvi, R., Sumbul, A., & Safiuddin. (2017). Siderophores: Augmentation of soil health and crop productivity. *Probiotics in agroecosystem*, 291–312.

Asif, M., Rafique, M. A., Munir, M., Anjum, M., & Arfan, A. (2024). Comparative study of crop and water productivity under drip and furrow irrigation systems for plastic tunnel grown off-season vegetables. *Science Letters, 12*(1), 20–26.

Awais, M., et al. (2017). Nitrogen and plant population change radiation capture and utilization capacity of sunflower in semi-arid environment. *Environmental Science and Pollution Research, 24*(21), 17511–17525.

Awais, M., et al. (2018). Potential impacts of climate change and adaptation strategies for sunflower in Pakistan. *Environmental Science and Pollution Research, 25*(14), 13719–13730.

Bardgett, R. D., & Van Der Putten, W. H. (2014). Belowground biodiversity and ecosystem functioning. *Nature, 515*(7528), 505–511.

Benbi, D. K., & Brar, K. (2021). Pyrogenic conversion of rice straw and wood to biochar increases aromaticity and carbon accumulation in soil. *Carbon Management, 12*(4), 385–397.

Benedict, C., Cogger, C. G., & Andrews, N. N. (2014). *Methods for successful cover crop management in your home garden.*

Bera, T., Collins, H., Alva, A., Purakayastha, T., & Patra, A. (2016). Biochar and manure effluent effects on soil biochemical properties under corn production. *Applied Soil Ecology, 107*, 360–367.

Brackin, R., Schmidt, S., Walter, D., Bhuiyan, S., Buckley, S., & Anderson, J. (2017). Soil biological health—What is it, and how can we improve it? In *39th conference of the Australian Society of Sugar Cane Technologists, ASSCT 2017.*

Brahmanand, P., & Singh, A. K. (2022). Precision irrigation water management-current status, scope and challenges. *Indian Journal of Fertilisers, 18*, 372–380.

Cayley, R., Taylor, D., VandenBerg, A., & Moore, D. (2002). Proterozoic–Early Palaeozoic rocks and the Tyennan Orogeny in central Victoria: The Selwyn Block and its tectonic implications. *Australian Journal of Earth Sciences, 49*(2), 225–254.

Choudhury, B., Ferraris, S., Ashton, R., Powlson, D., & Whalley, W. (2018). The effect of microbial activity on soil water diffusivity. *European Journal of Soil Science, 69*(3), 407–413.

Colica, G., Li, H., Rossi, F., Li, D., Liu, Y., & De Philippis, R. (2014). Microbial secreted exopolysaccharides affect the hydrological behavior of induced biological soil crusts in desert sandy soils. *Soil Biology and Biochemistry, 68*, 62–70.

Coombs, C., Lauzon, J. D., Deen, B., & Van Eerd, L. L. (2017). Legume cover crop management on nitrogen dynamics and yield in grain corn systems. *Field Crops Research, 201*, 75–85.

Deb, D., Kloft, M., Lässig, J., & Walsh, S. (2016). Variable effects of biochar and P solubilizing microbes on crop productivity in different soil conditions. *Agroecology and Sustainable Food Systems, 40*(2), 145–168.

Derpsch, R. (2001). Keynote: Frontiers in conservation tillage and advances in conservation practice. In D. E. Stott, R. H. Mohtar, & G. C. Steinhardt (Eds.), *Sustaining the global farm: Selected papers from the 10th international soil conservation organization meeting held may 24-29, 1999*. Purdue University Press.

Dias, T., Dukes, A., & Antunes, P. M. (2015). Accounting for soil biotic effects on soil health and crop productivity in the design of crop rotations. *Journal of the Science of Food and Agriculture, 95*(3), 447–454.

Fageria, N., & Baligar, V. (2008). Ameliorating soil acidity of tropical Oxisols by liming for sustainable crop production. *Advances in Agronomy, 99*, 345–399.

Fageria, N., Gheyi, H., & Moreira, A. (2011). Nutrient bioavailability in salt affected soils. *Journal of Plant Nutrition, 34*(7), 945–962.

Farid, H. U., et al. (2017). Site-specific aquifer characterization and identification of groundwater potential areas in district Rahim Yar Khan. *Pakistan. Polish Journal of Environmental Studies, 26*(1), 17–27.

Farid, H. U., et al. (2019a). Estimation of infiltration models parameters and their comparison to simulate the site-specific soil infiltration characteristics. *International Journal of Agricultural and Biological Engineering, 12*(3), 84–91.

Farid, H. U., et al. (2019b). Identification of influencing factors for optimal adoptability of high efficiency irrigation system (HEIS) in Punjab, Pakistan. *Sarhad Journal of Agriculture, 35*(2), 539–549.

Fernández García, I., Lecina, S., Ruiz-Sánchez, M. C., Vera, J., Conejero, W., Conesa, M. R., Domínguez, A., Pardo, J. J., Léllis, B. C., & Montesinos, P. (2020). Trends and challenges in irrigation scheduling in the semi-arid area of Spain. *Water, 12*(3), 785.

Finney, D. M., Murrell, E. G., White, C. M., Baraibar, B., Barbercheck, M. E., Bradley, B. A., Cornelisse, S., Hunter, M. C., Kaye, J. P., & Mortensen, D. A. (2017). Ecosystem services and disservices are bundled in simple and diverse cover cropping systems. *Agricultural & Environmental Letters, 2*(1), 170033.

Fischer, B. M., Manzoni, S., Morillas, L., Garcia, M., Johnson, M. S., & Lyon, S. W. (2019). Improving agricultural water use efficiency with biochar–A synthesis of biochar effects on water storage and fluxes across scales. *Science of the Total Environment, 657*, 853–862.

Gabriele, I., Race, M., Papirio, S., & Esposito, G. (2021). Phytoremediation of pyrene-contaminated soils: A critical review of the key factors affecting the fate of pyrene. *Journal of Environmental Management, 293*, 112805.

Gebretsadikan, T., Munro, P., Forge, T. A., Jones, M. D., & Nelson, L. M. (2023). Mulching improved soil fertility, plant growth and productivity, and postharvest deficit irrigation reduced water use in sweet cherry orchards in a semi-arid region. *Archives of Agronomy and Soil Science, 69*(8), 1419–1436.

Ghimire, R., Thapa, V. R., Acosta-Martinez, V., Schipanski, M., Slaughter, L. C., Fonte, S. J., Shukla, M. K., Bista, P., Angadi, S. V., & Mikha, M. M. (2023). Soil health assessment and management framework for water-limited environments: Examples from the Great Plains of the USA. *Soil Systems, 7*(1), 22.

Gobin, A., Jones, R., Kirkby, M., Campling, P., Govers, G., Kosmas, C., & Gentile, A.-R. (2004). Indicators for pan-European assessment and monitoring of soil erosion by water. *Environmental Science & Policy, 7*(1), 25–38.

Guo, M. (2021). Soil health assessment and management: Recent development in science and practices. *Soil Systems, 5*(4), 61.

Haghnazari, F., Shahgholi, H., & Feizi, M. (2015). Factors affecting the infiltration of agricultural soils. *International Journal of Agronomy and Agricultural Research, 6*(5), 21–35.

Hammad, H. M., et al. (2018). Offsetting land degradation through nitrogen and water management during maize cultivation under arid conditions. *Land Degradation and Development*, 1366–1375.

Hansen, A. T., Campbell, T., Cho, S. J., Czuba, J. A., Dalzell, B. J., Dolph, C. L., Hawthorne, P. L., Rabotyagov, S., Lang, Z., & Kumarasamy, K. (2021). Integrated assessment modeling reveals near-channel management as cost-effective to improve water quality in agricultural watersheds. *Proceedings of the National Academy of Sciences, 118*(28), e2024912118.

Hartmann, M., & Six, J. (2023). Soil structure and microbiome functions in agroecosystems. *Nature Reviews Earth & Environment, 4*(1), 4–18.

Helliwell, J., Miller, A., Whalley, W., Mooney, S., & Sturrock, C. (2014). Quantifying the impact of microbes on soil structural development and behaviour in wet soils. *Soil Biology and Biochemistry, 74*, 138–147.

Hussain, S., et al. (2022a). Using space–time scan statistic for studying the effects of COVID-19 in Punjab, Pakistan: A guideline for policy measures in regional agriculture. *Environmental Science and Pollution Research, 30*, 42495–42508.

Hussain, S., et al. (2022b). Spatiotemporal variation in land use land cover in the response to local climate change using multispectral remote sensing data. *Land, 11*, 595. https://doi.org/10.3390/land11050595

Hussain, S., et al. (2022c). Monitoring the dynamic changes in vegetation cover using Spatio-temporal remote sensing data from 1984 to 2020. *Atmosphere, 13*, 1609.

Hussain, S., et al. (2023a). Investigation of irrigation water requirement and evapotranspiration for water resource Management in Southern Punjab, Pakistan. *Sustainability, 15*, 1768.

Hussain, S., et al. (2023b). Relation of land surface temperature with different vegeation indices using multi-temporal remote sensing data in Sahiwal region, Pakistan. *Geoscience Letters, 15*, 1768.

Hussain, S., et al. (2024a). Assessment of future prediction of urban growth and climate change in district Multan, Pakistan using CA-Markov method. *Urban Climate, 53*, 101766.

Hussain, S., et al. (2024b). Agricultural land suitability analysis of southern Punjab, Pakistan using analytical hierarchy process (AHP) and multi-criteria decision analysis (MCDA) techniques. *Cogent Food & Agriulture., 10*(1), 2294540.

Hussain, S., et al. (2024c). Assessing the impact of land use land cover changes on soil moisture and vegetation cover in southern Punjab, Pakistan using multi-temporal satellite data. *Geology, Ecology, and Landscapes*. https://doi.org/10.1080/24749508.2024.2338970

Irfan, M., et al. (2022). Exploring the potential efect of Achnatherum splendens L. derived biochar treated with phosphoric acid on bioavailability of cadmium and wheat growth in contaminated soil. *Environmental Science and Pollution Research, 29*, 37676–37684.

ITPS. (2015). *Status of the world's soil resources (SWSR)—Main report.* Food and Agriculture Organization of the United Nations and Intergovernmental Technical Panel on Soils, p. 650.

Jarwar, A. H., Wang, X., Long Wang, L. Z., Zhaoyang, Q., Mangi, N., Pengjia, B., Jinjin, W., Ma, Q., & Shuli, F. (2019). Performance and evaluation of drip irrigation system, and its future advantages. *Journal of Biology, Agriculture and Healthcare, 9.*

Jayaraman, S., Naorem, A., Dalal, R. C., Sinha, N. K., Rao, C. S., Lal, R., Kundu, S., Prasad, J., & Singh, A. (2024). No-till farming and climate change mitigation: Lessons learnt from long-term no-till experiments and future perspectives. *Advances in Agronomy, 187*, 21–107.

Jiang, W., Meng, L., Liu, F., Sheng, Y., Chen, S., Yang, J., Mao, H., Zhang, J., Zhang, Z., & Ning, H. (2023). Distribution, source investigation, and risk assessment of topsoil heavy metals in areas with intensive anthropogenic activities using the positive matrix factorization (PMF) model coupled with self-organizing map (SOM). *Environmental Geochemistry and Health, 45*(8), 6353–6370.

Jordán, A., Zavala, L. M., & Gil, J. (2010). Effects of mulching on soil physical properties and runoff under semi-arid conditions in southern Spain. *Catena, 81*(1), 77–85.

Joshi, A., Arora, D., & Kumar, R. (2021). A review impact of soil erosion on agriculture. *Asian Journal of Multidimensional Research, 10*(10), 373–380.

Junhao, C., Pengpeng, C., Xiaodong, G., Qifang, Z., Yunjie, F., Xiaobo, G., Xining, Z., & Yuannong, L. (2022). Effects of plastic film residue and emitter flow rate on soil water infiltration and redistribution under different initial moisture content and dry bulk density. *Science of the Total Environment, 807*, 151381.

Kader, M., Senge, M., Mojid, M., & Ito, K. (2017). Recent advances in mulching materials and methods for modifying soil environment. *Soil and Tillage Research, 168*, 155–166.

Kalantari, Z., Ferreira, C. S. S., Pan, H., & Pereira, P. (2023). *Nature-based solutions to global environmental challenges* (Vol. 880, p. 163227). Elsevier.

Kelly, C., Schipanski, M. E., Tucker, A., Trujillo, W., Holman, J. D., Obour, A. K., Johnson, S., Brummer, J. E., Haag, L., & Fonte, S. J. (2021). Dryland cover crop soil health benefits are maintained with grazing in the US High and Central Plains. *Agriculture, Ecosystems & Environment, 313*, 107358.

Khaliq, M. A., et al. (2022). Assessment of heavy metal accumulation and health risks in okra (Abelmoschus Esculentus L.) and spinach (Spinacia Oleracea L.) fertigated with wastewater. *International Journal of Food Contamination, 09*, 11.

Khangura, R., Ferris, D., Wagg, C., & Bowyer, J. (2023). Regenerative agriculture—A literature review on the practices and mechanisms used to improve soil health. *Sustainability, 15*(3), 2338.

Khurshid, N., et al. (2023). Exogenously applied nicotinic acid alleviates drought stress by enhancing morpho-physiological traits and antioxidant defense mechanisms in wheat. *Ecotoxicology and Environmental Safety, 263*, 115350.

Kisaka, M. O., Shisanya, C., Cournac, L., Manlay, J. R., Gitari, H., & Muriuki, J. (2023). Integrating no-tillage with agroforestry augments soil quality indicators in Kenya's dry-land agroecosystems. *Soil and Tillage Research, 227*, 105586.

Kumar, A., Kadam, S., Arif, M., Meena, R., & Verma, T. (2020). Legumes an alternative land use options for sustaining soil health. *Agriculture & Food e-newsletter, 6.*

Kumar, N., Singh, H., Giri, K., Kumar, A., Joshi, A., Yadav, S., & Mishra, G. (2024). Physiological and molecular insights into the allelopathic effects on agroecosystems under changing environmental conditions. *Physiology and Molecular Biology of Plants, 30*(3), 417–433.

Labrière, N., Locatelli, B., Laumonier, Y., Freycon, V., & Bernoux, M. (2015). Soil erosion in the humid tropics: A systematic quantitative review. *Agriculture, Ecosystems & Environment, 203*, 127–139.

Lal, R. (2015). Soil carbon sequestration and aggregation by cover cropping. *Journal of Soil and Water Conservation, 70*(6), 329–339.

Leal Filho, W., Nagy, G. J., Setti, A. F. F., Sharifi, A., Donkor, F. K., Batista, K., & Djekic, I. (2023). Handling the impacts of climate change on soil biodiversity. *Science of the Total Environment, 869*, 161671.

Lehmann, J., Gaunt, J., & Rondon, M. (2006). Bio-char sequestration in terrestrial ecosystems–a review. *Mitigation and Adaptation Strategies for Global Change, 11*, 403–427.

Lehmann, J., Rillig, M. C., Thies, J., Masiello, C. A., Hockaday, W. C., & Crowley, D. (2011). Biochar effects on soil biota–a review. *Soil Biology and Biochemistry, 43*(9), 1812–1836.

Levidow, L., Zaccaria, D., Maia, R., Vivas, E., Todorovic, M., & Scardigno, A. (2014). Improving water-efficient irrigation: Prospects and difficulties of innovative practices. *Agricultural Water Management, 146*, 84–94.

Li, X., Qiao, L., Huang, Y., Li, D., Xu, M., Ge, T., Meersmans, J., & Zhang, W. (2023). Manuring improves soil health by sustaining multifunction at relatively high levels in subtropical area. *Agriculture, Ecosystems & Environment, 353*, 108539.

Lone, A., Najar, G., Ganie, M., Sofi, J., & Ali, T. (2015). 生物炭对土壤可持续健康的影响: 探索和热点综述. *Pedosphere, 25*(5), 639–653.

López-Vicente, M., Calvo-Seas, E., Álvarez, S., & Cerdà, A. (2020). Effectiveness of cover crops to reduce loss of soil organic matter in a rainfed vineyard. *Land, 9*(7), 230.

Manda, R. R., Addanki, V. A., Giabardo, A., Benjamin, J., Hossain, M. J., Khanna, S., Gaddam, M., Kumar, R., & Srivastava, S. (2023). Soil health management and microorganisms: Recent development. In *Detection, diagnosis and management of soil-borne phytopathogens* (pp. 437–493). Springer.

Mauget, S. A., Himanshu, S. K., Goebel, T. S., Ale, S., Lascano, R. J., & Gitz, D. C., III. (2021). Soil and soil organic carbon effects on simulated southern high plains dryland cotton production. *Soil and Tillage Research, 212*, 105040.

McDaniel, M., Tiemann, L., & Grandy, A. (2014). Does agricultural crop diversity enhance soil microbial biomass and organic matter dynamics? A meta-analysis. *Ecological Applications, 24*(3), 560–570.

Molden, D. (2013). *Water for food water for life: A comprehensive assessment of water management in agriculture*. Routledge.

Matos, C. D. C., Costa, M. D., Silva, I. R., & Silva, A. A. (2019). Competitive capacity and rhizosphere mineralization of organic matter during weed-soil microbiota interactions. *Planta Daninha, 37*, e019182676.

Mubeen, M., et al. (2021). Effect of PGPR and drought on spring maize (Zia mays L.). *Pakistan Journal of Botany, 53*(2), 731–739.

Muhammad, H., et al. (2023). A paradigm shift towards beneficial microbes enhancing the efficiency of organic and inorganic nitrogen sources for a sustainable environment. *Land, 12*(3), 680.

Murindangabo, Y. T., Kopecký, M., Konvalina, P., Ghorbani, M., Perná, K., Nguyen, T. G., Bernas, J., Baloch, S. B., Hoang, T. N., & Eze, F. O. (2023). Quantitative approaches in assessing soil organic matter dynamics for sustainable management. *Agronomy, 13*(7), 1776.

Murtaza, B., Murtaza, G., Sabir, M., Owens, G., Abbas, G., Imran, M., & Shah, G. M. (2017). Amelioration of saline–sodic soil with gypsum can increase yield and nitrogen use efficiency in rice–wheat cropping system. *Archives of Agronomy and Soil Science, 63*(9), 1267–1280.

Naeem, M. A., Khalid, M., Aon, M., Abbas, G., Amjad, M., Murtaza, B., Khan, W.-U.-D., & Ahmad, N. (2018). Combined application of biochar with compost and fertilizer improves soil properties and grain yield of maize. *Journal of Plant Nutrition, 41*(1), 112–122.

Nasim, W., et al. (2018). Future risk assessment by estimating historical heat wave trends with projected heat accumulation using SimCLIM climate model in Pakistan. *Atmospheric Research, 205*, 118–133.

Naz, S., et al. (2022). Modeling the impact of climate warming on potato phenology. *European Journal of Agronomy, 132*, 126404.

Naz, T., et al. (2023). Bio-fortification of two wheat cultivars with iron and zinc through their soil and foliar application in salt-factored soil: Growth, ionic, physiological, and biochemical modifications. *Journal of Plant Growth Regulation, 42*, 5727–5745.

Ning, T., Liu, Z., Hu, H., Li, G., & Kuzyakov, Y. (2022). Physical, chemical and biological subsoiling for sustainable agriculture. *Soil and Tillage Research, 223*, 105490.

Nyamangara, J., Bergström, L., Piha, M., & Giller, K. (2003). Fertilizer use efficiency and nitrate leaching in a tropical sandy soil. *Journal of Environmental Quality, 32*(2), 599–606.

Nyawade, S., Gitari, H. I., Karanja, N. N., Gachene, C. K., Schulte-Geldermann, E., Sharma, K., & Parker, M. L. (2020). Enhancing climate resilience of rain-fed potato through legume intercropping and silicon application. *Frontiers in Sustainable Food Systems, 4*, 566345.

Oguntunde, P. G., Abiodun, B. J., Ajayi, A. E., & Van De Giesen, N. (2008). Effects of charcoal production on soil physical properties in Ghana. *Journal of Plant Nutrition and Soil Science, 171*(4), 591–596.

Pedda Ghouse Peera, S. K., Debnath, S., & Maitra, S. (2020). Mulching: Materials, advantages and crop production. In S. Maitra, D. J. Gaikwad, & S. Tanmoy (Eds.), *Protected cultivation and smart agriculture* (pp. 55–66). New Delhi Publishers.

Pieri, C., Evers, G., Landers, J., O'Connell, P., & Terry, E. (2002). *No-till farming for sustainable rural development*. The International Bank for Reconstruction and Development.

Powlson, D. S., Whitmore, A. P., & Goulding, K. W. (2011). Soil carbon sequestration to mitigate climate change: A critical re-examination to identify the true and the false. *European Journal of Soil Science, 62*(1), 42–55.

Prabha, A. S., Thangakani, J. A., Devi, N. R., Dorothy, R., Nguyen, T. A., Kumaran, S. S., & Rajendran, S. (2022). Nanotechnology and sustainable agriculture. In *Nanosensors for smart agriculture* (pp. 25–39). Elsevier.

Price, A. J., & Norsworthy, J. K. (2013). Cover crops for weed management in southern reduced-tillage vegetable cropping systems. *Weed Technology, 27*(1), 212–217.

Prosdocimi, M., Tarolli, P., & Cerdà, A. (2016). Mulching practices for reducing soil water erosion: A review. *Earth-Science Reviews, 161*, 191–203.

Quintarelli, V., Radicetti, E., Allevato, E., Stazi, S. R., Haider, G., Abideen, Z., Bibi, S., Jamal, A., & Mancinelli, R. (2022). Cover crops for sustainable cropping systems: A review. *Agriculture, 12*(12), 2076.

Rehberger, E., West, P. C., Spillane, C., & McKeown, P. C. (2023). What climate and environmental benefits of regenerative agriculture practices? An evidence review. *Environmental Research.* Communications, *5*.

Rhodes, C. J. (2024). Trees—Protectors against a changing climate. *Ecological Civilization, 1*(2), 10002.

Romdhane, S., Spor, A., Busset, H., Falchetto, L., Martin, J., Bizouard, F., Bru, D., Breuil, M.-C., Philippot, L., & Cordeau, S. (2019). Cover crop management practices rather than composition of cover crop mixtures affect bacterial communities in no-till agroecosystems. *Frontiers in Microbiology, 10*, 1618.

Sabagh, A. E. L., et al. (2019). Drought and salinity stresses in barley: Consequences and mitigation strategies. *Australian Journal of Crop Science, 13*(06), 810–820.

Sainju, U. M., Lenssen, A. W., Allen, B. L., Jabro, J. D., & Stevens, W. B. (2021). Crop water and nitrogen productivity in response to long-term diversified crop rotations and management systems. *Agricultural Water Management, 257*, 107149.

Sardar, A., Shahid, M., Natasha, Khalid, S., Anwar, H., Tahir, M., Shah, G. M., & Mubeen, M. (2020). Risk assessment of heavy metal (loid)s via Spinacia oleracea ingestion after sewage water irrigation practices in Vehari District. *Environmental Science and Pollution Research, 27*, 39841–39851.

Schipanski, M. E., Barbercheck, M., Douglas, M. R., Finney, D. M., Haider, K., Kaye, J. P., Kemanian, A. R., Mortensen, D. A., Ryan, M. R., & Tooker, J. (2014). A framework for evaluating ecosystem services provided by cover crops in agroecosystems. *Agricultural Systems, 125*, 12–22.

Schlegel, A. J., Assefa, Y., Dumler, T. J., Haag, L. A., Stone, L. R., Halvorson, A. D., & Thompson, C. R. (2016). Limited irrigation of corn-based no-till crop rotations in west central Great Plains. *Agronomy Journal, 108*(3), 1132–1141.

Schlegel, A. J., Assefa, Y., Haag, L. A., Thompson, C. R., & Stone, L. R. (2019). Soil water and water use in long-term dryland crop rotations. *Agronomy Journal, 111*(5), 2590–2599.
Shah, K. K., Modi, B., Pandey, H. P., Subedi, A., Aryal, G., Pandey, M., & Shrestha, J. (2021). Diversified crop rotation: An approach for sustainable agriculture production. *Advances in Agriculture, 2021*, 1–9.
Shojaei, S., Ardakani, M. A. H., & Sodaiezadeh, H. (2019). Optimization of parameters affecting organic mulch test to control erosion. *Journal of Environmental Management, 249*, 109414.
Spanner, J., & Napolitano, G. (2015). *Healthy soils are the basis for healthy food production*. FAO.
Suliman, W., Harsh, J. B., Abu-Lail, N. I., Fortuna, A.-M., Dallmeyer, I., & Garcia-Pérez, M. (2017). The role of biochar porosity and surface functionality in augmenting hydrologic properties of a sandy soil. *Science of the Total Environment, 574*, 139–147.
Sun, L., Wang, S., Zhang, Y., Li, J., Wang, X., Wang, R., Lyu, W., Chen, N., & Wang, Q. (2018). Conservation agriculture based on crop rotation and tillage in the semi-arid loess plateau, China: Effects on crop yield and soil water use. *Agriculture, Ecosystems & Environment, 251*, 67–77.
Tahat, M. M., Alananbeh, K. M., Othman, Y. A., & Leskovar, D. I. (2020). Soil health and sustainable agriculture. *Sustainability, 12*(12), 4859.
Tariq, M., et al. (2008). The impact of climate warming and crop management on phenology of sunflower-based cropping systems in Punjab, Pakistan. *Agricultural and Forest Meteorology, 256–257*(2018), 270–282.
Ullah, I., et al. (2022). Comparative effects of biochar and NPK on wheat crops under different management systems. *Crop & Pasture Science*. Retrieved from https://www.publish.csiro.au/CP/CP21146/.
Van Zwieten, L., Kimber, S., Morris, S., Chan, K., Downie, A., Rust, J., Joseph, S., & Cowie, A. (2010). Effects of biochar from slow pyrolysis of papermill waste on agronomic performance and soil fertility. *Plant and Soil, 327*, 235–246.
Vanlauwe, B., Amede, T., Bationo, A., Bindraban, P., Breman, H., Cardinael, R., Couedel, A., Chivenge, P., Corbeels, M., & Dobermann, A. (2023). Fertilizer and soil health in Africa: The role of fertilizer in building soil health to sustain farming and address climate change.
Waleed, M., et al. (2022). Evaluating the efficiency of coarser to finer resolution multispectral satellites in mapping paddy rice fields using GEE implementation. *Scientific Reports, 12*, 13210.
Wang, J., Xiong, Z., & Kuzyakov, Y. (2016). Biochar stability in soil: Meta-analysis of decomposition and priming effects. *GCB Bioenergy, 8*(3), 512–523.
Wang, S., Gao, X., Yang, M., Zhang, L., Wang, X., Wu, P., & Zhao, X. (2022). The efficiency of organic C sequestration in deep soils is enhanced by drier climates. *Geoderma, 415*, 115774.
Wang, L., Lu, P., Feng, S., Hamel, C., Sun, D., Siddique, K. H., & Gan, G. Y. (2024). Strategies to improve soil health by optimizing the plant–soil–microbe–anthropogenic activity nexus. *Agriculture, Ecosystems & Environment, 359*, 108750.
Yuvaraj, M., Bose, K. S. C., Elavarasi, P., & Tawfik, E. (2021). *Soil salinity and its management* (Vol. 1). IntechOpen.
Zahed, Z., Mufti, S., Kumar, S. S., Wani, O. A., Mushtaq, F., Rasool, R., Babu, S., Abidi, I., Gaber, A., & Hossain, A. (2022). Organic and inorganic mulches combination improves the productivity, quality and profitability of rainfed potato in the temperate himalayan region. *Gesunde Pflanzen, 74*(4), 1109–1122.
Zaman, G., Murtaza, B., Imran, M., Shahid, M., Shah, G. M., Amjad, M., Naeem, M. A., Mubeen, M., & Murtaza, G. (2020). Utilization of bio-municipal solid waste improves saline-sodic soils and crop productivity in rice-wheat. *Compost Science & Utilization, 28*(1), 16–27.
Zimmerman, A. R. (2010). Abiotic and microbial oxidation of laboratory-produced black carbon (biochar). *Environmental Science & Technology, 44*(4), 1295–1301.

Water Harvesting Approaches for Irrigation

Muhammad Farooq Azhar, Ehsan Ali, Abida Aziz, Mazhar Ali, and Iqra Mubeen

Abstract The recognition of water as a critical resource for agriculture has brought attention to the pivotal role of water harvesting in mitigating the challenges posed by climate change and the increasing demand for irrigation. The agricultural sector's status as the world's largest consumer of freshwater underscores the urgency of sustainable water management practices. While water harvesting emerges as a time-honored practice with diverse strategies, its implementation is not without challenges. Issues such as labor, expenses, land degradation, and equitable access underscore the importance of tailoring water harvesting techniques to local socio-economic and environmental conditions. In addressing these challenges, it becomes imperative to explore alternatives and multifaceted solutions. Financial incentives can play a crucial role in promoting the adoption of appropriate water harvesting methods, fostering resilience in the face of climate uncertainties. Additionally, incorporating advanced technologies into irrigation water management is proposed as a strategic approach for long-term solutions. This involves optimizing water use by aligning availability with crop needs, thereby reducing waste and contributing to sustainable agricultural practices. In essence, the journey towards effective water harvesting and irrigation water management requires a holistic and adaptive

M. F. Azhar (✉) · E. Ali
Department of Forestry and Range Management, Bahauddin Zakariya University, Multan, Pakistan
e-mail: farooqazhar@bzu.edu.pk

A. Aziz
Department of Botany, The Women University, Multan, Pakistan

M. Ali
Cholistan Institute of Business Administration, CUVAS Bahawalpur, Bahawalpur, Pakistan

I. Mubeen
Department of Environmental Sciences, COMSATS University Islamabad, Vehari, Punjab, Pakistan

Department of Renewable Resources, University of Alberta, Edmonton, AB, Canada

M. Mubeen et al. (eds.), *Innovations in Agricultural Water Management*,
https://doi.org/10.1007/978-3-031-91883-4_6

approach. By acknowledging the site-specific nature of water collection methods and aligning strategies with local conditions, we can pave the way for a more resilient agricultural sector that not only meets the demands of today but also addresses the challenges of tomorrow, including those posed by climate change. As we navigate these complexities, a commitment to sustainable practices and innovative solutions will be paramount in ensuring the continued prosperity of agriculture and the ecosystems it sustains.

Keywords Water shortage · Water harvesting · Micro-basin methods · Macro-basin methods · Water management · Traditional approaches · Modern approaches · Sustainable agriculture

1 Water Harvesting and Its Importance

Water resource utilization in agriculture has advanced significantly during the last six decades, driven by population increase and the rising need for food production. The global population has increased from 2.5 billion in 1950 to 7 billion now, resulting in a doubling of irrigated land and a tripling of water withdrawals. This spike in water infrastructure development has been critical in fulfilling growing food demand (FAO, 2020). However, water scarcity and drought have grown common in many nations throughout the world in recent years. Inadequate rainfall and rising demand have resulted in a hazardous loss in water supplies, threatening the food security of many countries. Addressing these difficulties and finding sustainable solutions is critical to ensuring the availability of water resources for agriculture and mitigating the hazards associated with water shortage (WRI, 2021).

To fulfill the requirements of a rising and wealthier global population, global food production is expected to quadruple by 2050 (Beltran-Peña et al., 2020). Achieving this target necessitates not just decreasing food waste and adopting dietary changes, but also addressing agriculture's environmental implications (Gerten et al., 2020; Dawar et al., 2021; Ali et al., 2021; Bukhari et al., 2021; Rehman et al., 2021; Rasool et al., 2021). While agricultural improvements have helped to reduce poverty worldwide, they have also played a key influence in climate change, biodiversity loss, and the deterioration of land and water resources (Zhang et al., 2021).

Irrigation is an effective approach for agricultural intensification, particularly in areas with little rainfall (Rosa et al., 2020a, 2020b). While rainfed agriculture is dependent on natural rainfall to supply agricultural water, irrigated agriculture requires the artificial application of water via water infrastructure. Irrigation has been demonstrated to enhance agricultural yield greatly, making it critical for guaranteeing food security (World Bank, 2022). It is crucial to remember, however, that irrigation accounts for a sizable share of world water use and withdrawals. Finding sustainable techniques to irrigation and water resource management is therefore critical to mitigating these negative consequences while yet achieving food production objectives.

To fulfill rising global food demand without infringing on natural ecosystems, irrigation must be expanded over underperforming rainfed croplands (Droppers et al., 2021). Global warming is already lowering agricultural production and threatening global food security (Jägermeyr et al., 2021; Kummu et al., 2021). Rainfed agriculture is largely dependent on good weather conditions, accounting for 60% of world food production. As a result, rural populations that rely largely on rainfed agriculture will be more vulnerable to climate change consequences (Graves et al., 2019). With global warming aggravating water and heat stress events on rainfed croplands (Fitton et al., 2019), irrigation as a climate adaptation method is critical. This technique has the potential to minimize agricultural heat and water stress, reduce climatic variability, and attenuate extreme weather occurrences (IPCC, 2022; Nasim et al., 2016, 2017; Mubeen et al., 2013a, 2013b, 2013c, 2016; Wajid et al., 2014; Sultana et al., 2014; Khaliq et al., 2013).

According to the FAO (2022), the amount of irrigated land has more than quadrupled over the last 60 years and currently accounts for 22% of world croplands. There has been an additional 52 million ha increase in worldwide irrigated lands between 2000 and 2019, totaling 341 million ha. Population increase will fuel future irrigation expansion, with forecasts indicating that worldwide irrigated agriculture might quadruple by 2050, potentially reaching 800 million ha (Puy et al., 2020). As a result, this growth will increase the demand for irrigation water (Puy et al., 2021).

Water harvesting is a beneficial supplement to irrigation requirements, providing adaptive, flexible, simple-to-understand, and cost-effective solutions to productivity, climate adaption, and water security concerns. The fundamental advantage of water harvesting is its ability to increase water buffering capacity. This practice includes a variety of technologies, including traditional methods built via local knowledge as well as inventive new ones. Water collecting has long been used in rain-fed areas where other freshwater supplies are rare or nonexistent, allowing for a more sustainable way of life. It is critical for refilling depleted groundwater supplies and increasing resistance to climate change, especially during droughts and high rains. This chapter explores into numerous water collecting methods to address the world's rising demand for irrigation water.

2 Basic Definition of Water Harvesting

Water harvesting is the process of collecting rainwater from rainstorms within a defined catchment area. This water is subsequently collected and used for a variety of reasons, including human, animal, and agricultural consumption. The water can be used immediately for irrigation or saved for later use in aboveground ponds or subterranean reservoirs such as cisterns or shallow aquifers. Water harvesting systems are classed as farm systems (microsystems), roof systems (microsystems), valley floor systems (macrosystems), and nonvalley systems (macrosystems) based on the size of the catchment area (Yetik & Şen, 2020; Shah et al., 2021; Mehmood et al., 2021; Rasool et al., 2020; Hussain et al., 2020; Ali et al., 2020). The right method is chosen based on the area's unique water requirement and existing climatic conditions.

3 Components of a Water Harvesting System Model

Water harvesting is a traditional activity that includes three main components: a collecting system, a transportation system, and a storage system (Verbist et al., 2009).

3.1 Collection Systems

Water harvesting is the collecting of water from various sources after heavy rains or flooding. Catchments are classified into two technological systems in this practice: land surface harvesting and roof surface harvesting (Shakya & Thanju, 2013). When compared to other rainwater harvesting techniques, land surface harvesting is a straightforward and cost-effective way of collecting water from surfaced areas such as terraced landscapes, roads, and courtyards. Unpaved regions such as bare soil, meadows, cultivated land, rocky surfaces, natural slopes, and uncultivated areas can also be used as water collecting sites where floodwater runs into stream beds (Gould, 2015). Another typical option is roof surface collection, which involves collecting rainwater from rooftops using materials such as galvanized sheet metal and corrugated sheet metal and connecting it to a storage system (Shakya & Thanju, 2013; Mubeen et al., 2021; Rasool et al., 2020). The use of proper surface coatings in rainwater collecting regions is critical because it affects maximum water collection capacity, water quality, and overall efficiency (Ling & Benham, 2014). The quantity and quality of rainwater collected are influenced by the location, size, and material of the roof. Bamboo roofs, for example, produce lower-quality water, necessitating the usage of alternative materials like galvanized iron, aluminum, or cement. Cleaning the rainwater collection surface on a regular basis is necessary to remove dirt, leaves, and bird droppings (Pradhan & Sahoo, 2019).

3.2 Transportation Systems

The material used for the water conveyance system, which transports water from the collecting location to the storage area, is determined by the collection system. As components of the collecting system, many materials or systems, such as galvanized iron or polyvinyl chloride, may be used. In a roof collecting system, for example, the conveyance system in dwellings might include canals and ditches that route water to subterranean storage (Shakya & Thanju, 2013). Once collected, rainwater is delivered through transportation systems and either filtered or left unfiltered dependent on needs in rainwater collecting systems. It is then moved to storage facilities for safekeeping (Toyran, 2022).

Wider gutters are required in places with considerable rainfall than in areas with lower rainfall levels. It is typically suggested that gutters be at least 12.7 cm wide. The gutters should be large enough to accommodate drainage during periods of heavy rain. Standard gutter forms and sizes are widely available, but custom-made profiles set on a slope can also be used to enhance rainfall harvesting. Gutters and downpipes, like roof surfaces, should be made of chemically inert materials such as wood, plastic, metal, or fiberglass to avoid compromising water quality (Abdulla & Al-Shareef, 2009).

3.3 Storage System

Harvested water is often stored in one of two ways: in a tank or underground. However, in water harvesting approaches, the storage system is often the last component used (Shakya & Thanju, 2013). Because it is more expensive than other components, the storage tank is critical to the rainwater collecting system. To achieve an acceptable size at a fair cost, it requires careful design. The storage tank is intended to hold the necessary amount of water and can be situated above or underneath (Worm, 2006). The storage tanks' material, size, and position are governed by the planned use of the collected rainwater. The capacity of the tank should be proportionate to parameters such as expected monthly water consumption, location of usage, monthly rainfall, and collection area size (Stringer et al., 2012).

4 The State of the Water Resources at Global Level

Over the last four decades, worldwide water use has increased by around 1% per year (FAO, 2022). This increase is most noticeable in middle- and lower-income countries, particularly those with growing economies (Ritchie & Roser, 2017). This growing trend has been fueled by the interaction of population increase, socioeconomic development, and changing purchasing preferences. Notably, Northern America and Central Asia have the greatest per capita water extraction rates (FAO, 2022; Hussain et al., 2020). Globally, groundwater provides half of the water drawn for residential consumption and a quarter of the water extracted for irrigation, with depletion hotspots frequently identified in areas with intensive groundwater usage for irrigation or feeding major metropolitan centers (Connor & Miletto, 2022).

Predicting future patterns in water demand is difficult. Burek et al. (2016) project a 1% yearly rise in global water consumption, culminating in a 20–30% increase by 2050, with a margin of error surpassing 50%. This projection is dependent on the implementation of water-saving measures in towns, industry, and agriculture. Water availability per person varies greatly by area but has generally decreased as a result of population expansion. Global per capita internal renewable water resources (IRWRs) decreased by about 20% between 2000 and 2018, with more significant declines in Sub-Saharan Africa (41%), Central Asia (30%), Western Asia (29%), and Northern Africa (26%), while Europe experienced the smallest decline at 3% (FAO, 2022).

In 2017, global freshwater extraction reached around 3800 km^3/year, accounting for nearly 10% of available renewable water resources (UNESCO, 2021). This worldwide statistic, however, conceals regional issues associated with physical water stress (WRI, 2019), which takes into account water usage in proportion to availability. Between 2010 and 2015, 2.2–3.2 billion people, or 32–46% of the global population, suffered water stress for at least 1 month each year, with the majority living in Asia, mainly northeast China, India, and Pakistan (Vanham et al., 2021). According to projections, the worldwide urban population suffering water shortages might grow from 933 million in 2016 to 1.7–2.4 billion in 2050, with India bearing a disproportionate share of the burden (He et al., 2021; Saeed et al. 2013; Ali et al., 2013; Aslam et al., 2013; Khaliq et al., 2013; Aslam et al., 2013).

Currently, roughly 25% of worldwide croplands confront agricultural economic water shortages, with irrigation absent due to institutional and economic restrictions rather than hydrological limits (Rosa et al., 2020b). According to the World Bank (2016), water scarcity, worsened by climate change, may result in a loss of up to 6% of GDP for some regions by 2050, affecting agriculture, health, and earnings, as well as potentially generating migration and war (Fig. 1).

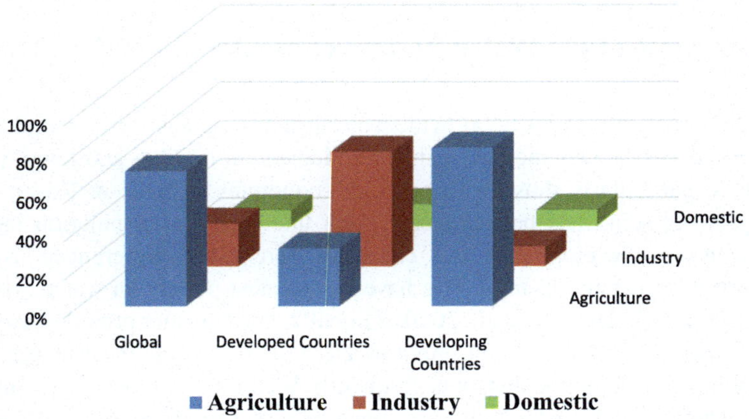

Fig. 1 Water use distribution in developed and developing countries

5 History of Water Harvesting

Water harvesting has been used since ancient times in varied locations worldwide, adapting to varying conditions in both the driest and wettest areas, as well as in communities ranging from the poorest to the wealthiest (Oweis et al., 2001). Water harvesting may be seen throughout history, beginning with the Roman roof basin system, when entire towns and dwellings were constructed to rely on rainwater as the major supply for drinking and domestic use as early as 2000 B.C. The use of tanks to collect runoff from hillsides for agricultural and residential reasons has been reported in Israel's Negev Desert, which gets low rainfall of 100 mm/year, with evidence of 200–2000 m^3 tanks in Egypt in use for around two millennia.

Rainwater harvesting has significant historical roots in places such as Asia, where techniques in Thailand date back over 2000 years, and Africa, where primitive earthen pots with rudimentary streams or roof eaves have been used for thousands of years. Notably, the Yerebatan Cistern in Istanbul, which measures 140 m long, 70 m broad, and has a capacity of 80,000 m^3, is the world's largest water gathering cistern. Farmers in the Palestinian region have built stone terraces throughout history to alleviate the impacts of heavy rainfall, increase soil organic matter content by reducing runoff and soil erosion, and protect soil water structures (Hammad et al., 2006). The enduring use of rainwater harvesting systems is evident globally, reflecting a practice that has persisted for millennia (Örs et al., 2011).

Throughout history, different civilizations, notably the Romans and Venetians, used inventive infrastructures to control water supply, such as aqueducts, reservoirs, and rain cisterns. Recognizing the importance of water in Muslim culture, fountains and baths were popular throughout the Ottoman Empire (1669–1898), with vast hydraulic facilities producing water. Although the function of cisterns in centralized places declined with the emergence of centralized water systems during the Ottoman period, their construction and usage in remote locations survived, with circular cisterns in rural settings. These circular cisterns are still used to supply the water demands of animals today (Selimoğlu & Yamaçli, 2022).

Evidence from prehistoric ages across the world highlights humanity's efforts to meet water needs for residential consumption, irrigation, and cattle reproduction by collecting and storing rainfall. Rainwater has traditionally been the principal supply for both potable and nonpotable purposes, making rainwater gathering critical to human life (Yannopoulos et al., 2019). Furthermore, rainwater has been used in desert places and locales with difficult water availability since ancient times (Selimoğlu & Yamaçli, 2022).

6 Categories of Water Harvesting Techniques

Various classifications exist for water harvesting methods, with the most commonly utilized system being based on the size of the basin—specifically, distinguishing between micro-basins and macro-basins.

6.1 Infiltration Trenches or Stormwater Harvesting

An infiltration trench is often made out of a reservoir filled with gravel aggregates or plastic lattice structures and coated with geotextile filter cloths (Wang & Guo, 2020). Infiltration chambers with prefabricated modular constructions are also used for temporary stormwater storage.

Infiltration trenches, which are distinguished by rectangular ditches filled with granular stones, serve as temporary subterranean reservoirs for rainwater runoff. By trapping finer particles, the granular filter inside retains the soil's natural permeability. These trenches, which are often long, narrow, and shallow, collect water from neighboring impermeable surfaces, allowing it to percolate and infiltrate through the subsoil and along the borders into the surrounding soils. The coarse stones in the trenches are more permeable than the underlying soil, allowing rainfall to percolate into the depression. Furthermore, water-percolating plants can be planted in the region, helping to replenish groundwater for potential reuse (Chahar et al., 2012).

6.2 Swales or Vegetative Open Channel

Grass swales are vegetated open channels meant to channel runoff and reduce it by infiltration into the underlying soils (Shafique et al., 2018). A vegetated ponding region supported by a porous media layer where plants may grow is typical of a bioretention cell (Zhang & Guo, 2014). These infiltration-focused stormwater best management strategies have a comparable storage structure that includes a depression storage layer, a soil layer, and a storage layer. Rainfall directly onto these operations and runoff from contributing catchment regions are the most common sources of inflows into these practices. These techniques' water losses include evapotranspiration and infiltration via their sides and bottoms.

6.3 Embankments or Basins (Micro- and Macro-water Harvesting)

6.3.1 Micro-basin Methods

6.3.2 Negarim

The approach entails the construction of tiny diamond or rectangular grid earth embankments surrounded by artificial embankments known as dikes, which are particularly useful in small flow catchments (Oweis & Hachum, 2006). A Negarim may

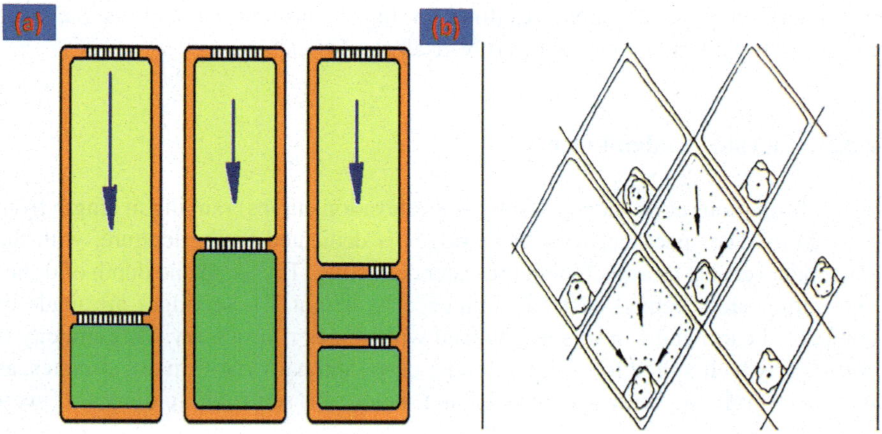

Fig. 2 Schematic diagram of micro-basin methods (**a**) Negarim and (**b**) Mesqat

be built on practically any slope; however, slopes greater than 10% should be avoided since they may cause soil erosion and demand nonideal increases in embankment height. As a result, this approach is advised for areas with an annual rainfall of 150–500 mm (Fig. 2).

6.3.3 Mesqat

A mesqat is a catchment region that may serve numerous cultivated areas (target areas) by redirecting surplus runoff water from one cultivated area to another, effectively channeling water to places in need (Yetik & Şen, 2020).

The mesqat system, (Fig. 2) measuring around 500 m^2, includes a water catchment basin bordered by dikes known as a "meskat," and a planting area of roughly 250 m^2 known as a "manka." The entire mesqat system is surrounded by a 20 cm high embankment with spillways that allow for the controlled flow of runoff into the manka plots while enabling surplus water to depart without creating erosion. This micro-catchment technology, which is only used for tree cultivation, is widely used in Tunisia, especially for the cultivation of olive trees on 300,000 ha (Örs et al., 2011).

6.3.4 Elevated Terraces

This approach is especially developed to prevent soil erosion and is built on steep terrain with slopes ranging from 20 to 50%. Cultivated terraces are deliberately designed to be level, with stone walls to restrict runoff and successfully regulate erosion. These terraces include simple drainage systems to enable the safe evacuation of excess water. The terrace method is appropriate for places with annual rainfall ranging from 200 to 600 mm. While manual building is possible, heavy machinery may be required, resulting in higher installation and maintenance expenses as well as labor demands (Oweis et al., 2012).

6.3.5 Elevated Embankments

These formations correspond to isohypses (elevation curves) ranging in length from 5 to 20 m. The space between these ridges is dedicated for agriculture, with the remainder serving as a catchment for water collecting. The slope and depth of drainage immediately behind each hill influence its height. These ridges are made of compacted earth and can be strengthened with stones if necessary. This strategy is particularly well-suited for farming fodder crops and hardwoods in dry climates, as well as sorghum, millet, cowpeas, and beans in semi-arid climates (Oweis et al., 2012).

6.3.6 Small Pits

The "Zay system" consists of shallow pits with depths ranging from 5 to 15 cm and diameters ranging from 0.3 to 2.0 m (Wright, 1985). It comprises a process in which fertilizers and herbs are blended with a portion of the soil, improving the fertility and structure of the pit soil. The leftover soil is used to build a tiny embankment at the pit's bottom. This technology, which combines the Zay system with embankments to slow surface runoff, is most commonly used for growing annual crops, particularly grains like millet, maize, and sorghum (Prinz 2002). This approach is usually used on flat terrain or slopes of less than 5% in areas with annual rainfall ranging from 350 to 600 mm. On level ground, small pits serve the purpose of moisture conservation rather than water harvesting (Oweis et al., 2012).

6.3.7 Eyebrow-Shaped Terraces

Semicircular constructions supported by stones pointing in the direction of the slope form eyebrow-shaped terraces that serve as micro-basins specifically for the water demands of trees or bushes. These structures, often known as "platform terraces," keep runoff at a consistent level. The basin dimensions are usually 5–50 m^2, with a planting area of 1–5 m^2. This approach is applicable to slopes ranging from 1 to 50%, with higher inclines required the strengthening of embankments using stone material. Eyebrow-shaped terraces are appropriate in areas with annual rainfall ranging from 200 to 600 mm (Örs et al., 2011).

6.3.8 Surface or Runoff Flow

Rainfall that falls on the ground is channeled through diverter pipes, allowing it to be collected and stored in either underground or aboveground tanks, basins, or canals. Runoff harvesting may be used on a variety of ground surfaces, such as asphalt, pavement, parking lots, and soil. Rainwater collected from open ground areas can be used without treatment to irrigate green spaces or as flushing water for home use, although it may be more polluted than water collected from roofs. Alternatively, it can be filtered and used for groundwater replenishment (Gupta & Harvesting-pub, 2006). In comparison to rooftop rainwater collection, ground harvesting systems have the benefit of collecting water from a broader surface area. Surface rain harvesting transports and directs collected rainfall held in regions such as dams down streams and rivers, adding to water supply demands during dry seasons (NWRM, 2023).

6.3.9 Roof Top Flow

Rainwater collected from rooftops is an important supply of drinking water, especially in rural regions where tap water is scarce (Worm, 2006). A significant amount of the rainfall, ranging between 80 and 85%, can be properly gathered and stored. A well-designed system may provide a family's water needs for an entire year by collecting and storing water through rooftop systems in areas with less than 200 mm of annual rainfall (Morgan, 1990). The system's operating premise is simple: it uses the natural slope of roofs to funnel water into storage tanks. Furthermore, there is a recent trend of storing and using rainwater collected from greenhouse roofs for agricultural irrigation once it has been filtered (Fig. 3).

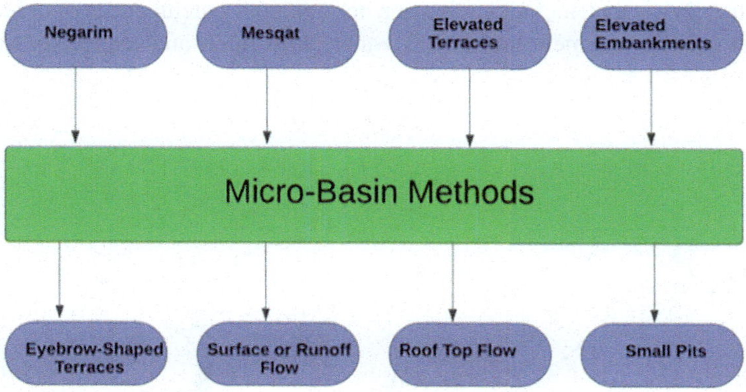

Fig. 3 Micro-basin methods

6.4 Macro-Basin and Floodwater Methods

6.4.1 Small Farm Reserves

Farmers in valleys build small ponds to absorb and hold surface runoff if suitable sites are available (Oweis et al., 2012). These reservoirs have capacities ranging from 1000 to 500,000 m^3. These reservoirs, however, have high installation and maintenance expenses. Proper planning, including projects and designs, is required, as is continual cleaning to remove accumulated debris such as soil and plants from ponds as runoff runs through. Because the water storage is exposed, it is prone to evaporation and leakage. To reduce these losses, it is critical to transfer stored water to plants as soon as possible (Oweis et al., 2012).

6.4.2 Hillside Channel System

Hillside canal systems use tiny conveyance channels to transport water from long slopes to agricultural regions at the hill's base. This approach is appropriate for places with annual rainfall ranging between 200 and 600 mm and slopes exceeding 10%. Aside from the need to handle surplus water, the buildings must be strong and need precise planning (Örs et al., 2011).

6.4.3 Jessour System

In Arabic, "Jessour" refers to walls built perpendicular to the slope in moderately steep valleys in southern Tunisia as shown in (Fig. 4). This approach entails the construction of miniature dams made of earth, rock, or sand baskets and placed at the base of the hill or along seasonal stream channels. The Jessour system is a hydraulic device that includes a barrier, terrace, and catchment area. The barrier helps to contain sediment and runoff water, with main and lateral spillways for

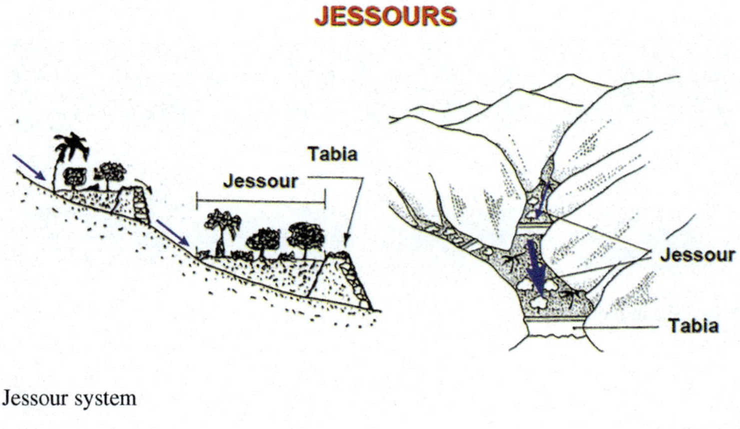

Fig. 4 Jessour system

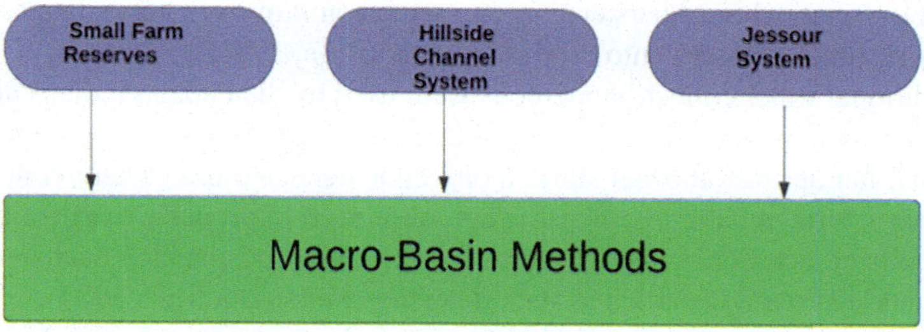

Fig. 5 Macro-basin methods

regulated water release. Through the deposition of silt, the terrace emerges throughout time as a dedicated place for flora. Fruit trees such as olives, figs, almonds, and dates, as well as legume crops like as peas, chickpeas, lentils, and broad beans, are commonly grown near the barrier (Khouri et al., 1995) (Fig. 5).

7 Water Harvesting and Management Approaches

7.1 *Traditional Approach*

As evidenced by archaeological discoveries in Jordan, Syria, Iraq, the Negev, and the Arabian Peninsula, notably Yemen, the Middle East has obvious historical relevance in the evolution of ancient water collecting systems. Traces of early water gathering structures dating back over 9000 years have been unearthed in Jordan. Simple water collection structures may be traced back to 4500 BC in Southern Mesopotamia (Bruins et al., 1986). The famed runoff-irrigation systems are most notably found in Israel's semi-arid to dry Negev desert, with historical origins reaching back to the tenth century BC, when the Israelites of that era introduced them (Adato, 1987).

A system dating back at least 1000 years directed floodwater to irrigate huge regions in North Yemen, potentially supporting a population of up to 300,000 people (Adato, 1987). Notably, farmers in this region continue to employ floodwater irrigation, making it one of the few sites on the planet where runoff agriculture has been practiced regularly since early times (Bamatraf, 1994).

Ancient water harvesting systems in Pakistan's Baluchistan province include the "Khuskaba" system, which involves bunds across the land slope to improve infiltration, and the "Sailaba" system, which captures floods in natural water courses using earthen bunds (Oosterbaan, 1983).

In India, the ancient "tank" system collects rainfall by bunding or excavation and serves as the agricultural backbone in dry and semi-arid regions. Large bunds known as "Khadin" were built in West Rajasthan as early as the fifteenth century, allowing wheat and chickpeas to be grown utilizing conserved moisture following the monsoon season (Kolarkar et al., 1983). A comparable method known as "Ahar" was created in Bihar, where Ahars erected in succession improved groundwater quality, converting brackish water into drinkable water (Rain, 1983).

Traditional water collection methods were used by the Papago Indians and other indigenous communities in the Sonoran Desert. Papago fields were grown on alluvial flats, fan aprons, and fan skirts along ephemeral washes, where considerable catchments formed naturally. Brush weirs were used to properly disperse floodwaters. In other places, crops were watered using gravity-fed channels such as arroyos, with water diverted via earth and stick diversion weirs (Doolittle, 1984).

Extensive water collecting techniques have been used in North Africa from Roman times, as revealed by the UNESCO Libyan Valleys team's archaeological investigation, highlighting the important role of runoff irrigation in the prosperity of the "granary of the Roman empire" (Gilbertson, 1986). Sudan has a long history of using "Haffirs" for household and animal consumption, pasture development, and paddy production (Rain, 1983).

7.2 Modern Approach

To fulfill the possibility for future irrigation growth, more water storage is required to overcome temporal mismatches between water supply and agricultural water demand. In a warmer environment, considerable water storage infrastructure is required to ensure an adequate water supply for sustaining present irrigation growth possibilities (Rosa et al., 2020a). The paucity of such infrastructure is projected to remain a significant contributor to agricultural economic water scarcity (Rosa et al., 2020b). The increase in water storage facilities poses new social, economic, and ecological issues for future food systems, creating a considerable knowledge vacuum in assessing the economic and environmental effects of increasing irrigation. The construction of huge dams and reservoirs has been a distinguishing component of infrastructure development during the twentieth century (Molle et al., 2009), and these man-made reservoirs serve an important role as a key water supply for irrigation (Biemans et al., 2011).

The adoption of a "hard-path" water governance model, defined by massive, centralized, and capital-intensive irrigation projects, has had significant consequences for ecosystems, coastal processes, and riparian people' livelihoods (Albert et al., 2021). The "hydraulic mission" of the twentieth century affected nearly all major rivers (Molle et al., 2009), with just around 23% of the world's rivers retaining continuous flow to the ocean, making them a focal focus for global conservation efforts (Grill et al., 2019). The hard path approach to water governance also poses systemic issues. Water storage expansion for irrigation can result in "reservoir effects," which are vicious cycles in which increased storage leads to disproportionate growth in demand and reliance on reservoir storage, fostering unsustainable water usage and increased vulnerability to climatic extremes (Di Baldassarre et al., 2018). The increased need for irrigation water storage may lead to the building of more dams, increasing rivalry for water storage between irrigated agriculture and hydropower generation (Zeng et al., 2017). Furthermore, the management of surface water flows for irrigation through dammed reservoirs adds to enhanced evapotranspiration rates and influences climatic conditions (Jaramillo & Destouni, 2015). Soft-path water harvesting methods are an alternative to hard-path approaches (Gleick, 2003). These options include water collection using tiny check dams, regulated aquifer recharging through the building of artificial streams and ponds, and the implementation of local catchment systems. Managed aquifer recharge is a viable strategy for decreasing flood risks and storing extra water for agriculture (He et al., 2021). Significantly, because soft-path irrigation systems reduce both the capital and operational costs associated with hard-path irrigation systems, they are better positioned to help smallholder farms dealing with agricultural economic water shortages and energy difficulties (Burney et al., 2013).

8 Conclusions

This chapter digs into a thorough examination of various approaches and techniques for improving agricultural water supplies. The varied variety of ways covered, from historic methods to cutting-edge modern ones, emphasizes the necessity and relevance of sustainable water harvesting and management in the context of irrigation for agriculture. This chapter not only discusses the importance, components, history, and many micro- and macro-level ways of water harvesting, but it also highlights traditional and modern approaches to water harvesting and management. As we face the problems of water scarcity and climate change, the ideas offered in this chapter will be a significant resource for academics, policymakers, and practitioners working toward a more resilient and water-efficient agricultural future. It is recommended that financial incentives can play an important role in encouraging the use of appropriate water collecting systems and creating resilience in the face of climate uncertainty. In addition, introducing modern technology into irrigation water management is advocated as a long-term solution strategy.

References

Abdulla, F. A., & Al-Shareef, A. W. (2009). Roof rainwater harvesting systems for household water supply in Jordan. *Desalination, 243*(1–3), 195–207.

Adato, M. (1987). *Runoff agriculture. For arid and semiarid lands*. Prospects based on research at the desert runoff farms in the Negev Desert of Israel. The desert runoff farms unit of the Hebrew University of Jerusalem and the Jacob Blaustein Institute for Desert Research, Ben-Gurion University of the Negev. CINADCO.

Albert, J. S., Destouni, G., Duke-Sylvester, S. M., Magurran, A. E., Oberdorff, T., Reis, R. E., et al. (2021). Scientists' warning to humanity on the freshwater biodiversity crisis. *Ambio, 50*(1), 85–94.

Ali, A., et al. (2013). Comparison of yield performance of soybean varieties under semiarid conditions. *Journal of Animal and Plant Sciences, 23*(3), 828–832.

Ali, S., et al. (2020). Assessment of arsenic exposure by drinking well water and associated carcinogenic risk in peri-urban areas of Vehari, Pakistan. *Environmental Geochemistry and Health, 42*, 121–133.

Ali, S., et al. (2021). Future climatic changes, extreme events, related uncertainties, and policy recommendations in the Hindu Kush sub-regions of Pakistan. *Theoretical and Applied Climatology, 143*, 193–209.

Aslam, M., et al. (2013). Drought stress, its effect on maize production and development of drought tolerance through potassium application. *Journal Cercetări Agronomiceîn Moldova (Agronomical Research in Moldavia), Iaşi—Romania, XLVI*(2), 99–114.

Bamatraf, A. M. (1994). *Water harvesting and conservation systems in Yemen.*

Beltran-Peña, A., Rosa, L., & D'Odorico, P. (2020). Global food self-sufficiency in the 21st century under sustainable intensification of agriculture environ. *Research Letters, 15*, 095004.

Biemans, H., Haddeland, I., Kabat, P., Ludwig, F., Hutjes, R. W., Heinke, J., et al. (2011). Impact of reservoirs on river discharge and irrigation water supply during the 20th century. *Water Resources Research, 47*(3).

Bruins, H. J., Evenari, M., & Nessler, U. (1986). Rainwater-harvesting agriculture for food production in arid zones: The challenge of the African famine. *Applied Geography, 6*(1), 13–32.

Bukhari, M. A., et al. (2021). Silicon mitigates drought stress in wheat (Triticum aestivum L.) through improving photosynthetic pigments. *Biochemical and Yield Characters. Silicon, 13*, 4757–4772.

Burek, P., Satoh, Y., Fischer, G., Kahil, M. T., Scherzer, A., Tramberend, S., et al. (2016). *Water futures and solution-fast track initiative.*

Burney, J. A., Naylor, R. L., & Postel, S. L. (2013). The case for distributed irrigation as a development priority in sub-Saharan Africa. *Proceedings of the National Academy of Sciences, 110*(31), 12513–12517.

Chahar, B. R., Graillot, D., & Gaur, S. (2012). Storm-water management through infiltration trenches. *Journal of Irrigation and Drainage Engineering, 138*(3), 274–281.

Connor, R., & Miletto, M. (2022). *The United Nations world water development report 2022: Groundwater: Making the invisible visible; executive summary.*

Dawar, M., et al. (2021). The effect of biochar and nitrogen inhibitor on ammonia and nitrous oxide emissions and wheat productivity. *Journal of Plant Growth Regulation, 40*, 2465–2475.

Di Baldassarre, G., Wanders, N., AghaKouchak, A., Kuil, L., Rangecroft, S., Veldkamp, T. I., et al. (2018). Water shortages worsened by reservoir effects. *Nature Sustainability, 1*(11), 617–622.

Doolittle, W. E. (1984). Agricultural change as an incremental process. *Annals of the Association of American Geographers, 74*(1), 124–137.

Droppers, B., Supit, I., Van Vliet, M. T., & Ludwig, F. (2021). Worldwide water constraints on attainable irrigated production for major crops. *Environmental Research Letters, 16*(5), 055016.

FAO. (2020). *L'eau, L'Agriculture and L'Alimentation.* L'Utilisation de L'Eau en Agriculture. Retrieved 2 January, 2020, from https://www.fao.org/.

FAO (Food and Agriculture Organization of the United Nations). (2022). *The state of the world's land and water resources for food and agriculture: Systems at breaking point. Main report.* FAO. https://doi.org/10.4060/cb9910en

Fitton, N., Alexander, P., Arnell, N., Bajzelj, B., Calvin, K., Doelman, J., et al. (2019). The vulnerabilities of agricultural land and food production to future water scarcity. *Global Environmental Change, 58*, 101944.

Gerten, D., Heck, V., Jägermeyr, J., Bodirsky, B. L., Fetzer, I., Jalava, M., et al. (2020). Feeding ten billion people is possible within four terrestrial planetary boundaries. *Nature Sustainability, 3*(3), 200–208.

Gilbertson, D. D. (1986). Runoff (floodwater) farming and rural water supply in arid lands. *Applied Geography, 6*(1), 5–11.

Gleick, P. H. (2003). Global freshwater resources: Soft-path solutions for the 21st century. *Science, 302*(5650), 1524–1528.

Gould, J. (2015). Rainwater harvesting for domestic supply. *Rainwater Harvesting for Agriculture and Water Supply*, 235–268.

Graves, A., Rosa, L., Nouhou, A. M., Maina, F., & Adoum, D. (2019). Avert catastrophe now in Africa's Sahel. *Nature, 575*(7782), 282–286.

Grill, G., Lehner, B., Thieme, M., Geenen, B., Tickner, D., Antonelli, F., et al. (2019). Mapping the world's free-flowing rivers. *Nature, 569*(7755), 215–221.

Gupta, A. K., & Harvesting-pub, R. (2006). *Indian railways institute of civil engineering*. Pune.

Hammad, A. H. A., Børresen, T., & Haugen, L. E. (2006). Effects of rain characteristics and terracing on runoff and erosion under the Mediterranean. *Soil and Tillage Research, 87*(1), 39–47.

He, C., Liu, Z., Wu, J., Pan, X., Fang, Z., Li, J., & Bryan, B. A. (2021). Future global urban water scarcity and potential solutions. *Nature Communications, 12*(1), 4667.

Hussain, S., et al. (2020). Study of land cover/land use changes using RS and GIS: A case study of Multan district, Pakistan. *Environmental Monitoring & Assessment, 192*(2), 1–15.

IPCC. (2022). Climate change 2022: Impacts, adaptation, and vulnerability. In H. O. Pörtner et al. (Eds.), *Contribution of working group II to the sixth assessment report of the intergovernmental panel on climate change*. Cambridge University Press. Retrieved from www.ipcc.ch/report/ar6/wg2/about/how-to-cite-this-report.

Jägermeyr, J., Müller, C., Ruane, A. C., Elliott, J., Balkovic, J., Castillo, O., et al. (2021). Climate impacts on global agriculture emerge earlier in new generation of climate and crop models. *Nature Food, 2*(11), 873–885.

Jaramillo, F., & Destouni, G. (2015). Local flow regulation and irrigation raise global human water consumption and footprint. *Science, 350*(6265), 1248–1251.

Khaliq, T., et al. (2013). A study of thermal changes on physiology and paddy yield of fine rice cultivars under subtropical conditions. *Asian Journal of Agriculture and Biology, 1*(4), 208–216.

Khouri, J., Salih, A. M. A., Abdulrazzak, M., Kallel, M. R., & Adam, H. (1995). *Rainfall water Management in the Arab Region: State of the art report* (p. 147). UNESCO Regional Office for Science and Technology for the Arab States.

Kolarkar, A. S., Murthy, K. N. K., & Singh, N. (1983). Khadin—A method of harvesting water for agriculture in the Thar Desert. *Journal of Arid Environments, 6*(1), 59–66.

Kummu, M., Heino, M., Taka, M., Varis, O., & Viviroli, D. (2021). Climate change risks pushing one-third of global food production outside the safe climatic space. *One Earth, 4*(5), 720–729.

Ling, E., & Benham, B. L. (2014). Rainwater harvesting systems.

Mehmood, A., et al. (2021). Exploring the potential of moringa leaf extract as bio stimulant for improving yield and quality of black cumin oil. *Scientific Reports, 11*, 24217.

Molle, F., Mollinga, P. P., & Wester, P. (2009). Hydraulic bureaucracies and the hydraulic mission: Flows of water, flows of power. *Water Alternatives, 2*(3), 328–349.

Morgan, P. Rural water supplies and sanitation; Macmillan Publishers Ltd.: , 1990; p. 358. ISBN 0333485696.

Mubeen, M., et al. (2013a). Evaluating CSM-CERES-maize model for irrigation scheduling in semi-arid conditions of Punjab, Pakistan. *International Journal of Agriculture and Biology, 15*, 1–10.

Mubeen, M., et al. (2013b). Evaluating different irrigation scheduling criteria for autumn-sown maize under semi-arid environment. *Pakistan Journal of Botany, 45*(4), 1293–1298.

Mubeen, M., et al. (2013c). Effect of growth stage-based irrigation schedules on biomass accumulation and resource use efficiency of wheat cultivars. *American Journal of Plant Sciences, 4*, 1435–1442.

Mubeen, M., et al. (2016). Application of CSM-CERES-maize model in optimizing irrigated conditions. *Outlook on Agriculture., 45*(3), 173–184.

Mubeen, M., et al. (2021). Effect of PGPR and drought on spring maize (Zia mays L.). *Pakistan Journal of Botany, 53*(2), 731–739.

Nasim, W., et al. (2016). Correlation studies on nitrogen for sunflower crop across the agroclimatic variability. *Environmental Science and Pollution Research, 23*, 3658–3670. https://doi.org/10.1007/2Fs11356-015-5613-1

Nasim, W., et al. (2017). Response of sunflower (Helianthus annuus L.) hybrids to nitrogen application grown under different agro-environments. *Journal of Plant Nutrition, 40*(1), 82–92.

Natural Water Retention Measures, European Commission (NWRM). (2023). Retieved 26 March, 2023, from http://nwrm.eu.
Oosterbaan, R. J. (1983). Modern interferences in traditional water resources in Baluchistan. *ILRI Annual Report, 1982*, 23–34.
Örs, İ., Safi, S., Ünlükara, A., & Yürekli, K. (2011). Water harvesting techniques, structures and their impacts. *TABAD, Tarım Bilimleri Araştırma Dergisi, 4*(2), 65–71.
Oweis, T., Prinz, D., & Hachum, A. (2001). *Water harvesting: Indigenous knowledge for the future of the drier environments*. ICARDA.
Oweis, T., & Hachum, A. (2006). Water harvesting and supplemental irrigation for improved water productivity of dry farming systems in West Asia and North Africa. *Agricultural Water Management, 80*(1–3), 57–73.
Oweis, T. Y., Prinz, D., & Hachum, A. Y. (2012). Rainwater harvesting for agriculture in the dry areas. CRC press.
Pradhan, R., & Sahoo, J. (2019). *Smart rainwater management: New technologies and innovation, smart urban development, vito bobek* (p. 10). Intech Open.
Prinz, D., & Malik, A. H. (2002). Runoff farming. Institute of Water Resources Management, Hydraulic and Rural Engineering: Germany.
Puy, A., Lo Piano, S., & Saltelli, A. (2020). Current models underestimate future irrigated areas. *Geophysical Research Letters, 47*(8), e2020GL087360.
Puy, A., Borgonovo, E., Lo Piano, S., Levin, S. A., & Saltelli, A. (2021). Irrigated areas drive irrigation water withdrawals. *Nature Communications, 12*(1), 4525.
Rain, U. N. E. P. (1983). *Stormwater harvesting in rural areas* (p. 5). Water Resources Series.
Rasool, A., et al. (2020). Microbial diversity response in thallium polluted riverbank soils of the Lanmuchang. *Ecotoxicology and Environmental Safety, 187*, 109854.
Rasool, A., et al. (2021). Microbial diversity and community structure in alpine stream soil. *Geomicrobiology Journal, 38*(3), 210–219.
Rehman, M. H., et al. (2021). Influence of semi-arid environment on radiation use efficiency and other growth attributes of lentil crop. *Environmental Science and Pollution Research, 28*, 13697–13711.
Ritchie, H., & Roser, M. (2017). *Water use and stress*. Our World in Data.
Rosa, L., Chiarelli, D. D., Sangiorgio, M., Beltran-Peña, A. A., Rulli, M. C., D'Odorico, P., & Fung, I. (2020a). Potential for sustainable irrigation expansion in a 3 C warmer climate. *Proceedings of the National Academy of Sciences, 117*(47), 29526–29534.
Rosa, L., Chiarelli, D. D., Rulli, M. C., Dell'Angelo, J., & D'Odorico, P. (2020b). Global agricultural economic water scarcity. *Science Advances, 6*(18), eaaz6031.
Selimoğlu, P., & Yamaçli, R. (2022). Sürdürülebilir Yağmur Suyu Hasadı Üzerine Yapısal Bir İnceleme. *Sürdürülebilir Mühendislik Uygulamaları ve Teknolojik Gelişmeler Dergisi, 5*(2), 210–231.
Shafique, M., Kim, R., & Kyung-Ho, K. (2018). Evaluating the capability of grass swale for the rainfall runoff reduction from an urban parking lot, Seoul, Korea. *International Journal of Environmental Research and Public Health, 15*(3), 537.
Shah, M. Z., et al. (2021). Synthesis of silver nanoparticles using Plantago lanceolata extract and assessing their antibacterial and antioxidant activities. *Scientific Reports, 11*, 20754.
Shakya, B., & Thanju, J. P. (2013). Technical guidelines for installation of rainwater harvesting system and its operation. *Hydro Nepal: Journal of Water, Energy and Environment, 12*, 45–51.
Stringer, A., Vogel, J., Lay, J., & Nash, K. (2012). Design of rainwater harvesting systems in Oklahoma. *Oklahoma Cooperative Extension Service.*
Sultana, S. R., et al. (2014). Normalized difference vegetation index as a tool for wheat yield estimation: A case study from Faisalabad, Pakistan. *The Scientific World Journal, 2014*, 725326.
Toyran, C. (2022). *Rain Harvesting potential in urban areas and its use in landscape: Case study of Buyukcekmece District, Istanbul*. Master's Thesis, Yıldız Technical University Graduate School of Natural and Applied Sciences, Istanbul, Turkey, p. 152.
UNESCO. (2021). *The United Nations world water development report 2021*. Valuing Water. United Nations.

Vanham, D., Alfieri, L., Flörke, M., Grimaldi, S., Lorini, V., de Roo, A., & Feyen, L. (2021). The number of people exposed to water stress in relation to how much water is reserved for the environment: A global modelling study. *The Lancet Planetary Health, 5*(11), e766–e774.

Verbist, K., Cornelis, W. M., Gabriels, D., Alaerts, K., & Soto, G. (2009). Using an inverse modelling approach to evaluate the water retention in a simple water harvesting technique. *Hydrology and Earth System Sciences, 13*(10), 1979–1992.

Wajid, A., et al. (2014). Modeling growth, development and seed-cotton yield for varying nitrogen increments and planting dates using DSSAT. *Pakistan Journal of Agricultural Sciences, 51*, 639–647.

Wang, J., & Guo, Y. (2020). Dynamic water balance of infiltration-based stormwater best management practices. *Journal of Hydrology, 589*, 125174.

World Bank. (2016). *High and dry: Climate change, water, and the economy*. The World Bank.

World Bank. (2022). *Water in agriculture*. Retrieved March, 2022, from www.worldbank.org/en/topic/water-in-agriculture#1.

World Resources Institute (WRI). (2021). *Data, water stress by country*. Retrieved 2 January, 2020, from https://www.wri.org.

Worm, J. (2006). *AD43E rainwater harvesting for domestic use (No. 43)*. Agromisa Foundation.

WRI (World Resources Institute). (2019). *WRI Aqueduct website*. Retrieved from www.wri.org/aqueduct.

Wright, P. (1985). Soil and water conservation as a starting point for rural forestry: The Oxfam project in Ouahigouya, Burkina Faso. *Rural Africana, 23*, 79–86.

Yannopoulos, S., Giannopoulou, I., & Kaiafa-Saropoulou, M. (2019). Investigation of the current situation and prospects for the development of rainwater harvesting as a tool to confront water scarcity worldwide. *Water, 11*(10), 2168.

Yetik, A. K., & Şen, B. (2020). Importance and techniques of water harvesting systems. *Turkish Journal of Agriculture-Food Science and Technology, 8*(1), 46–53.

Zeng, R., Cai, X., Ringler, C., & Zhu, T. (2017). Hydropower versus irrigation—An analysis of global patterns. *Environmental Research Letters, 12*(3), 034006.

Zhang, S., & Guo, Y. (2014). Stormwater capture efficiency of bioretention systems. *Water Resources Management, 28*, 149–168.

Zhang, X., Yao, G., Vishwakarma, S., Dalin, C., Komarek, A. M., Kanter, D. R., et al. (2021). Quantitative assessment of agricultural sustainability reveals divergent priorities among nations. *One Earth, 4*(9), 1262–1277.

Terrace Farming for Better Water Management in Agriculture

Rafi Qamar, Sundas Ashraf, Farrukh Tanzeel Abbas, Hafiz Muhammad Rashad Javeed, Muhammad Ather Nadeem, Atique-ur-Rehman, Muhammad Yaseen, Usman Saleem, Bilal Ahmad Khan, Touqeer Abbas, Farhan Saeed, and Zulqarnain Haider

Abstract Terrace farming is the oldest methods of farming that has recently been brought back and reinvented as a sustainable solution for effective water management in farming. This chapter unfolds the benefits and skills of farming in terraces, emphasizing how the method raises the efficiency in water, minimizes soil erosion,

R. Qamar · S. Ashraf · F. T. Abbas · M. A. Nadeem
Department of Agronomy, College of Agriculture, University of Sargodha, Sargodha, Pakistan
e-mail: rafi.qamar@uos.edu.pk; ather.nadeem@uos.edu.pk

H. M. R. Javeed (✉) · F. Saeed
Department of Environmental Sciences, COMSATS University Islamabad, Vehari, Pakistan
e-mail: rashadjaveed@cuivehari.edu.pk; farhansaeed@cuivehari.edu.pk

Atique-ur-Rehman
Department of Agronomy, Bahauddin Zakariya University, Multan, Pakistan
e-mail: dr.atique@bzu.edu.pk

M. Yaseen
Department of Agricultural Extension and Rural Studies, College of Agriculture, University of Sargodha, Sargodha, Pakistan
e-mail: yaseen.baksh@uos.edu.pk

U. Saleem
Deptartment of Plant Breeding and Genetics, University of Sargodha, Sargodha, Pakistan
e-mail: usman.saleem@uos.edu.pk

B. A. Khan
Department of Soil and Crop Sciences, Texas A&M University, College Station, TX, USA
e-mail: bilalahmadkhan2022@tamu.edu

T. Abbas
College of Horticulture and Agronomy, University of Nebraska, Lincoln, NE, USA
e-mail: tabbas3@unl.edu

Z. Haider
Department of Agricultural Economics, College of Agriculture, University of Sargodha, Sargodha, Pakistan
e-mail: zulqarnain.haider@uos.edu.pk

M. Mubeen et al. (eds.), *Innovations in Agricultural Water Management*,
https://doi.org/10.1007/978-3-031-91883-4_7

and increases productivity on crops. By creating flat or gently sloping land surfaces, terracing allows better retention of water, reduced runoff, and greater infiltration, making water conservation strategies very effective. Terrace farming is among the earliest forms of agriculture that were ever conceptualized and offered a sustainable solution for efficient water management in modern-day farming. This chapter elaborates the principles and merits of terrace farming, along with the scope for it to become a technique for efficient water use, reduced soil erosion, and better productivity in crop harvesting. By merging gently sloping and flat lands, terracing ensures that more water is retained, lessening loss through runoff and aiding in the recharging of water aquifers to conserve water effectively. It discusses design and construction of terraces, the selection and management of crops to use on terraces, and how to link the other conservation agriculture practices with terracing. Terrace farming offers benefits in a variety of areas: better water harvesting, decreased land degradation, and enhanced biodiversity. This chapter clearly introduces terrace farming as an important intervention in sustainable agriculture and water resource management. It spells out practical ways of adopting and promoting the terrace farming practices for the benefit of farmers, administrators, and interested scholars. By presenting the benefits and application of land-terrace farming, this chapter is an effort in tandem with the global motive to alleviate water scarcity and soil degradation, leading further to food insecurity, and henceforth, a move in support of the transition toward a more sustainable and resilient agricultural system.

Keywords Conservation tillage · Water management · Terracing · Contorting · Slope · Erosion

1 Introduction

A slanting block of land divided into a series of regularly receding flat surfaces or platforms, resembling steps, is called a terrace, to enhance the productivity of agriculture. This method of land modification is called terracing (Opilo, 2023). It is a common practice to carry on agriculture in hills and mountainous regions where fully formed steps of the terrace are at one's disposal (Bertolino & Corrado, 2021). Terraced fields can be cultivated with irrigated crops, like rice, and have the added advantage of checking soil erosion and surface run-off (Mishra et al., 2020). The rice terraces of the Philippine Cordilleras are acknowledged as a UNESCO World Heritage Site because they highly serve this significant technique (Bairagi et al., 2021). Extensive labor was invested in terrace farming to avoid erosion, prevent loss of water, and gain more planting areas on different landforms. This is usually a low-level, flat ridge of earth built across a slope, with the runoff water channel located just above the ridge (do Prado et al., 2022; Khan et al., 2019; Amin et al., 2018a, 2018b; Rahman et al., 2018; Nasim et al., 2018a, 2018b). Terraces are usually built with some grade to allow water caught in the channel to move slowly down to the outlet of the terrace. Level terraces are made for use in cases where rainfall is scarce and soil that absorbs the same exists. That is, in parts of Africa and the Andes of South America, terrace farming had been a common practice for centuries, together with the Mediterranean region, China, Japan, the Philippines, and other parts of Oceania and Southeast Asia. Terrace farming is a technique used on slopes to minimize the flow of water and restrict soil erosion (Tamura et al., 2021).

But there are more variants than using it in the mountains. While others resemble graduated steps, some are relatively level on gently sloping or wavy lands; this shows that terracing applies to a variety of elevations in the fields (Wohl, 2020). Soil type and steepness of slope determine the orientation of terraces. It is especially unsuitable on slopes that slide. Bench, contour, and parallel are the common forms of terracing systems (Osman & Osman, 2018). By their names, one may easily deduce their layout.

- One of the Cordillera rice terraces in the Philippines is included in the UNESCO World Heritage List. Quite definitely before 1000 AD, the Wari culture and other peoples of the south-central Andes invented terraced farming, hundreds of years before the Inka.
- Terrace farming is an extensive practice in countries where rice is grown in Asia, for instance, Vietnam, the Philippines, and Indonesia.
- Places with hilly topography, such as the Canary Islands, are very common places to practice terrace farming.
- Potato and maize are also grown in terrace farming, along with rice.
- Rice, wheat, and barley are the three considerable crops grown through terrace farming.
- As rice does not thrive where the land is elevated to a great extent, it would be possible to grow other crops where rice cannot be cultivated.
- Terraces can be made on sloping land also; quite probably the Hanging Gardens of Babylon were built on a series of stepped terraces similar to those of a ziggurat.

A terrace is an earth embankment with a channel cut across the slope at regular vertical intervals and with a sufficient horizontal slope. These are constructed to reduce surface water runoff, reduce erosion, and maximize moisture retention for crop growth (Singh, 2023). It is hence a water–soil conservation measure. Without them, it would be nearly impossible to farm hillsides in mountainous regions; in arid areas, the terraces enhance plant water storage and groundwater recharging (Acharya, 2021). There are three main types of terraces: diversion, retention, and bench-style terraces. The diversion terrace essentially intercepts overland flow and diverts the water down the slope to some suitable outlet (Kameri-Mbote et al., 2021). A small gradient of 1:250 to the contour is generally adopted for the development of diversion terraces. Again, diversion terraces have been broadly classified into Nichols, Magnum, broad-based, and narrow-based types. Magnum-type diversion terrace is constructed using earth from both sides of the embankment (Awulachew et al., 2009; Hammad et al., 2018; Nasim et al., 2018a, 2018b; Ali et al., 2018a, 2018b; Tariq et al., 2018). Only the Nichols-type diversion terrace is formed when soil from the embankment's upslope side is taken. The wide-based type diversion terraces are constructed with a 15 m wide embankment and channel.

The banks have steeper slopes that are impracticable for cultivation, and the narrow based type diversion terraces are only 3–4 m wide (Yesuph, 2020). The retention terraces are the flat terraces mainly used on hillsides for storing surface water to conserve it. Bench terraces are platform-like agriculture implemented along a slope for steep slope cultivation (Lingard, 2020). Internationally, they are in everyday use and constructed mostly on slopes that are less than 10%. Other types of bench terraces include level bench terraces, outward-sloping bench terraces, and inward-sloping bench terraces (Singh, 2023). A very huge labor investment is required both during installation and upkeep of a terrace, and poor maintenance can lead to catastrophic consequences. Neglected terraces may lead to mudslides, the formation of deep gulleys, and increased soil erosion, especially on sandy soils or very steep land (Plekhov, 2022). It also causes a loss in the quality of the soil due to the leaching process of terrace farming. During the monsoon and rainy seasons, the processing of terrace farming can be harmful as it supports the oversaturation of the soil (Ishak, 2022; Nasim et al., 2016, 2017; Mubeen et al., 2013a, 2013b, 2013c, 2016; Wajid et al., 2014; Sultana et al., 2014; Khaliq et al., 2013). Besides seasonal changes, bringing droughts, floods, and other extreme weather events previously uncommon in this area, maintenance itself has become very difficult for the historical terraces due to climate change. Flooding water may create runoffs and landslides if not monitored continuously (Marino et al., 2020). There are many ways to farm various crops, and raise animals, and forest species on terraces, either separately or together. Crop selection should be based on a set of factors that include, but are not limited to, the following: crop adaptation to the environment, low input requirement, and increased nutrition and/or income (Talabi et al., 2022). Tools and practices should be economical, light, scalable, and effective at reducing female drudgery. Since most terracing works are done in remote hills and mountains, strategies that are easy to implement and cost less to labor and transport need due attention in order to reduce the farmers' dependency on organizations or service providers (Romeo et al., 2021). Finally, that for the case of the terrace farmers, unless there is some obvious and direct financial benefit of the selected interventions in comparison with the current status, then they are unlikely to take up any new initiative.

2 Terrace Farming

Terrace farming, otherwise known as terracing, is one common means of subsistence for many farmers who reside in hilly or mountainous regions around the world. With this technique, the steep terrain is changed into a sequence of short, progressively descending stairs (Avni, 2022). These steps, which are typically 2–3 m wide and 50–80 m long, aim to assist with a range of crop growth, including field, horticultural, and fodder crops as well as other crops that require specific management techniques like irrigation (Minhas et al., 2020). Terrace farming can be utilized separately or alongside agroforestry systems. By using this technique, farmers can increase their means of subsistence and livelihood by growing a variety of crops and making the most of the difficult growing conditions (Achmad et al., 2022;

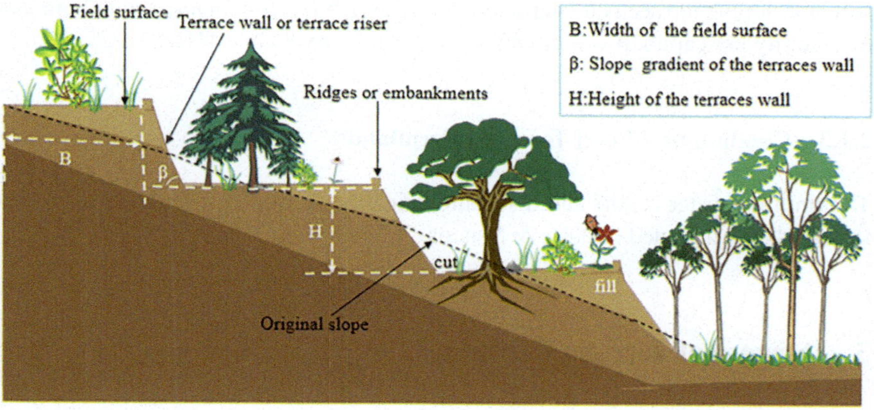

Fig. 1 Terrace conservation and agricultural productivity. (Source: Deng et al., 2021)

Wang et al., 2019; Fahad et al., 2019; Shah et al., 2019; Shahzad et al., 2018). As shown in (Fig. 1), ridges and channels are used to create a terrace across the slope of a field in order to store runoff, collect sediment, and decrease erosion which help to preserve water (Deng et al., 2021). To make sure that a suitable drainage system is in place, terraces can be built in fields. Cut and fill sections are used to create terraces (Singh, 2023). In hilly regions, large-scale crop cultivation is made feasible by expanding arable land through filling in existing areas. Field water and runoff are significantly intercepted by the ridges and embankments (Rao et al., 2022). However, terraces can also collapse; the chance of this happening increases with the height of the terrace's wall.

2.1 *Types of Terrace*

Here are three types of terrace farming:

2.1.1 Bench Terraced Farmlands

This is the most common kind of mountain farmland. It accumulates rainwater for irrigation due to its projections and level platform. This kind of farmland requires farmers to provide adequate drainage for irrigation (Maina, 2020).

2.1.2 Slopping Terrace Farmlands

Farmers don't level the land for farming; instead, riser dikes are formed. Over time, it ploughs and eventually becomes flattered. The natural slope causes variations in the distance between riser dikes (Xiao et al., 2024). In terms of saving water and soil, the sloping terrace is less efficient than bench terracing. In areas with high land availability per capita, it is more adaptive (Cucchiaro et al., 2021).

2.1.3 Combination Level Terraced Farmlands

To minimize water runoff and distribute irrigation or rainfall evenly across fields, farmlands are flattened (Duarte & Mateos, 2022).

3 Purpose of Terrace Farming

The general objective of terracing is usually runoff and soil loss reduction, but it also reduces the river peak discharge rates and improves infiltration, raising the soil moisture content. Soil conversation thus aims at the physical and chemical preservation of soil while keeping it in place (Deng et al., 2021). Water is the principal agent in causing erosion, particularly by surface runoff, which is amplified on sloping ground (Bettoni et al., 2023). Soil erosion becomes serious in hilly areas where runoff water achieves great velocities. We construct terraces in those hills that catch the flow of the water and greatly slow it down (Strahler, 2020) (Table 1). The levels of the terrace store the amount of water needed for the level itself and the excess drains down to the lower levels due to drainage outlets and gradients (Phillips, 2022). Areas which have steep hills and heavy rainfall also support mud slides and landslides which make the entire area unusable for any form of activity forever (Lalbiakzuali, 2020). By combining food crops with tree species they even help prevent such events in such regions. As noted earlier, the chemical properties of soil are equally important as its physical integrity (Wilpert, 2022; Ali et al., 2013; Aslam et al., 2013; Khaliq et al., 2013). It is one of the environment-friendly farming practices since it mainly uses organic fertilizers and uses fewer artificial fertilizers. Thus, soil fertility and quality continue to increase. Water flow helps in maintaining soil cover and fertility because they do not wash away manure and topsoil (Esilaba et al., 2021).

3.1 Systems of Terrace Farming

There are three types of farming practice, depending upon the type and magnitude of disturbance caused to the slopes for building the terraces: bench, contour, and parallel terracing (Ramser, 2020).

Table 1 The data which shows the reduction in runoff and soil erosion in different area according to the adoption of different terraces designs

Terrace type	Location	Meteorological factor	Slope range	Soil loss	Water loss	NPK loss	References
Arid terrace	Jianyang, Sichuan Province, Southwestern China	Annual rainfall 700–800 mm	5–15%	34.41%	21.5%	–	Zhang et al. (2008)
Dryland terrace	Southern Ningxia, Northwest China	Precipitation use efficiency (PUE) 6.3–15.3%	–	0.2–1.9 $t \times hm^{-2}$	0.7–3.2 $m^3 \times hm^{-2}$	N loss 42.2–73.3% P loss 40%	Hou et al. (2012)
Grassed terrace	Jhon River Valley, New Brunswick, Canada	Annual rainfall 768.9 mm	2–14%	20–30 t/ha/year	19.5%	–	Chow et al. (1999) Kramer et al. (1999)
Bench terrace	West Java, Indonesia	Humid tropical climate	–	11–30 t ha^{-1} $year^{-1}$	3.0–3.9%	–	Van Dijk et al. (2005)
Bench terrace	Dehradun, India and Northern plains, Sidney, Montana	Rainfall 7400 mm Sub-humid climate	0.5–1.5%	90%	80%	High	Sharda et al. (2002) Black (1968)
Parallel terrace	Luanhe River Basin, China	Arid- and semi-arid region	Level range 10 cm (4 in.)	88.2%	78.9%	Less	Liu et al. (2021)
Paddy terrace	Northern Vietnam	Annual rainfall 1560 mm	Area of slope 250 (Fa)	1722 kg ha^{-1} $year^{-1}$	88%	–	Mai et al. (2013)
Agriculture terrace	Pokhare Khola, Nepal	Annual rainfall 1202 mm	–	18–28%	11%	N loss 5.6 kg ha^{-1} $year^{-1}$ P loss 0.02 kg ha^{-1} $year^{-1}$	Tiwari et al. (2009)

3.1.1 Bench Terracing

In bench terracing, the hills are divided into wide, uniformly spaced flatlands, which prevent water from flowing downhill and removing topsoil (Le, 2022). If enhancing bench terraces on a hillside, you begin at the top and use an approach called "cutting and filling" to create broad, level spaces that slope down the hillside to the appropriate width (Eren, 2022) as shown in (Fig. 2). The recently built terraces can be stabilized and strengthened over time with the use of additional materials such as stones, border plants, and bio-engineering techniques (Acharya, 2020). This means that due to the challenging terrain, there will be a lot of manual earthworks involved, which will take a lot of time and labor. They are even more successful at catching the flowing water when they have bunds on the sides to keep the water stagnant, which is essential for growing rice (Walia et al., 2024). They are unsuitable for cultivating plants like potatoes, which struggle to thrive in wet conditions (Chiankem, 2022).

3.1.2 Contour Terracing

Required landscaping and earthwork are much less than those of the bench terracing systems because the terraces or growing flatlands follow the natural contour of the hill (Brown et al., 2021). Since they have been built with each terrace level at the same height that follows the relief, they are not spaced equally. Their constant intra-level gradient allows holding inside the water for a long time (Reimer, 2022). Bunds and drainage are some instances of features laid on each level in the same fashion as bench terraces (Kurdekar et al., 2023; Rahman et al., 2019; Ahmad et al., 2019; Sabagh et al., 2019; Danish et al., 2019; Iqbal et al., 2019). Very often, a cut of the grassed waterway is put right across the slope at points which receive precipitation to take the accumulated water straight down (Sadeghi & Eslamian, 2021).

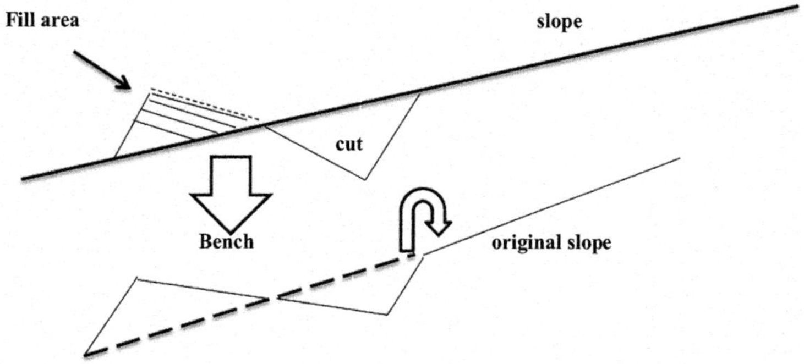

Fig. 2 Cut and fill procedure of bench terrace

3.1.3 Parallel Terracing

Creating something parallel requires much more work than merely utilizing what we already have. The development of terraces on sloping terrain implies the same standards. To construct terraces that are parallel to each other, we will have to drastically change the terrain by taking out a large amount of material from the top and filling it in at the base (Cucchiaro et al., 2021).

3.1.3.1 Examples of Terrace Farming

Here are some of the more well-known examples of terrace farming:

1. **The Rice Terraces of Banaue**

For more than 2000 years, people have farmed rice in terraces in a few secluded villages among some of the Philippine mountain ranges known as Cordillera; as such, it is now a UNESCO World Heritage Site (Zerrudo, 2022). Apart from the economic role which the terraces play for the people, they are also part of the region's tradition and heritage (Fig. 3).

2. **Machu Picchu**

One of the Seven Wonders of the World, Machu Picchu is a site located in the Andes Mountains of South America that comprises nothing but terraces (Mardon et al., 2021). Not only did they provide a site upon which such a highly developed civilization, complete with countless temples and homes, could flourish but also they made farming possible in very challenging terrain (Angelakıs et al., 2020). The ingenuity of the terrace irrigation systems remains, to this day, amazing to scientists, geologists, and farmers alike because the systems completely harnessed rainfall in areas where other irrigation methods were impractical (Pueppke, 2021). Contrary to common types of terraces, the terraces at Machu Picchu, as depicted by Fig. 4, rely a great deal on walls and rocks to support the levels (Fig. 4). Even to this day, the terraces at Machu Picchu remain one of the most beautiful creations ever made (Kavoori, 2020).

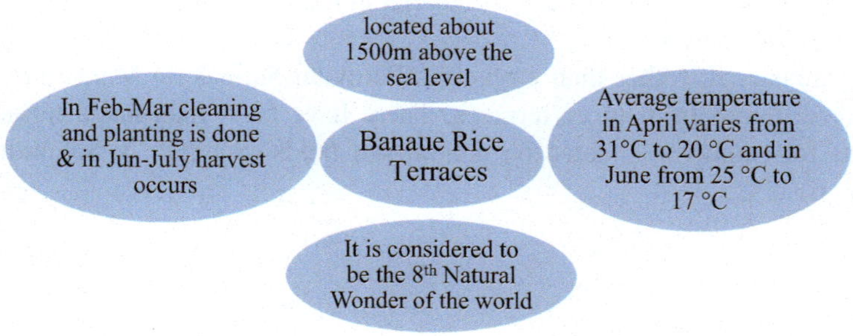

Fig. 3 Schematic diagram of rice terrace of Banaue

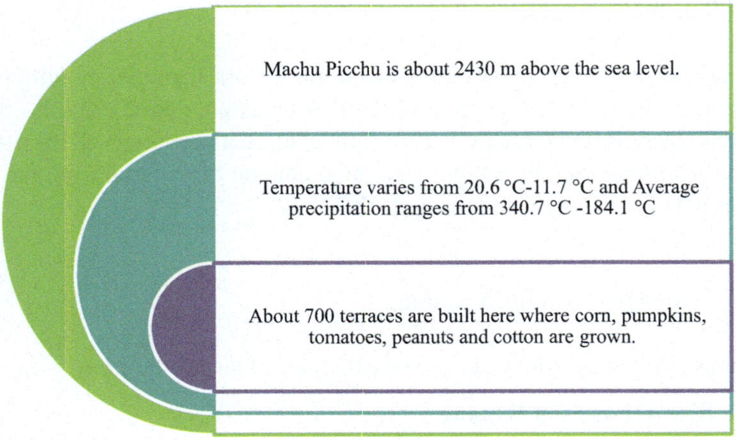

Fig. 4 Schematic diagram of Machu Picchu

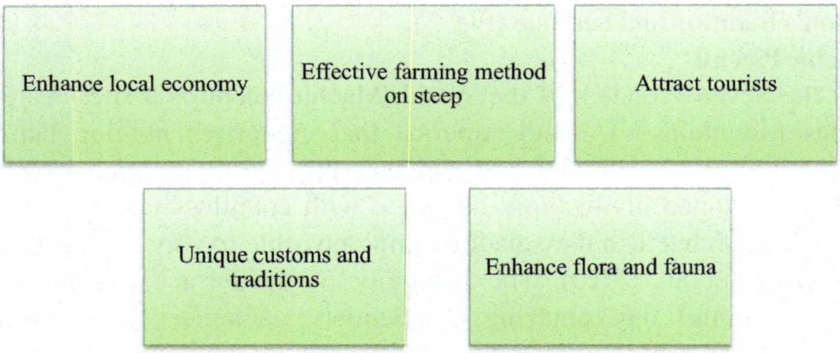

Fig. 5 Schematic sketch shows the importance of Sa pa Terrace

3. **Sa Pa Terraces**

Vietnam's Sa Pa terraces, which can produce significant amounts of crops like rice, maize, and vegetables, are also essential to the tourism industry in the area (Nguyen et al., 2022). It is also beneficial in the following ways (Fig. 5).

4. **Salinas De Maras**

As they make salt rather than crops, the Peruvian Salinas de Maras terraces are quite different from other terraces. There have been man-made terraces like these for centuries, situated on natural salt fields (Schmidt & Schmidt, 2022) (Fig. 6).

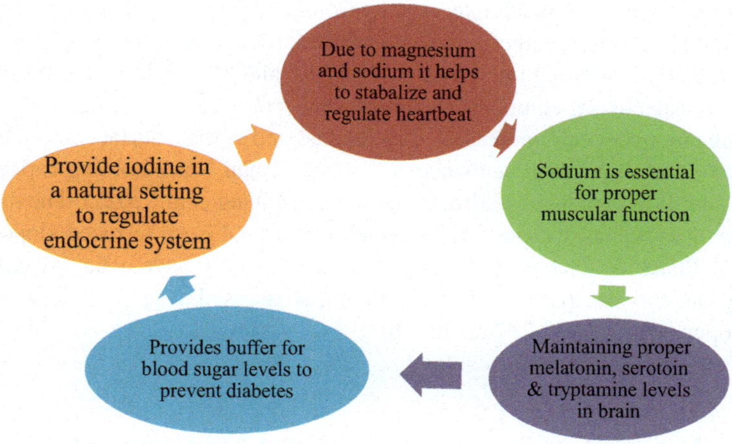

Fig. 6 Schematic sketch of Salinas De Maras

4 Types of Crops Grown in Terrace Farming

Many crops, including grains, natural products, vegetables, flowers, pulses, oilseeds, millets, buckwheat, saffron, dark cumin, grain amaranth, and flavors, are grown on terraces (Sen et al., 2023). Apples, subtropical crops, and mild natural products which include dry fruits and nuts are the main crops grown for natural products (Malik et al., 2020).

Cropping pattern:

- The cropping pattern that is suitable for terrace farming is as follows for several crops;
  Maize and cowpea
  Ginger, soybeans and maize
  Soybean and millet
  Pea and mustard
  Pea and wheat

Advantages:

- Increases the capacity of legume intercrops to fix nitrogen, raises soil N accumulation, increases yield, and decreases insect, disease, and weed issues (Kumawat et al., 2022).

5 Terracing and Sustainable Irrigation

Water conservation through terracing is based on two main reasons: first, terracing directly affects the micro-topography of hill slopes and generates a series of micro-watersheds within slope channels or over the entire slopes (Uwacu et al., 2021).

These modifications may alter specific hydrological pathways, leading to high concentration, divergence, and efficiency in rainwater harvesting (Doost et al., 2024). It is indicated that terracing reduces runoff by 92.6 and 80% in humid and sub-humid climates, respectively, compared to natural slopes (Uwacu et al., 2021). Second, the terracing practice would increase soil roughness and vertical surface relief, decrease overland flow connectivity, enhance soil moisture and water-holding capacity, and finally alter how raindrops infiltrate into the soil (Deng et al., 2021). The mean soil moisture may increase from 15.7% on slopes to 29.4% on terraced slopes in the Yun-Gui Plateau's dryland (Wei et al., 2016). Terraces have a far greater water-holding capacity compared to slopes, which was assessed to be 5.0–6.2 times higher than slopes (Wang et al., 2024a, 2024b).

6 Ecological Benefits of Terrace Water Management

6.1 Reduced Runoff and Improved Water Conservation

In several ways, terracing minimizes runoff and sediment loss. First, it reduces the length and gradient of the slope, which significantly reduces the amount of runoff and sediment yield (Chen et al., 2022). This involves breaking up steep slopes into an artificial sequence of relatively level surfaces. Because reshaped slopes alter particular hydrological pathways, reduce hydrological connectivity, and increase catchment area, they can effectively intercept rainfall and reduce flood peak discharge (Sidle, 2021) as illustrated in (Fig. 3). In addition, rainfall should also be harvested and channeled toward controlled outlets at a controlled velocity and surface runoff from the terraced fields should be intercepted to promote infiltration to prevent soil erosion and water loss (Deng et al., 2021). Second, terracing enhances the surface roughness and vertical surface relief. It improves, according to, infiltration, soil moisture, and the ability of the soil to retain more water. That is to say, terracing can increase the content of soil moisture by 4.24–12.9%, or even by as much as 5.0–6.2 times compared with sloped land (Wang et al., 2024a, 2024b). Through its potential to hold water, terracing enables the growth of vegetation in otherwise arid ecosystems. The third and most common benefit of terracing would generally be the reduction in runoff by constructing ridges or embankments (Deng et al., 2021).

6.2 Reducing Erosion and Preserving Soil

Soil erosion goes together with runoff. Part of the rainwater infiltrates the ground during huge storms and becomes runoff. Together with it comes the increase in quantity, velocity, and erosivity of runoff (Blanco & Lal, 2023). The critical runoff velocity is related to the slope gradient and the size of the soil particles, which is the

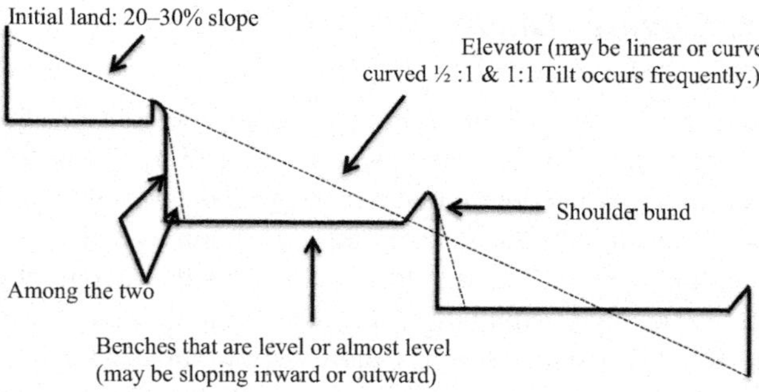

Fig. 7 Terrace for soil and water erosion control

speed at which soil particles move about the surface (Chen et al., 2022). Hence, terraces are constructed to intercept and store overland runoff so that, in due course of time, the formation of sediment is reduced. When more than 40% of the land under terracing covers the total land area on a mountain slope (Deng et al., 2021) as shown in Fig. 7, soil erosion can be reduced by 43–70%. This controls erosion both in soil and water. Some of the ways through which terracing improves erosion control effectiveness include the following (Mukai et al., 2021). First, it reduces the erosive forces by cutting down runoff velocity and total amount: Second, soil erosion is usually stopped in terraced fields with features such as upslope contour hedgerows, vegetative filter strips, and grass barriers (Darboux et al., 2023). Besides, it improves the level of canopy cover and promotes plant growth for two pivotal factors that reduce soil erosion. Moreover, the increased surface roughness of the root and litter layer decreases soil erosion in areas experiencing splash, rill, and inter-rill erosion (Wang, 2023).

6.3 *Increase Agricultural Productivity*

Increasing population pressures in areas of low arable land are thus attributed to current terracing trends worldwide (Socci et al., 2019). Terracing makes a steep slope relatively level, hence increasing the area farmed and allowing more intensive farming practices compared to sloping fields (Rutebuka et al., 2021). Therefore, this type of creation of arable land generally through slope land terrace transformation can increase by 20–40%, thus contributing to increasing grain yield (Deng et al., 2021). Ploughing and other cultivation practices facilitate the creation of pedo-environments, and modified profiles with huge amounts of nutrients and organic matter. This reason makes terraced soils, in most cases, hold more nutrients and organic matter compared to agricultural plots without terraces (Chen et al., 2021). In addition, terracing prevents water loss, thus adding moisture into the soil, hence improving crop resistance to droughts, thereby increasing crop yield (Suna et al., 2023).

6.4 Improve Cultural Landscape

Culturally landscaped terracing can be built to clearly demonstrate humanity's—and especially visiting adjective's—intelligent understanding of our dependence on crucial natural ecosystems and our mutual reliance on a natural environment (Tupan, 2023). A variety of disciplines have experienced a stimulating increase in research on terraced landscapes within the last decade that included geography, landscape architecture, rural sociology, ethnology, and other spatial studies (Lawson et al., 2022). They are not only essential for the beautiful view but also for spiritual growth and recognized as a part of the heritage of a place as a "cultural landscape." Examples of some famous ones are agricultural terraces in Asia, Central America, South America, and also Arab gardens in Spain (Nowicka, 2022). Their beautiful views with the beautiful, pleasing, neat, organized, and sustainable landscapes capture the attention of thousands of visitors every year (Cervera Alonso de Medina, 2022). As such, terraces consequently generate a good amount of money for the residents. As an example, in the scenario of China's Longji terraces, 70.8% of the income of local families owes itself to tourism.

6.5 Protect Natural Habitat

In many of the degraded or water-constrained ecosystems, afforestation or reforestation efforts are hard to carry through successfully without the other harvester inputs, more so because of the low site quality and adverse climatic conditions. Field terracing can increase regional landscape heterogeneity and thus provide habitats and facilitate the symbiosis of organisms as an added measure in preventing erosion, while conserving biodiversity as crucial for habitat restoration and improvement (Deng et al., 2021).

6.6 Sustainable Biodiversity Development

Terracing is done primarily for two reasons to increase biodiversity. First, the increased availability of water and nutrients in terraced fields makes better plant growth possible (Deng et al., 2021). Second, terracing would enhance the environment in which the different species can grow, thus contributing to an increased biodiversity. As many as 189 plant species belonging to 81 families were found in the terraced fields of Toho village, Japan (Kimura, 2024). It is also proven that, with similar environmental conditions, the biomass in terraced lands is 32–70% more than in non-terraced fields (Deng et al., 2021).

6.7 *Environmental Benefits*

Going by the impact of agricultural terraces on relief in hilly and mountainous areas, they bring a canopy of environmental benefits that would somewhat neutralize the impacts of climate change (Arora & Bhatt, 2023).

- Participation in the rainfall-enhanced
- Soil erosion control
- Reduced runoff
- Storage of biomass
- Tamed hot summer temperatures
- Reduced risk from fires and flooding
- Protection from landslides and major movements
- Habitat and biodiversity protection

6.8 *Socioeconomic Benefits*

One clear advantage of all types of agricultural terraces is that they increase the amount of arable land and improve agricultural productivity per unit of plan metric land area (Avni, 2022). Some studies have reported cases of crop yield increase, with the extreme case reaching as high as 270% in Tanzania. These increases range from almost zero to over 60% (Yengoh & Ardö, 2020). These productive agricultural systems both allowed and required greater population density, which is now decreasing due to competition from less labor-intensive agriculture in Europe's lowlands (Hayami & Ruttan, 2020). As noted in Fig. 8, terraces reconfigure the topography and foster plant growth, reducing runoff and sediments.

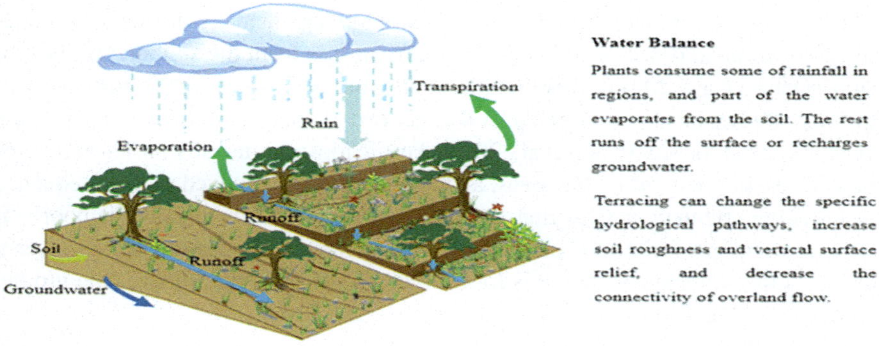

Fig. 8 The relationship between terracing and soil health. (Source: Deng et al., 2021)

7 Disadvantages of Terrace Farming

7.1 Terracing Impact on Hydrology

While terracing is synonymous with conservation benefits for soil and water, it creates hydrological problems in areas that have limited water resources (Rutebuka et al., 2021). Indeed, there are some notable alterations in the water cycle, such as soil infiltration, rainwater collection, irrigation, and evapotranspiration (Riedel & Weber, 2020). All of these changes can affect the amount of water available for human consumption. Large-scale terracing, through runoff acquisition and course change, enhances water retention in fields and reduces outflow by upsetting hydrological connectivity (McLain, 2022). Through improvement in soil conditions and encouraging the growth of vegetation, terracing increases crop water consumption while reducing runoff proportionately to the surface vegetation cover (Abiye, 2022). It builds concerns about the amount of water involved by the recently planted vegetation in areas that lie seasonally dry. Evaporation from terraces, as in the semi-arid northwest Loess Plateau, has an impact on water circulation, thereby influencing the downstream flow of the Yellow River (Zhang et al., 2021). This not only causes a remarkable decline in the freshwater resource quantity available but also has an impact on the hydrological cycle, causing the weakening of the large cycle and reducing the river runoff, changing the regional dynamic balance of water resources (Mishra, 2023). These negative effects are generally the result of bad planning, bad management, and abandonment in agriculture, which is problematic for the stability of the ecosystems, or the man–land relationship (Ustaoglu & Williams, 2022) (Fig. 9).

7.2 Decline in Soil Quality

The main causes of deterioration in newly built terraced land are the loss of productive topsoil, breakdown of soil structure, nutrient loss from the soil, and water erosion (Deng et al., 2021). Topographical alteration of slopes in the formation of terraces implies an alteration in the soil structure, which has a greater impact on the vegetation and the soil (Chen et al., 2021). Fertile topsoil is usually buried at the tips of terraces, thus extending flat areas, and may have to be removed as a consequence of terracing (Bennett & Chapline, 2020). This upside–down subsoil has poor soil structure and poor nutrient conditions, which further exacerbate these adverse effects. Indeed, local vegetation is lost in the process, just like it leaves the ground bare and produces loose materials Bustos et al., 2023).

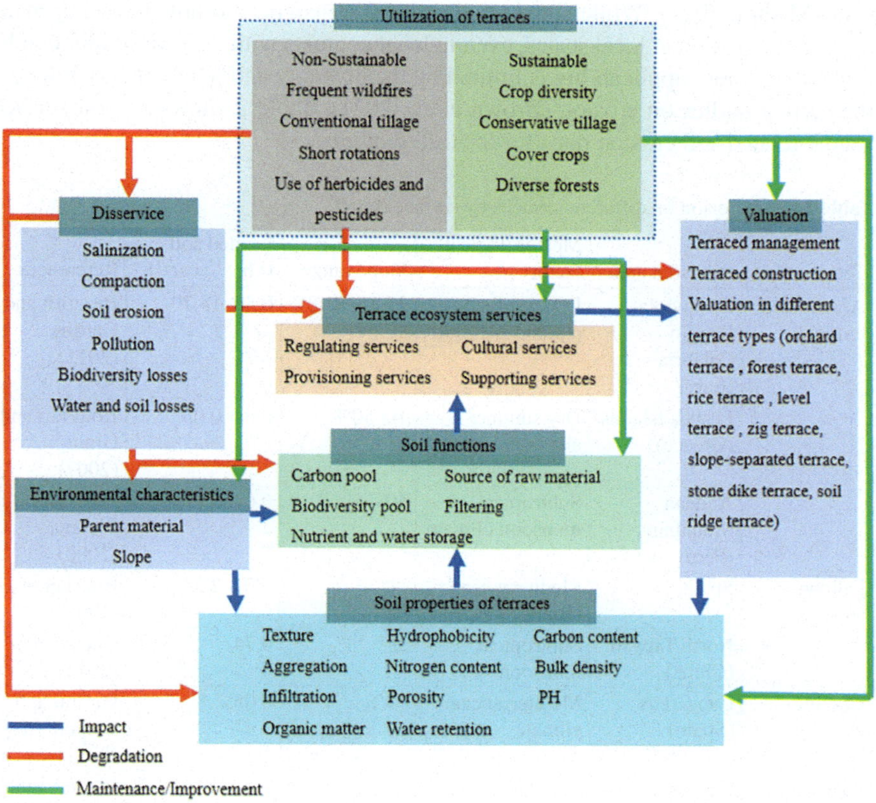

Fig. 9 Disadvantages of terrace irrigation management. (Source: Deng et al., 2021)

8 Factors Affecting the Terrace Irrigation Management

Various variables affecting the management of irrigation on terraces have been studied till date. Terracing is, undeniably, one of the established methods of soil and water conservation in the mountainous regions of the world (Kamamia, 2023). However, it is still limited to topography, the physical properties of the soil, climate, use of the land, culture, demography, and socioeconomic status (Bhattarai et al., 2020). In particular, topography, soil texture, land use, climate, cultivation practice, and the structure of the terraces turn out to be variables that considerably affect the impacts of terracing on soil and water conservation (Rutebuka et al., 2021). A clear understanding of all of the effects of how terracing would affect processes related to water erosion is critical in the management of soil loss and water loss. However, different methods have been used to assess how terracing affects water erosion, which makes it difficult to draw useful comparisons (Deng et al., 2021). Given the ways through which topography, land use, and terrace structure affect water erosion, building, managing, and maintaining terraces in a correct way are more urgent than ever (Mishra, 2023). While some studies provide insight into how terracing manages water erosion at local scales, even in local contexts with very dissimilar conditions, their wider applicability is limited by the diversity and inherent variability of the earlier studies on erosion (Brown et al., 2021). Table 2 shows several surface runoff-related soil erosion rates for the different terraces.

Table 2 Soil erosion on different terraces by surface runoff

Terrace type	Research area	Meteorological factor	Slope range	Eroded soil (t ha^{-1} year^{-1})	References
Abandoned terraces	Costaviola-Reggio Calabria (Italy)	Dry summer climate	29.65–81.8%	0.50–18.70	Bazzoffi and Gardin (2011)
	Lesbos Islands (Greece)	Dry summer climate	10–50%	0.04–0.05	Koulouri and Giourga (2007)
	Andean Mountains (Perú)	Subtropical monsoon climate	50–60°	0.01	Inbar and Llerena (2000)
Fallow terraces	Spain	Mediterranean climate		0.53–2.22	Ruecker et al. (1998)
	North Taiwan (China)	Subtropical monsoon	25°	0.74	Chen et al. (2012)
Grazing terraces	Cap Creus (Spain)	Mediterranean climate	18%	0.05	Pardini and Gispert (2012)
Forest terraces	Cap Creus (Spain)	Mediterranean climate	18%	0.01–0.35	Pardini and Gispert (2012)
	Cap Creus (Spain)	Dry climate	15%	0.03	Pardini and Gispert (2012)
Rice terraces	Northern Vietnam	Tropical monsoon	3–8°	0.16–1.77	Mai et al. (2013)
	North Taiwan (China)	Subtropical monsoon	25°	0.77	Chen et al. (2012)
Mixed crops terrace	Yimeng Mountains (China)	Continental monsoon	12°	27.00	Zhang et al. (2014)
	Baiquan County (China)	Temperate region	3–10°	25.04	Zhang et al. (2014)
	Lesbos Islands (Greece)	Mediterranean climate	10–50%	0.03	Koulouri and Giourga (2007)
	West Java (Indonesia)	Warm and humid climate	15–20°	30.0–34.0	Van Dijk (2002)
	West Java (Indonesia)	Warm and humid climate	15–20°	80.00	Van Dijk (2002)

9 Opportunities

Terrace land can be used in one or more of the following ways to intensify agroecologically:

9.1 Terraces for Providing Inexpensive Techniques and Materials for Sustainable Farming

Table 3 summarizes the problems faced by farmers with the traditional approaches, the benefits seen by adopting the use of terracing farming, low-cost approaches and tools that offer opportunities to enhance the intensity of the Terrace Cropping Systems while concomitantly increasing the sustainability and/or labor intensification (Mello et al., 2023). Opposite slopes and conservation tillage practices on hillsides use soil moisture for germination, thereby enhancing yields of terraces: contour ridging, tied-ridging, and mulch-ripping, or zero/minimum till on hillsides (Njiru, 2023). These increased germination and yields, no-till tied-ridging and ripping into maize residues in Zimbabwe reduced surface runoff by some 90% increased infiltration and produced a higher grain and biomass yield than the conventional untilled flat and flat treatments (Kaushik & Grichar, 2024). Grain and biomass yields were highest on the no-till tied-ridging treatment, likely due to the more substantial root depth and density of the longer roots (Wainaina, 2023). You can build ridges and tie ridges using readily available equipment such as this mold-board plough designed to be pulled by an animal. Similarly compared with the conventional practice, the application of plastic film plus straw mulch increased grain yield by 35% and WUE by 25% in winter wheat in Loess Plateau, China (Lin

Table 3 Challenges faced by farmers in conventional farming and opportunities in terrace farming as a solution for regarding hurdles

Challenges (in conventional farming)	Opportunities (in terrace farming)
Limited lands for intensive agriculture	Deploying barriers, edges, and vertical slopes
Farm operations are challenging to mechanize	Farm tools are lightweight and inexpensive
A spike in female work and employment	Implementing methods and instruments that decrease the amount of work that women do in agriculture
Inadequate service accessibility	There is an animal supporting network available
Loss of soil and erosion	Practice are done to conserve resources such as mulching, catch crops, and cover crops
Lack of resources	Low-cost innovations
Scarcity of employees	Tools and methods that reduce labor costs
Poor revenue and production	Practice to increase utilization of resources as well agricultural output

et al., 2019). In the highlands of Kenya, which are water-scarce, their use conjoined with minimum tillage using living vegetative barriers made of leucaena trees reduced competition for water between companion crops and the barriers.

9.2 Using the Slopes as Diet, Feed, and Medical Sources

Because it can produce a wide variability of crops, from small herbs to big trees, terrace agriculture may have opportunities for diversification and intensification (Caporali & Caporali, 2021). Crops and cropping systems are selected by farmers based on many different factors, like the climate, type of soil, topography, land holdings, needs of the farmers, cultural preference, availability of agricultural inputs including seeds and fertilizers, and, last but not least, local market opportunities (Giller et al., 2021). In Nepal, for instance, the most important crops cultivated on terraces include finger millet (*Eleusine coracana* L.), rice (*Oryza sativa* L.), and maize (*Zea mays* L.) (Chapagain & Raizada, 2017). Other crops that can possibly be grown are field peas, common beans, wheat, and wild and underutilized legumes based on interest by the farmers and seasonality of cultivation (Abobatta et al., 2021).

9.3 Using Micro-climates to Increase and Diversify Agricultural Production

The different climatic conditions prevalent in hills and mountains make the production of a wide range of horticultural as well as agronomic crops possible (Kumar et al., 2021). For instance, in Nepal, there are four distinct agroecological zones for which off-season fruits, vegetables, and other cash crops can easily be grown during the entire year (Gyawali et al., 2022). All these agroclimatic zones have small, niche-based microclimatic pockets that offer very great opportunities for the cultivation of many financially high-value food crops, fibers, fruits, medicinal plants, and trees (Kumar et al., 2021).

9.4 Rice Field Applications of the Anabaena-Azolla Symbiosis

On the other hand, photo-insensitive and tolerant Azolla strains like A. microphylla show promise (Chapagain & Raizada, 2017). Application of either fresh biomass of Azolla fronds at the rate of 200 kg ha^{-1} or frond-based spore inoculum of A. microphylla at 2.5 kg ha^{-1} in transplanted rice fields can produce a fairly thick layer of about 15–25 tons of Azolla (Chapagain & Raizada, 2017). The symbiosis has been reported to increase the grain yield by 36–38% over a sole rice system by contributing nitrogen of 40–60 kg ha^{-1} and organic matter to the soil following decomposition (Amanullah et al., 2019). Alternate wetting and drying (AWD) or intermittent irrigation combined with Azolla has been observed to increase the productivity of crop for producing water-efficient rice in areas where water becomes scarce (Shahid et al., 2021).

10 Conclusion and Future Prospect

Theoretically, research on terraces lags behind practical accomplishments. Under current technical conditions and databases, it is both possible and required to use high-resolution remote sensing image data and spatial analysis tools for determining the real state of terraced fields and their spatial boundaries. In the future, we suggest that some theoretical work should be tackled in parallel with the development of a terrace-monitoring network. We recommend that future research on the major following issues and related challenges be taken into consideration to provide a theoretical and scientific basis for the selection and optimal application of terracing techniques: (1) All the functions and effects of terraces at medium and permanent timescales, at different scales. (2) Research in the functioning of the terracing system for tiny watershed ecosystems. (3) Functional qualities of terraces and their suitability for different territories. Further, terracing will be adapted to local conditions, and interrelations between a person and an environment will be strengthened.

Combined effects of various factors on the terraced fields include variables that describe land use and geographical features of the environment. In contrast, very few studies join the different types of terraces with specific purposes; fewer still focus on how terracing is going to affect vegetation cover and water circulation. Current research, in relation, does not reveal the mechanism of action of the terraces. Effective treatment strategies and research associated with pollution related to terraces are also not available. We would recommend that future research has a close focus on the examination of the overall impact the terraces have, considering the purpose and mechanisms associated, and come up with feasible solutions for associated problems.

10.1 Investigations on Terrace Maintenance

In that sense, the terraced fields have become a typical example of how the design of human ecosystems and the relationship between humans and the environment are related. Human beings form an integral constituent in decision-making, management, and creation as far as terracing is concerned. Improper use and subsequent abandonment of terraces may result in environmental problems and threaten the socioeconomic conditions and food security of the area. Proper management and maintenance of the terraces in perpetuity should therefore back the provision of services by ecosystems. In the light of changing national policies, demographic pressures, and the state of socioeconomic development, we therefore propose that future studies on the subject consider the following: (1) Maintenance and repair of the terraces; (2) Quantifiable goals and activities for maintaining the terraces in perpetuity. The first three are protection from construction-related risks and compensation. (3) Legislation of terracing-based environmental legislation. (4) Provision of special funds, and monetary incentives for terracing to incentivize improved farming practices. (5) Knowledge provision from universities and policy framers regarding constructing terraces, to the local farmers about sustainable land management. It is in these areas that research can help in the integrated and more equal development of the economy, society, and ecology in the future.

References

Abiye, W. (2022). Soil and water conservation nexus agricultural productivity in Ethiopia. *Advances in Agriculture, 2022*(1), 8611733.

Abobatta, W. F., El-Hashash, E. F., & Hegab, R. H. (2021). Challenges and opportunities for the global cultivation and adaption of legumes. *Journal Of Applied Biology & Biotechnology, 8*(5), 160–172.

Acharya, M. S. (2020). Bamboo crib wall a sustainable soil bioengineering method to Stabiliz e Slopes in Nepal. *Journal of Development Innovations, 4*(1), 99–118.

Acharya, C. L. (2021). Soil and water conservation in the Himalayan region: Must to increase and stabilize crop yields. *Journal of Soil and Water Conservation, 20*(2), 119–130.

Achmad, B., Siarudin, M., Widiyanto, A., Diniyati, D., Sudomo, A., et al. (2022). Traditional subsistence farming of smallholder agroforestry systems in Indonesia: A review. *Sustainability, 14*(14), 8631.

Ahmad, A., Shahid, M., et al. (2019). Residues of endosulfan in cotton growing area of Vehari, Pakistan: An assessment of knowledge and awareness of pesticide use and health risks. *Environmental Science and Pollution Research, 26*, 20079–20091.

Ali, A., et al. (2013). Comparison of yield performance of soybean varieties under semiarid conditions. *Journal of Animal and Plant Sciences, 23*(3), 828–832.

Ali, S., et al. (2018a). Effects of sucrose and growth regulators on the microtuberization of potato germplasm. *Pakistan Journal of Botany, 50*(2), 763–768.

Ali, S., et al. (2018b). In vitro effects of gibberellic acid on morphogenesis of CIP potato explants and acclimatization of plantlets in field. *In Vitro Cellular & Developmental Biology, 54*(1), 104–111.

Amanullah, Khalid, S., Imran, Khan, H. A., Arif, M., Altawaha, A. R., et al. (2019). Organic matter management in cereals based system: Symbiosis for improving crop productivity and soil health. In R. Lal & R. Francaviglia (Eds.), *Sustainable agriculture reviews* (Sustainable agriculture reviews) (Vol. 29, pp. 67–92). Springer.

Amin, A., et al. (2018a). Evaluation and analysis of temperature for historical (1996–2015) and projected (2030–2060) climates in Pakistan using SimCLIM climate model: Ensemble application. *Atmospheric Research, 213*, 422–436.

Amin, A., et al. (2018b). Regional climate assessment of precipitation and temperature in southern Punjab (Pakistan) using SimCLIM climate model for different temporal scales. *Theoretical Applied Climatology, 131*, 121–131.

Angelakıs, A. N., Zaccaria, D., Krasilnikoff, J., Salgot, M., Bazza, M., Roccaro, P., et al. (2020). Irrigation of world agricultural lands: Evolution through the millennia. *Water, 12*(5), 1285.

Arora, S., & Bhatt, R. (2023). Resource conservation techniques for sustaining crop production in Rainfed foothills under changing climate. In *Advances in crop production and climate change* (pp. 367–405). CRC Press.

Aslam, M., et al. (2013). Drought stress, its effect on maize production and development of drought tolerance through potassium application. *Journal Cercetări Agronomiceîn Moldova (Agronomical Research in Moldavia), Iaşi—Romania, XLVI*(2), 99–114.
Avni, Y. (2022). The emergence of terrace farming in the arid zone of the Levant—Past perspectives and future implications. *Land, 11*(10), 1798.
Awulachew, S. B., Lemperiere, P., & Tulu, T. (2009). *Training manual on agricultural water management.* International Livestock Research Institute.
Bairagi, S., Custodio, M. C., Durand-Morat, A., & Demont, M. (2021). Preserving cultural heritage through the valorization of cordillera heirloom rice in The Philippines. *Agriculture and Human Values, 38*(1), 257–270.
Bazzoffi, P., & Gardin, L. (2011). Effectiveness of the GAEC standard of cross compliance retain terraces on soil erosion control. *Italian Journal of Agronomy, 6*, 43–51.
Bennett, H. H., & Chapline, W. R. (2020). Soil erosion a national menace. In *Environmental geomorphology and landscape conservation* (pp. 57–83). Routledge.
Bertolino, M. A., & Corrado, F. (2021). Rethinking terraces and dry-stone walls in the Alps for sustainable development: The case of Mombarone/alto Eporediese in Piedmont region (Italy). *Sustainability, 13*(21), 12122.
Bettoni, M., Maerker, M., Bosino, A., Conedera, M., Simoncelli, L., & Vogel, S. (2023). Land use effects on surface runoff and soil erosion in a southern alpine valley. *Geoderma, 435*, 116505.
Bhattarai, K., Yousef, M., Greife, A., & Naraharisetti, S. S. (2020). Influence of topography on sustainable land management: An analysis of socioeconomic and ecodemographic conditions of Nepal. *Agriculture, 10*(6), 224.
Black, A. L. (1968). Conservation bench terraces in Montana. *Transactions of the ASAE, 11*(3), 393–0395.
Blanco, H., & Lal, R. (2023). Water erosion. In *Soil conservation and management* (pp. 23–51). Springer Nature Switzerland.
Brown, A. G., Fallu, D., Walsh, K., Cucchiaro, S., Tarolli, P., Zhao, P., et al. (2021). Ending the Cinderella status of terraces and lynchets in Europe: The geomorphology of agricultural terraces and implications for ecosystem services and climate adaptation. *Geomorphology, 379*, 107579.
Bustos, B., Engel-Di Mauro, S., García-López, G., Milanez, F., & Ojeda, D. (Eds.). (2023). *Routledge handbook of Latin America and the environment.* Taylor & Francis.
Caporali, F., & Caporali, F. (2021). *Sustainable agriculture through ecological intensification* (pp. 123–258). Bridging the Ecological Gaps.
Cervera Alonso de Medina, M. (2022). *A landscape architectural guide to the United Nations 17 Sustainable Development Goals*. Col· legi d'Arquitectes de Catalunya (COAC).
Chapagain, T., & Raizada, M. N. (2017). Agronomic challenges and opportunities for smallholder terrace agriculture in developing countries. *Frontiers in Plant Science, 8*, 331.
Chen, S. K., Liu, C. W., & Chen, Y. R. (2012). Assessing soil erosion in a terraced paddy field using experimental measurements and universal soil loss equation. *Catena, 95*, 131–141.
Chen, D., Wei, W., & Chen, L. (2021). Effects of terracing on soil properties in three key mountainous regions of China. *Geography and Sustainability, 2*(3), 195–206.
Chen, T., Shu, J., Han, L., Tian, G., Yang, G., & Lv, J. (2022). Modeling the effects of topography and slope gradient of an artificially formed slope on runoff, sediment yield, water and soil loss of sandy soil. *Catena, 212*, 106060.
Chiankem, B. N. (2022). *Impacts and adaptation of small-scale potatoes farmers to climate change in the Western highlands of Cameroon, A case study of the impacts and adaptation of small-scale potato farmers to climate change in the Western highlands of Cameroon*. Master's thesis, The University of Bergen.
Chow, T. L., Rees, H. W., & Daigle, J. L. (1999). Effectiveness of terraces/grassed waterway systems for soil and water conservation: A field evaluation. *Journal of Soil and Water Conservation, 54*(3), 577–583.
Cucchiaro, S., Paliaga, G., Fallu, D. J., Pears, B. R., Walsh, K., Zhao, P., et al. (2021). Volume estimation of soil stored in agricultural terrace systems: A geomorphometric approach. *Catena, 207*, 105687.

Danish, S., et al. (2019). Allevation of Cr toxicity in maize by Fe fortification and Cr tolerant ACC deaminase producing PGPR. *Ecotoxicology & Environmental Safety, 185*, 109706.

Darboux, F., Räsänen, T., Johannsen, L., Schmaltz, E., Moussa, R., Pellegrini, S., & Deproost, P. (2023). *Guideline on the current implementation of erosion measures and other connectivity elements depending on scale and modelling approach.* Doctoral dissertation, EJP Soil.

Deng, C., Zhang, G., Liu, Y., Nie, X., Li, Z., Liu, J., & Zhu, D. (2021). Advantages and disadvantages of terracing: A comprehensive review. *International Soil and Water Conservation Research, 9*(3), 344–359.

do Prado, A. H., de Almeida, R. P., Galeazzi, C. P., Sacek, V., & Schlunegger, F. (2022). Climate changes and the formation of fluvial terraces in Central Amazonia inferred from landscape evolution modeling. *Earth Surface Dynamics, 10*(3), 457–471.

Doost, Z. H., Alsuwaiyan, M., & Yaseen, Z. M. (2024). Runoff management based water harvesting for better water resources sustainability: A comprehensive review. *Knowledge-Based Engineering and Sciences, 5*(1), 1–45.

Duarte, A. C., & Mateos, L. (2022). How changes in cropping intensity affect water usage in an irrigated Mediterranean catchment. *Agricultural Water Management, 260*, 107274.

Eren, G. (2022). *Make place for thy Lydian kings: Monumental urban terraces of iron age Sardis.* Doctoral dissertation, Boston University.

Esilaba, A. O., Miriti, J., Gitari, J., Mangale, N., Kathuku-Gitonga, A. N., Njiru, E., et al. (2021). *KCEP-CRAL integrated soil fertility and water management extension manual.* Kenya Agricultural and Livestock Research Organization.

Fahad, S., et al. (2019). Suppressing photorespiration for the improvement in photosynthesis and crop yields: A review on the role of S-allantoin as a nitrogen source. *Journal of Environmental Management, 237*, 644–651.

Giller, K. E., Delaune, T., Silva, J. V., Descheemaeker, K., van de Ven, G., Schut, A. G., et al. (2021). The future of farming: Who will produce our food? *Food Security, 13*(5), 1073–1099.

Gyawali, P., Khanal, S., & Bhandari, S. (2022). Need to take precedence for off seasonal vegetable farming? Issues in context of Nepal. *Turkish Journal of Agriculture-Food Science and Technology, 10*(12), 2495–2503.

Hammad, H. M., et al. (2018). Uptake and toxicological effects of pharmaceutical active compounds on maize. *Agriculture, Ecosystem and Environment, 258*, 143–148.

Hayami, Y., & Ruttan, V. W. (2020). Population growth and agricultural productivity. In *Technological prospects and population trends* (pp. 11–69). Routledge.

Hou, X. Q., Li, R., Han, Q. F., Jia, Z. K., Wang, W., Yan, B., & Yang, B. P. (2012). Effects of strip planting and fallow rotation on the soil and water loss and water use efficiency of slope farmland. *Ying Yong Sheng tai xue bao= The Journal of Applied Ecology, 23*(8), 2191–2198.

Inbar, M., & Llerena, C. A. (2000). Erosion processes in high mountain agricultural terraces in Peru. *Mountain Research and Development, 20*(1), 72–79.

Iqbal, J., A., et al. (2019). Purple nutsedge (Cyperus rotundus) control through interference by summer crops. *International Journal of Agriculture Biology, 21*, 1083–1088.

Ishak, C. F. (2022). SOIL CONSTRAINTS FOR CROP PRODUCTION IN SOUTH-EAST ASIA. Soil Constraints on crop. *Production, 318.*

Kamamia, A. W. (2023). *Using hydrological models and digital soil mapping for the assessment and management of catchments: A case study of the Nyangores and Ruiru catchments in Kenya (East Africa).*

Kameri-Mbote, P., Kamunge, I., & Yatich, J. K. (2021). Country report for Kenya. In *African soil protection law* (pp. 177–294). Nomos Verlagsgesellschaft mbH & KG.

Kaushik, P., & Grichar, W. J. (2024). *New prospects of maize.* BoD–Books on Demand.

Kavoori, A. (2020). The inkas married the earth: Excavating machu picchu. *ISLE: Interdisciplinary Studies in Literature and Environment, 27*(3), 648–660.

Khaliq, T., et al. (2013). A study of thermal changes on physiology and paddy yield of fine rice cultivars under subtropical conditions. *Asian Journal of Agriculture and Biology, 1*(4), 208–216.

Khan, A., et al. (2019). Managing tillage operation and manure to restore soil carbon stocks in wheat–maize cropping system. *Agronomy Journal, 111*(5), 2600–2609.

Kimura, S. (2024). *Theorizing post-disaster literature in Japan: Revisiting the literary and cultural landscape after the triple disasters*. Rowman & Littlefield.

Koulouri, M., & Giourga, C. (2007). Land abandonment and slope gradient as key factors of soil erosion in Mediterranean terraced lands. *Catena, 69*(3), 274–281.

Kramer, L. A., Burkart, M. R., Meek, D. W., Jaquis, R. J., & James, D. E. (1999). Field-scale watershed evaluations on deep-loess soils: II. Hydrologic responses to different agricultural land management systems. *Journal of Soil and Water Conservation, 54*(4), 705–710.

Kumar, P., Sharma, P. K., Kumar, P., Sharma, M., & Butail, N. P. (2021). Agricultural sustainability in Indian Himalayan region: Constraints and potentials. *Indian Journal Of Ecology, 48*(3), 649–661.

Kumawat, A., Bamboriya, S. D., Meena, R. S., Yadav, D., Kumar, A., Kumar, S., et al. (2022). Legume-based inter-cropping to achieve the crop, soil, and environmental health security. In *Advances in legumes for sustainable intensification* (pp. 307–328). Academic Press.

Kurdekar, A. K., Ashoka, P., Dambale, A. S., & Zaman, M. I. (2023). Modern techniques in soil and water conservation. *ADVANCED FARMING TECHNOLOGY, 1*.

Lalbiakzuali, C. (2020). *Landslide susceptibility mapping of Lunglei town and its vicinity in Mizoram*. Doctoral dissertation, Mizoram University.

Lawson, K., Bavaj, R., & Struck, B. (2022). A guide to spatial history: Areas, aspects, and avenues of research. *Olsokhagen*.

Le, D. H. (2022). *Water erosion process on sloping cropland in Central Vietnam—A case study in A Luoi district, Thua Thien Hue province*.

Lin, W., Liu, W., Zhou, S., & Liu, C. (2019). Influence of plastic film mulch on maize water use efficiency in the loess plateau of China. *Agricultural Water Management, 224*, 105710.

Liu, S., Qin, T., Lv, X., Shi, X., Dong, B., Wang, J., & Liu, C. (2021). Experimental study of runoff and sediment yield affected by ridge direction and width of sloping farmland. *Frontiers in Earth Science, 9*, 694773.

Lingard, A. (2020). *The fresh and the salt: The story of the Solway*. Birlinn Ltd.

Mai, V. T., van Keulen, H., Hessel, R., Ritsema, C., Roetter, R., & Phien, T. (2013). Influence of paddy rice terraces on soil erosion of a small watershed in a hilly area of Northern Vietnam. *Paddy and Water Environment, 11*, 285–298.

Maina, J. W. (2020). *Rainfall and temperature trends, farming patterns and rainwater harvesting Technologies in Kieni Sub-County*. Doctoral dissertation, University of Nairobi.

Malik, A. R., Mushtaq, R., Bhat, Z. A., & Ganie, M. A. (2020). Diversity of fruits. In *Emerging Technologies for Shelf-Life Enhancement of fruits* (pp. 1–20). Apple Academic Press.

Mardon, A., Alad, S., Barara, G., Chana, I., Falade, I., Harun, R., et al. (2021). *Machu Picchu?*.

Marino, P., Comegna, L., Damiano, E., Olivares, L., & Greco, R. (2020). Monitoring the hydrological balance of a landslide-prone slope covered by pyroclastic deposits over limestone fractured bedrock. *Water, 12*(12), 3309.

McLain, K. (2022). *Hydrological controls on flow conveyance losses on the middle Rio Grande*. New Mexico Institute of Mining and Technology.

Mello, I., Roloff, G., Laurent, F., Gonzalez, E., & Kassam, A. (2023). Sustainable land management with conservation agriculture for Rainfed production: The case of Paraná III watershed (Itaipu dam) in Brazil. In *Rainfed systems intensification and scaling of water and soil management: Four case studies of development in family farming* (pp. 99–126).

Minhas, P. S., Ramos, T. B., Ben-Gal, A., & Pereira, L. S. (2020). Coping with salinity in irrigated agriculture: Crop evapotranspiration and water management issues. *Agricultural Water Management, 227*, 105832.

Mishra, R. K. (2023). Fresh water availability and its global challenge. *British Journal of Multidisciplinary and Advanced Studies, 4*(3), 1–78.

Mishra, P. K., Rai, A., & Rai, S. C. (2020). *Indigenous knowledge of terrace management for soil and water conservation in the Sikkim Himalaya, India*.

Mubeen, M., et al. (2013a). Evaluating CSM-CERES-maize model for irrigation scheduling in semi-arid conditions of Punjab, Pakistan. *International Journal of Agriculture and Biology, 15*, 1–10.

Mubeen, M., et al. (2013b). Evaluating different irrigation scheduling criteria for autumn-sown maize under semi-arid environment. *Pakistan Journal of Botany, 45*(4), 1293–1298.

Mubeen, M., et al. (2013c). Effect of growth stage-based irrigation schedules on biomass accumulation and resource use efficiency of wheat cultivars. *American Journal of Plant Sciences, 4*, 1435–1442.

Mubeen, M., et al. (2016). Application of CSM-CERES-maize model in optimizing irrigated conditions. *Outlook on Agriculture., 45*(3), 173–184.

Mukai, S., Billi, P., Haregeweyn, N., & Hordofa, T. (2021). Long-term effectiveness of indigenous and introduced soil and water conservation measures in soil loss and slope gradient reductions in the semi-arid Ethiopian lowlands. *Geoderma, 382*, 114757.

Nasim, W., et al. (2016). Correlation studies on nitrogen for sunflower crop across the agroclimatic variability. *Environmental Science and Pollution Research, 23*, 3658–3670. https://doi.org/10.1007/2Fs11356-015-5613-1

Nasim, W., et al. (2017). Response of sunflower (Helianthus annuus L.) hybrids to nitrogen application grown under different agro-environments. *Journal of Plant Nutrition, 40*(1), 82–92.

Nasim, W., et al. (2018a). Future risk assessment by estimating historical heat wave trends with projected heat accumulation using SimCLIM climate model in Pakistan. *Atmospheric Research, 205*, 118–133.

Nasim, W., et al. (2018b). Radiation efficiency and nitrogen fertilizer impacts on sunflower crop in contrasting environments. *Environmental Science and Pollution Research, 25*(2), 1822–1836.

Nguyen, A. T., Turner, S., & Kalacska, M. (2022). Challenging slopes: Ethnic minority livelihoods, state visions, and land-use land cover change in Vietnam's northern mountainous borderlands. *Environment, Development and Sustainability, 24*(2), 2412–2431.

Njiru, E. N. (2023). *Effect of Fanya Juu terraces with varying ditch dimensions on selected soil properties and crop yields in semi-arid eastern Kenya*. Doctoral dissertation, University of Nairobi.

Nowicka, K. (2022). The heritage given: Cultural landscape and heritage of the vistula delta mennonites as perceived by the contemporary residents of the region. *Sustainability, 14*(2), 915.

Opilo, B. N. (2023). IMPACT OF SUSTAINABLE LAND MANAGEMENT TECHNOLOGIES ON FLOOD RISK REDUCTION IN WESTERN REGION, KENYA.

Osman, K. T., & Osman, K. T. (2018). Soils on steep slopes. In *Management of soil problems* (pp. 185–217).

Pardini, G., & Gispert, M. A. (2012). Soil quality assessment through a multi-approach analysis in soils of abandoned terraced land in NE Spain. *Cuadernos de Investigación Geográfica, 38*(2), 7–30.

Phillips, J. D. (2022). Store and pour: Evolution of flow systems in landscapes. *Catena, 216*, 106357.

Plekhov, D. (2022). *Terraced landscapes: The historical ecology of agricultural infrastructure*. Doctoral dissertation, Brown University.

Pueppke, S. G. (2021). Ancient WEF: Water–energy–food nexus in the distant past. *Water, 13*(7), 925.

Rahman, M. H. R., et al. (2018). Multi-model projections of future climate and climate change impacts uncertainty assessment for cotton production in Pakistan. *Agricultural and Forest Meteorology, 253–254*, 94–113.

Rahman, H. M., et al. (2019). Application of CSM-CROPGRO-cotton model for cultivars and optimum planting dates. *Field Crops Research, 238*, 139–152.

Ramser, C. E. (2020). Prevention of the erosion of farm lands by terracing. In *Environmental geomorphology and landscape conservation* (pp. 299–313). Routledge.

Rao, B. K., Annapurna, S., Rani, B. R., Rao, Z. S., Sunitha, K., SchinDutt, M., et al. (2022). *Soil and water conservation techniques in Rainfed areas [E-book]*. National Institute of Agricultural Extension Management (MANAGE) & Water and Land Management Training and Research Institute (WALAMTARI), Hyderabad.

Reimer, K. (2022). *A longitudinal study of the natural gas pipeline industry in British Columbia*. Doctoral dissertation, University of British Columbia.

Riedel, T., & Weber, T. K. (2020). The influence of global change on Europe's water cycle and groundwater recharge. *Hydrogeology Journal, 28*(6), 1939–1959.

Romeo, R., Manuelli, S., Geringer, M., & Barchiesi, V. (2021). *Mountain farming systems–seeds for the future: Sustainable agricultural practices for resilient mountain livelihoods*. Food & Agriculture Org.

Ruecker, G., Schad, P., Alcubilla, M. M., & Ferrer, C. (1998). Natural regeneration of degraded soils and site changes on abandoned agricultural terraces in Mediterranean Spain. *Land Degradation & Development, 9*(2), 179–188.

Rutebuka, J., Uwimanzi, A. M., Nkundwakazi, O., Kagabo, D. M., Mbonigaba, J. J. M., Vermeir, P., & Verdoodt, A. (2021). Effectiveness of terracing techniques for controlling soil erosion by water in Rwanda. *Journal of Environmental Management, 277*, 111369.

Sabagh, A. E. L., et al. (2019). Drought and salinity stresses in barley: Consequences and mitigation strategies. *Australian Journal of Crop Science, 13*(6), 810–820.

Sadeghi, N., & Eslamian, S. (2021). Road water harvesting. In *Handbook of water harvesting and conservation: Basic concepts and fundamentals* (pp. 109–126).

Schmidt, C., & Schmidt, C. (2022). Landscape explorations between given and acquired resilience. In *Landscape resilience: Basics, case studies, practical recommendations* (pp. 139–201).

Sen, C. T., Saberi, H., & Bhattacharyya, S. (2023). *The Bloomsbury handbook of Indian cuisine*.

Shah, G. M., et al. (2019). Composting of mu municipal solid waste by different methods improved the growth of vegetables and reduced the health risks of cadmium and lead. *Environmental Science and Pollution Research, 26*(06), 5463–5474.

Shahid, M., Munda, S., Khanam, R., Chatterjee, D., Kumar, U., Satapathy, B. S., et al. (2021). Climate resilient rice production system: Natural resources management approach. *Oryza, 58*, 143–167.

Shahzad, K., et al. (2018). Tillage and biochar effects on wheat productivity under arid conditions. *Crop Science, 59*, 1–9.

Sharda, V. N., Juyal, G. P., & Singh, P. N. (2002). Hydrologic and sedimentologic behavior of a conservation bench terrace system in a sub–humid climate. *Transactions of the ASAE, 45*(5), 1433.

Sidle, R. C. (2021). Strategies for smarter catchment hydrology models: Incorporating scaling and better process representation. *Geoscience Letters, 8*(1), 24.

Singh, R. (2023). Terrace. In *Soil and water conservation structures design* (pp. 63–93). Springer Nature Singapore.

Socci, P., Errico, A., Castelli, G., Penna, D., & Preti, F. (2019). Terracing: From agriculture to multiple ecosystem services. In *Oxford research encyclopedia of environmental science*.

Strahler, A. N. (2020). The nature of induced erosion and aggradation. In *Environmental geomorphology and landscape conservation* (pp. 18–35). Routledge.

Sultana, S. R., et al. (2014). Normalized difference vegetation index as a tool for wheat yield estimation: A case study from Faisalabad, Pakistan. *The Scientific World Journal, 2014*, 725326.

Suna, T., Kumari, A., Paramaguru, P. K., & Kushwaha, N. L. (2023). Enhancing agricultural water productivity using deficit irrigation practices in water-scarce regions. In *Enhancing resilience of dryland agriculture under changing climate: Interdisciplinary and convergence approaches* (pp. 177–206). Springer Nature Singapore.

Talabi, A. O., Vikram, P., Thushar, S., Rahman, H., Ahmadzai, H., Nhamo, N., et al. (2022). Orphan crops: A best fit for dietary enrichment and diversification in highly deteriorated marginal environments. *Frontiers in Plant Science, 13*, 839704.

Tamura, K., Miura, H., Kaneko, S., Sano, T., & Kubotera, H. (2021). Soil-forming factors. In *The Soils of Japan* (pp. 13–51).

Tariq, M., et al. (2018). The impact of climate warming and crop management on phenology of sunflower-based cropping systems in Punjab, Pakistan. *Agricultural and Forest Meteorology, 256-257*, 270–282.

Tiwari, K. R., Sitaula, B. K., Bajracharya, R. M., & BØrresen, T. (2009). Runoff and soil loss responses to rainfall, land use, terracing and management practices in the Middle Mountains of Nepal. *Acta Agriculturae Scandinavica Section B–Soil and Plant Science, 59*(3), 197–207.

Tupan, M. A. (2023). *Phenomenology and cultural difference in high modernism*. Cambridge Scholars Publishing.
Ustaoglu, E., & Williams, B. (2022). Institutional settings and effects on agricultural land conversion: A global and spatial analysis of European regions. *Land, 12*(1), 47.
Uwacu, R. A., Habanabakize, E., Adamowski, J., & Schwinghamer, T. D. (2021). Using radical terraces for erosion control and water quality improvement in Rwanda: A case study in Sebeya catchment. *Environmental Development, 39*, 100649.
van Dijk, A. I. J. M. (2002). *Water and sediment dynamics in bench-terraced agricultural steeplands in West Java, Indonesia* (p. 384). AIJM van Dijk.
van Dijk, A. I. J. M., Bruijnzeel, L. A., Vertessy, R. A., & Ruijter, J. (2005). Runoff and sediment generation on bench - terraced hillsides: Measurements and up - scaling of a field - based model. *Hydrological Processes: An International Journal, 19*(8), 1667–1685.
Wainaina, M. W. (2023). *Effect of tied ridges on yield and yield components of maize, sorghum and dry bean under different agronomic management practices in Central Kenya*. Doctoral dissertation, University of Nairobi.
Wajid, A., et al. (2014). Modeling growth, development and seed-cotton yield for varying nitrogen increments and planting dates using DSSAT. *Pakistan Journal of Agricultural Sciences, 51*, 639–647.
Walia, S. S., Kaur, K., & Kaur, T. (2024). Soil and water conservation techniques in rainfed areas. In *Rainfed agriculture and watershed management* (pp. 115–124). Springer Nature Singapore.
Wang, L. (2023). *Advances in soil erosion research*. Scientific Research Publishing, Inc.
Wang, D., Fahad, S., et al. (2019). Morphological acclimation to agronomic manipulation in leaf dispersion and orientation to promote "Ideotype" breeding: Evidence from 3D visual modeling of "super" rice. *Plant Physiology and Biochemistry, 135*, 499–510.
Wang, M., Rong, L., Li, Y., Huang, J., Jiao, Y., & Wei, X. (2024a). Drainage of paddy terraces impacts structures and soil properties in the globally important agricultural heritage of Hani Paddy terraces, China. *International Soil and Water Conservation Research, 12*(1), 64–76.
Wang, Q., Yang, X., Liu, X., & Furuya, K. (2024b). Rice terrace experience in Japan: An ode to the beauty of seasonality and nostalgia. *Land, 13*(1), 64.
Wei, W., Chen, D., Wang, L., Daryanto, S., Chen, L., Yu, Y., et al. (2016). Global synthesis of the classifications, distributions, benefits and issues of terracing. *Earth-Science Reviews, 159*, 388–403.
Wilpert, K. V. (2022). Forest soils—What's their peculiarity? *Soil Systems, 6*(1), 5.
Wohl, E. (2020). *Rivers in the landscape*. Wiley.
Xiao, Y., Cao, H., Pan, H., Luo, G., & Zhu, D. (2024). Incipient motion of sands near the pipe tip. *Acta Geotechnica, 19*(6), 4107–4119.
Yengoh, G. T., & Ardö, J. (2020). Climate change and the future heat stress challenges among smallholder farmers in East Africa. *Atmosphere, 11*(7), 753.
Yesuph, A. Y. (2020). *Prospects of sustainable land management amidst interlocking challenges in the upper Beshillo catchments, northeastern highlands of Ethiopia*. Doctoral dissertation, University of South Africa. Retrieved from http://hdl.handle. net/10500/26493.
Zerrudo, E. B. (2022). Reinforcing Philippine cultural landscape values through the schools of living tradition (SLTs): Mimicry to memory. In *The Routledge handbook of cultural landscape heritage in the Asia-Pacific* (pp. 341–355). Routledge.
Zhang, J. H., Su, Z. A., & Liu, G. C. (2008). Effects of terracing and agroforestry on soil and water loss in hilly areas of the Sichuan Basin, China. *Journal of Mountain Science, 5*, 241–248.
Zhang, Y., Long, Y., An, J., Yu, X., & Wang, X. (2014). Spatial patterns of 137Cs inventories and soil erosion from earth-banked terraces in the Yimeng Mountains, China. *Journal of Environmental Radioactivity, 136*, 1–9.
Zhang, S., Zhao, G., Mu, X., Tian, P., Gao, P., & Sun, W. (2021). Changes in streamflow regimes and their responses to different soil and water conservation measures in the loess plateau watersheds, China. *Hydrological Processes, 35*(10), e14401.

Virtual Water Trade: The Need for Water-Intensive Crops

Muhammad Asad ur Rehman Naseer, Shoaib Akhtar, Ahmad Abdul Wahab, and Shakeel Ahmad

Abstract One of the most important global issues of the twenty-first century is the critical importance of water for millions of people whose livelihoods depend on its wise use. In many regions around the world, water demand already exceeds supply, a phenomenon exacerbated by the world's growing population. What matters is the planning, development, management, and trade of water—vital for natural resources, basic human needs, and national wealth. In recent years, water demand has increased due to the production of a variety of goods—ranging from food, fiber, and shelter. Over the years, China has emerged as a virtual water net importer, while countries like India and Pakistan are still dependent on water-intensive production techniques. These countries continue to export agricultural commodities posing a significant risk to water sustainability. This chapter covers the water marketing reality as a hidden cost in the agricultural production system and international trade. It is important to understand the political economy in the unsustainability of food and water resources and its influence on other factors related to food system dysfunction. Several case studies are also presented in this chapter highlighting the crucial role of water in agriculture to achieve food security. It explores how the international trade policy can improve water scarcity in water-stressed regions by a tradeoff between water-intensive and water-efficient crops. Finally, this chapter encourages a balanced approach to virtual water trade, highlighting the integration of sustainable water trade with technological innovation and the adoption of realistic water trading in agriculture. The authors call for collaborative efforts among governments, policymakers, and stakeholders to tackle water scarcity effectively and ensure water security for a more sustainable future.

M. A. u. R. Naseer · S. Akhtar
Department of Agricultural Business and Marketing, Bahauddin Zakariya University, Multan, Pakistan
e-mail: asadrehman@bzu.edu.pk; shoaibakhtar@bzu.edu.pk

A. A. Wahab · S. Ahmad (✉)
Institute of Agronomy, FAS&T, Bahauddin Zakariya University, Multan, Pakistan
e-mail: shakeelahmad@bzu.edu.pk

M. Mubeen et al. (eds.), *Innovations in Agricultural Water Management*,
https://doi.org/10.1007/978-3-031-91883-4_8

Keywords Virtual water trade · Water-intensive crops · Water scarcity · Global food security · Sustainable water management

1 Introduction

Water is a crucial resource that is essential for sustaining life on our planet. Nevertheless, the distribution of this resource is not uniform worldwide, resulting in significant inequalities in both access to and utilization of water. The notion of virtual water trade pertains to the movement of water-intensive commodities among different regions, effectively shifting the water utilized in their production from one location to another (Ma et al., 2020). This chapter presents case studies and findings that emphasize the vital role of water-intensive crops in ensuring food security and meeting global demands (Ahmed et al., 2024; Wahab et al., 2024; Raza et al., 2021; Ahmad et al., 2008, 2014, 2015; Ahmad and Hasanuzzaman, 2012). It also explores the benefits and drawbacks of trading water for the production of these crops (Dang et al., 2015). This practice facilitates countries to combat local water scarcity issues, encourage the adoption of more sustainable water practices, and ensure the efficient allocation of water resources at a global level. The concept of virtual water trade highlights the interconnection of water resource management across the globe and it also highlights the importance of strategic trade policies to address the growing issue of water scarcity.

Virtual water trade is even more necessary than mere water saving. It helps to boost food security and economic stability by enabling water-scarce countries not to over-exhaust their limited supply of freshwater on essential goods (Xing & Chen, 2023). Water consumption is a big problem in growing agricultural products. A previous study suggested that countries such as China, India, and Pakistan have managed their water resources by outsourcing the use of virtual-imported billions of liters of domestic freshwater a year in imported high-water merchandize (Zheng et al., 2022; Hammad et al., 2018a, 2018b; Awais et al., 2018a, 2018b; Rasool et al., 2018; Fahad et al., 2018). This improves the local food supplies and also contributes to development by enabling countries with comparative advantage in other goods so that they can specialize themselves as producers.

The agricultural sector uses a huge amount of water, especially for the crops such as rice, wheat cotton, and sugarcane that require more volume of water to cultivate. As global demand for these crops grows further with increasing population growth and changing dietary habits, this exerts extreme pressure on water reserves (Vallino et al., 2021). Moreover, the benefit of virtual water has enormous environmental and social repercussions. Such practice further encourages global water saving as it locates productions in areas of high water availability, leading to a reduction in the environmental burdens created by depleting ground and surface waters in arid environments. From a societal standpoint, it contributes to the prevention of conflicts over water resources by reducing competition and promoting equitable division of waters (Ingrao et al., 2023; Afzal et al., 2018; Amin et al., 2018c; Khan et al., 2018; Hashmi et al., 2018; Adnan et al., 2018; Awais et al., 2017a, 2017b; Porkka, 2011). Policymakers have an opportunity to develop policies that can balance the water demand locally with resource management globally, by using and understanding virtual water exchange. But it can help build something more enduring and effective for the critical water issues facing us in the twenty-first century of increasing scarcity.

1.1 *Importance of Water in the Twenty-First Century*

This most important substance for bacteria, plants, and animals is equally so beneficial to us human beings being that water makes up the foundation of everything we do in life on earth. The importance of water is felt in the twenty-first century among many challenges we face. This era characterized by rapid population growth, urbanization, and climate change combined with the growing demands in agriculture has put unprecedented pressure on global water resources.

1.1.1 Growing Global Water Demand

There is an emerging global challenge of increasing water demand, which arises from factors like population growth, urbanization, and economic progress. Over the next three decades, amid demands from a global population estimated to grow to 9.7 billion by 2050, there will be an additional demand for freshwater resources of up to 50%. As water demand increases—the vast majority of which will be for agriculture and hydropower—water stress worsens, leading to scarcity and growing competition over common pool resources.

Of all sectors, agriculture is one of the biggest drivers behind growing world water demand and accounts for around 70% of global freshwater withdrawals. With the growth of the global population, food production requires more water, and therefore water needs increase as well. Industrial processes, energy production, and domestic use also add to the growing thirst. Unchecked, the situation could result in great human suffering with severe water stress leading to famine and possibly further collapse of a fragile economy.

Effective management of water-stressed areas improved agricultural production and sustainable industrial development are some adaptations required to climate hazards associated with the growing global demand for water. Advances in water-saving technology—such as low-flow fixtures and efficient irrigation systems—can minimize the waste of freshwater resources. If we come together to solve the growing scarcity of freshwater, then maybe there is hope that beneficial agreements can help us all in bettering our chances for centuries ahead.

1.2 Implications of Water Scarcity

There are major environmental and economic impacts that occur due to water scarcity. This in turn is seeing the loss of aquatic ecosystems and biodiversity, while wetlands that perform significant ecosystem services such as water purification and flood control systems are also being lost (Koebele & Simpson, 2023). Reduced water flow also amplifies pollution in bodies of water, thereby threatening wildlife and local ecosystems. Economically, agriculture suffers from lower crop yields and increased production costs which endangers food security for the wellbeing of farmers. Shemer et al. (2023) further stated that water-intensive industries for manufacturing, cooling, and processing experience failures or delays in operation resulting in higher costs leading to reduced economic yield and ultimately affecting the employment rate In addition, the water shortage causes a negative impact on energy production—especially in hydroelectric and thermoelectric plants—which leads to lack of energy supply with increasing costs.

Water scarcity carries significant public health effects, as the limited access to potable water increases vulnerability to diseases transmitted through contaminated drinking water such as cholera, dysentery, and typhoid. It also causes the transfer of transmittable diseases from one individual to another, thereby leading to nightmare public health tragedies affecting primarily vulnerable populations like children, elders, and poor communities (Rhue et al., 2023; Sabagh et al., 2019; Farid et al., 2019a, 2019b; Tariq et al., 2008; Awais et al., 2017a, 2017b, 2018a, 2018b; Amin et al., 2017a, 2017b, 2018a, 2018b, 2018c; Hammad et al., 2018a, 2018b; Nasim et al., 2018, 2016). On the societal front, competition for increasingly scarce resources causes unrest and conflict between communities perhaps even regions or nations. Urban services and infrastructure could come under increased pressure as people move in search of better water availability for their families. What is more, the onus of fetching water—as often experienced by women and children in underdeveloped areas—may discourage children particularly girls from attending school thus translating into a hindrance for them from acquiring an education leading to a good future (Nesar et al., 2022).

2 Understanding Virtual Water Trade

As with physical water trade, virtual water trade allows for resources to be transferred from regions that are rich in them into others where the needs remain unmet. This makes it possible for countries without much water capacity to meet their needs by importing items that take lots of water in assembly, crude as crops and textiles, to locations with numerous available. Understanding and improving virtual water trade should help alleviate regional water stress, increase food security, and adopt more sustainable agricultural practices between countries. It also highlights the interconnectedness of global water supplies and emphasizes that international policymaking will be needed to control water usage sustainably.

Virtual water trade has implications far beyond water conservation, by allowing the free transfer of a portion of the earth's freshwater resources to those regions sharing in virtual terms and where it is used more soundly, acting as an enhancement tool for social progress and environmental steadiness. Countries can reduce their water footprint by importing instead of producing water-intensive goods, saving precious freshwater resources for other important uses. Furthermore, through this virtual water trade the impacts of climate change on stagnation against changing climatic patterns to availability of food and water can be reduced. This understanding of virtual water trade, in turn, enables policymakers to develop a set of suitable strategies—between local agricultural reliance and global resource governance that can potentially manage the situation by responding better to both host countries' desire for greater resources with increasing investment opportunities.

2.1 *Concept of Virtual Water*

Virtual water is the hidden flow of water if food or other commodities are traded from one place to another. Virtual water is a term, first coined by Professor John Allan in the early 1990s, which embeds how much water has been used within the production of goods. This is called virtual water including the water used in agriculture to grow crops, breed livestock, or manufacture products. On their export, the water utilized during production accumulates at the importing region—even though much of the physical water is within the boundaries of the exporting country. Virtual water trade allows regions deficient in water to purchase resources that are inherently high-water use rather than strip their irreplaceable supply. Such trade offers a solution to water scarcity at local levels and, when conducted internationally, increases the overall efficiency of global water use by directing resources toward areas in which they are most abundant. For example, in dry regions that may endanger scarce water reserves if more grains are grown there instead of importing them or meat from countries with a surplus of abundant water. Awareness of the virtual water content in things helps improve a country's decisions and proper interventions on water management, agricultural practice, or trade; reinforcing a more sustainable and equitable use of planetary resources.

2.2 *Significance in Global Virtual Water Trade*

The value of virtual water trade in the broader context of global trading lies within its capacity to improve the efficiency and sustainability with which water resources are used across a worldwide scale. Exchanging agricultural goods and textiles that demand a lot of water allows countries confronted with scarcity to avoid depleting their limited resources by importing such products instead. Beyond just easing local water pressure, this strategy has a broader objective to help enhance the global efficiency of use by facilitating production in regions where there are relatively more plentiful and less stressed resources. Through virtual water trade, optimal allocation of water provision between regions enhances sustainable management strategies for the dwindling global freshwater resources. As shown in Fig. 1, annual water flows in agricultural trade over this century from three different studies (Abbas et al., 2017; Fahad et al., 2017; Saud & e al., 2017; Zia et al., 2017; Amin et al., 2017a, 2017b, 2017c; Dalin et al., 2017; Graham et al., 2020; Hoekstra & Mekonnen, 2012). The time water fossil embodied in agricultural trade (in billion cubic meters)—and their split between virtual green, blue, and goal nonrenewable groundwater exports from 2010 to the end of this century—is presented here. The virtual green water exports are represented by the green line, and here we can see a sharp upward trajectory—within one century they increase rapidly toward levels that would be unsustainable. Blue virtual water exports in blue are increasing as well, but significantly more slowly than green water. Nonrenewable virtual groundwater exports, in purple, have remained low and roughly constant since 1992 due to falling levels of nonrenewable resources.

In addition, changing the hands of the virtual water is important in terms of food security and the stability of economies. It enables countries with scarce water to access the goods they need without tapping their water reserves too heavily and ensuring more reliable supplies of food and other critical products. This type of commerce could neutralize the impact droughts, and for that matter, any other climate issues, have on those countries revolving around elusive water channels. Economically, it supports the global market by enabling countries to produce products that suit their comparative advantage through virtual water trade and reinforcement of trade between nations thus creating economic interconnections. Thus,

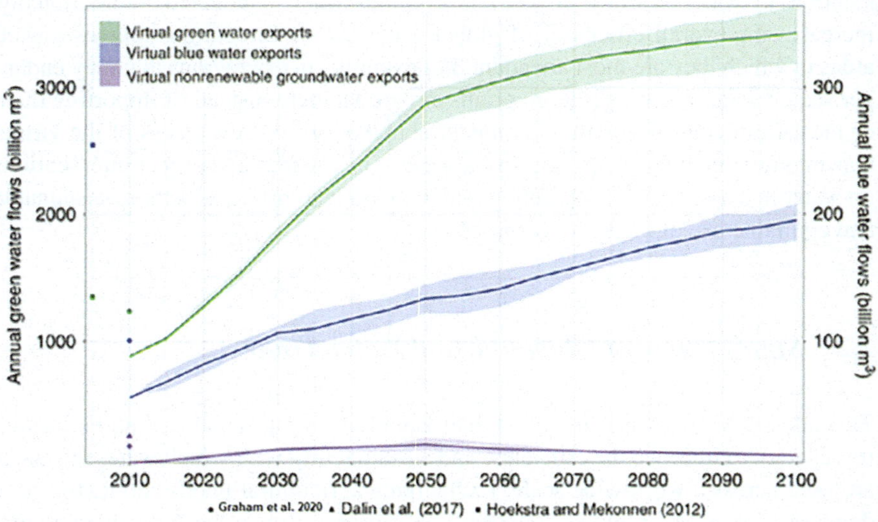

Fig. 1 Annual green water, blue water, and groundwater flows in agricultural trade. (Source: Adapted from Graham et al. (2020))

virtual water trade is one of the important instruments to improve sustainability, adaptability, and cooperation regarding growing global water threats.

2.3 *Historical Perspectives: China, India, and Pakistan*

Such demand could be satiated by the imported virtual water originating from multiple countries for better managing the available scarce resource of any country and also rooted in historical agricultural traditions experienced much before, often coupled with regions known to face some degree of scarcity regarding access to freshwater. China, India, and Pakistan are important examples due to their high agricultural yield while being endowed with a significant water resource. These nations are the pure example of virtual water trading since these nations have long been involved in activities that are consistent with the basic tenets of it.

2.3.1 China

The historical virtual water strategies of China, which deal with its immense amount of internal water resources, were based on the historical trading pattern. Throughout the past centuries, China has witnessed an asymmetrical distribution of water resources with a dry north and a wet south, which adds to this aforementioned disparity (Zhang et al., 2021). China follows the old system where agricultural ventures in water-deficient zones are transferred to regions that have plenty of water, thus keeping up with resourcefully and also fighting for access by restriction (Shakeel et al., 2017; Jabran et al., 2017; Amin et al., 2017a, 2017b, 2017c; Mirza et al., 2017; Hammad et al., 2017; Sarpong et al., 2020). In recent years, the trade of virtual water has become a more important issue due to China's economic restructuring and participation in the global economy (Dalin et al., 2017). Given that China has limited water reservoirs, it can import loads of soybeans and wheat (water-intensive crops) from abroad to preserve its storage with a view to more valuable agricultural or industrial products cultivated there (Zheng et al., 2022). Such a strategic fact was fundamentally an essential element in the configuration of water scarcity management inside, principally directed at the growth of population and economic expectations.

2.3.2 India

The agricultural sector in India, which uses nearly 90% of the country's freshwater resources, has a long tradition of trading embedded virtual water. With large variations in water accessibility across areas, India has been forced to import high-water-use crops to suit its national-scale hydrobalance (Harris et al., 2020). The Green Revolution in the 1960s and 1970s led to a fivefold increase in water use, effectively increasing agricultural production. Substance goods that India has decided to import—such as oilseeds, pulses, and some water-intensive items—were identified following the need of its local water reservoirs (Eliazer Nelson et al., 2019). During the 1960s and 1970s in the era of the green revolution in India, a fivefold increase in water use was seen, effectively increasing agricultural production.

2.3.3 Pakistan

Pakistan has a history of nearly 70 years of water shortage, given its worst-ever arid climate and reliance on the Indus River for irrigation but also implementing trade policies (Zahra et al., 2023). India has long been battling water reserves with a near-exclusive focus on rice, cotton, and other dry crops. However, due to the simple fact that this method is unfeasible from a long-term perspective for Pakistan (as is obvious); therefore, there has been an evolution in Pakistan's approach and virtual water trade options have started coming alive. This may help Pakistan to save the pressure of its water reservoirs by way of importing wheat and other significant cereal plants (Ullah et al., 2023). Furthermore, the established Indus Water Treaty with India, which dictates the apportionment of river waters between the two countries, serves as a resource management mechanism that is in line with virtual water principles by facilitating enhanced water distribution and utilization according to regional requirements.

3 Water-Intensive Cropping and Food Security

Water is considered the most vital natural resource for sustaining life. However, around 80% of the global population faces significant risks to their water security (Vörösmarty et al., 2010; Yu, 2016). The agricultural sectors will be the initial victims of water shortage, as the demand for water from non-agricultural sources rises and water scarcity becomes more severe (Hanjra & Qureshi, 2010; Li et al., 2021).

Water-intensive agricultural methods have substantial consequences for worldwide food security. Several studies have shown that farming systems that depend too much on crops that need a lot of water can lead to water waste and groundwater loss, especially in dry places like Pakistan, India, and Telangana, as well as in countries that already have a lot of water problems/shortage, like Jordan and Egypt (Christoforidou et al., 2023; Kuchimanchi et al., 2023; Rashid et al., 2004). The effects of increased agricultural activity on water resources highlight the necessity for sustainable farming methods and governmental measures to alleviate water scarcity and guarantee enduring food security (Christoforidou et al., 2023) (Kuchimanchi et al., 2023). Crops that are ignored and grown in less ideal conditions have a significant effect on the production of crops through diversifying the production system of cropping. The negative effect of the environment declines by promoting the sustainability process and enriching the diet with different techniques (Farooq et al.,

2023). The selection of water-efficient crops and various agroecological principles to fulfill worldwide food requirements while saving resources for the future.

Du et al. (2015) answered that improvement in agricultural water productivity can be seen by efficient utilization of scarce water resources. It can also be made more efficient by implementing "water-saving agriculture," which is a tool that includes effective distribution of water use in agriculture, water transference through the soil-plant-atmosphere range, and increasing water availability for yield formation. Various methods of water utilization in agriculture essential to ensure food security are shown in Fig. 2. In this figure the notations used are W_o is the ratio of water diverted from reservoir to farm; W_f is the water at the farm gate; W_t is the water taken by the crop; W_r is the water retained by roots; m_{as} is mass of CO_2; m_{bm} is the plant biomass produced; Y is the crop yield; ET is evapotranspiration; and SPAC stands for soil–plant–atmosphere continuum.

In current scenarios, the availability of sufficient and nutritious food for all is necessary and a prominent issue, but it holds a prominent standing in global goals set for the year 2030 (Mechiche-Alami et al., 2021). The water resources are limited due to increasing population, changes in dietary patterns, climate changes, and urbanization. Due to environmental and economic constraints, the issue of water shortage is becoming increasingly urgent (Fito et al., 2017). Given that agriculture is the greatest consumer of water globally, water shortage has a significant effect on agricultural outputs, and that causes food insecurity. The significance of advocating

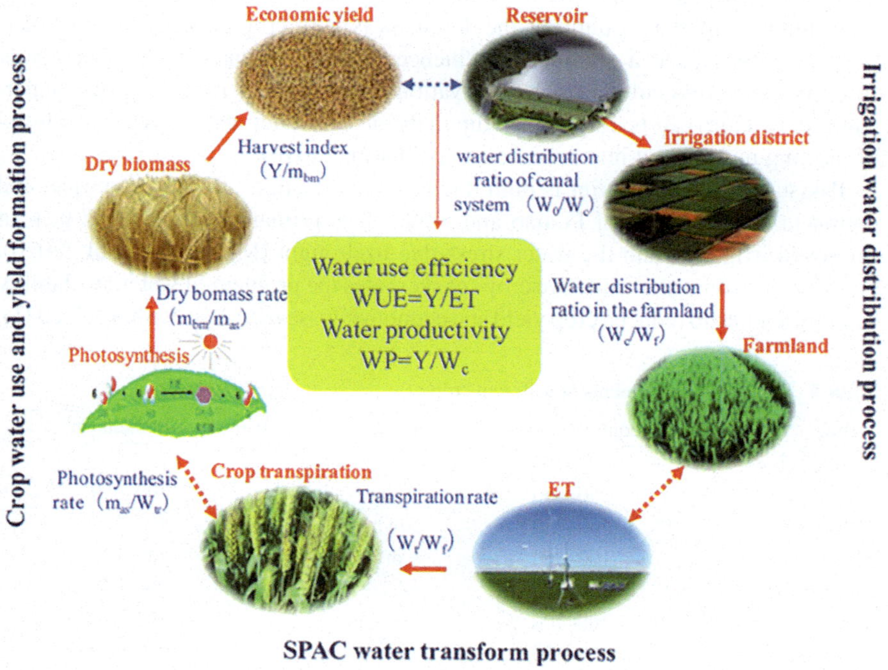

Fig. 2 Multiple methods of agricultural water use for food security. (Source: Du et al. (2015))

for selections of water-efficient crops and ecological methods to fulfill worldwide food requirements while preserving water for future use.

3.1 Role of Water-Intensive Crops

He believes that crops like cotton, sugarcane, and rice, which are more water dependent than other food-grain fields, hold great importance in global agriculture due to their critical productivity gain and economic values. Table 1 illustrates water requirements for different crops. Rice is a staple and primary source of calories for 50% of the world's population and has been widely recognized as a cereal at the interface with water management leading to different methods of technological adoption enhancing production (Cesari de Maria et al., 2017). Water is required to produce bioethanol and sugar from sugarcane. Ethiopia further illustrates the high water footprint of the production of bioethanol from sugarcane, with weight averages of water footprint of 217.69 m^3/ton cane and 3106.7 L/L, respectively. This, in turn, suggests a shortage of surface water (Fito et al., 2017). Martinelli et al. (2013) reported that water utilized for sugarcane ethanol production was 1.53 cm^3 of water/ton of sugarcane in Brazil with an average usage rate of up to 18 L used per liter of ethanol. Many of these crops need regular watering and a reliable water source to produce any useful yield. The result is these crops are a vital security blanket for food, sustenance, and economy in multiple geographies.

In India, rainfed agriculture is the dominant type of crop production and water-intensive crops have a significant influence on both irrigated-outlandish matter resources and cropland yield. The cultivation of water-intensive crops (rice, sugarcane, and cotton) exerts a significant dip in the groundwater table thus sustainability of farming must be put into practice as per Sharma (2016).

Pakistan's water footprint is high due to the cultivation of sugarcane, rice, and cotton in the provinces of Punjab and Sindh. It is primarily caused by a lack of improved irrigation and the water stress due to drought (Muzammil et al., 2020). High reliance on water-intensive crops exacerbates the problem of water availability scarcity leading to reduced crop yield and economic losses (Zeshan & Shakeel, 2021).

Table 1 Irrigation requirements of different crops

Crops	Water requirement (mm)	Crops	Water requirement (mm)
Sunflower	600–1000		
Barley	250–350	Bajra	400–650
Sugarcane	1500–2500	Soybean	450–750
Gram	100–250	Cotton	700–1300
Mustard	300–450	Maize	500–800
Potato	300–700	Sorghum	450–650
Groundnut	500–700	Paddy	900–1500
Wheat	450–650	Alfalfa	800–1600

Source: Ouda et al. (2016) and Rai et al. (2017)

3.2 *Meeting Global Food Demands*

The world food demands are met in water-intensive agriculture systems by the transfer of the current system into sustainable intensification. There is a need to enhance the division of crops within existing farmland to better production efficiency and reduce water consumption (Davis et al., 2017; Droppers et al., 2021; Joseph et al., 2023). Sustainable intensification increases crop productivity while reducing the negative impacts on the environment. This approach highlights the actual consumption of resources and implementation of proper management practices (Davis et al., 2016; Shrestha & Subedi, 2019). It is important to limit the water quantity used for irrigation and increase the amount that can be naturally restored. There is one more thing that carefully regulates nutrient application to make certain sustainable levels for production without causing harm to water natural resources (Droppers et al., 2021). By redesigning worldwide crop allocations according to their water usage, it is easy to provide sustenance to millions more individuals while simultaneously increasing water for future purposes (Davis et al., 2017).

Implementation of new cropping patterns and the use of advanced irrigation techniques (drip and sprinkler irrigation) can successfully lessen crop water requirements, providing a more sustainable approach to water utilization in agriculture (Muzammil et al., 2020). Water resource management is essential for focusing on the difficulties of water-intensive crops. This is critical for providing food security and encouraging sustainable agricultural practices under water-intensive cropping (Javed & Ahmad, 2020).

4 Advantages and Disadvantages of Virtual Water Trade

Virtual water trade supports numerous advantages, including water scarcity mitigations in countries with inadequate water resources by enabling them to require a significant water quantity from other areas. In China, interprovincial trade plays a significant role in contributing to economic development. Many developed provinces of China get advantages of this trade in the form of economic advantages (Zheng et al., 2022). Nevertheless, virtual water trade can affect disparities, as developing regions face net virtual water exports and encounter difficult economic consequences in their trade with more developed regions (Zheng et al., 2022). Moreover, commercial virtual water exchange at a global scale may reinforce asymmetries in which exporting countries with higher levels of water scarcity export their surplus to importing ones characterized by lower values of the drought index, hence, underlining economic disparities and underscoring the necessity for redistributive schemes aimed by them (Vallino et al., 2021). Virtual water commerce has the potential to alleviate water scarcity and stimulate economic growth. However, it also shows problems in terms of economic and water resource inequalities, which require adequate management.

4.1 Alleviating Water Scarcity

To overcome water scarcity, a range of sustainable strategies can be utilized to hold the risks and problems associated with this serious problem. It is essential to employ alternative water sources such as desalination processes and sewage reclamation methods, optimize crop mixtures to minimize water usage in irrigated agriculture, and create innovative solutions like degradable radiating cooling mulch to maintain soil moisture and address water scarcity for agriculture (Richter et al., 2023; Shemer et al., 2023). There are a few more techniques to overcome this scarcity: providing knowledge about water scarcity, improving water infrastructures (storage and shifting), developing water catchment and harvesting techniques, advancement in agriculture and irrigation practices (crop rotation, drip irrigation, etc.), minimizing water pollution, novel technologies (treatment of wastewater), and establishment of water cooperation between countries.

4.2 Risks and Challenges

Regional consumption can have a substantial impact on water demand in remote locations, and the trading of virtual water can exacerbate water scarcity in regions that are already experiencing a shortage of water. Water scarcity presents significant hazards to the overall economy as it is a crucial resource for various industries including agriculture, power generation, mining, and utilities. The risk of water scarcity in water-reliant industries may be transferred to other sectors of the economy through trade networks resulting in economic loss. In this manner, the Water Shortage Risk (WSR) primarily affects the whole supply chain of the economy (Dalin et al., 2012; Lenzen et al., 2013).

In the United States of America (USA), urban water scarcity has been caused by both households at the local level and across the supply chain in the production process. Therefore, management of water resources focuses on water scarcity as well as on potential risks linked to upstream supply chain commerce (Djehdian et al., 2019).

4.3 Sustainable Approaches

The solution for water scarcity can be managed by adopting a sustainable approach that deals with many facets. Sustainable development goals are directly related to water security and include issues such as the availability of water, demand for water resources, abatement concerning environmental quality degradation (in particular regarding quantity), economic prioritization of uses through least-cost methods in policy formulation, and stakeholder involvement when deciding how decisions should be reached on use allocation choices among competing users including plans implemented over space-time scales covering years or 10-year increments. Determining who will benefit from resource comments about arrangements requiring changes adoption involving stakeholders most affected while improving capacity-building enhance implementation align with socio-cultural factors. Furthermore, the resolution of water scarcity requires rational pricing for water, public awareness campaigns, and programs by the government bodies along with an efficient structure to govern it or agreements on a global level (Padder & Bashir, 2023).

Additionally, sustainable measures are needed to overcome the sociocultural aspects of water scarcity and these include community participation, consumer empowerment in self-management, and appropriate modes of dissemination (Emile et al., 2022; Wutich et al., 2022). An interdisciplinary and holistic approach should be adopted while taking into account a few factors, to deal with water scarcity issues along with being assertive toward having a secure water future. This way, long-standing actions of sustainable water management could be effectively materialized.

5 Balancing Virtual Water Trade

Balanced virtual water trade integrates sustainability to maximize the efficiency of water use and equitable distribution. Given the fact that end-use change can result in water saving and is effective in sustainable management of water resources, implementing water-efficient functional nexus with respective aspects such as agriculture-based best practices for high crops' virtual water reduction or wise practice to improve efficient and sufficient utilization. The use of High-Efficiency Irrigation Systems (HEIS) such as drip and sprinkler irrigation technologies can reserve water and provide crops with the required amount of water to grow. It can similarly be made more sustainable with the aid of crop selections that are better suited to local climate and soil conditions, reducing water requirements further. Mitigating water scarcity involves sustainably managing the limited resources. Water is not only a fundamental life source but also an asset to enrich economies helping the nations grow faster and alleviating poverty (FAO, 2017). These practices can be embraced by nations to enhance their water efficiency and make the virtual water trade network more equitable as well as sustainable.

In addition, technological inventions also promote practicing virtual water trade appropriately to balance the pitfall by improving usage efficiency and productivity of water. Progress in technologies like remote sensing, IoT-based soil moisture sensors, and AI-driven water management systems is resulting in real-time information about the conditions of both the soil and crops helping to provide exact amounts of required water. One example is the use of remote sensing technologies to track crop health status as well as soil moisture, which enables farmers to alter irrigation schedules and prevent depletion due to water shortage while not affecting their yield

(Obaideen et al., 2022). Similarly, the introduction of genetically modified crops enhancing efficiency could protect high yields under limited supply and would suit especially those regions that are scarce in access to water (Blum, 2012). Virtual water trade can become more effective, and cope with the challenges brought through water scarcity by a combination of technological advances in our ability to grow food and efficient use of this form of virtual wealth. This chapter discusses sustainable practices, technological innovations, and pragmatic ways of optimizing virtual water trade.

5.1 *Integrating Sustainable Practices*

Embedding sustainable practices into virtual water trade implies seeking the minimum returns per area in agricultural and industrial processes. Sustainability practices include advanced irrigation systems, smart crop selection (for water stress), and applying best practices for waste recycling. FAO reports that proper water management can decrease the size of agriculture's water footprint—the biggest user and consumer of global freshwater resources—with considerable efficiency gains. Drip and sprinkler systems are two examples of precision irrigation technologies, indeed both minimize water loss and maximize crop yield per unit of applied water (FAO, 2017). Furthermore, using suitable crops in the right region can also decrease over-irrigation. For example, in arid areas drought-resistant or low-water crops can be given high priority (Muhammad et al., 2024).

Further can also improve the water efficiency in their process/technology adaption. Industrial operations use water recycling and treatment systems that reduce the virtual water trade network, thereby lowering their high water consumption (Yasmeen et al., 2023).

5.2 *Technological Innovations in Agriculture*

Water-saving productivity produced through technological advancements can optimize the efficiency of water use in agriculture, which enables a great deal of virtual water trade. New developments such as remote sensing, Internet of Things (IoT-enabled) soil moisture sensors, and water management systems powered by artificial intelligence provide immediate insights regarding the state of crop health and quality to ensure exacting application of this support. Remote sensing technology can provide real-time data over entire agricultural fields that measure crop health, water stress, and soil moisture levels. When this data is fed into AI, it can tweak watering schedules to use less water but still give good yields (Shahab et al., 2024). Moreover, IoT-enabled systems are capable of automatic irrigation applications when that is needed and in the right volumes. Biotechnological advances in producing genetically modified crops to improve water use efficiency are also a part of agriculture that can help us reduce our large footprints. These crops are able to withstand low precipitations of water which enable high production yields and are commonly grown in parts that are often faced with problems if the scarcity of available natural sweet waters where irrigation facilities do not function (Munaweera et al., 2022).

5.3 *Realistic Approaches to Water Trading*

Realistic water trading solutions require the development of supporting policies and frameworks that provide an equitable process for reallocating limited water supply. Water trading can be used for efficient allocation of water among competing users, to support the likelihood that consumers conserve a scarce resource and in dealing with regional scarcity problems. Charging prices are suitable to the true cost of water including environmental and social costs leads people to execute more financially feasible consumption (Wheeler et al., 2017). Also, the implementation of water markets acts as a crucial tool by which waterfront rights could be exchanged between users to streamline and improve the equal distribution of water (Razzaq et al., 2022). But these mechanisms must be such that they should not lead to over-exploitation of water resources and also the rights of marginalized communities. Policies should also incentivize investments in water-efficient technology and infrastructure, ensuring a sustainable future (Ahmad et al., 2020).

6 Collaborative Solutions for Water Security

This is to emphasize that water security cannot be achieved by the governments, stakeholders, and in collaboration with communities. In successive sections, different programs and engagement strategies can be seen to guarantee water management is sustainable.

6.1 *Government and Policy Initiatives*

Water securities rely particularly on governments and policy initiatives. Integrated Water Resources Management (IWRM) policies may be one way to do this by encouraging not only a good mix of trade-offs between sectors and regions but also agreements favorable for environmental flows. Water policies at the national and regional levels should further concentrate on improving water governance, investing in infrastructure, as well as increasingly promoting more efficient use of water. It has been suggested that governments could enact regulations for sustainable water use, clean the polluted and contaminated sources of drinking water, or play a role in managing regional resources fairly (Razzaq et al., 2019). Transnational water management also requires international cooperation. Given so, the sharing of water resources among countries can be accomplished through agreements and treaties that will serve as safeguards to potential conflicts thereby ensuring regional stability.

6.2 Stakeholder Engagement

In the field of water management, it is crucial to involve all relevant stakeholders, from local communities and industries through non-governmental organizations. Water management decisions benefit from involving stakeholders in the designs of policies and interventions to facilitate both their acceptance and effectiveness. Community-based water management frameworks can provide local populations agency to actively manage their waters (Ahmad et al., 2020). This can result in more viable and culturally sensitive remedies. Companies can further conserve water while helping local efforts to science by implementing corporate water stewardship practices. PPPs can also promote investment in water infrastructure and technology to enhance the access, and quality of supply for all kinds of users.

6.3 Ensuring a Sustainable Future

Taking into account current and future use is essential to prevent over-extraction, so monitoring extraction will be our new sustainable feature of water for the ages. Strategies that can operate in different climate conditions and also handle changes in population growth, and geopolitical situations as changing water demands are fundamental so it is crucial to get the better combination (Ahmadov, 2020). Research and development of new water-saving technologies—as well as innovative practices in using existing technology more efficiently—is essential for managing future droughts. There should also be education and public awareness programs to promote water conservation among the general masses. At the same time, natural ecosystems that supply critical water-related services (such as wetlands and forests) must also be conserved to preserve a healthy water cycle by safeguarding long-term water supplies.

6.3.1 Long-Term and Short-Term Sustainability

Over the long term, countries with water abundant resources must be linked to virtual water markets globally. At a generalized level, virtual water exporters are those countries with more than 75% of their population exposed to low levels—less than zero-point-five megameters per annum (mm/year) sometimes categorized as effectively no—water stress and highest overall productivity in terms of volume or value

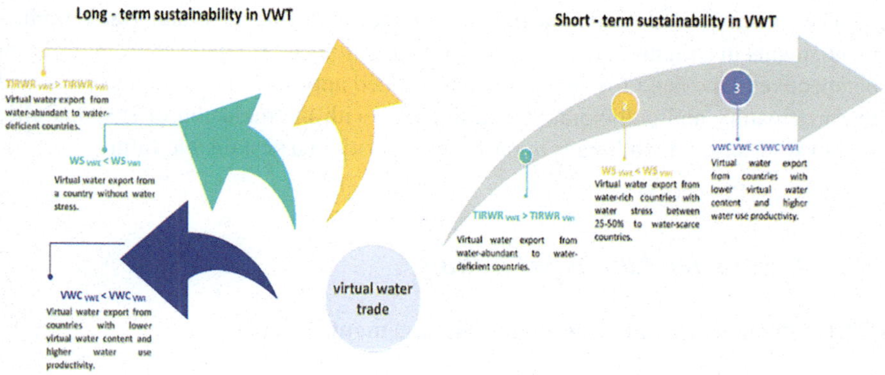

Fig. 3 Long-run and short-run sustainability in virtual water markets. (Source: Hekmatnia et al. (2024))

or financial cost production efficiencies, therefore, it may be claimed the sustainable use of freshwater resources. It will also help to reduce the water scarcity stress associated with other countries. The short-term goal is to heal the critical water stress in water-scarce nations. Resultantly, short-term water-rich countries are included having water stress levels in the range of 25–50% in virtual water markets as shown in Fig. 3.

7 Conclusion

This chapter has explored various dimensions of water security through the lens of virtual water trade, technological innovations, sustainable practices, and collaborative solutions. As we conclude, it is essential to synthesize the key points discussed and offer actionable recommendations for future efforts.

7.1 *Summary of Key Points*

This chapter has tried to provide a comprehensive overview of the importance of water in the twenty-first century, emphasizing the growing global water demand and the severe implications of water scarcity. It also introduces the concept of virtual water trade and discusses its immense implications in world trade, with a brief history for countries like China, India, and Pakistan. Based on case studies that exemplify the water shortages and, consequently, food security in those regions with scarce resources for agricultural activities, this chapter discusses virtual water trade advantages and disadvantages (alleviating water scarcity while showing risks and issues). Those risks need to be mitigated by working in a more sustainable way, the report said. The optimization of water use was discussed through the balance between virtual water trade, sustainable agricultural practices, and water trading mechanisms in a relative way to technical innovations or efficiencies from different perspectives. The need for an adaptation-oriented approach recommended was tied to participatory and collaborative action along with the inclusion of sustainability within solutions, which hinges upon water security in a sustainable future.

7.2 Recommendations for Action

7.2.1 Promote Sustainable Water Management Practices

We need regulations to sustainably manage water that are insistently put in place by governments and industries alike. It uses precision irrigation, water recycling, and drought-tolerant plants. These efforts can drastically diminish water footprints and reinforce the sustainability of trade in virtual waters.

7.2.2 Invest in Technological Innovations

Agri.Tech: it only makes sense, as agriculture directly impacts the food we eat which is vital for our sustenance. Biological Genomics, Remote Sensing, IoT-Based Water Management, Biotechnological Innovations, and more to be explored. The properties and characteristics of water resources determine obvious biophysical elements in the ecological evolution process, which is where many smart technologies perform their role to maximize utilization efficiency for sustainable ecosystem management.

7.2.3 Develop and Implement Effective Water Trading Policies

Water trading frameworks should be designed by policymakers to promote the efficiency and equity of water allocation. True water costs, including environmental and social externality costs, as well as transparent pricing mechanisms that account for these should encourage sustainable water resource management.

7.2.4 Enhance Stakeholder Engagement

This is because the involvement of local stakeholders in decision-making related to water management is essential. Engaging in a public process ultimately results in water management strategies more efficient and well-accepted public–private partnerships can help finance the critical infrastructure and technologies that are essential in order to make this happen.

7.2.5 Protect and Restore Ecosystems

Wetlands and forest areas are crucial to the natural environment, especially because they play a major part in maintaining the water cycle. These ecosystems also protect water resources over time and foster biodiversity. Any activities that help to restore habitat should be prioritized in order to encourage and improve both the quality of life for desert terrestrial species (restoration projects) and ecosystem services associated with water regulation.

7.3 Call to Collaborative Efforts

Water security for a growing population and in an era of climate change will be possible only through the unprecedented mobilization of all sectors. Also, the recommendations as defined above should be implemented in a collaborative way. Sustainable water management solutions require coordinated collective action by governments, industries, and communities at all levels as individuals to develop sustainable practices of water management. A culture of innovation and cooperation when it comes to water scarcity will help us ensure sustainable consumption or utilization of this very important finite resource for the welfare of not only discount seniors but also future generations. This shared effort will bolster global food security, economic stability, and environmental sustainability in addition to enhancing water security. Water security is a crucial issue that needs to be urgently and continuously addressed. Together, and with smart solutions, we can develop water security for a human society functioning within planetary boundaries (Chawla et al., 2018; Farid et al., 2017).

References

Abbas, G., et al. (2017). Quantification the impacts of climate change and crop management on phenology of maize-based cropping system in Punjab, Pakistan. *Agricultural Forest Meteorology, 247*, 42–55.

Adnan, M., et al. (2018). Phosphate-solubilizing bacteria nullify the antagonistic effect of soil calcification on bioavailability of phosphorus. *Scientific Reports, 2018*, 07–22653.

Afzal, M., et al. (2018). Current status and future possibilities of molecular genetics techniques in Brassica napis. *Biotechnology Letters, 40*(3), 479–492.

Ahmad, & Hasanuzzaman. (2012). Integrated effect of plant density, N rates and irrigation regimes on the biomass production, N content, PAR use efficiencies and water productivity of rice under irrigated semiarid environment. *Notulae Botanicae Horti Agrobotanici Cluj-Napoca, 40*(1), 201–211.

Ahmad, S., et al. (2008). Water and radiation use efficiencies of transplanted rice (Oryza sativa L.) at different plant densities and irrigation regimes under semi-arid environment. *Pakistan Journal of Botany, 40*(1), 199–209.

Ahmad, S., et al. (2014). Response of cotton crop to exogenous application of glycinebetaine under sufficient and scarce water conditions. *Brazilian Journal of Botany, 37*(4), 407–415.

Ahmad, S., et al. (2015). Radiation, water and nitrogen use efficiencies of Gossypium hirsutum L. *Turkish Journal of Agriculture and Forestry, 39*(5), 825–837.

Ahmad, B., Pham, H. D., Ashfaq, M., Memon, J. A., Bano, R., Dahri, Z. H., Mustafa, R. N., Baig, I. A., Naseer, M. A., & u. R. (2020). Impact of institutional features on the overall performance assessment of participatory irrigation management: Farmers' response from Pakistan. *Water, 12*(2), 497. https://www.mdpi.com/2073-4441/12/2/497.

Ahmed, M., et al. (2024). Simulation of source sink portioning in wheat under varying nitrogen regimes using DSSAT-CERES wheat model. *Agricultural Water Management, 303*, 109028.

Ahmadov, E. (2020). Water resources management to achieve sustainable development in Azerbaijan. *Sustainable Futures, 2*, 100030. https://doi.org/10.1016/j.sftr.2020.100030

Amin, A., et al. (2017a). Comparaison of future and base precipitation anomalies by SimCLIM statistical projection through ensemble approach in Pakistan. *Atmospheric Research, 194*, 214–225.

Amin, A., et al. (2017b). Optimizing the phosphorus use in cotton by using CSM-CROPGRO-cotton model for semi-arid climate of Vehari-Punjab, Pakistan. *Environmental Science and Pollution Research, 24*(6), 5811–5823.

Amin, A., et al. (2017c). Comparison of future and base precipitation anomalies by SimCLIM statistical projection through ensemble approach in Pakistan. *Atmospheric Research, 294*, 214–225.

Amin, A., et al. (2018a). Evaluation and analysis of temperature for historical (1996-2015) and projected (2030-2060) climates in Pakistan using SimCLIM climate model: Ensemble application. *Atmospheric Research, 213*, 422–436.

Amin, A., et al. (2018b). Regional climate assessment of precipitation and temperature in southern Punjab (Pakistan) using SimCLIM climate model for different temporal scales. *Theoretical and Applied Climatology, 131*(1–2), 121–131.

Amin, A., et al. (2018c). Simulated CSM-CROPGRO-cotton yield under projected future climate by SimCLIM for southern Punjab, Pakistan. *Agricultural Systems, 167*, 213–222.

Awais, M., et al. (2017a). Nitrogen and plant population change radiation capture and utilization capacity of sunflower in semi-arid environment. *Environmental Science and Pollution Research, 24*(21), 17511–17525.

Awais, M., et al. (2017b). Nitrogen and plant population change radiation capture and utilization capacity of sunflower arid environment. *Environmental Science and Pollution Research, 24*(21), 17511–17525.

Awais, M., et al. (2018a). Potential impacts of climate change and adaptation strategies for sunflower in Pakistan. *Environmental Science and Pollution Research, 25*(14), 13719–13730.

Awais, M., et al. (2018b). Potential impacts of climate change climate change impact assessment and adaptation strategies for sunflower in Pakistan. *Environmental Science and Pollution Research, 25*(14), 13719–13730.

Blum, A. (2012). *Plant breeding for water-limited environments*. Springer.

Cesari de Maria, S., Bischetti, G. B., Chiaradia, E. A., Facchi, A., Miniotti, E. F., Rienzner, M., Romani, M., Tenni, D., & Gandolfi, C. (2017). The role of water management and environmental factors on field irrigation requirements and water productivity of rice. *Irrigation Science, 35*(1), 11–26. https://doi.org/10.1007/s00271-016-0519-3

Chawla, M. S., et al. (2018). Air pollution and climate warming on the earth: A review. *Applied Sciences and Business Economics, 5*(1), 1–10.

Christoforidou, M., Borghuis, G., Seijger, C., van Halsema, G. E., & Hellegers, P. (2023). Food security under water scarcity: A comparative analysis of Egypt and Jordan. *Food Security, 15*(1), 171–185. https://doi.org/10.1007/s12571-022-01310-y

Dalin, C., Konar, M., Hanasaki, N., Rinaldo, A., & Rodriguez-Iturbe, I. (2012). Evolution of the global virtual water trade network. *Proceedings of the National Academy of Sciences, 109*(16), 5989–5994.

Dalin, C., Wada, Y., Kastner, T., & Puma, M. J. (2017). Groundwater depletion embedded in international food trade. *Nature, 543*(7647), 700–704. https://doi.org/10.1038/nature21403

Dang, Q., Lin, X., & Konar, M. (2015). Agricultural virtual water flows within the United States. *Water Resources Research, 51*(2), 973–986.

Davis, K. F., Gephart, J. A., Emery, K. A., Leach, A. M., Galloway, J. N., & D'Odorico, P. (2016). Meeting future food demand with current agricultural resources. *Global Environmental Change, 39*, 125–132. https://doi.org/10.1016/j.gloenvcha.2016.05.004

Davis, K. F., Rulli, M. C., Seveso, A., & D'Odorico, P. (2017). Increased food production and reduced water use through optimized crop distribution. *Nature Geoscience, 10*(12), 919–924.

Djehdian, L. A., Chini, C. M., Marston, L., Konar, M., & Stillwell, A. S. (2019). Exposure of urban food–energy–water (FEW) systems to water scarcity. *Sustainable Cities and Society, 50*, 101621. https://doi.org/10.1016/j.scs.2019.101621

Droppers, B., Supit, I., Van Vliet, M. T., & Ludwig, F. (2021). Worldwide water constraints on attainable irrigated production for major crops. *Environmental Research Letters, 16*(5), 055016.

Du, T., Kang, S., Zhang, J., & Davies, W. J. (2015). Deficit irrigation and sustainable water-resource strategies in agriculture for China's food security. *Journal of Experimental Botany, 66*(8), 2253–2269. https://doi.org/10.1093/jxb/erv034

Eliazer Nelson, A. R. L., Ravichandran, K., & Antony, U. (2019). The impact of the Green revolution on indigenous crops of India. *Journal of Ethnic Foods, 6*(1), 8. https://doi.org/10.1186/s42779-019-0011-9

Emile, R., Clammer, J. R., Jayaswal, P., & Sharma, P. (2022). Addressing water scarcity in developing country contexts: A socio-cultural approach. *Humanities and Social Sciences Communications, 9*(1), 144. https://doi.org/10.1057/s41599-022-01140-5

Fahad, S., et al. (2017). Crop production under drought and heat stress: Plant responses and management options. *Frontier in Plant Sciences, 08*, 1147.

Fahad, S., et al. (2018). Consequences of high temperature under changing climate optima for rice pollen characteristics concept and perspectives. *Archives of Agronomy and Soil Science, 64*(11), 1473–1488.

FAO. (2017). *The state of food and agriculture 2017. Leveraging food systems for inclusive rural transformation.* Food and Agriculture Organization (FAO). Retrieved from https://www.fao.org/family-farming/detail/en/c/1043688/.

Farid, H. U., et al. (2017). Site-specific aquifer characterization and identification of groundwater potential areas in district Rahim Yar Khan, Pakistan. *Polish Journal of Environmental Studies, 26*(1), 17–27.

Farid, H. U., et al. (2019a). Estimation of infiltration models parameters and their comparison to simulate the site-specific soil infiltration characteristics. *International Journal of Agricultural and Biological Engineering, 12*(3), 84–91.

Farid, H. U., et al. (2019b). Identification of influencing factors for optimal adoptability of high efficiency irrigation system (HEIS) in Punjab, Pakistan. *Sarhad Journal of Agriculture, 35*(2), 539–549.

Farooq, M., Rehman, A., Li, X., & Siddique, K. H. (2023). Neglected and underutilized crops and global food security. In *Neglected and underutilized crops* (pp. 3–19). Elsevier.

Fito, J., Tefera, N., Demeku, S., & Kloos, H. (2017). Water footprint as an emerging environmental tool for assessing sustainable water use of the bioethanol distillery at Metahara sugarcane farm, Oromiya region. *Ethiopia. Water Conservation Science and Engineering, 2*(4), 165–176. https://doi.org/10.1007/s41101-017-0038-y

Graham, N. T., Hejazi, M. I., Kim, S. H., Davies, E. G. R., Edmonds, J. A., & Miralles-Wilhelm, F. (2020). Future changes in the trading of virtual water. *Nature Communications, 11*(1), 3632. https://doi.org/10.1038/s41467-020-17400-4

Hammad, H. M., et al. (2017). Maize plant nitrogen uptake dynamics at limited irrigation water and nitrogen. *Environmental Science & Pollution Research, 24*(3), 2549–2557.

Hammad, H. M., et al. (2018a). Offsetting land degradation through nitrogen and water management during maize cultivation under arid conditions. *Land Degradation and Development, 2018*, 1366–1375.

Hammad, H. M., et al. (2018b). Offsetting land degradation through nitrogen and water management during maize cultivation under arid conditions. *Land Degradation and Development, 29*(5), 1366–1375.

Hanjra, M. A., & Qureshi, M. E. (2010). Global water crisis and future food security in an era of climate change. *Food Policy, 35*(5), 365–377. https://doi.org/10.1016/j.foodpol.2010.05.006

Harris, F., Dalin, C., Cuevas, S., Lakshmikantha, N. R., Adhya, T., Joy, E. J., Scheelbeek, P. F., Kayatz, B., Nicholas, O., & Shankar, B. (2020). Trading water: Virtual water flows through interstate cereal trade in India. *Environmental Research Letters, 15*(12), 125005.

Hashmi, M. Z., et al. (2018). PCB118-induced cell proliferation mediated by oxidative stress and MAPK signaling pathway in HELF cells. *Dose-Response, 2018*(16), 1559325817751525.

Hekmatnia, M., Ardakani, A. F., Isanezhad, A., & Monibi, H. (2024). A novel classification of virtual water trade for the sustainability of global freshwater resources. *Environment, Development and Sustainability, 26*(3), 7377–7408. https://doi.org/10.1007/s10668-023-03012-7

Hoekstra, A. Y., & Mekonnen, M. M. (2012). The water footprint of humanity. *Proceedings of the National Academy of Sciences of the United States of America, 109*(9), 3232–3237. https://doi.org/10.1073/pnas.1109936109

Ingrao, C., Strippoli, R., Lagioia, G., & Huisingh, D. (2023). Water scarcity in agriculture: An overview of causes, impacts and approaches for reducing the risks. *Heliyon, 9*(8), e18507. https://doi.org/10.1016/j.heliyon.2023.e18507

Jabran, K., et al. (2017). Water-saving technologies affect the grain characteristics and recovery of fine-grain rice cultivars. *Environmental Science and Pollution Research, 24*, 12971–12981.

Javed, M. A., & Ahmad, S. R. (2020). A decision support system for crop water requirement estimation using advanced geospatial techniques. *Pakistan Journal of Agricultural Sciences, 57*(4), 981–991.

Joseph, J., Tramberend, S., Luna Gonzalez, D., Fischer, G., & Kahil, T. (2023). *Modeling sustainable intensification of agriculture under resource constraints*. EGU General Assembly Conference Abstracts.

Khan, N., et al. (2018). Interaction between PGPR and PGR for water conservation and plant growth under drought condition. *Biologia, 73*(11), 1083–1098.

Koebele, E. A., & Simpson, K. (2023). 4. The impacts of water scarcity on the security of democracies. In *Handbook on democracy and security* (pp. 45–63). Edward Elgar Publishing.

Kuchimanchi, B. R., Ripoll-Bosch, R., Steenstra, F. A., Thomas, R., & Oosting, S. J. (2023). The impact of intensive farming systems on groundwater availability in dryland environments: A watershed level study from Telangana, India. *Current Research in Environmental Sustainability, 5*, 100198. https://doi.org/10.1016/j.crsust.2022.100198

Lenzen, M., Moran, D., Bhaduri, A., Kanemoto, K., Bekchanov, M., Geschke, A., & Foran, B. (2013). International trade of scarce water. *Ecological Economics, 94*, 78–85. https://doi.org/10.1016/j.ecolecon.2013.06.018

Li, M., Singh, V. P., Fu, Q., Liu, D., Li, T., & Zhou, Y. (2021). Optimization of agricultural water–food–energy nexus in a random environment: An integrated modelling approach. *Stochastic Environmental Research and Risk Assessment, 35*(1), 3–19. https://doi.org/10.1007/s00477-019-01672-4

Ma, T., Sun, S., Fu, G., Hall, J. W., Ni, Y., He, L., Yi, J., Zhao, N., Du, Y., & Pei, T. (2020). Pollution exacerbates China's water scarcity and its regional inequality. *Nature Communications, 11*(1), 650.

Martinelli, L. A., Filoso, S., de Barros Aranha, C., Ferraz, S. F., Andrade, T. M., Ravagnani, E. d. C., Della Coletta, L., & de Camargo, P. B. (2013). *Water use in sugar and ethanol industry in the state of São Paulo (Southeast Brazil)*.

Mechiche-Alami, A., Yagoubi, J., & Nicholas, K. A. (2021). Agricultural land acquisitions unlikely to address the food security needs of African countries. *World Development, 141*, 105384. https://doi.org/10.1016/j.worlddev.2020.105384

Mirza, N., et al. (2017). Constitutional tolerance and chlorophyll fluorescence of Boehmeria nivea L in response to the antimony (Sb) and arsenic (As) co-contamination. *Toxicological and Environmental Chemistry, 99*(2), 265–272.

Muhammad, M., Waheed, A., Wahab, A., Majeed, M., Nazim, M., Liu, Y.-H., Li, L., & Li, W.-J. (2024). Soil salinity and drought tolerance: An evaluation of plant growth, productivity, microbial diversity, and amelioration strategies. *Plant Stress, 11*, 100319. https://doi.org/10.1016/j.stress.2023.100319

Munaweera, T. I. K., Jayawardana, N. U., Rajaratnam, R., & Dissanayake, N. (2022). Modern plant biotechnology as a strategy in addressing climate change and attaining food security. *Agriculture & Food Security, 11*(1), 26. https://doi.org/10.1186/s40066-022-00369-2

Muzammil, M., Zahid, A., & Breuer, L. (2020). Water resources management strategies for irrigated agriculture in the Indus Basin of Pakistan. *Water, 12*(5), 1429. https://www.mdpi.com/2073-4441/12/5/1429.

Nasim, W., et al. (2016). Correlation studies on nitrogen for sunflower crop across the agroclimatic variability. *Environmental Science and Pollution Research, 23*, 3658–3670.

Nasim, W., et al. (2018). Future risk assessment by estimating historical heat wave trends with projected heat accumulation using SimCLIM climate model in Pakistan. *Atmospheric Research, 205*(2018), 118–133.

Nesar, A. B., Mahmud, T., & Hossain, F. (2022). Simulating the behaviour and displacement of women in water-stressed areas. In *2022 IEEE world AI IoT congress (AIIoT).*

Obaideen, K., Yousef, B. A. A., AlMallahi, M. N., Tan, Y. C., Mahmoud, M., Jaber, H., & Ramadan, M. (2022). An overview of smart irrigation systems using IoT. *Energy Nexus, 7*, 100124. https://doi.org/10.1016/j.nexus.2022.100124

Ouda, S., El-Latif, K. A., & Khalil, F. (2016). Water requirements for major crops. In S. Ouda (Ed.), *Major crops and water scarcity in Egypt: Irrigation water management under changing climate* (pp. 25–32). Springer International Publishing. https://doi.org/10.1007/978-3-319-21771-0_2

Padder, F. A., & Bashir, A. (2023). Scarcity of water in the twenty-first century: Problems and potential remedies. *MEDALION JOURNAL: Medical Research, Nursing, Health and Midwife Participation, 4*(1), 1–5.

Porkka, M. (2011). The role of virtual water trade in physical water scarcity: Case Central Asia.

Rai, R. K., Singh, V. P., & Upadhyay, A. (2017). Chapter 6—Estimating irrigation design parameters. In R. K. Rai, V. P. Singh, & A. Upadhyay (Eds.), *Planning and evaluation of irrigation projects* (pp. 243–282). Academic Press. https://doi.org/10.1016/B978-0-12-811748-4.00006-6

Rashid, A., Ryan, J., & Chaudhry, M. A. (2004). Challenges and strategies for dryland agriculture in Pakistan. In *Challenges and strategies of dryland agriculture* (pp. 359–371). https://doi.org/10.2135/cssaspecpub32.c22.

Rasool, A., et al. (2018). A review of global outlook on fluoride contamination in groundwater with prominence on the Pakistan current situation. *Environmental Geochemistry and Health, 40*(4), 1265–1281.

Raza, M. A., et al. (2021). Land productivity and water use efficiency of maize soybean strip intercropping systems in semi-arid areas. *Journal of Cleaner Production, 308*, 127282.

Razzaq, A., Qing, P., Naseer, M. A. u. R., Abid, M., Anwar, M., & Javed, I. (2019). Can the informal groundwater markets improve water use efficiency and equity? Evidence from a semi-arid region of Pakistan. *Science of the Total Environment, 666*, 849–857. https://doi.org/10.1016/j.scitotenv.2019.02.266

Razzaq, A., Xiao, M., Zhou, Y., Liu, H., Abbas, A., Liang, W., & Naseer, M. A. u. R. (2022). Impact of participation in groundwater market on farmland, income, and water access: Evidence from Pakistan. *Water, 14*(12), 1832. https://www.mdpi.com/2073-4441/14/12/1832.

Rhue, S. J., Torrico, G., Amuzie, C., Collins, S. M., Lemaitre, A., Workman, C. L., Rosinger, A. Y., Pearson, A. L., Piperata, B. A., & Wutich, A. (2023). The effects of household water insecurity on child health and well-being. *Wiley Interdisciplinary Reviews: Water, 10*(6), e1666.

Richter, B. D., Ao, Y., Lamsal, G., Wei, D., Amaya, M., Marston, L., & Davis, K. F. (2023). Alleviating water scarcity by optimizing crop mixes. *Nature Water, 1*(12), 1035–1047. https://doi.org/10.1038/s44221-023-00155-9

Sabagh, A. E. L., et al. (2019). Drought and salinity stresses in barley: Consequences and mitigation strategies. *Australian Journal of Crop Science, 13*(06), 810–820.

Sarpong, K. A., Xu, W., Mensah-Akoto, J., Neequaye, J. K., Dadzie, A. A., & Frimpong, O. (2020). Waterscape, state and situation of China's water resources. *Journal of Geoscience and Environment Protection, 8*(10), 26. https://doi.org/10.4236/gep.2020.810003

Saud, S., & e al. (2017). Effects of nitrogen supply on water stress and recovery mechanisms in Kentucky bluegrass plants. *Frontier in Plant Sciences, 8*, 983.

Shahab, H., Iqbal, M., Sohaib, A., Ullah Khan, F., & Waqas, M. (2024). IoT-based agriculture management techniques for sustainable farming: A comprehensive review. *Computers and Electronics in Agriculture, 220*, 108851. https://doi.org/10.1016/j.compag.2024.108851

Shakeel, M., et al. (2017). Environment polluting conventional chemical control compared to an environmentally friendly IPM approach for control of diamondback moth, Plutella xylostella (L), in China: A review. *Environmental Science and Pollution Research, 24*, 14537–14550.

Sharma, C. P. (2016). *Overdraft in India's water banks: Studying the effect of production of water intensive crops on groundwater depletion*. Georgetown University.

Shemer, H., Wald, S., & Semiat, R. (2023). Challenges and solutions for global water scarcity. *Membranes, 13*(6), 612.

Shrestha, J., & Subedi, S. (2019). Improving crop productivity through sustainable intensification. *South Asian Research Journal of Agriculture and Fisheries, 1*(1), 8–11.

Tariq, M., et al. (2008). The impact of climate warming and crop management on phenology of sunflower-based cropping systems in Punjab, Pakistan. *Agricultural and Forest Meteorology, 256–257*(2018), 270–282.

Ullah, I., Zeng, X.-M., Hina, S., Syed, S., Ma, X., Iyakaremye, V., Yin, J., & Singh, V. P. (2023). Recent and projected changes in water scarcity and unprecedented drought events over southern Pakistan. *Frontiers in Earth Science, 11*, 1113554.

Vallino, E., Ridolfi, L., & Laio, F. (2021). Trade of economically and physically scarce virtual water in the global food network. *Scientific Reports, 11*(1), 22806.

Vörösmarty, C. J., McIntyre, P. B., Gessner, M. O., Dudgeon, D., Prusevich, A., Green, P., Glidden, S., Bunn, S. E., Sullivan, C. A., Liermann, C. R., & Davies, P. M. (2010). Global threats to human water security and river biodiversity. *Nature, 467*(7315), 555–561. https://doi.org/10.1038/nature09440

Wahab, A. A., et al. (2024). Evaluating the interactive impact of nitrogen levels and cultivars on yield traits, nitrogen use efficiency, water use efficiency and benefit cost ratio of ratoon rice in Punjab, Pakistan. *International Journal of Plant Production, 18*, 97–107.

Wheeler, S. A., Loch, A., Crase, L., Young, M., & Grafton, R. Q. (2017). Developing a water market readiness assessment framework. *Journal of Hydrology, 552*, 807–820. https://doi.org/10.1016/j.jhydrol.2017.07.010

Wutich, A., Beresford, M., Montoya, T., Radonic, L., & Workman, C. (2022). Water security and scarcity. In *Oxford research encyclopedia of anthropology*. https://doi.org/10.1093/acrefore/9780190854584.013.475.

Xing, L., & Chen, W. (2023). Structural characteristics and evolutionary drivers of global virtual water trade networks: A stochastic actor-oriented model for 2000–2015. *International Journal of Environmental Research and Public Health, 20*(4), 3234.

Yasmeen, R., Hao, G., Ye, Y., Shah, W. U. H., & Tang, C. (2023). The synergy of water resource agglomeration and innovative conservation technologies on provincial and regional water usage efficiency in China: A super SBM-DEA approach. *Water, 15*(19), 3524. https://www.mdpi.com/2073-4441/15/19/3524.

Yu, H.-L. (2016). Water, climate, and health. *Stochastic Environmental Research and Risk Assessment, 30*(8), 2041–2042. https://doi.org/10.1007/s00477-016-1359-7

Zahra, S. M., Shahid, M. A., Misaal, M. A., Zaman, M., Imran, M., Azam, S., & Hussain, F. (2023). Sustainable water Management in Indus Basin and Vulnerability due to climate change. *Environmental Sciences Proceedings, 25*(1), 36.

Zeshan, M., & Shakeel, M. (2021). Water crisis in Pakistan: A dynamic CGE-water model (preprint). *Research Square*. https://doi.org/10.21203/rs.3.rs-337680/v1.

Zhang, Y., Zhang, J. H., & Tian, Q. (2021). Virtual water trade in the service sector: China's inbound tourism as a case study. *International Journal of Environmental Research and Public Health, 18*(4). https://doi.org/10.3390/ijerph18041769

Zheng, J., Wu, S., Li, L., Li, S., Li, Q., & An, Q. (2022). Revealing virtual water transfers and imbalanced economic benefits hidden in China's interprovincial trade. *Water, 14*(11), 1677.

Zia, Z., et al. (2017). Effect of water management and silicon on germination, growth, phosphorus and arsenic uptake in rice. *Ecotoxicology and Environmental Safety, 144*, 11–18.

Risks of Wastewater Use in Agriculture

Sobia Parveen, Mazhar Ali, Muhammad Mubeen, Maham Fatima, Muhammad Razaq, Wajid Nasim Jatoi, and Ghulam Mustafa Shah

Abstract The effluent use in agriculture offers numerous benefits, but it also presents significant dangers to public health, particularly when untreated effluent is utilized during crop irrigation. However, in some cases, farmers frequently have no choice but to use untreated sewage as there is no treatment for wastewater and freshwater is either insufficient or prohibitively expensive. Chemical and microbial hazards provide the most serious threats to public health. Wastewater use in farming can potentially pose environmental problems, such as groundwater and soil degradation. However, once effectively planned, managed, and implemented, sewage irrigation can provide several environmental, agricultural, and water resource benefits. Furthermore, given these benefits and risks, many countries trying to improve agricultural wastewater usage, which can minimize risks, particularly those to public health. Additionally different physical, chemical, and biological approaches are used to improve the risks of wastewater reuse in agriculture.

Keywords Wastewater · Benefits · Risks · Physical · chemical · and biological approaches · Wastewater treatment process

---

S. Parveen (✉) · M. Ali · M. Mubeen
Department of Biotechnology, COMSATS University Islamabad, Vehari Campus, Vehari, Pakistan

M. Fatima
Department of Environmental and Polymer Engineering, Inha University, Incheon, South Korea

M. Razaq
Department of Entomology, Faculty of Agricultural Sciences and Technology, Bahauddin Zakariya University, Multan, Pakistan

W. N. Jatoi
Department of Agronomy, Faculty of Agriculture and Environment, The Islamia University of Bahawalpur (IUB), Bahawalpur, Punjab, Pakistan

G. M. Shah
College of Engineering (Key Laboratory for Clean Renewable Energy Utilization Technology, Ministry of Agriculture), China Agricultural University, Beijing, People's Republic of China

Department of Environmental Sciences, COMSATS University Islamabad, Vehari, Pakistan

M. Mubeen et al. (eds.), *Innovations in Agricultural Water Management*,
https://doi.org/10.1007/978-3-031-91883-4_9

1 Introduction

The fast expansion of the human population has put an enormous strain on the world's limited freshwater supplies. Freshwater is an essential element of life, and its demand in numerous areas has steadily increased in the previous decades. The Food and Agricultural Organization reported an almost 6.3-fold rise in annual water withdrawal from 600 km^3 $year^{-1}$ at the turn of the century to over 3800 km^3 $year^{-1}$ at the start of the twenty-first century (Boliko, 2019). Increased freshwater withdrawal, which contributes to worldwide freshwater crises, produces massive amounts of wastewater. However, as the globe's population grows, the disparity between availability and demand for water widens and reaches frightening proportions in certain regions of the world, posing a danger to the survival of humans. Global scientists are working on innovative strategies for water conservation. However, one promising method is the recycling of wastewater from cities for different agricultural objectives (Jhansi & Mishra, 2013). Overall, wastewater refers to waste fluids generated by households, commercial sources, and industries due to daily consumption and production actions. Municipal treatment facilities are designed to handle untreated wastewater, resulting in a liquid waste of acceptable quality that can released into the natural waterways with minimal impact on environment and health. Wastewater disposal is a big issue for communities, especially in large urban regions with little acreage for treatment and disposal. However, in developing and developed countries, most typical approach is to utilize municipal waste (treated and untreated) for land. Furthermore, in developed nations with stringent environmental regulations, most of the effluent is cleaned before being utilized for seed, fiber, and feed crop irrigation. Other significant usage of wastewater consists of groundwater recharging, landscaping (playgrounds, golf courses, freeways, parks, and schoolyards), construction, dust control, improvement of wildlife habitat, and aquaculture (Feng et al., 2024; Hussain et al., 2023a, 2023b, 2023c, 2024; Hammad et al., 2023; Abdullah et al., 2023; Fito & Van Hulle, 2021).

Untreated wastewater is commonly utilized for aquaculture and agriculture, and this practice is common in nations like India, Mexico, and China. However, inefficient or inadequate treatment services, combined with the costly treatment facilities for wastewater in developing nations, resulted in the releasing untreated liquid waste into the surrounding environment and other water bodies such as streams, rivers, and lakes (Bunting & Edwards, 2018).

Irrigation agriculture consumes 90% of available freshwater in developing nations, leading to a push for wastewater reuse (Samian et al., 2015). The reuse of wastewater is a proven method for cultivating field crops and vegetables in dry places with insufficient resources for freshwater (Jaramillo & Restrepo, 2017). Thus, wastewater can be seen as a resource and a challenge. The nitrogen content in wastewater can be widely utilized for irrigation and other ecological services, benefiting the farming towns, community, and society. Additionally, reusing wastewater has negative consequences for individuals and natural systems, which must be discovered and evaluated (Wijaya & Soedjono, 2018; Hammad et al., 2010a, 2010b; Nasim et al., 2011, 2012; Nasim & Bano, 2012).

Wastewater use in agriculture is far more frequent than many people realize. Currently, about 20 million ha of fertile land are reported to be irrigated with wastewater, and actual unreported use is much greater. This practice is widespread in developing-world cities and peri-urban areas, where low financial resources and institutional capacities present challenges in maintaining and establishing wastewater collection and treatment infrastructure. Wastewater usage in agriculture has several advantages, supplying nutrients and water for agricultural cultivation, supplying towns with food, and lowering demand for limited freshwater sources. However, wastewater is a cause of pollution that can harm the health of users, customers, and the environment if proper precautions are not taken. Soil health and water scarcity have an impact on many human activities, such as food production, agricultural irrigation, ecosystem preservation, and drinking water supply (Shahriar et al., 2021). However, water stress affects 20% of the Mediterranean population, with over 50% experiencing it during summer (Farid et al., 2023; Haseeb et al., 2023; Alsuhaibani et al., 2023; Yang et al., 2023; Muhammad et al., 2023; De Roo et al., 2021). Furthermore, agriculture requires 70% of freshwater (Meffe et al., 2021). An estimated 20 million ha of agricultural land are irrigated worldwide using non-conventional water supplies (Rusănescu et al., 2022). In approximately 3000 B.C., the concept of reusing wastewater for irrigation first emerged, and while there are no specific global statistics reuse of wastewater, it is estimated that a minimum of 7% of the land under irrigation utilizes liquid effluent (Wajid et al., 2010; Usman et al., 2010; Ahmad et al., 2009; Usman et al., 2009; Wajid et al., 2007). However, it is important to consider that wastewater contains heavy metals, organic pollutants, and salts that can contaminate soil, crops, and groundwater. Furthermore, wastewater reuse in agriculture may pose dangers to the environment and human health (Verma et al., 2023). The possible growth of bacteria resistant to popular antibiotics, which can withstand their inhibitory effects, is another adverse consequence of reusing wastewater (Bougnom & Piddock, 2017).

2 Characteristics of Wastewater

2.1 Sources of Wastewater

Municipal liquid waste mainly consists of stormwater, industrial wastewater, and groundwater seepage entering the sewage system. However, effluent discharged from residential, business, and institutional buildings is called domestic wastewater. The effluent released by manufacturing and food processing facilities is known as industrial wastewater. In Faisalabad, industrial wastewater discharges consist of municipal wastewater in certain areas. Unlike in some modern cities with separate systems, the network of municipal sewage also provides a stormwater (Khan et al.,

2022a, 2022b). Additionally, seepage of groundwater occurs due to sewerage system issues, leading to an increase in the sewage volume that needs to be disposed of. Domestic wastewater that enters municipal wastewater systems follows a daily cycle. This flow is modest early in the morning, with a first peak in the late morning and a second high in the evening, after dinner. Nonetheless, the ratio of average flow to peak flow is expected to fluctuate inversely with the community's size and the length of the sewage system. In general, peak flows may also occur on festive occasions, business hours, religious rites, tourist seasons, and in university campuses, among other things. Industrial runoff discharges closely abide by the processing guidelines of local industries, which depend on the number of shifts and process involved, and the industry's water consumption (Gonzalez-Serrano et al., 2005).

Modifications from defined schedules are limited and only occur during shift transitions. Seasonal product processing might also cause flow changes. As a result, seasonal changes in industrial wastewater outputs become more important. In urban areas where wastewater from municipalities accounts for a large portion of total effluents, variations in industrial wastewater discharge are likely critical for water cycle administration (Rolls et al., 2012).

3 Composition of Wastewater

However, specifically makeup of wastewater may vary per community, all of the municipal wastewater includes the following groups of constituents.

- Organic and inorganic matter
- Nutrients (potassium, phosphorous, and nitrogen)
- Toxic chemicals and pathogens (Table 1)

4 Use of Wastewater in Agriculture

According to the Berlin/Rome (19 January 2017) report, while increasing water constraints and food demand, now is the moment to stop considering wastewater as trash and start managing it as a valuable asset for agriculture. However, liquid waste can be safely used to enhance crop productivity, either indirectly by replenishing aquifers or directly through irrigation, because this requires careful control of health concerns through suitable treatment or utilization. Furthermore, untreated waste frequently contains bacteria and pathogens, antibiotic residues, chemical pollutants, and other hazards that put food chain employees' health at risk. Numerous technologies and methodologies are employed globally to treat, utilize, and manage agricultural waste, many of which are particular to the base of natural resources like crops, and farming systems (Obi et al., 2016; Rasool et al., 2016; Alghabari et al., 2016; Fahad et al., 2016a, 2016b, 2016c, 2016d).

Table 1 Impact of different pollutants/components in agriculture

Components	Standards	Effect
Food nutrients	K, N, P	1. Excess nitrogen cause economic loss, delayed growing maturity and season 2. Increased amount of P, and N causes eutrophication
Pathogens	Helminth eggs, viruses, and bacteria, etc.	Communicable diseases caused by these components
Degradable organics	COD, BOD	1. These components causes decrease in oxygen-dissolved content 2. Adverse effect on environment and increase humus build-up
Submerged inorganic compounds	Cl, B, Na, Mg, Ca	These cause salinity, and phytotoxicity
Concentration of hydrogen ion	pH	1. Related to industrial wastewater 2. Negative effect on the growth of plants due to alkalinity and acidity 3. Positive impact on soil fauna and flora
Heavy metals	Cd, Zn, Ni, As, Pb, Hg	1. Build-up in aquatic animals 2. Toxic towards animals, and plants 3. May render wastewater inappropriate for irrigation

In Egypt, where water resources are few and wastewater is often highly contaminated, built wetlands are emerging as a feasible and economically viable treatment option. Similarly, wastewater in Tunisia is frequently employed in agroforestry system, which aids both production of wood and anti-desertification effect (Abdel-Shafy & Aly, 2002). In Central Mexico, farmers have used municipal wastewater for crop irrigation. In the past, ecological mechanisms helped to lower health concerns. Additionally, some authorities have developed crop limitations, permitting few crops to be safely raised using sewage, and they have developed waste treatment opportunities to improve the overall process (Valdes Ramos et al., 2019; Shahid et al., 2014; Masood et al., 2013; Nasim et al., 2012).

In Jordan, 25% of total water utilization is from recycled sources. However, treatment and regulated aquifer recharge are typical practices in the United States, particularly in the Western states. According to De Silva, infrastructure and management systems and infrastructure for reusing, treating, and reclaiming wastewater can create jobs in addition to addressing the issue of water shortages, minimizing environmental contamination, and assisting with agricultural production (Da Silva et al., 2020).

4.1 *Categories and Types of Wastewater Use*

The wastewater from diverse urban activities, ranging in quality from raw to diluted, is used for agricultural irrigation and comes from diverse sources (Qadir et al., 2010). The prevalent varieties are as follows:

1. **Urban wastewater** is a mixture of one or more of the following categories:
   - Residential effluent, which is composed of gray water (cooking and bathing wastewater) and black water (urine, excreta, and associated sludge).
   - Sewage from businesses and organizations, such as medical facilities.
   - Industrial and other urban wastewater.
2. **Treated wastewater**: Treated effluent has been purified by a wastewater treatment facility and underwent one or more chemical, biological, and physical processes to minimize contaminants by hazardous compounds (Chaturvedi, 2022).
3. **Recycled or reclaimed water**: Water is purified wastewater that can be legally used under restricted conditions for beneficial uses like irrigation.
4. **Gray water**: It is very appropriate for reuse. It is generated by residences not linked to a drainage system and can be processed and used to irrigate trees and home gardens. Gray water is a vital element of water conservation. It accounts for 50–80% of domestic wastewater and holds considerable promise as a cost-effective and resource-saving part of integrated water source management in arid regions (Matos et al., 2012).

As with the many types of wastewater mentioned above, there are also diverse methods for utilizing sewage/wastewater:

- Direct utilization of untreated sewage: Untreated effluent from a sewage outlet can be disposed of directly on land for cultivation.
- Direct use of treated wastewater: It happens when wastewater is treated before being utilized for irrigation, agriculture, and recycling.
- Indirect usage of untreated or treated urban wastewater happens when farmers use that water from a river for agriculture. However, that happens when cities do not have a robust sewage collecting network and drainage systems that release stored sewage (wastewater) into rivers.

4.2 Benefits of Wastewater Usage

Despite its high level of use, wastewater's potential as a resource is sometimes overestimated. However, instead of being a source of difficulties, properly managed wastewater shows a positive impact on the environment, economy, and society for social fairness and increasing food security. Wastewater can provide essential nutrients for plants. However, by utilizing these minerals and fertilizers, the need for artificial fertilizers can be decreased. Furthermore, this effort not only reduces the environmental impact of production and use of artificial fertilizers, but also boosts farmers' incomes. Farmers benefit from faster growth cycles, higher yield, and productivity while decreasing their need for extra water sources and artificial fertilizers (Corcoran et al., 2010; Gillani et al., 2017; Jabran et al., 2017; Ahmad et al., 2017; Nasim et al., 2017). Additionally, wastewater is valuable because of its increased availability, making it an effective resource for farmers in urban areas where alternative water sources are restricted. However, improving nutrition and food supply in dry and food-insecure areas may also have considerable health benefits. Additionally, there is no systematic global evaluation of the beneficial health effects of wastewater application in agriculture, and the balance of advantages and dangers may vary greatly depending on the context. For instance, subsistence farmers, who could benefit greatly from better food nutrition and security, are also at the greatest risk due to negative health consequences, particularly when untreated wastewater is utilized for irrigation. In contrast, in areas where alternative water resources are least, treatment quality is excellent, and farming and food processing techniques are advanced, the possible benefits are expected to outweigh the hazards (Raschid-Sally et al., 2005; Mahmood et al., 2017; Awais et al., 2017; Nasim et al., 2016a, 2016b; Mubeen et al., 2016).

Additionally, there are many cases where farmers have no choice but to use low-quality water resources (for example, in areas where reliable water sources are scarce and municipal wastewater discharge into the environment damages water bodies) or where the farmers are not aware that they are consuming wastewater (for example, when farmers live downstream of urban areas where wastewater released into open water). However, planned wastewater usage for irrigation is becoming a growing resource due to the potential benefits, particularly in peri-urban and urban agriculture. This trend is driving its use in developed and developing nations, especially in water-scarce places, where alternative resources are insufficient (Buechler et al., 2005).

5 Risks of Wastewater Utilization in Agriculture

The usage of wastewater creates both obstacles and possibilities. Its utilization can benefit farmers, society, and towns by properly utilizing its nutrient content in productive activities and agriculture (Hettiarachchi et al., 2018). Wastewater has a significant impact on crop production by raising output (Ofori et al., 2021), accelerating the development of microorganisms, optimizing soil performance, and improving its physical makeup.

The European Council and Parliaments are promoting the efficient utilization of effluents to conserve water supplies and establish a sustainable economy. However, it also aims to safeguard human health by ensuring proper collection, disposal, and treatment of industrial and urban wastewater. The fact that sewage muck is a nutrient source; however, it can be converted to biogas for commercial purposes (Elalami et al., 2019).

The amount of nitrogen delivered by wastewater boost fertility of soil and are potentially suitable for plants (lettuce, tomato, and eggplants) with short cropping cycles (Cirelli et al., 2012). However, lettuce fed with purified waste exhibited larger macronutrients (Mg, K, Ca, P, S, and N), larger fresh weights (during first cycle and second cycle are +50% and +100%), and larger surface area of roots than treated with drinking water (filtered). Furthermore, tomato plant supplied with treated sewage produced greater quantity (20%) than crops sustained with freshwater, increasing the amount of commercially viable products (Cirelli et al., 2012). While reusing water has several advantages, scientists have found several environmental and human health risks and difficulties. Larger nutrients in wastewater can benefit both plant growth and herbivorous insects. However, it is unclear whether vigorous plants can offset the adverse impact on pest population. So, pesticides are expected to be used more frequently in this scenario.

Urban sewage treatment plants often release effluents holding a variety of organic compounds, microbiological pathogens, and hazardous metals (Salgot et al., 2006). Such effluents pose several possible risks due to the existence of soil pollutants. In this regard, apparently there are serious worries concerning the possible threats toward human health, such as poisoning of drinking water, and the buildup of hazardous compounds in edible goods.

5.1 Crops

In general, wastewater (untreated and treated) is widely used in agriculture due to its high nutritional content and ability to provide all of the nutrients required for the growth of crops. Many crops produce greater yields by using wastewater irrigation system, which reduces the requirement for synthetic fertilizers, thus causing overall financial savings for cultivators. However, whole nitrogen content given using wastewater irrigation to the crop exceeds the necessary dosage of nitrogen for maximum yields, can boost vegetative growth while delaying maturity and ripening, while in severe cases, causes yield loss (Chojnacka et al., 2020).

The consumption of untreated local wastewater is common in different nations and raises several issues. Nevertheless, the significant amount of plant-based nutrients encourages farmers to utilize untreated effluents since it lowers fertilizers expenses. Most crops, especially those cultivated in the urban agricultural sector, require particular quantities of NPK to achieve maximum yield. When the suggested dosage of NPK exceeds, crop production and development may suffer. For instance, urea plant discharges are abundant fertilizers in liquid form; however, in concentrated amount, they hurt corn and rice yields (Kesari et al., 2021). Plant health hazards affect nutritional value and lead to lower crop yields, particularly when sewage water utilized for irrigation exhibits greater amounts of boron (B), heavy metals, sodium, nitrogen, and other industrial pollutants. Plant health dangers can decrease by refraining from combining industrial wastewater with local effluent; however, regardless of the case, the five parameters, including N, B, salinity levels, sodium adsorption ratio, and pH, should be continually monitored during the crop period. However, if any parameters cross the acceptable limit, suitable action is taken to fix the problem to prevent adverse impacts on crop food quality and productivity. Several investigations, however, have discovered pharmaceutical waste for non-potable wastewater reuse, including influent and effluent, using a Triple

Quadrupole Mass Spectrometer (TQD-MS) coupled with ESI (electrospray ionization system) (Khan et al., 2022a, 2022b).

5.2 *Public Health*

Parasites, bacteria, and viruses are among the pathogenic microorganisms found in wastewater. Human parasites, such as helminth eggs and protozoa, are effective in this regard because they are the most difficult to treat and are linked to a variety of severe gastrointestinal disorders in developing and developed countries. Additionally, the discharge of untreated sewage water for irrigation undoubtedly holds a significant danger to one's health. However, the danger level varies by age group. Untreated sewage irrigation increases the hookworms (Feenstra et al., 2000) and childhood cases of ascariasis.

Moreover, heavy metallic elements in sewage offer a health concern if consumed in significant proportions and may prove hazardous. As an idea, heavy metal consumption risk and crop uptake might not be a problem because plants can't withstand excessive amounts of contaminants and expire prior to their health risks (Table 2). Saravanan and Mollinga (2011) conducted an extensive investigation on the health impacts of viruses, but there is a lack of thorough investigation that

Table 2 Impact of different risk factors on environment and humans

Risk factors	Impact on agriculture, and humans	Description
Microbes and pathogens	Skin infection, increased diarrheal diseases in children, intestinal parasitic	Contamination of different crops with different bacterial pathogens, viruses, parasites
Heavy metals	Carcinogenic hazards toward crop toxicity and humans	Accumulation of different heavy metals in crops and soils
High level of salinity	Reduced crop production, and soil degradation	High salinization effect on yield of crops
Organic contaminants	Spread of pollutants of rising concern	Organic pollutants in cultivated crops
Different risks and their advantages, and disadvantages in wastewater re-usage		
Risks	**Advantages**	**Disadvantages**
Adverse effect of organic matter, heavy metals, and nitrates on ground water	1. Increasing the economic effectiveness of interventions in wastewater irrigation and disposal 2. Replenishment of aquifers by water seeping (natural solution)	Wastewater is produced constantly; however, sewage irrigation is typically restricted to the period of growth
Negative impact of pathogenic germs on human health	By utilization of wastewater nutrient like phosphate, nitrogen, etc. • Decreased synthetic fertilizers usage • Soil properties like yield and fertility are improved	Some compounds may be found in effluent in such large quantities that are potentially hazardous to plants or cause harm to the environment

assesses the heavy metals impacts and the true threats to human health (Saravanan & Mollinga, 2011).

However, QMRA (Quantitative Microbial Risk Assessment) is a model-based method that predicts the possible human health hazards associated with exposure to various pathogens (e.g., protozoa, pathogenic viruses, and bacteria), particularly through water and food. This structured approach combines data and information with mathematical models to analyze the dissemination and exposure of microbial pathogens, highlighting human health hazards. QMRA is used to estimate infection/ risks for various groups affected by wastewater reuse (Haas et al., 2014).

5.3 Groundwater and Soil Resources

Wastewater treatment has the potential to have a lasting effect on groundwater quality due to excess salts and nutrients seeping below the level of plant roots. However, the real impact is determined by several parameters, which include soil drainage, groundwater quality, depth of water table, and the degree of wastewater irrigation. For example, groundwater quality controls the amount of harshness from nitrate degradation. If the groundwater becomes blackish, then nitrate leaching could be less because the water holds no considerable value for usage (Quist-Jensen et al., 2015).

Many underdeveloped countries rely heavily on groundwater as their primary source of potable water. As a result, the risk of groundwater contamination must be assessed before engaging in a wastewater irrigation system. In addition to nitrate and salt accumulation, wastewater irrigation can introduce harmful viruses and bacteria into groundwater under specific conditions (Hanjra, 2000).

The impact of wastewater on the soil used for agriculture is due to the existence of high levels of nutrients (phosphorus and nitrogen), excessive solids in solution, and other elements like heavy metals, all of which are deposited in the soil over time. Salts in wastewater can be collected in the root zone, posing a risk to crop yield and soil health. The leaking of such salts under the root zone could contaminate the groundwater and soil. Prolonged use of sodium-rich and saline wastewater can degrade soil structure of soil and reduces efficiency. However, this could lead to developing UN-sustainability in the future (Elgallal et al., 2016). The main problem of sodicity and salinity in soil can be resolved by using synthetic or natural soil additives. Wastewater-induced salinity can negatively affect crop output by suppressing general growth in the pre-seedling stage, causing nutritional imbalance, and suppressing growth through harmful ions. Moreover, wastewater irrigation can transfer heavy metals into soils and contaminate crops, damaging soil fauna and flora. Some heavy metals are absorbed in the soil, whereas others, especially Cu and Cd, are transferred by soil fauna like earthworms (Hussain et al., 2021).

5.4 Ecological Impact

When water drains from wastewater irrigation systems flows into small aquatic environments and lakes then phosphates in the form of orthophosphate are present in it in produce eutrophication. It disrupts the plant-microbe populations of lakes and rivers. This can hurt other superior types of aquatic organisms and affect on water birds existence. As long as such water bodies satisfy the requirements of local communities, the ecological consequences can be converted into quantifiable economic benefits. For example, organic material overload results in lower dissolved oxygen and causes fluctuations in the composition of aquatic life, like a reduction in fisheries and fish deaths (Aloe et al., 2014).

6 Approaches Used for Regulation of Wastewater Re-usage in Agriculture

The composition and source of wastewater, storage, transport, treatment, agricultural practice and irrigation technology, groundwater, and climatic conditions affect the chemical risks associated with water reuse. It is essential to examine the potential exposure paths during the reuse of water for agricultural irrigation (Qadir et al., 2010).

Various risk management systems are implemented, notably the Australian Rules for Water Reusing are WHO's Water superiority, Hazard Analysis and Critical Control Point (HACCP) approach, and Sanitation Safety Management (Recharge, 2009). The WHO security planning strategy permits the investigation of potential system hazards, risk assessment, and development of monitoring and measuring criteria in collaboration with key stakeholders. Portugal, by the latest ISO Guidelines, offers a qualitative risk evaluation and a multi-barrier technology for reusing water to evaluate the risks to the environment and human health. This includes determining the risk levels of surface water and groundwater to set pollutant thresholds (Morgado et al., 2019).

6.1 Physical Approaches

Screening is the initial step in any wastewater treatment process. This procedure primarily comprises the removal of floating and non-biodegradable materials that frequently make their way into wastewater treatment plants such as rags, tins, plastics, papers, containers, and wood. However, the removal of these substances protects the downstream equipment and plants from pipe clogs and potential damage. The wastewater screening is often characterized as fine or coarse screening (Asthana et al., 2017).

The Grit Chamber is a sedimentation chamber located at the top of a plant that treats wastewater for the removal of eggshells, sand, coffee beans, and other items that might clog waterways or cause scratches on pumps. In some bigger plants, grease, and fats are eliminated by passing wastewater via a tiny container where skimmers capture the fat that floats to the surface. Air blowers at the tank's base can also help retrieve the fat as a foam. Most plants still operate primary clarifiers equipped with mechanized surface skimmers to remove grease and fat (Shammas & Wang, 2008; Kinsemolu and Akinsemolu, 2020).

The flotation method is also often utilized in industrial sewage treatment plants to remove oils, grease, fats, and other suspended elements from wastewater. They are known as dissolved air flotation (DAF) units. Specifically, DAF units are employed to remove oil from the wastewater of chemical plants, oil refineries, plants of natural gas extraction, and other industrial advantages (Fig. 1). Flow equalization is a method for improving the performance of developed systems for treating wastewater. Flow equalization smooths operational aspects such as temperature, pollutant levels, and flow over time (often 24 h), reducing the downstream consequences of these variables (Palaniandy et al., 2017).

Wastewater clarifiers, also known as sedimentation tanks, are used before or after biological treatment for the removal of heavy sludge particles by separating them from liquid state. When used before biological treatment, primary treatment can help substantially lowering the BOD levels, minimizing the load supplied into the aeration lakes. These basic sedimentation reservoirs are planned to handle the increased loading rate while maintaining a short retention period for water (Hirom & Devi, 2022).

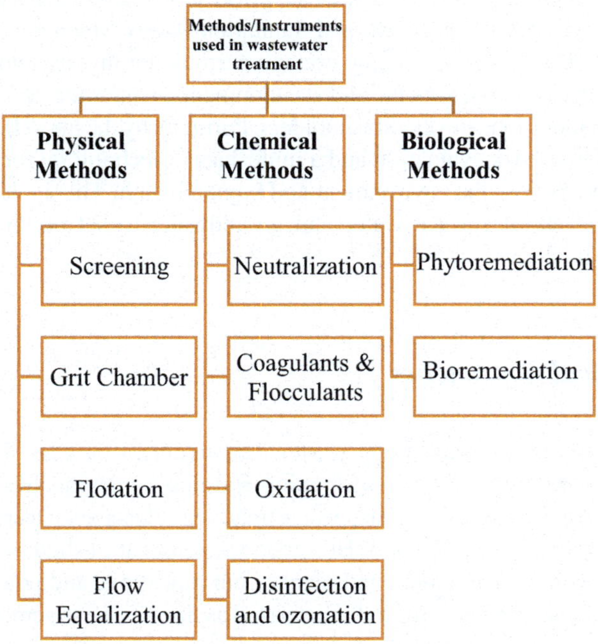

Fig. 1 Methods for treatment of wastewater

6.2 Chemical Approaches

- **Neutralization**:Neutralization adjusts the wastewater pH to fit the needs of individual processing units. Neutralization can be used to remediate acid wastewater holding metals. The procedure entails adjusting the pH of the acidic wastewater by adding a basic reagent. Through this approach, the solution's pH is adjusted to the ideal level for precipitating metallic substances in the form of hydroxides. This stage comes prior to the primary phase of water treatment, to achieve the overall treatment goals (Ibrahim, 2008).
- **Coagulants and Flocculants**
  Flocculation and coagulation processes are used to soften lime, remove particles from water, clarify it, and thicken sludge through wastewater treatment. However, coagulants neutralize the negatively charged particles, which destabilizes the binding forces that hold colloids apart. Moreover, coagulants are positively charged compounds that, when mixed with water, neutralize the charge. However, organic or inorganic coagulants or a mixture is frequently utilized to treat wastewater for the removal of suspended particles. Flocculants bind destabilized molecules, causing them to form clusters and leave the solution (Bratby, 2016).
- **Oxidation**
  Oxidation lowers the biological oxygen need of wastewater and may lessen the toxicity of certain contaminants. Secondary treatment turns certain pollutants into water, carbon dioxide, and biosolids. Chemical oxidation is frequently utilized for disinfection (Bartolomeu et al., 2018).
- **Ozonation**
  Ozone is utilized to disinfect drinkable water, removal of effluent by treatment systems, and degradation of inorganic or organic contaminants in sewage (Tripathi & Hussain, 2022).
- **Disinfection**
  Disinfection in sewage treatment aims to drastically lower the amount of microorganisms in the water before it is discharged into the environment for future use in bathing, drinking, and irrigation (Fig. 1). The efficacy of disinfection is calculated by the overall condition of the water being disinfected, the kind of disinfection utilized, the disinfectant prescription (time and concentration), along with other environmental factors (Ngwenya et al., 2012).

6.3 Biological Approaches

The traditional approach regarding wastewater treatment involves treatment plants for sewage that utilize physicochemical processes (like activated sludge, trickling filters, and high-rate stabilizing ponds) and anaerobic methods (upflow anaerobic wastewater blanket power plants). However, because of greater operating costs, high electrical supply, equipment corrosion, and the inability to reuse treated water, these traditional techniques are not more viable or efficient than biological techniques (Patel et al., 2021). Biological sewage treatment solutions are more affordable and environmentally friendly. They frequently use two methods: phytoremediation (utilization of green plants to degrade, remove, and detoxify contaminants), and bioremediation (usage of organisms to transform, convert, degrade, and detoxify pollutants). In typical wastewater treatment facilities, there is a danger of operational failure owing to poor maintenance, resulting in significant financial losses (Fig. 1). However, microbe- and plant-based eco-technology, on the other hand, poses no dangers, produces no residues, and requires no energy (Chandra et al., 2015; Kumar et al., 2011).

7 Conclusion

The deterioration of river water quality as a result of human and industrial activity has necessitated the establishment of effective water management strategies. Conventional wastewater treatment technologies rely on physicochemical processes and have restrictions due to the high cost of operation and implementation. Plant-based innovations are solar-powered, ecologically friendly, and affordable as they need no energy and are relatively inexpensive to build and operate. After the biological procedure, sewage can be utilized for irrigation and gardening. As a result, treating wastewater by cultivating effective aquatic algae and plants in a plant-based system might be utilized not just to replace traditional processes, but also to control and regulate river water contamination.

8 Future Prospects

More investigation is required to develop efficient management systems for reuse of wastewater in agriculture for the purpose of evaluating different treatment process and their monitoring for better water quality.

References

Abdel-Shafy, H. I., & Aly, R. O. (2002). Water issue in Egypt: Resources, pollution and protection endeavors. *Central European Journal of Occupational and Environmental Medicine, 8*(1), 3–21.

Abdullah, H., et al. (2023). Investigating on-site energy consumption patterns using vertical electrical sounding (VES) and GIS techniques. *Sustainability, 2023*(15), 16782. https://doi.org/10.3390/su152416782

Ahmad, A., et al. (2009). Seasonal growth, radiation interception, its conversion efficiency and aerial dry biomass production of Oryza sativa L. under diverse agro-environments in Pakistan. *Pakistan Journal of Botany, 41*(3), 1241–1257.

Ahmad, S., et al. (2017). Quantification of climate warming and crop management impacts on cotton phenology. *Plants, 6*(7), 1–16.
Akinsemolu, A. A. (2020). *The principles of green and sustainability science*. Springer.
Alghabari, F., et al. (2016). Gibberellin-sensitive Rht alleles confer tolerance to heat and drought stresses in wheat at booting stage. *Journal of Cereal Science, 70*, 72–78.
Aloe, A. K., Bouraoui, F., Grizzetti, B., Bidoglio, G., & Pistocchi, A. (2014). *Managing nitrogen and phosphorus loads to water bodies: Characterisation and solutions towards macro-regional integrated nutrient management*. Joint Research Centre of the European Commission.
Alsuhaibani, A. M., et al. (2023). Green buildings model: Impact of rigid polyurethane foam on indoor environment and sustainable development in energy sector. *Heliyon, 9*, 14451.
Asthana, M., Kumar, A., & Sharma, B. S. (2017). Wastewater treatment. In *Principles and applications of environmental biotechnology for a sustainable future* (pp. 173–232).
Awais, M., et al. (2017). Modeling the water and nitrogen productivity of sunflower using OILCROP-SUN model in Pakistan. *Field Crops Research, 205*, 67–77.
Bartolomeu, M., Neves, M. G. P. M. S., Faustino, M. A. F., & Almeida, A. (2018). Wastewater chemical contaminants: Remediation by advanced oxidation processes. *Photochemical & Photobiological Sciences, 17*(11), 1573–1598.
Boliko, M. C. (2019). FAO and the situation of food security and nutrition in the world. *Journal of Nutritional Science and Vitaminology, 65*(Supplement), S4–S8.
Bougnom, B. P., & Piddock, L. J. (2017). *Wastewater for urban agriculture: A significant factor in dissemination of antibiotic resistance*.
Bratby, J. (2016). *Coagulation and flocculation in water and wastewater treatment*. IWA Publishing.
Buechler, S., Mekala, G. D., & Keraita, B. (2005). Wastewater use for urban and peri-urban agriculture. *Cities farming for the future, urban agriculture for green and productive cities*, 244–260.
Bunting, S. W., & Edwards, P. (2018). Global prospects for safe wastewater reuse through aquaculture. In *Wastewater management through aquaculture* (pp. 55–72).
Chandra, R., Saxena, G., & Kumar, V. (2015). Phytoremediation of environmental pollutants: An eco-sustainable green technology to environmental management. In *Advances in biodegradation and bioremediation of industrial waste* (pp. 1–29).
Chaturvedi, N. K. (2022). Comparison of available treatment techniques for hazardous aniline-based organic contaminants. *Applied Water Science, 12*(7), 173.
Chojnacka, K., Witek-Krowiak, A., Moustakas, K., Skrzypczak, D., Mikula, K., & Loizidou, M. (2020). A transition from conventional irrigation to fertigation with reclaimed wastewater: Prospects and challenges. *Renewable and Sustainable Energy Reviews, 130*, 109959.
Cirelli, M., Moulin, E., Panci, P., Serpico, P. D., & Viana, A. (2012). Gamma ray constraints on decaying dark matter. *Physical Review D—Particles, Fields, Gravitation, and Cosmology, 86*(8), 083506.
Corcoran, J., Winter, M. J., & Tyler, C. R. (2010). Pharmaceuticals in the aquatic environment: A critical review of the evidence for health effects in fish. *Critical Reviews in Toxicology, 40*(4), 287–304.
Da Silva, T. L., Sánchez-Román, R. M., Queluz, J. G. T., & Pletsch, T. A. (2020). Treatment options for the direct reuse of reclaimed water in developing countries. In *Advances in chemical pollution, environmental management and protection* (Vol. 6, pp. 1–31). Elsevier.
De Roo, A., Trichakis, I., Bisselink, B., Gelati, E., Pistocchi, A., & Gawlik, B. (2021). The water-energy-food-ecosystem nexus in the Mediterranean: Current issues and future challenges. *Frontiers in Climate, 3*, 782553.
Elalami, D., Carrere, H., Monlau, F., Abdelouahdi, K., Oukarroum, A., & Barakat, A. (2019). Pretreatment and co-digestion of wastewater sludge for biogas production: Recent research advances and trends. *Renewable and Sustainable Energy Reviews, 114*, 109287.
Elgallal, M., Fletcher, L., & Evans, B. (2016). Assessment of potential risks associated with chemicals in wastewater used for irrigation in arid and semiarid zones: A review. *Agricultural Water Management, 177*, 419–431.

Fahad, S., et al. (2016a). Exogenously applied plant growth regulators enhance the morpho-physiological growth and yield of rice under high temperature. *Frontier in Plant Sciences, 07*, 1250.

Fahad, S., et al. (2016b). Responses of rapid viscoanalyzer profile and other rice grain qualities to exogenously applied plant growth regulators under high day and high night temperatures. *Plos One, 11*(7), 01–13.

Fahad, S., et al. (2016c). Exogenously applied plant growth regulators affect heat-stressed rice pollens. *Journal of Agronomy and Crop Science, 202*, 139–150.

Fahad, S., et al. (2016d). A combined application of biochar and phosphorus alleviates heat-induced adversities on physiological, agronomical and quality attributes of rice. *Plant Physiology and Biochemistry, 103*, 191–198.

Farid, H. M., et al. (2023). Groundwater quality risk assessment using hydro-chemical and geospatial analysis. *Environment, Development and Sustainability, 25*, 8343–8365.

Feenstra, R. C., Hanson, G. H., & Swenson, D. L. (2000). Offshore assembly from the United States: Production characteristics of the 9802 program. In *The impact of international trade on wages* (pp. 85–125). University of Chicago Press.

Feng, L., et al. (2024). Seasonal dynamics in land surface temperature in response to land use land cover changes using google earth engine. *IEEE Journal of Selected Topics in Applied Earth Observations and Remote Sensing, 17*, 17983.

Fito, J., & Van Hulle, S. W. (2021). Wastewater reclamation and reuse potentials in agriculture: Towards environmental sustainability. *Environment, Development and Sustainability, 23*(3), 2949–2972.

Gillani, R. A., et al. (2017). Biosorption analysis of Agrobacterium tumefacians 12b3 against Cr (1 III) and Pb(II) by 2 kinetic modeling and equilibrium studies. *Desalination and Water Treatment, 67*, 206–214.

Gonzalez-Serrano, E., Rodriguez-Mirasol, J., Cordero, T., Koussis, A. D., & Rodriguez, J. J. (2005). Cost of reclaimed municipal wastewater for applications in seasonally stressed semi-arid regions. *Journal of Water Supply: Research and Technology—AQUA, 54*(6), 355–369.

Haas, C. N., Rose, J. B., & Gerba, C. P. (2014). *Quantitative microbial risk assessment*. Wiley.

Hammad, H. M., et al. (2010a). Optimizing water and nitrogen requirement in maize under semi arid conditions of Pakistan. *Pakistan Journal of Botany, 43*(6), 2919–2923.

Hammad, H. M., et al. (2010b). Organic farming in wheat crop under arid conditions of Pakistan. *Pakistan Journal of Agricultural Sciences, 48*(2), 97–102.

Hammad, H. M., et al. (2023). Impacts of COVID-19 pandemic on environment, society, and food security. *Environmental Science and Pollution Research, 30*, 99261–99272.

Hanjra, M. A. (2000). *Socio-economic and environmental effects of wastewater irrigation: State of the art, problem areas and policy alternatives.*

Haseeb, M., et al. (2023). Quantifying irrigation water demand and supply gap using remote sensing and GIS in Multan, Pakistan. *Environmental Monitoring and Assessment, 195*, 990.

Hettiarachchi, H., Ryu, S., Caucci, S., & Silva, R. (2018). Municipal solid waste management in Latin America and the Caribbean: Issues and potential solutions from the governance perspective. *Recycling, 3*(2), 19.

Hirom, K., & Devi, T. T. (2022). Application of computational fluid dynamics in sedimentation tank design and its recent developments: A review. *Water, Air, & Soil Pollution, 233*(1), 22.

Hussain, M. M., Farooqi, Z. U. R., Mohy-Ud-Din, W., Younas, F., Shahzad, M. T. T., Ghani, M. U. M. U., et al. (2021). Application of Bioremediation as sustainable approach to remediate heavy metal and pesticide polluted environments. *Plant and Environment, 2*(01), 62–92.

Hussain, S., et al. (2023a). Assessment of land use/land cover changes and its effect on land surface temperature using remote sensing techniques in Southern Punjab, Pakistan. *Environmental Science and Pollution Research, 30*, 99202–99218.

Hussain, S., et al. (2023b). Relation of land surface temperature with different vegetation indices using multi-temporal remote sensing data in Sahiwal region, Pakistan. *Geoscience Letters, 10*, 33.

Hussain, S., et al. (2023c). Investigation of Irrigation Water Requirement and Evapotranspiration for Water Resource Management in Southern Punjab, Pakistan. *Sustainability, 15*, 1768.

Hussain, S., et al. (2024). Assessing the impact of land use land cover changes on soil moisture and vegetation cover in Southern Punjab, Pakistan using multi-temporal satellite data. *Geology, Ecology, and Landscapes*. https://doi.org/10.1080/24749508.2024.2338970

Ibrahim, R. (2008). *Practical modelling and control implementation studies on a pH neutralization process pilot plant*. Doctoral dissertation, University of Glasgow.

Jabran, K., et al. (2017). Growth and physiology of basmati rice under conventional and water-saving production systems. *Archives of Agronomy and Soil Science, 63*(10), 1465–1476.

Jaramillo, M. F., & Restrepo, I. (2017). Wastewater reuse in agriculture: A review about its limitations and benefits. *Sustainability, 9*(10), 1734.

Jhansi, S. C., & Mishra, S. K. (2013). Wastewater treatment and reuse: Sustainability options. *Consilience, 10*, 1–15.

Kesari, K. K., Soni, R., Jamal, Q. M. S., Tripathi, P., Lal, J. A., Jha, N. K., et al. (2021). Wastewater treatment and reuse: A review of its applications and health implications. *Water, Air, & Soil Pollution, 232*, 1–28.

Khan, M. M., Siddiqi, S. A., Farooque, A. A., Iqbal, Q., Shahid, S. A., Akram, M. T., et al. (2022a). Towards sustainable application of wastewater in agriculture: A review on reusability and risk assessment. *Agronomy, 12*(6), 1397.

Khan, R. U., Hamayun, M., Altaf, A. A., Kausar, S., Razzaq, Z., & Javaid, T. (2022b). Assessment and removal of heavy metals and other ions from the industrial wastewater of Faisalabad, Pakistan. *Processes, 10*(11), 2165.

Kumar, A., Bisht, B. S., Joshi, V. D., & Dhewa, T. (2011). Review on bioremediation of polluted environment: A management tool. *International Journal of Environmental Sciences, 1*(6), 1079–1093.

Mahmood, F., et al. (2017). Economic and environmental impacts of introducing grain legumes in farming systems of Midi-Pyrénées region (France): A simulation approach. *International Journal of Plant Production, 01*(1), 65–87.

Masood, N., et al. (2013). Whitefly (Aleurolobus barodensis mask) spatio-temporal trends in semi-arid agro-ecological zones of Pakistan. *Pakistan Journal of Agricultural Sciences, 50*(1), 69–74.

Matos, C., Sampaio, A., & Bentes, I. (2012). Greywater use in irrigation: Characteristics, advantages and concerns. In *Irrigation-water management, pollution and alternative strategies* (pp. 159–184).

Meffe, R., de Santiago-Martín, A., Teijón, G., Hernández, V. M., López-Heras, I., Nozal, L., & de Bustamante, I. (2021). Pharmaceutical and transformation products during unplanned water reuse: Insights into natural attenuation, plant uptake and human health impact under field conditions. *Environment International, 157*, 106835.

Morgado, L., Silva, F. J. G., & Fonseca, L. M. (2019). Mapping occupational health and safety management systems in Portugal: Outlook for ISO 45001: 2018 adoption. *Procedia Manufacturing, 38*, 755–764.

Mubeen, M., et al. (2016). Application of CSM-CERES-Maize model in optimizing irrigated conditions. *Outlook on Agriculture, 45*(3), 01–12.

Muhammad, H., et al. (2023). A paradigm shift towards beneficial microbes enhancing the efficiency of organic and inorganic nitrogen sources for a sustainable environment. *Land, 12*, 680.

Nasim, W., & Bano, A. (2012). Impact of nitrogen and plant growth promoting rhizobacteria on yield and yield components of sunflower in glasshouse environment. *Journal of Crop Science and Biotechnology, 15*(4), 319–324.

Nasim, W., et al. (2011). Nitrogen effects on growth and development of sunflower hybrids under agro-climatic conditions of Multan. *Pakistan Journal of Botany, 43*(4), 2083–2092.

Nasim, W., et al. (2012). Effect of nitrogen on growth and yield of sunflower under semiarid conditions of Pakistan. *Pakistan Journal of Botany, 44*(2), 639–648.

Nasim, W., et al. (2016a). Correlation studies on nitrogen for sunflower crop across the agroclimatic variability. *Environmental Science and Pollution Research, 23*(4), 3658–3670.

Nasim, W., et al. (2016b). Modelling climate change impacts and adaptation strategies for sunflower in Punjab-Pakistan. *Outlook on Agriculture, 45*(1), 39–45.
Nasim, W., et al. (2017). Response of sunflower hybrids to nitrogen application grown under different Agro-environments. *Journal of Plant Nutrition, 40*(1), 82–92.
Ngwenya, N., Ncube, E. J., & Parsons, J. (2012). Recent advances in drinking water disinfection: Successes and challenges. In *Reviews of environmental contamination and toxicology* (pp. 111–170).
Obi, F. O., Ugwuishiwu, B. O., & Nwakaire, J. N. (2016). Agricultural waste concept, generation, utilization and management. *Nigerian Journal of Technology, 35*(4), 957–964.
Ofori, S., Puškáčová, A., Růžičková, I., & Wanner, J. (2021). Treated wastewater reuse for irrigation: Pros and cons. *Science of the Total Environment, 760*, 144026.
Palaniandy, P., Adlan, M. N., Aziz, H. A., Murshed, M. F., & Hung, Y. T. (2017). Dissolved air flotation (DAF) for wastewater treatment. In *Waste treatment in the service and utility industries* (pp. 145–182). CRC Press.
Patel, A., Arkatkar, A., Singh, S., Rabbani, A., Medina, J. D. S., Ong, E. S., et al. (2021). Physicochemical and biological treatment strategies for converting municipal wastewater and its residue to resources. *Chemosphere, 282*, 130881.
Qadir, M., Wichelns, D., Raschid-Sally, L., McCornick, P. G., Drechsel, P., Bahri, A., & Minhas, P. S. (2010). The challenges of wastewater irrigation in developing countries. *Agricultural Water Management, 97*(4), 561–568.
Quist-Jensen, C. A., Macedonio, F., & Drioli, E. (2015). Membrane technology for water production in agriculture: Desalination and wastewater reuse. *Desalination, 364*, 17–32.
Raschid-Sally, L., Carr, R., & Buechler, S. (2005). Managing wastewater agriculture to improve livelihoods and environmental quality in poor countries. *Irrigation and Drainage: The Journal of the International Commission on Irrigation and Drainage, 54*(S1), S11–S22.
Rasool, A., et al. (2016). Arsanic and heavy metal contaminations in the tube well water of Punjab, Pakistan and risk assessment: A case study. *Ecological Enginee., 95*, 90–100.
Recharge, M. A. (2009). *Australian guidelines for water recycling*.
Rolls, R. J., Leigh, C., & Sheldon, F. (2012). Mechanistic effects of low-flow hydrology on riverine ecosystems: Ecological principles and consequences of alteration. *Freshwater Science, 31*(4), 1163–1186.
Rusănescu, C. O., Rusănescu, M., Voicu, G., Paraschiv, G., Biriş, S. Ş., & Popescu, I. N. (2022). The recovery of vermicompost sewage sludge in agriculture. *Agronomy, 12*(11), 2653.
Salgot, M., Huertas, E., Weber, S., Dott, W., & Hollender, J. (2006). Wastewater reuse and risk: Definition of key objectives. *Desalination, 187*(1–3), 29–40.
Samian, M., Mahdei, K. N., Saadi, H., & Movahedi, R. (2015). Identifying factors affecting optimal management of agricultural water. *Journal of the Saudi Society of Agricultural Sciences, 14*(1), 11–18.
Saravanan, V. S., & Mollinga, P. P. (2011). *The environment and human health: An agenda for research*.
Shahid, M., et al. (2014). EDTA-enhanced phytoremediation of heavy 1 metals: A review. *Soil and Sediments Contamination, 23*, 389–416.
Shahriar, A., Tan, J., Sharma, P., Hanigan, D., Verburg, P., Pagilla, K., & Yang, Y. (2021). Modeling the fate and human health impacts of pharmaceuticals and personal care products in reclaimed wastewater irrigation for agriculture. *Environmental Pollution, 276*, 116532.
Shammas, N. K., & Wang, L. K. (2008). 10 hazardous waste deep-well. In *Advances in hazardous industrial waste treatment* (p. 385).
Tripathi, S., & Hussain, T. (2022). Water and wastewater treatment through ozone-based technologies. In *Development in wastewater treatment research and processes* (pp. 139–172). Elsevier.
Usman, M., et al. (2009). Development and application of crop water stress index for scheduling irrigation in cotton (Gossypium hersutumn L.) under semiarid environment. Journal of Food. *Agriculture and Environment, 07*(3&4), 386–391.

Usman, M., et al. (2010). Lower and upper baselines for crop water stress index and yield of Gossypium hirsutum L. under variable irrigation regimes in irrigated semiarid environment. *Pakistan Journal of Botany, 42*(4), 2541–2550.

Valdes Ramos, A., Aguilera Gonzalez, E. N., Tobón Echeverri, G., Samaniego Moreno, L., Díaz Jiménez, L., & Carlos Hernández, S. (2019). Potential uses of treated municipal wastewater in a semiarid region of Mexico. *Sustainability, 11*(8), 2217.

Verma, N., Rachamalla, M., Kumar, P. S., & Dua, K. (2023). Assessment and impact of metal toxicity on wildlife and human health. In *Metals in water* (pp. 93–110). Elsevier.

Wajid, A., et al. (2007). Maize response to yield under semi arid conditions of Pakistan. *Pakistan Journal of Agricultural Sciences, 44*(2), 217–220.

Wajid, A., et al. (2010). Quantification of growth, yield and radiation use efficiency of promising cotton cultivars at varying nitrogen levels. *Pakistan Journal of Botany, 42*(3), 1703–1711.

Wijaya, I. M. W., & Soedjono, E. S. (2018). Domestic wastewater in Indonesia: Challenge in the future related to nitrogen content. *Geomate Journal, 15*(47), 32–41.

Yang, X., et al. (2023). 2023 quantitative evaluation of soil water and wind erosion rates in Pakistan. *Remote Sensing, 15*, 2404.

Wastewater Evaluation and Characterisation Before Irrigation

Marryam Qamar, Muhammad Mubeen, Samra Tariq, Faiz Rabbani, Sana Khalid, Muhammad Habib ur Rahman, and Muhammad Azhar Ehsan

Abstract Global population density and freshwater resources show uneven distribution patterns around the globe. This has forced farmers to use wastewater for the irrigation of food crops. Most industrialised nations purify their wastewater before using it to irrigate crops for food, fibre and other uses. This practice presents both positive and negative effects with respect to agricultural use, as well as in the context of environmental contamination and toxicology. This chapter highlights the harmful and beneficial impacts of wastewater irrigation. Cu, Cr, Mn, Fe, Pb, Zn and Ni are some of the harmful metals found in it. Treated wastewater reuse has been expanding at a rate of between 10% and 29% annually in Europe, the United States and China, and as much as 41% in Australia. The chemical pollutants found in untreated wastewater may be rather diverse and include substances such as fertilisers, heavy metals, textiles, leather, paper, sewage, food processing waste, pesticides and chemical waste from industrial effluent The direct use of wastewater for irrigation has serious health consequences. This chapter also highlights the irrigation water criteria, list of parameters and methods adopted for analysis of water quality and sampling methods of water for analysis. This chapter also highlights the Private/Government laboratories, institutes and organisations which provide financial and

M. Qamar · S. Tariq · F. Rabbani · S. Khalid
Department of Environmental Sciences, COMSATS University Islamabad, Vehari Campus, Vehari, Pakistan

M. Mubeen (✉)
Department of Biotechnology, COMSATS University Islamabad, Vehari Campus, Vehari, Pakistan
e-mail: muhammadmubeen@cuivehari.edu.pk

M. Habib ur Rahman
Crop Science Group, Institute of Crop Science and Resource Conservation (INRES), University of Bonn, Bonn, Germany

Department of Agronomy, MNS-University of Agriculture, Multan, Multan, Pakistan

M. A. Ehsan
International Research Institute for Climate and Society (IRI), Columbia Climate School, Palisades, NY, USA

M. Mubeen et al. (eds.), *Innovations in Agricultural Water Management*,
https://doi.org/10.1007/978-3-031-91883-4_10

technical support to universities and research institutes for collaborative research like Pakistan Council of Research in Water Resources, Pakistan Environmental Protection Agency, Institute of Soil and Environmental Sciences, University of Agriculture Faisalabad, Pakistan, Centre of Excellence in Analytical Chemistry, University of Sindh, Jamshoro, Pakistan, Institute of Environmental Studies, University of Karachi, Department of Soil and Environmental Science, University of Peshawar, Peshawar and IWMI. In order to fulfil the criteria for irrigation water quality and to guarantee the safety of crops and the environment, it is necessary to treat wastewater in order to eliminate pollutants including pathogens, organic materials, nutrients and trace elements. This chapter delves into the several treatment procedures, including physical, chemical and biological approaches, and highlights how successful they are in creating irrigation-quality water. It goes on to discuss how treated wastewater may improve soil fertility, increase agricultural production in areas with water scarcity and decrease the demand for freshwater. All things considered, a sustainable method of managing water resources is to use treated wastewater for irrigation. This helps with both water conservation and the sustainability of agricultural practices.

Keywords Wastewater irrigation · Evaluation · Characterization · Heavy metals · Sustainability · Freshwater resources

1 Introduction

The term 'wastewater' is used to describe the liquid byproducts of many human activities, including those in homes, hospitals, restaurants and businesses. Most industrialised nations purify their wastewater before using it to irrigate crops for food, fibre and other uses (Silva, 2023). The need for freshwater is increasing at a rapid pace, and one of the most pressing needs right now is wastewater reclamation or reuse (Khan et al., 2022). Based on many studies, agriculture is the primary user of water on a global scale (102%). About 70% of the freshwater that originates from rivers and subterranean water sources is used for irrigation (Scanlon et al., 2023). As an example of the water problem in irrigation, 40% of the world's population lives in basins that are severely water-stressed (Singh, 2021). So, instead of using freshwater, farms may benefit from reusing wastewater (Khan et al., 2022).

Irrigated agriculture using untreated wastewater might have beneficial and bad effects on the ecosystem (Pratap et al., 2021). In contrast to dumping treated municipal wastewater into untreated surface or groundwater, which is harmful to the environment, treating wastewater before irrigation is strongly advocated (Minhas et al., 2022). Despite its usefulness as a plant nutrition and organic matter source, wastewater may include harmful chemical compounds and bacteria that harm both humans and the environment (Pratap et al., 2021). As a result, there are several areas where access to high-quality water for irrigation is under risk. Plant nutrients and organic debris may be found in wastewater, making it a useful resource (Kurniawan et al., 2021).

The presence of harmful bacteria and other chemicals might make it a health and environmental hazard, however. There are several known dangers associated with wastewater reuse. Some of these dangers are immediate (like microbial infections), while others develop over time and become more severe (like saline effects on soil) as the practice continues (Kesari et al., 2021).Consequently, a plethora of standards outlining quality requirements and best practices for the reuse of treated wastewater (effluents) for irrigation purposes have been established (Revitt et al., 2021). Recent studies found many health risks related to wastewater reuse for agriculture irrigation. There was an increased danger when farmers used industrial effluent for irrigation, either alone or in combination with water from homes (Mahjoub et al., 2022). Because heavy metals are both non-biodegradable and have a prolonged biological half-life, pathogen contamination with heavy metals increases risk factors (Chen et al., 2022). Cu, Cr, Mn, Fe, Pb, Zn and Ni are some of the harmful metals found in it (Enyoh et al., 2022). The heavy metals build up in the topsoil, which is 20 cm deep, and they reach plants and animals via plant roots, leafy greens and polluted air (Hossain et al., 2021). It is crucial to evaluate the potential health risks of using such wastewater for irrigation purposes, particularly in adults (Shahid et al., 2021).

In the world, millions of small-scale farmers rely on treated wastewater irrigation to sustain their livelihoods and increase agricultural productivity. There is a large range of values, from 1.6% to 6.6%, for the worldwide reuse of treated wastewater for agricultural applications (Mishra et al., 2023). The World Health Organization estimates that over 10% of the world's population relies on agricultural goods grown by wastewater irrigation (Khalid et al., 2018). Treated wastewater reuse has been expanding at a rate of between 10% and 29% annually in Europe, the United States and China, and as much as 41% in Australia (Kesari et al., 2021). Among Asian countries, China's projected 1.3 million hectares (ha) of wastewater reuse puts it ahead of Vietnam, India and Pakistan (Hashem & Qi, 2021). According to recent estimates, only about 376% of India's urban wastewater is being treated (Singh et al., 2021). Israel uses treated wastewater for agricultural land irrigation more than any other country, using 90% of recovered water (Mordechay et al., 2021).

2 Wastewater Impact on Agriculture

It is widely recognised that the agricultural sector is the biggest consumer of water, responsible for over 70% of the world's total consumption (Motoshita et al., 2023). A significant source for irrigation is shown by the fact that almost 20 million ha of land throughout the globe are watered using waste-water (Gökçeku et al., 2023). The majority of wastewater utilised for irrigation purposes remains untreated. Irrigation using partly treated or untreated wastewater is common in underdeveloped nations (Yahaya et al., 2023). The chemical pollutants found in untreated wastewater may be rather diverse and include substances such as fertilisers, heavy metals, textiles, leather, paper, sewage, food processing waste, pesticides and chemical waste from industrial effluent. The direct use of wastewater for irrigation has serious health consequences, according to the World Health Organization (WHO)

(Pacific, 2023). A large range of organic chemicals are also present in wastewater. Some of these substances are carcinogenic or poisonous, and they may damage an embryo (Shetty et al., 2023).

According to many studies, there are negative health concerns linked to microbiological dangers, also known as infectious pathogens, if wastewater or grey water is reused (Fernandes et al., 2023). Scientific investigations have shown that helminth infection illnesses, which may be transmitted by wastewater exposure, have been associated with anaemia and stunted physical and mental growth (Albareedy, 2023). There is a vast array of organic chemicals found in wastewater as well. There are a few of them that are poisonous or carcinogenic that may damage an embryo. Figure 1 shows the flow of untreated wastewater utilised for irrigation and the impacts on human health (Dickin et al., 2016).

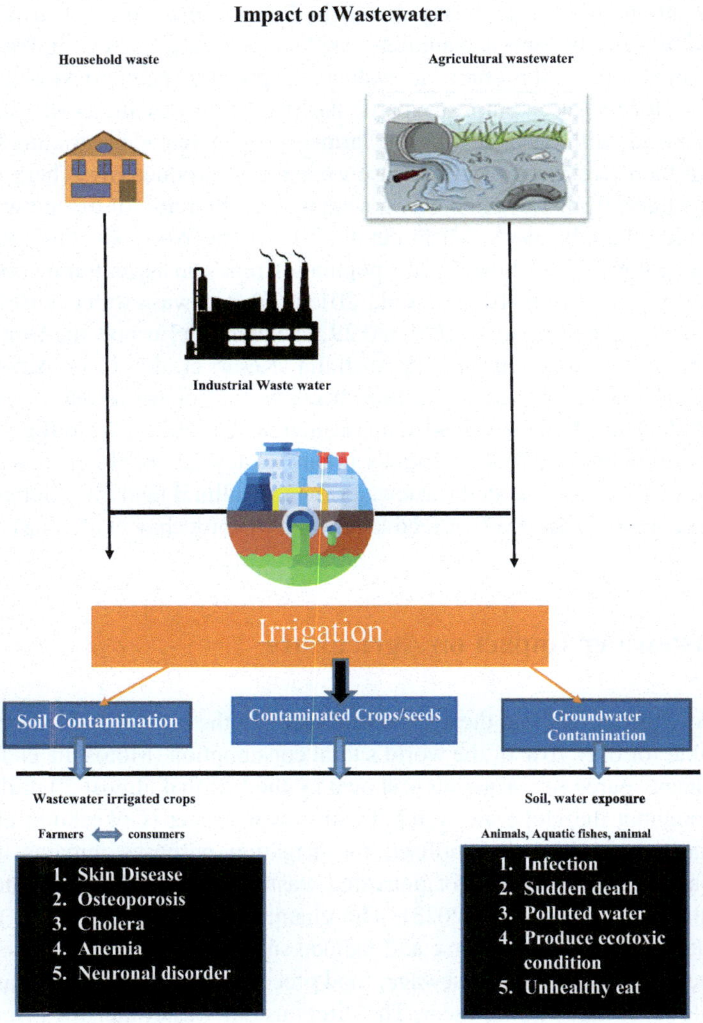

Fig. 1 Impact of wastewater

3 Irrigation Water Criteria

Watercourses used for irrigation must have a systematic water quality monitoring and analysis programme, particularly if treated wastewater is discharged into these watercourses. It is necessary to measure the water quality characteristics, including hydrogen ion concentration, salinity, organic matter and trace elements (Mishra et al., 2023).

3.1 Salinity

Soil salinity is an essential component of agricultural water quality as it may make nutrient absorption more difficult for crops and increase the toxicity of some ions (Mezlini et al., 2024). In order to ensure the safe reuse of wastewater in agricultural settings, the FAO has set salinity standards. In 1985, Westcot and Ayers revised these recommendations and added restrictions for applications ranging from none to mild to severe.

3.2 Essential Elements

The growth of crops may be affected by the nitrogen (N) and phosphorus (P) that are necessary for plant development and growth. Groundwater pollution and eutrophication in coastal regions and lakes are both caused by irrigation water that contains too many nutrients (Sharma et al., 2024).

3.3 Natural Substances

The biological oxygen demand (BOD) is a measure of the index of organic matter. Aerobic activity consumes dissolved oxygen to finish degrading organic materials at high BOD levels, creating an anaerobic condition; oxides in the soil, such Fe^{3+}, Mn^{5+}, and SO_4^-, use oxygen during decomposition to reduce the oxidation-reduction potential. The organic acids, manganese and iron that are produced might ultimately hinder the paddy rice's ability to absorb nutrients (Das, 2014).

3.4 The Hydrogen Ion Level

pH is a measure of acidity or alkalinity that allows for a rapid evaluation of water quality. The typical pH range for water used in agriculture is between 6.5 and 8.4. Water with an abnormally high or low pH might cause problems for trickle irrigation systems.

3.5 *Elements of Trace*

While trace elements in water are essential for plant development, excessive amounts of these elements may be toxic to plants. The author found that copper (Cu) may lead to chlorosis of the leaves and stunted root development (Xu et al., 2024). The effects of zinc (Zn) and arsenic (As) on stem chlorosis and root development inhibition are substantial (Hafeez et al., 2023). Crop yields may be reduced when aluminium (Al) is present in soils that are acidic. Typically, lead (Pb) and cyanide (CN) are subject to stringent regulations because, when dissolved in water or soil, they may build in crops over time, posing health risks to humans (Rahman et al., 2018). Limits on trace element levels in irrigation water have been established by the FAO's standards for agricultural water quality (Anyango et al., 2024).

4 List of Parameters and Methods Adopted for Water Quality Analysis

Serial numbers	Parameters	Units	Methods/instrument
1	Potential of hydrogen (pH)	–	pH meter
2	Chemical oxygen demand (COD)	mg/L	Open reflux technique
3	Total suspended solids (TSS)	mg/L	Calculation formula
4	Sodium (Na)	mg/L	Flame photometer (PFP-7, Jenway, UK)
5	Magnesium (Mg)	me/L	Atomic absorption, spectrophotometer (Analytik Jena: NovAA 400P, Germany)
6	Carbonate and bicarbonate	me/L	Titration
7	Sodium adsorption ratio (SAR)	me/L	Calculation formula
8	Electrical conductivity (EC)	μS/cm	EC meter
9	Biochemical oxygen demand (BOD)	mg/L	5 day BOD test
10	Total solids (TS)	mg/L	Gravimetric determination method
11	Total dissolved solids (TDS)	mg/L	TDS meter (EZ9901)
12	Calcium (Ca)	me/L	Flame photometer (PFP-7, Jenway, UK)
13	Potassium (K)	mg/L	Flame photometer (PFP-7, Jenway, UK)
14	Residual sodium carbonate (RSC)	me/L	Calculation formula
15	Heavy metals of Zn, Cd, Ni, Cu	mg/L	Atomic absorption, spectrophotometer (Analytik Jena: NovAA 400P, Germany)

Raza et al. (2022)

5 Laboratory Determinations Needed to Evaluate Common Irrigation Water Quality Problems

Table below describes the laboratory findings and calculations necessary to follow the instructions, along with the symbols used. Publications such as USDA Handbook 60, Rhoades and Clark, FAO Soils Bulletin 10 and Standard Methods for Examination of Waters and Wastewaters provide analytical methodologies for laboratory determinations. Use the approach that makes the most logical in context with the resources (money, time, and equipment) at your disposal. It is deemed sufficient when the analytical accuracy is within ±5%.

Water parameter	Symbol	Unit	Range in irrigation water
Salinity			
Salt content			
Electrical conductivity	EC_w	dS/m	0–3
(or)			
Total dissolved solids	TDS	mg/L	0–2000
Cations and anions			
Calcium	Ca^{2+}	me/L	0–20
Magnesium	Mg^{2+}	me/L	0–5
Sodium	Na^+	me/L	0–40
Carbonate	CO_3^-	me/L	0–0.1
Bicarbonate	HCO_3^-	me/L	0–10
Chloride	Cl^-	me/L	0–30
Sulphate	SO_4^-	me/L	0–20
Nutrients			
Nitrate–nitrogen	NO_3–N	mg/L	0–10
Ammonium–nitrogen	NH_4–N	mg/L	0–5
Phosphate–phosphorus	PO_4–P	mg/L	0–2
Potassium	K^+	mg/L	0–2
Miscellaneous			
Boron	B	mg/L	0–2
Acid/basicity	pH	1–14	6.0–8.5
Sodium adsorption ratio	SAR		0–15

Raza et al. (2022)

6 Contaminants Trends in Wastewater

Parameters	Min	Max	Mean	Std. dev.	Permissible limit
Temperature (°C)	28.00	29.00	28.667	0.577	≤3 °C
pH of wastewater (pH)	9.20	9.50	9.333	0.152	6–10
Chemical oxygen demand (COD)	400	416	408	8	150

Parameters	Min	Max	Mean	Std. dev.	Permissible limit
Total suspended solids (TSS)	4870	4890	4880	0.01	150
Sodium (Na)	2.95	2.97	2.96	1.00	
Magnesium (Mg)	116	124	119.33	4.163	
Carbonate (CO_3^-)	7.00	7.20	7.0667	0.115	
Bicarbonate (HCO_3^-)	6.60	8	7.16	0.425	
Sodium adsorption ratio (SAR)	1.4191	1.5045	1.4641	0.0429	
Electrical conductivity (EC)	2.45	2.50	2.47	0.02	
Biochemical oxygen demand (BOD)	303	305	304	1.00	80
Total solids (TS)	19,167	19,171	19,169	2.223E-03	150
Total dissolved solids (TDS)	14,290	14,298	14,296	2.222E-03	3500
Calcium (Ca)	34.0	35.0	37.667	1.5275	
Potassium (K)	30.0	33.0	31.33	1.5275	
Residual sodium carbonate (RSC)	2.2167	2.8667	2.5056	0.3310	
Zinc (Zn)	4.4550	4.4590	4.4570	2.000E-03	5
Cadmium (Cd)	0.0546	0.0550	0.0548	2.082E-04	0.1
Nickle (Ni)	0.0244	0.0248	0.0246	2.000E-04	1
Copper (Cu)	0.1540	0.1544	0.1542	2.000E-04	1

The wastewater samples had an average BOD level of 304 mg/L. According to PNEWS, the EPA of Pakistan has established a 50 mg/L standard for BOD. Water with a sufficient BOD level is clean, but water with a high BOD value is polluting. Bacterial oxygen demand (BOD) is a measure of the amount of dissolved oxygen that bacteria need to aerobically degrade organic compounds. The biological oxygen demand (BOD) was used as a measure for the quantity of organic contaminants in the majority of aquatic environments (Wojnárovits et al., 2024). An average of 408 mg/L was the COD level. Because it shows how sewage discharges could affect ecosystems, it is an important measure of water quality. The concentration of contaminant in wastewater is measured by the COD. According to recent studies, it is defined as the concentration of oxygen needed to completely oxidise organic materials in acidic environments using a powerful oxidising agent such as potassium dichromate, resulting in the production of CO_2, NH_3, and H_2O (Reineke & Schlömann, 2023). Wastewater samples had a pH of 9.33, which is higher than the World Health Organization's recommended range of 6.5–8.5 (Lemessa et al., 2023) but lower than the limit authorised by Pakistan's environmental protection agency.

An essential metric for determining the acidity or basicity of effluent is pH. Soil metal fixation is enhanced by slightly alkaline (7.2–7.6) wastewater, according to research (Tahir et al., 2023). Similarly, it was shown that macrophytes thrive in environments with a pH between 6 and 9 (Srivastava et al., 2008). The sewage wastewater had an alkaline pH since the samples were collected from a residential neighbourhood's sewage treatment facility. This means that the wastewater may include waste products like urine, soap and detergent that are created by home activities. The samples of wastewater had an average temperature of 28.67 °C, which is within the 20–32 °C range recommended by the World Health Organization (Raza et al., 2022). The electrical conductivity (EC) of water is a measure of its electrical conductivity that is directly related to the concentration of ions in the water. The wastewater's electrical conductivity (EC) is 3450 μS/cm, which is higher than the 400–600 μS/cm range that the World Health Organization (WHO) recommends (South Wairarapa District Council, 2017).

Sodium absorption ratio, residual sodium carbonate, bicarbonates and carbonates were 7.06, 7.16, 1.46, and 2.50 me/L, respectively. Among the nutrients found in wastewater, the amounts of calcium (37.6 me/L), potassium (31.33 mg/L), sodium (2.96 mg/L) and magnesium (19.33 mg/L) are the greatest (Raza et al., 2022). Potassium is present in soil in significant amounts (about 3% of the lithosphere), but its insoluble chemical form makes it useless for plant growth without the use of fertilisers (Singh, 2020).

A certain amount of potassium 185 kg/ha is needed to cultivate certain crops. For this criterion, the reclamation of sodic soil is quite important. The sodium adsorption ratio and residual sodium carbonate were both 1.46 me/L. Wastewater with different concentrations of potassium, calcium, magnesium and sodium might affect the saturation of the soil's base. The relationship between sodium concentration and Ca plus Mg levels may be shown by the sodium adsorption ratio (SAR) (Salem et al., 2024).

Metals in wastewater represent a threat to ecological systems and human health since they are poisonous, bioaccumulate, persist and biomagnify throughout food chains (Habib et al., 2024). As a result, metal analysis is necessary before wastewater is used in agriculture. Enzymes rely on copper, which is also a micronutrient in soil and irrigation water and an essential element for haemoglobin synthesis (Husen, 2024). On the other hand, severe copper poisoning may cause gastrointestinal distress, nausea, vomiting, abdominal pain and perhaps death (Lubal, 2024). As a result of air deposition, corrosion of plumbing fittings, industrial processes and animal manure, the water had 0.154 mg/L of copper (Ullah et al., 2024).

According to PNEWS, the zinc content in this research was 4.45 mg/L, which is below the permitted range of 5 ppm. In addition, Younas and Younas found that an overabundance of zinc may lead to anaemia, soreness in the muscles and joints, sudden kidney failure and pancreatitis. Researchers found that there was 0.058 mg/L of cadmium in the wastewater sample. Prostate, kidney, lung, cardiovascular, and skeletal issues, as well as malignancies of the stomach, lungs, kidneys, and prostate, may develop in children exposed to cadmium for long periods of time (Budi et al., 2024). In Fig. 2, a wastewater treatment flowchart showing the several processes that enhance the wastewater's quality from its source all the way to its final destination, which is used for irrigation (Kesari et al., 2021).

7 Organisational Setup and Responsibilities

7.1 National Organisation(s)

7.1.1 Pakistan Council of Research in Water Resources

Organisation type: Public Sector Research Organisation
Roles and responsibilities:

1. To monitor the water quality and to establish a permanent water quality-monitoring network
2. To provide financial and technical support to universities and research institutes for collaborative research

7.1.2 Pakistan Environmental Protection Agency (Plus Provincial Chapters)

Organisation type: Government Organisation for National Policy Drafting
Roles and responsibilities:

1. Strictly enforce national environmental quality standards and self-monitoring and reporting system
2. Devise and implement master plans for the treatment of municipal and industrial wastewater in urban and rural areas

7.2 Regional (Local) Organisation(s)

7.2.1 Institute of Soil and Environmental Sciences, University of Agriculture Faisalabad, Pakistan

Organisation type: University/Research Institute
Roles and responsibilities:

1. Monitoring effect of wastewater application in agricultural lands where mostly vegetables are grown
2. Risk Assessments of metal-contaminated soils
3. Pakistan role of organic and inorganic amendments in immobilisation of heavy metals in wastewater irrigated soils
4. Predict the bioavailability of trace and toxic metals in soil and transfer to the human food chain

7.2.2 Centre of Excellence in Analytical Chemistry, University of Sindh, Jamshoro, Pakistan

Organisation type: University/Research Centre
Roles and responsibilities:

1. Determine effect of sewage sludge application in agricultural lands where mostly vegetables are grown

2. Predict the bioavailability of trace and toxic metals in soil and transfer to the human food chain

7.2.3 Centre of Excellence in Water Resources Engineering, University of Engineering and Technology Lahore

Organisation type: University/Research Institute
Roles and responsibilities:

1. Evaluate environmental impacts of the untreated wastewater irrigation

2. Quantify the extent of groundwater contamination using wastewater irrigation and to evaluate their effects on human health of the local community

7.2.4 Institute of Environmental Studies, University of Karachi

Organisation type: University/Research Institute
Roles and responsibilities:

1. Determine the effect of treated wastewater and equivalent basal fertiliser on growth parameters, chlorophyll and nutrient contents of grain crops and vegetables

7.2.5 Department of Soil and Environmental Science, University of Peshawar, Peshawar

Organisation type: University/Research Institute
Roles and responsibilities:

1. Evaluate concentration of metals in sewage and canal water used for irrigation

2. Use of constructed wetland for the removal of heavy metals from industrial wastewater

7.2.6 Institute of Soil Chemistry, Ayyub Agriculture Research Institute Faisalabad

Organisation type: Research Institute
Roles and responsibilities:

1. Evaluate hazardous effects of sewage water on the environment: Focus on heavy metals and chemical composition of soil and vegetables
2. Long-term disposal of city sewage effluent on agricultural lands

7.2.7 Department of Soil Science, University of Arid Agriculture Rawalpindi

Organisation type: University/Research Institute
Roles and responsibilities:

1. Assessing quality of sewage for irrigation
2. Heavy metals, soil salinity/sodicity and soil fertility status of soils and plants of some sewage-irrigated farms
3. Heavy metal contents of vegetables irrigated by sewage/tube well water

7.3 Other Organisation(s)

7.3.1 IWMI

Organisation type: International Institute/Research Centre
Roles and responsibilities:

1. Health risks of irrigation with untreated urban wastewater: a valuable resource for agriculture
2. The use of untreated wastewater in agriculture: Strategies for Health Risk Management

(Murtaza & Zia, 2012)

7.4 Sampling Methods

There are several methods of sampling

1. Systematic
2. Random
3. Judgmental (nonstatistical)

4. Stratified
5. Haphazard

(Ilyas & Ashfaq, 2020)

References

Albareedy, A. (2023). Water security and its impact on the liver and gut health in Africa. In *Impact of climate change on health in Africa: A focus on liver and gastrointestinal tract* (pp. 195–214). Springer.

Anyango, G. W., Bhowmick, G. D., & Bhattacharya, N. S. (2024). A critical review of irrigation water quality index and water quality management practices in micro-irrigation for efficient policy making. *Desalination and Water Treatment, 100304*.

Budi, H. S., Catalan Opulencia, M. J., Afra, A., Abdelbasset, W. K., Abdullaev, D., Majdi, A., Taherian, M., Ekrami, H. A., & Mohammadi, M. J. (2024). Source, toxicity and carcinogenic health risk assessment of heavy metals. *Reviews on Environmental Health, 39*(1), 77–90.

Chen, R., Tu, H., & Chen, T. (2022). Potential application of living microorganisms in the detoxification of heavy metals. *Foods, 11*(13), 1905.

Das, S. K. (2014). Role of micronutrient in rice cultivation and management strategy in organic agriculture—A reappraisal. *Agricultural Sciences, 5*(9), 765–769.

Dickin, S. K., Schuster-Wallace, C. J., Qadir, M., & Pizzacalla, K. (2016). A review of health risks and pathways for exposure to wastewater use in agriculture. *Environmental Health Perspectives, 124*(7), 900–909.

Enyoh, C. E., Ohiagu, F. O., Verla, A. W., & Verla, E. N. (2022). A chemometric review of heavy metals (Zn, Cd, Pb, Fe, Cu, Ni and Mn) in top soils of Imo state, Southeastern Nigeria. *International Journal of Environmental Analytical Chemistry, 102*(18), 6151–6176.

Fernandes, L. S., Galvão, A., Santos, R., & Monteiro, S. (2023). Impact of water reuse on agricultural practices and human health. *Environmental Research, 216*, 114762.

Gökçeku, H., Kassem, Y., & Seman, H. (2023). Review on waste water reuse for irrigation towards achieving environmental sustainability. *International Journal of Engineering and Applied Physics, 3*(1), 678–688.

Habib, S. S., Naz, S., Fazio, F., Cravana, C., Ullah, M., Rind, K. H., Attaullah, S., Filiciotto, F., & Khayyam, K. (2024). Assessment and bioaccumulation of heavy metals in water, fish (wild and farmed) and associated human health risk. *Biological Trace Element Research, 202*(2), 725–735.

Hafeez, A., Rasheed, R., Ashraf, M. A., Qureshi, F. F., Hussain, I., & Iqbal, M. (2023). Effect of heavy metals on growth, physiological and biochemical responses of plants. In *Plants and their interaction to environmental pollution* (pp. 139–159). Elsevier.

Hashem, M. S., & Qi, X. (2021). Treated wastewater irrigation—A review. *Water, 13*(11), 1527.

Hossain, M. M., Chowdhury, M. A., Hasan, M. J., Rashid, M. H. A., Acter, T., Khan, M. N., Mahatabuddin, S., & Uddin, N. (2021). Heavy metal pollution in the soil-vegetable system of Tannery Estate. *Environmental Nanotechnology, Monitoring & Management, 16*, 100557.

Husen, A. (Ed.). (2024). *Essential minerals in plant-soil systems: Coordination, signaling, and interaction under adverse situations*. Elsevier.

Ilyas, Z., & Ashfaq, M. (2020). *Water sampling, testing techniques and water borne diseases*.

Kesari, K. K., Soni, R., Jamal, Q. M. S., Tripathi, P., Lal, J. A., Jha, N. K., Siddiqui, M. H., Kumar, P., Tripathi, V., & Ruokolainen, J. (2021). Wastewater treatment and reuse: A review of its applications and health implications. *Water, Air, & Soil Pollution, 232*, 1–28.

Khalid, S., Shahid, M., Natasha, Bibi, I., Sarwar, T., Shah, A. H., & Niazi, N. K. (2018). A review of environmental contamination and health risk assessment of wastewater use for crop irrigation with a focus on low and high-income countries. *International Journal of Environmental Research and Public Health, 15*(5), 895.

Khan, M. M., Siddiqi, S. A., Farooque, A. A., Iqbal, Q., Shahid, S. A., Akram, M. T., Rahman, S., Al-Busaidi, W., & Khan, I. (2022). Towards sustainable application of wastewater in agriculture: A review on reusability and risk assessment. *Agronomy, 12*(6), 1397.

Kurniawan, S. B., Ahmad, A., Said, N. S. M., Imron, M. F., Abdullah, S. R. S., Othman, A. R., et al. (2021). Macrophytes as wastewater treatment agents: Nutrient uptake and potential of produced biomass utilization toward circular economy initiatives. *Science of the Total Environment, 790*, 148219.

Lemessa, F., Simane, B., Seyoum, A., & Gebresenbet, G. (2023). Assessment of the impact of industrial wastewater on the water quality of rivers around the Bole Lemi Industrial Park (BLIP), Ethiopia. *Sustainability, 15*(5), 4290.

Lubal, M. J. (2024). Health effects of heavy metal contamination in drinking water. *Uttar Pradesh Journal of Zoology, 45*(10), 16–25.

Mahjoub, O., Mauffret, A., Michel, C., & Chmingui, W. (2022). Use of groundwater and reclaimed water for agricultural irrigation: Farmers' practices and attitudes and related environmental and health risks. *Chemosphere, 295*, 133945.

Mezlini, W., Amor, R. B., Beneduci, A., Romdhane, I. B., Shammas, M. I., Almazroui, M., & Attia, R. (2024). Effects of irrigation water quality on soil physico-chemical proprieties: Case study in North-West of Tunisia. *Earth Systems and Environment, 8*(4), 1541–1561.

Minhas, P. S., Saha, J. K., Dotaniya, M. L., Sarkar, A., & Saha, M. (2022). Wastewater irrigation in India: Current status, impacts and response options. *Science of the Total Environment, 808*, 152001.

Mishra, S., Kumar, R., & Kumar, M. (2023a). Use of treated sewage or wastewater as an irrigation water for agricultural purposes-Environmental, health, and economic impacts. *Total Environment Research Themes, 6*, 100051.

Mordechay, E. B., Mordehay, V., Tarchitzky, J., & Chefetz, B. (2021). Pharmaceuticals in edible crops irrigated with reclaimed wastewater: Evidence from a large survey in Israel. *Journal of Hazardous Materials, 416*, 126184.

Motoshita, M., Pfister, S., Sasaki, T., Nansai, K., Hashimoto, S., Yokoi, R., Islam, K., & Finkbeiner, M. (2023). Responsibility for sustainable water consumption in the global supply chains. *Resources, Conservation and Recycling, 196*, 107055.

Murtaza, G., & Zia, M. H. (2012, May). Wastewater production, treatment and use in Pakistan. In *Second regional workshop of the project 'safe use of wastewater in agriculture* (pp. 16–18). University of Agriculture.

Pacific, W. (2023). *World Health Organization WHO Regional websites*. World.

Pratap, B., Kumar, S., Purchase, D., Bharagava, R. N., & Dutta, V. (2021). Practice of wastewater irrigation and its impacts on human health and environment: A state of the art. *International Journal of Environmental Science and Technology*, 1–16.

Rahman, M. A., Lee, S. H., Ji, H. C., Kabir, A. H., Jones, C. S., & Lee, K. W. (2018). Importance of mineral nutrition for mitigating aluminum toxicity in plants on acidic soils: Current status and opportunities. *International Journal of Molecular Sciences, 19*(10), 3073.

Raza, H., Haq, T., Imran, M., Ikram, N. A., & Shoukat, M. B. (2022). Wastewater contaminants characterization and evaluation from Wali Muhammad distributary, Multan, Punjab. *Journal of Innovative Sciences, 8*(2), 243–249.

Reineke, W., & Schlömann, M. (2023). Biological waste water treatment. In *Environmental microbiology* (pp. 467–491). Berlin, Heidelberg, Springer Berlin Heidelberg.

Revitt, D. M., Lundy, L., & Fatta-Kassinos, D. (2021). Development of a qualitative approach to assessing risks associated with the use of treated wastewater in agricultural irrigation. *Journal of Hazardous Materials, 406*, 124286.

Salem, M. A., Sharif, O. A., Alshatory, M. M., & Assad, M. E. H. (2024). Evaluation of groundwater quality and its suitability for irrigation of ALSHATI Agricultural Project, Libya. *Ned University Journal of Research, 21*(2), 19–35.

Scanlon, B. R., Fakhreddine, S., Rateb, A., de Graaf, I., Famiglietti, J., Gleeson, T., Grafton, R. Q., Jobbagy, E., Kebede, S., Kolusu, S. R., Konikow, L. F., Di Long, M. M., Schmied, H. M., Mukherjee, A., MacDonald, A., Reedy, R. C., Shamsudduha, M., Simmons, C. T., Sun, A.,

Taylor, R. G., Villholth, K. G., Vörösmarty, C. J., & Zheng, C. (2023). Global water resources and the role of groundwater in a resilient water future. *Nature Reviews Earth & Environment, 4*(2), 87–101.
Shahid, M., Khalid, S., Niazi, N. K., Murtaza, B., Ahmad, N., Farooq, A., Zakir, A., Imran, M., & Abbas, G. (2021). Health risks of arsenic buildup in soil and food crops after wastewater irrigation. *Science of the Total Environment, 772*, 145266.
Sharma, K., Rajan, S., & Nayak, S. K. (2024). Water pollution: Primary sources and associated human health hazards with special emphasis on rural areas. In *Water resources management for rural development* (pp. 3–14). Elsevier.
Shetty, S. S., Deepthi, D., Harshitha, S., Sonkusare, S., Naik, P. B., & Madhyastha, H. (2023). Environmental pollutants and their effects on human health. *Heliyon, 9*(9).
Silva, J. A. (2023). Wastewater treatment and reuse for sustainable water resources management: A systematic literature review. *Sustainability, 15*(14), 10940.
Singh, V. (2020). *Environmental chemistry: Water and soil pollution*. Horizon Books (A Division of Ignited Minds Edutech P Ltd).
Singh, S. (2021). *Crisis of water and water in crisis: Some reflections from India*. Reflections on 21st Century Human Habitats in India: Felicitation Volume in Honour of Professor MH Qureshi, pp. 143–166.
Singh, M., Karmakar, R., Ganguli, S., & Mitra Ghosh, M. (2021). Report of antibiotic resistance in urban and rural wastewaters from West Bengal, India.
South Wairarapa District Council. (2017). *Ecological survey of Donald Creek and Otauira Stream*, 2016.
Srivastava, J., Gupta, A., & Chandra, H. (2008). Managing water quality with aquatic macrophytes. *Reviews in Environmental Science and Bio/Technology, 7*, 255–266.
Tahir, M. S., Sagir, M., & Tahir, M. B. (2023). *Advances in water and wastewater treatment technology* (pp. 1–14). Springer.
Ullah, S., Shah, S. S. A., Altaf, M., Hossain, I., El Sayed, M. E., Kallel, M., El-Bahy, Z. M., ur Rehman, A., Najam, T., & Nazir, M. A. (2024). Activated carbon derived from biomass for wastewater treatment: Synthesis, application and future challenges. *Journal of Analytical and Applied Pyrolysis, 106480*.
Wojnárovits, L., Homlok, R., Kovács, K., Tegze, A., & Takács, E. (2024). Wastewater characterization: Chemical oxygen demand or total organic carbon content measurement? *Molecules, 29*(2), 405.
Xu, E., Liu, Y., Gu, D., Zhan, X., Li, J., Zhou, K., Zhang, P., & Zou, Y. (2024). Molecular mechanisms of plant responses to copper: From deficiency to excess. *International Journal of Molecular Sciences, 25*(13), 6993.
Yahaya, S. M., Mahmud, A. A., & Abdu, N. (2023). The use of wastewater for irrigation: Pros and cons for human health in developing countries. *Total Environment Research Themes, 6*, 100044.

Modern Irrigation Practices in the Use of Wastewater

Tayyaba Mumtaz, Maham Tariq, Mazhar Ali, Muhammad Mubeen, Muhammad Adnan Shahid, and Nasir Masood

Abstract The increasing demand for agricultural water by utilizing real-time sensor data, modern technologies such variable rate irrigation (VRI) and precision irrigation (PI) enable the effective and precise application of treated wastewater. These developments reduce environmental risks such as runoff and groundwater pollution, maximize water use, and enhance fertilizer delivery. Emitter clogging is a frequent issue in drip irrigation, which is vital in the campaign for worldwide water scarcity. It reduces crop yields by conflicting in irrigation uniformity and system performance. This problem is exacerbated by the use of salt water in fertigation, highlighting the need for innovative methods to prevent blockages.

Irrigation management is being transformed by automation, which is being propelled by developments in artificial intelligence, robots, sensors, and cloud computing. With web dashboards, farmers can now remotely operate water pumps and keep an eye on variables like soil moisture content and flow rates. In conclusion, modern wastewater method of irrigation enables safe, efficient, and environmentally friendly water management, which is crucial to maintaining in natural resources and sustainable agriculture.

Keywords Drip irrigation · Chemical clogging · Sensors · Internet of Things (IoT) · Fertigation

---

T. Mumtaz · M. Tariq
Department of Environmental Sciences, COMSATS University Islamabad, Vehari Campus, Vehari, Pakistan

M. Ali · M. Mubeen (✉)
Department of Biotechnology, COMSATS University Islamabad, Vehari Campus, Vehari, Pakistan
e-mail: mazharali@cuivehari.edu.pk; muhammadmubeen@cuivehari.edu.pk

M. A. Shahid
College of Life Sciences and Agriculture, University of New Hampshire, Durham, NH, USA

N. Masood
Memorial University of Newfoundland, St. John's, NL, Canada
e-mail: nmasood@grenfell.mun.ca

M. Mubeen et al. (eds.), *Innovations in Agricultural Water Management*,
https://doi.org/10.1007/978-3-031-91883-4_11

1 Introduction

Water recycling from ancient to the present is also the tale of how water recycling changed over time, especially in the late 1800s and early 1900s when enhanced wastewater treatment techniques emerged. This could be attributed to a number of factors, including the world's limited supply of freshwater, population increase, the rise of megacities, climate change, and rapid technological advancements.

In an era where the sustainable management of water resources has become a paramount concern, modern agricultural practices are undergoing a transformative shift towards efficient water usage. Among these, the utilization of wastewater for irrigation has emerged as a novel and innovative solution to address both water scarcity and environmental challenges. The judicious application of treated wastewater in agricultural settings not only provides an alternative water source but also offers an opportunity to recycle nutrients and reduce pollution. This integration of modern irrigation practices with the use of wastewater exemplifies a convergence of technology, environmental awareness, and agricultural productivity (El Sabagh et al., 2020d; Akram et al., 2019a, b; Fahad et al., 2019; Amanet et al., 2019; Sami ul Din et al., 2019; Ali et al., 2019; Tariq et al., 2019; Zahoor et al., 2019). However, its adoption is not without complexities, as it requires a delicate balance between harnessing the benefits of wastewater while mitigating potential health and ecological risks. This exploration delves into the contemporary landscape of modern irrigation practices involving wastewater, uncovering its benefits, challenges, and the broader implications it holds for sustainable agriculture and water resource management. This approach involves treating and reusing treated wastewater from various sources, such as households, industries, and municipalities, to irrigate crops and enhance agricultural productivity (Hussain et al., 2021a, b, 2022a, b, c; Khaliq, 2022; Hammad et al., 2022; Tahir et al., 2022; El Sabagh et al., 2022).

However, the adoption of wastewater irrigation also raises important considerations. Proper treatment and filtration processes are crucial to ensure that harmful pathogens, chemicals, and heavy metals are removed from the wastewater, preventing potential health risks to both crops and consumers. Careful planning is necessary to avoid soil and groundwater contamination, as well as to manage the buildup of salts and pollutants in the soil over time.

To successfully implement modern wastewater irrigation, a comprehensive approach is required. This includes advanced wastewater treatment technologies, efficient distribution systems, and monitoring mechanisms to assess water quality, soil conditions, and crop health. Local regulations and guidelines play a pivotal role in governing the safe use of wastewater in agriculture. Treated wastewater, also known as reclaimed water, can provide a new source of irrigation water, alleviating the pressure on traditional freshwater supplies (Akram et al., 2017a, b, 2018a, b; Hasnain et al., 2017; Hashmi et al., 2017; Ahmad et al., 2015).

2 Wastewater Treatment and Reclamation

Wastewater treatment processes aim to remove contaminants from sewage and industrial effluents, making the water safe for reuse. Reclaimed water, also known as recycled water, is the treated wastewater that meets specific quality standards for nonpotable uses like irrigation, industrial processes, and groundwater recharge. Advanced treatment technologies, including membrane filtration, reverse osmosis, and ultraviolet (UV) disinfection, enhance the quality of reclaimed water.

3 On-Farm Irrigation Techniques

3.1 Drip Irrigation

At the forefront of water-efficient irrigation practices stands drip irrigation, a game-changing technique that revolutionizes how water is delivered to plants. Unlike traditional surface irrigation methods, which can result in significant water losses due to evaporation and runoff, drip irrigation precisely delivers water directly to the root zones of plants through an intricate network of pipes, tubes, and emitters. This pinpoint accuracy significantly reduces the risk of water wastage and minimizes the potential for contamination, making it an ideal candidate for wastewater utilization. When considering the integration of wastewater into drip irrigation, proper treatment becomes paramount. Wastewater, after undergoing advanced filtration, disinfection, and nutrient removal processes, can meet the quality standards necessary for safe agricultural use. The synergy between drip irrigation and treated wastewater ensures efficient water use and optimal nutrient delivery, contributing to sustainable crop growth while conserving freshwater resources (Shahzad et al., 2022; Din et al., 2022; Balal et al., 2022; Abbas et al., 2020). Micro-irrigation is an effective plan for managing a regional water scarcity. In the past decade, both the global irrigated area and the use of micro-irrigation systems have grown, with micro-irrigation seeing a significant 43% increase over the last 5 years, according to the International Commission on Irrigation and Drainage (ICID). This expansion underscores the growing challenge of emitter contamination, as discussed in Shi et al.

3.1.1 Chemical Clogging Method

Drip irrigation is celebrated worldwide for its precise control and outstanding water efficiency, making it one of the most popular micro-irrigation techniques. This technology delivers fertilizers using irrigation systems like drip or sprinkler systems, revolutionizing conventional fertilization methods. It frequently involves fertigation. Fertigation dramatically lowers the amount of water and fertilizer used.

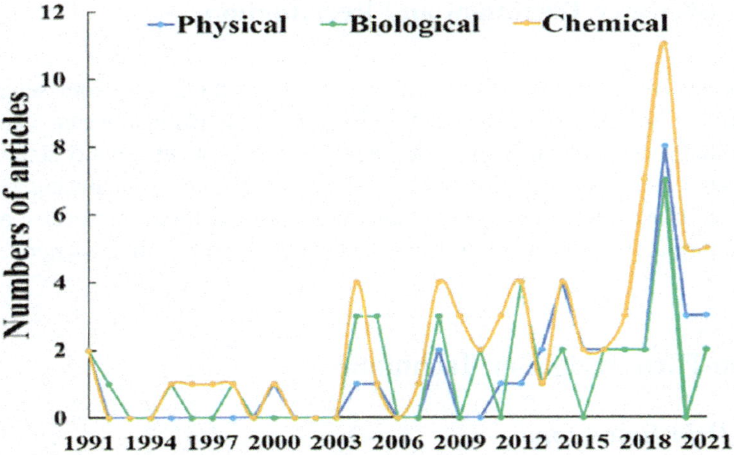

Fig. 1 Research articles on drip irrigation emitter clogging published between 1991 and 2021

However, because of the precipitate evolution caused by using saline water which has a high concentration of salt it can cause chemical blockage. Likewise, increased use of fertigation can lead to reactions between fertilizer nutrient ions and irrigation water ions, possibly producing precipitates that are insoluble and exacerbate chemical blockage in emitters. In addition, a major cause of emitter obstructions is the use of salted water for irrigation and fertigation (Mehmood et al., 2021; El Sabagh et al., 2021; Javeed et al., 2021a, b; Ali et al., 2021; Mubeen et al., 2021). When we searched for "Drip Irrigation and Clogging" in the Web of Science core collection database between 1991 and 2021, 294 items were found, as seen in Fig. 1. For in-depth examination, the references were divided into three themes: "Chemical clogging," "Biological clogging," and "Physical clogging." The results reveal that chemical clogging has emerged as a particularly pressing issue, particularly since 2018. This is likely attributed to the growing use of saline water for irrigation and the widespread implementation of drip fertigation techniques worldwide (Hussain et al., 2020; El Sabagh et al., 2020a, b, c; Ahmed et al., 2020).

3.1.2 Reasons and Techniques for Assessing Drip Irrigation Emitter Blockage

3.1.2.1 Types of Clogged Emitters and Their Causes

Particles of solid matter, chemical precipitates, bacteria, and other elements in irrigation water build up in the emitters or laterals of drip irrigation systems. Emitter clogging results from it. The accumulation reduces the flow rate and uniformity of the irrigation. A difficult and inevitable problem with agricultural practices is clogging. To the ideas provided in the International Organization for

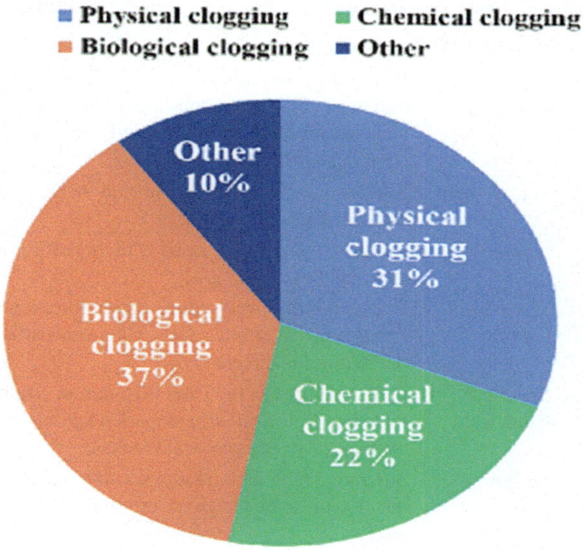

Fig. 2 The forms and likelihood of emitter blockage in drip irrigation, derived from data from an FAO consultant's statistics

Standardization (ISO), blockage is indicated when the actual flow rate of an emitter is less than 75% of its intended capacity. Based on facts supplied by a Food and Agriculture Organization consultant, there are three different types of clogging that can occur in drip irrigation systems: physical clogging (31%), chemical clogging (22%), and biological clogging (37%) with various forms making up the remaining 10% (Fig. 2). Furthermore, there is a greater chance of chemical blockage in emitters as fertigation becomes more widely used.

Physical Clogging

When filtration equipment is unable to remove organic or inorganic suspended debris from irrigation water, such as algae, plastic pieces, sand, silt, and clay particles, physical blockage develops. It also involves subsurface drip irrigation system obstruction brought on by negative pressure mud suction. Even with the use of high-quality water, solid particles are the most frequent source of physical clogging and frequently result in emitter obstructions. Table 1 illustrates how different water sources contribute to this problem. Physical clogging in drip irrigation systems is hard to avoid, and the more particles there are and their size, the worse the clogging gets.

Table 1 Solid particles under various water sources are the cause of emitter clogging

Factors	Water source	Main conclusion
Concentration of solid particles	Not applicable	When drip irrigation water's dissolved solids are less than 500 mg/L, clogging issues are minimal. However, when the concentration exceeds 2000 mg/L, the risk of severe clogging significantly increases
	Turbid water	Emitter clogging grew worse with increasing sediment content, especially at 1.25 g/L
Size of solid particles	Sewage	Emitter blockage worsened with increasing sediment content, especially over 1.25 g/L
	Water from rivers	Even with improved sedimentation filtration technology, the direct cause of emitter obstruction was particles smaller than 0.076 mm in diameter
	Sediment-laden water	Particles between 0.05 and 0.10 mm in size have the capacity to gather within the drip irrigation laterals' lumen and flow channels, causing physical blockage
	Turbid water	The range of 0.03–0.04 mm was the sensitive particle size at which emitter blockage occurred

Biological Clogging

Slime accumulation can result from the growth and reproduction of bacteria, zooplankton, algae, and other organisms in an environment that drip irrigation systems can foster. This slime has the potential to combine with mineral particles and clog emitters biologically. Furthermore, biological blockage in subsurface drip irrigation refers to a decrease in irrigation flow brought on by plant root invasion. In systems where the water has larger concentrations of microorganisms and organic materials, biological blockage is more likely. Table 2 lists the components that lead to biological blockage of emitters and highlights the fact that, whereas particulate matter is the primary substance involved, clogging of emitters does not originate from particulate matter. The persistent adsorption of suspended particulate matter can result from the viscosity of the emitter walls being increased by the secretions of microorganisms. This procedure is thought to be a major factor in emitter blockage. In actuality, the methods used for system operation, water quality, and active microbial species can all have a direct impact on the establishment of biofilms. By secreting viscous chemicals to build biofilms, microorganisms contribute significantly to emitter obstruction. Chlorine's potent oxidation can significantly reduce biological blockage by killing or suppressing the growth of bacteria and other bacteria and also the development of clumps and slime. Using hydraulic shear force, another popular technique for elimination of and cleaning biofilm from the lateral and emitter is lateral flushing. A combination of lateral flushing and chlorination could efficiently lower the microbial load. Based on gene sequencing, biological blockages have recently been demonstrated to be resolved by electromagnetic field treatment. It is projected that this treatments procedure is going to be effective, chemical-free, and antibiological.

Table 2 The primary factors leading to biological clogging of emitters

Variables	Source of water	Principal findings
Microbes and particulate matter	Drainage	The emitter became clogged due to the accumulation of solid particles and bacteria from the main body of biofilm on the inner wall of the emitter flow channel, which continuously adhered to one another and accumulated suspended particles
Protozoa and particulate substances	Freshwater + treated sewage effluent	The primary blocking materials were biomass from in-line proliferating protozoa and gelatinous aggregates of fine organic and inorganic particles
Microbial community, moss, and bacteria	A mixture of freshwater, secondary sewage, and winter storm runoff is present in the reservoir	Microbial communities, moss, and bacteria were found to release viscous compounds that were closely associated with clogged emitters
Organic materials and microorganisms	Recycled and potable water	The presence of adsorbed microbial floc, which is made up of bacteria and organic materials, was responsible for 90% of the blockage of emitters produced by sand particles

Chemical Clogging

When materials that dissolve in water, like hydroxides, carbonates, phosphates, sulfates, and silicates, Fe_3O_4, Ca_3O, Mg_3O, and sulfides form chemical precipitates under specific circumstances, it can lead to chemical clogging. Using fertigation, salt water, or groundwater can cause this kind of blockage. Water temperature, pH, ion concentration, and system pressure are some of the variables that affect chemical blockage. Emitter blockage is frequently caused by chemical precipitation, especially when using saline water and drip fertigation. The soluble materials that cause chemical blockage of emitters are listed in Table 3. The chemical blocking mechanisms are multifaceted and intricate.

Factors Contributing to Chemical Clogging

- **Salinity**: Wastewater often contains elevated levels of salts due to the presence of dissolved minerals. These salts can build up in the soil over time, leading to a decrease in soil permeability. As a result, water infiltration is hindered, and plants may suffer from water stress and reduced nutrient uptake.
- **Mineral precipitation**: Certain minerals present in wastewater, such as calcium, magnesium, and iron, can undergo precipitation reactions under certain conditions. These precipitates can accumulate in the narrow passages of irrigation pipes and emitters, gradually reducing their diameter and causing clogs.
- **Organic matter decomposition**: Organic materials present in wastewater can decompose over time. This decomposition can produce byproducts that, when combined with other substances, contribute to the formation of deposits that block irrigation components.

Table 3 Soluble materials in water sources causing blockages in emitters

Induce factors	Main results
Acidity level	When the pH of the water is more than 8, severe blockage is more likely to happen
Fe^{2+}	If the concentration of Fe^{2+} exceeds 1.5 g/m^3, it may severely damage the irrigation system
HPO_4^{2-}andPO_4^{3-}	Fertilizers containing phosphorus might cause chemical precipitation inside of emitters
Ca^{2+},Mg^{2+}andHCO_3^-	Emitter clogging is generally rare when the Ca + Mg (water hardness) is less than 150 mg/L. However, the risk of clogging rises considerably when hardness levels exceed 300 mg/L, increased levels of HCO_3^-, Mg^{2+}, and Ca^{2+} in irrigation water, especially during fertigation, further increase the risk. Additionally, the likelihood of scale formation grows when the irrigation water's pH is higher than 7.5 and its bicarbonate levels exceed 5 mmol/L
Mn^{2+}	If Mn^{2+} concentrations exceed 0.1 mg/L, clogging will occur, and blockage will get worse if they exceed than 1.5 mg/L
H_2S	When H^+ concentrations surpass 0.5 mg/L, clogging will happen, and when H^+ levels surpass 2 mg/L, the issue will get worse

- **Microbial growth**: Pathogens and microorganisms present in wastewater can multiply within irrigation systems, forming biofilms that adhere to surfaces. These biofilms can trap particles and promote the accumulation of deposits, leading to clogging.

Strategies for Managing Chemical Clogging

- **Pre-treatment**: The first line of defense against chemical clogging involves treating the wastewater before it enters the irrigation system. Filtration processes, sedimentation tanks, and chemical treatments can be used to remove or reduce contaminants, including suspended solids and some dissolved substances.
- **Crop selection**: Opting for crops that have higher tolerance to salinity and other contaminants can help mitigate the negative impacts of chemical clogging. Some plants are better equipped to handle water with elevated salt levels, thereby reducing the risk of water stress and yield reduction.
- **Irrigation techniques**: Drip irrigation, which delivers water directly to the root zone of plants, is less susceptible to clogging compared to other methods like sprinkler irrigation. Using filtered water in drip systems can further reduce the risk of clogs.
- **Regular maintenance**: Implementing a routine maintenance schedule is crucial. Cleaning filters, unclogging emitters, and inspecting the entire irrigation system can prevent the gradual buildup of deposits and maintain system efficiency.
- **Soil management**: Improving soil structure through practices like adding organic matter and applying soil conditioners can enhance water infiltration. Well-structured soil is less prone to clogging, as water can move through it more freely.

- **Water management**: Balancing the use of wastewater with other water sources can help dilute contaminants, reducing the overall risk of clogging. Proper scheduling of irrigation events can also prevent over-accumulation of salts and minerals.

Compound Clogging

Filtration systems cannot completely remove salts, organic materials, suspended particles, and microbes from water. The quality of water significantly affects the nature of clogged emitters in irrigation systems.

While saline water is commonly associated with chemical clogging, the presence of quartz and silicate in clogging materials suggests that use saline water for irrigation might result in both chemical and physical blockages. In water sources with high sediment levels, the size and concentration of sediment particles are major contributors to emitter blockage. Additionally, research has shown that biofilm formation on sediment surfaces exacerbates emitter clogging. This suggests that emitter clogging in high-sediment water sources involves not just physical clogging but also biological clogging.

Compound clogging is a phenomenon that occurs when various substances present in irrigation water accumulate and form deposits within the irrigation system components. These substances can include minerals, salts, organic matter, algae, and bacteria. When wastewater is used for irrigation, it often contains higher concentrations of these compounds compared to freshwater. As the water flows through the irrigation pipes, filters, and emitters, these compounds can precipitate and adhere to the inner surfaces, leading to clogs. These clogs can restrict water flow, increase pressure within the system, and ultimately reduce the efficiency of irrigation.

Factors influencing compound clogging include the quality of the wastewater, the type of irrigation system used, and the climatic conditions. Different types of irrigation systems, such as drip irrigation or sprinkler irrigation, can have varying susceptibility to clogging. Proper system design, maintenance, and filtration are essential to mitigate the effects of compound clogging.

3.2 Subsurface Irrigation

Subsurface irrigation is a revolutionary concept that reimagines the way water is delivered to crops. By infiltrating water directly below the soil surface, near the root zones of plants, this technique effectively addresses the limitations of surface irrigation methods. Surface evaporation losses are minimized as water bypasses the exposed soil surface, resulting in more efficient water utilization. Moreover, subsurface irrigation reduces the risk of crop contamination from wastewater, as the foliage remains largely untouched by water.

The integration of treated wastewater into subsurface irrigation systems enhances its impact. Treated wastewater, which has undergone stringent purification processes, can be confidently channeled into subsurface irrigation. The result is a dual advantage: efficient water uses and a reduced risk of pathogen transmission. This approach promotes not only agricultural productivity but also safeguards human health by minimizing the potential for exposure to contaminants.

3.3 Biofiltration and Phyto-remediation

Biofiltration systems present a holistic approach to wastewater treatment and irrigation. These systems harness the inherent capabilities of vegetation and microorganisms to purify wastewater. In a biofiltration setup, wastewater is directed through specially designed areas where plants absorb nutrients and contaminants, while microorganisms in the soil break down organic matter and neutralize pathogens. As the treated water permeates through the soil, pollutants are captured and rendered harmless, resulting in water that is safe for agricultural use.

The integration of biofiltration and phyto-remediation techniques into irrigation practices offers a compelling two-fold benefit. First, the treated wastewater contributes to efficient water utilization, conserving precious freshwater resources. Second, the process enriches the soil with nutrients derived from the breakdown of contaminants, thereby enhancing soil fertility and supporting sustainable crop growth. This approach underscores the interconnectedness of water management, soil health, and crop productivity.

4 Technological Innovations

4.1 Sensor-Based Irrigation

Irrigation management has advanced significantly as a result of the fusion of technology and agriculture, especially with sensor-based systems. This innovation relies heavily on soil moisture sensors, which provide up-to-date information on the water content of soil. This enables farmers to decide on irrigation volume and timing with knowledge. Soil moisture sensors improve water efficiency and promote healthy plant growth by guarding against both overwatering and underwatering incidents.

In addition to soil moisture sensors, environmental sensors also detect things like temperature, humidity, and light intensity. A thorough grasp of the crop's growth environment is possible through the integration of data from many sensors. With the help of this combined data, farmers can improve their irrigation techniques to make sure that plants are growing in ideal conditions and with right amounts of water.

4.2 Smart Irrigation Controllers

Smart irrigation controllers exemplify the transformative impact of data-driven decision-making in agriculture. These controllers amalgamate real-time weather forecasts, soil moisture measurements, and plant water requirements to dynamically adjust irrigation schedules. When rain is anticipated, the controller intelligently postpones irrigation, preventing unnecessary water application.

The influence of smart irrigation controllers extends beyond water conservation. By preventing overwatering, which can lead to root diseases and nutrient leaching, these controllers enhance plant health and productivity. Moreover, they empower farmers to manage their irrigation systems with greater precision, making adjustments that align with the ever-changing dynamics of weather and plant growth.

5 Weather-Based Irrigation Controllers (WBICs)

Weather-based irrigation controllers, also known as smart irrigation controllers or ET-based controllers, are innovative technologies that automatically adjust irrigation schedules based on real-time weather and environmental data. These controllers gather information on temperature, humidity, wind speed, and solar radiation to calculate evapotranspiration (ET) the rate at which water evaporates from the soil and transpires from plants. By using this data, they accurately determine the amount of water needed for irrigation, reducing the risk of overwatering and promoting water conservation.

Inefficient irrigation practices can lead to water waste, increased energy consumption, and diminished crop yields. Weather-based irrigation controllers (WBICs) address these problems by automatically adjusting irrigation schedules based on real-time weather conditions and environmental factors. These controllers collect data on temperature, humidity, wind speed, solar radiation, and rainfall to calculate evapotranspiration (ET), which indicate the rate at which water is lost from the soil through evaporation and plant transpiration.

By integrating ET data and weather forecasts, WBICs provide valuable insights into the water needs of crops. This facilitates accurate irrigation scheduling, making sure that crops get the precise amount of water they need while avoiding overwatering. As a result, water wastage is minimized, and crop health and yields are optimized. Advanced WBICs may also integrate soil moisture sensors, allowing for even more accurate irrigation decisions based on the actual moisture content of the soil.

6 Fertigation Using Treated Wastewater

Fertigation involves applying nutrients, particularly biogenic compounds like nitrogen and phosphorus. The presence of NPK (nitrogen, phosphorus, and potassium) in secondary effluent acts as a fertilizer, providing essential macronutrients that support plant growth.

7 Variable Rate Irrigation (VRI)

Variable rate irrigation (VRI) is an advanced irrigation method that optimizes the use of treated wastewater, enhancing both efficiency and sustainability. By incorporating VRI into wastewater irrigation systems, several key benefits are realized:

VRI systems use real-time data from sensors to apply treated wastewater in precise amounts tailored to different areas of a field. This approach ensures that each zone receives the optimal amount of water based on its specific needs, improving overall water use efficiency. Treated wastewater often contains valuable nutrients like nitrogen and phosphorus. VRI allows for precise adjustment of irrigation rates to match the nutrient needs of various crops and soil conditions, thereby maximizing nutrient uptake and reducing the risk of nutrient runoff. Advanced VRI systems integrate with monitoring technologies to track soil moisture, wastewater quality, and crop conditions. Automated controls use this data to dynamically adjust irrigation schedules and volumes, optimizing the application of treated wastewater and minimizing water waste.

8 An Automated Wastewater Irrigation System Based on IoT Framework

An automated farming framework enabled by the Internet of Things (IoT) consists of a network of sensors spread over the farm. These sensors connect to a central administration hub, which is outfitted with unique hardware and software, via a bidirectional channel (Habib ur Rehman et al., 2020). The quality, sustainability, transparency, and cost-effectiveness of this type of agriculture are all rising with the goal of enhancing crop production while minimizing labor costs. In this regard, we propose a system that collects real-time environmental data using Wireless Sensor Networks (WSN), thereby optimizing irrigation timing. The Kingdom of Saudi Arabia, with a coastline reaching 1750 km, this vast country, which spans 2.25 million km^2, is bordered to the north by Jordan, Iraq, and Kuwait; to the east by the Gulf, Bahrain, and the United Arab Emirates; to the south by Oman and Yemen; and to the west by the Red Sea. Developing self-sufficiency in poultry and dairy production is the main objective in Shabana Habib and Saudi Arabia.

Soil moisture data is collected through controlled farming procedures, indoor temperature profiles of covered agricultural sheds, rainfall, and intra-field wind speeds. This data is stored on a computer server that provides support for climate prediction, soil condition analysis, and disturbance analysis (Habib ur Rehman et al., 2020).

A fuzzy controller was used to manage irrigation by analyzing the moisture index, water stress, and ongoing monitoring of crop water fluctuation and soil moisture.

This system automates irrigation through flooding methods and allows remote observation of weather patterns and agricultural variables through an application, enabling farmers to control water gates in canals from a far. Additionally, a method was developed to evaluate root stress in hydroponic crops by regulating environmental leaf temperatures and light intensity between 28 and 32 °C. This method classifies root conditions into natural, stressed, and rotten categories based on observed patterns. By incorporating IoT technologies, the system lowered the reaction time from 5 min to less than 60 s and increased detection accuracy from 85% to 95%, outperforming the Easy Leaf Area application, which measures reductions in leaf area (Habib ur Rehman et al., 2020). Different methods of estimating irrigation needs have been used by various irrigation systems for both soil and soilless crops. Effective irrigation management requires monitoring and controlling factors that are essential for running drip irrigation, ventilation systems, micro-sprinklers, and soil moisture, evapotranspiration, and water stress index. For example, using temperature and humidity ventilation, and heating systems readings to regulate the environment in agricultural sheds by switching on the appropriate irrigation.

Similarly, Mohamed and Poyen estimate irrigation needs by measuring daily evapotranspiration. In order to improve water productivity and conservation, Mohamed employs the Penman–Monteith method, while Poyen integrates the ability to automatically choose between the Hargreaves, Samni, Kharufa, and Penman–Monteith methods based on climatic and geographical factors. Both methods make use of modern technologies and intelligent systems. These illustrations illustrate the diverse ways in which the term "smart" is applied in research and prototypes, showcasing various capabilities and features.

The research community has created a number of taxonomies that classify smart technologies according to their attributes in order to address the difficulty of implementing a one-size-fits-all strategy. Describe two key phases in their framework: design and implementation. Each phase involves distinct scenarios where services need to be customized accordingly. The Agro-IoT project, which operated from 2015 to 2020, utilized a six-layer architecture. This structure included sensors, actuators, and wireless nodes, along with a network layer featuring communication protocols. It also comprised an intermediate software layer, an application layer for forecasting and data analysis, and a user layer for sharing data with farmers, subject matter experts, and the supply chain. The project's objectives were early disease detection, crop variety identification, and real-time monitoring of small-scale greenhouses, optimized irrigation, and efficient use of pesticides and fertilizers.

In a similar vein, solar precision agriculture uses an IoT architecture with four layers. This consists of an application layer that gives researchers, experts, and farmers information, a network layer with IoT nodes and base stations, a decision layer with server services, workstations, and knowledge bases, and a sensory layer (Habib ur Rehman et al., 2020).

This study's irrigation solution goes beyond simply managing irrigation systems. In order to find leaks and malfunctions and to increase efficiency and lower costs, it also inspects every part, including sprinklers and water pipes. The system gathers data from the field in real time using Wireless Sensor Networks (WSN) in order to evaluate the state in affairs. Artificial intelligence algorithms use this data in conjunction with weather forecasts, evapotranspiration rates from soil and other surfaces, and farm parameters to calculate the precise quantity of water required for each area of the farm and modify the irrigation controller accordingly.

9 Advanced Communication Systems

Six key components make up an Internet the objects irrigation system: data collection, data monitoring, communication, data processing, storage, and visualization (Fig. 3).

9.1 Data Collection

To collect data from soil moisture sensors, rainfall sensors, ambient humidity sensors, temperature sensors, and other devices, a variety of sensors is needed. Each sensor is configured with specific threshold values based on the kind of sensor and

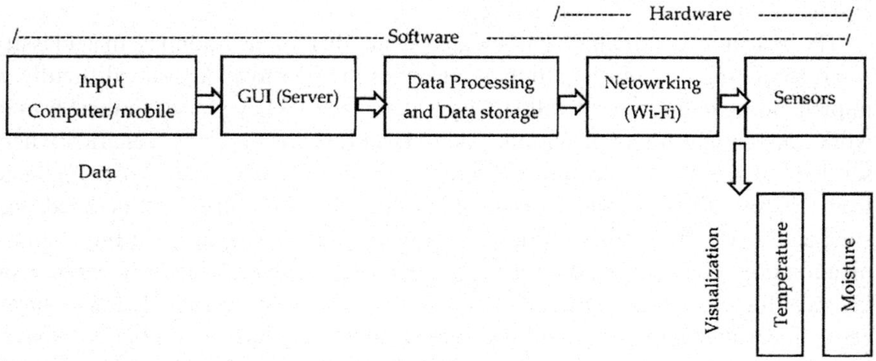

Fig. 3 Fertigation scheme

the crop being watched used (Habib ur Rehman et al., 2020). The sensors collect data at regular intervals, and these measurements are transmitted via Wi-Fi modules for processing according to the parameters of each measurement.

9.2 Data Surveillance and Analysis

To manage and track the sensor information temperature, moisture, and rainfall, a supervisory system is required. It also needs to track crop cycles and other related data. This data is displayed using a Graphical User Interface (GUI) that retrieves the information stored in the database (García et al., 2020). A statistical analysis of water consumption by crops, categorized by their respective dates and sections, has been conducted based on the monitoring data. All data is accessed using web services (WS), which are accessible to authorized personnel on the Internet.

9.3 Communication Network Management

The sensors are linked to development boards for Arduino microcontrollers, which connect to the database via ESP8266 Wi-Fi modules. Because the solution depends on cloud connectivity for data management and transmission, the whole system runs on the cloud. For security reasons, the sensor is not directly connected to the database. Alternatively, a web service (WS) is used to insert data into the database. The two main pieces of information delivered by the Wi-Fi module are: first, the sensor value; and second, the identification number (ID). A unique identity and IP address are used to assign each sensor a unique ID, which allows for tracking the data source. To decide if irrigation is required, this data is processed and if so, the schedule is updated accordingly.

One of the irrigation system's electronic switches is controlled by an Arduino using the Wi-Fi module. Thus, requests cannot be sent to the Wi-Fi module because it lacks a public IP address and cannot be accessed directly from the cloud. Rather, the Wi-Fi module keeps checking the web service for updates. In keeping with the power-on time specified in the WS, the valve is opened. Easy access from the Internet is made possible by the Wi-Fi module's static IP address, which is assigned each time it connects to an Internet gateway (Habib ur Rehman et al., 2020). An algorithm determines the ideal water amount based on the field's moisture levels, and irrigation occurs only when moisture is low. This strategy allowed us to save about 25% of our budget. Previous projects used collected data to increase garden or farm productivity and decrease water usage. These concepts hold theoretical value but may face practical implementation challenges. In this case study, a web application was used for data visualization, though alternative methods could also be considered (Habib ur Rehman et al., 2020). Despite these advancements, our previous systems have struggled with reliability issues, largely due to the chosen sensors. Additionally, they need testing in more extensive scenarios. To enhance results and ensure the solution's effectiveness across the wireless network, it will be crucial for maintain its functionality in a number various settings (García et al., 2020).

9.4 Data Processing

Sensor data, provided as integer values for soil moisture, temperature, and rainfall, is used to guide crop management decisions. If soil moisture levels drop below what the crop needs, this is when the valve will open. The water temperature affects the amount of water is needed for sprinkler irrigation. Furthermore, the system will automatically halt plant watering when it detects rainfall. Based on the sensor readings, the valve will either open or stay closed.

9.5 Information Storage

Crop types, sensors, moisture levels, and statistical data have to be stored in databases. Information about crops, their growth cycles, and irrigation schedules are managed using the application's database structure. Along with the sensors connected to each portion, it also stores data about the sections that get monitored. By utilizing a "smart irrigation" system, the system makes use of IoT technology to determine the exact amount of water needed for every crop during its growth cycle. The administrator will have the ability to manage variables, such as sensor data, and control crop irrigation to ensure that the data is presented effectively. Each crop type has specific irrigation configurations for both visualization and user interaction. If direct configuration is not feasible, the system allows for adjustments based on a predefined setup.

9.6 Visualization

It will be possible for the administrator to control variables like sensor data and oversee crop irrigation to ensure accurate data presentation (Fig. 4). Each crop type comes with tailored irrigation configurations for visualization and user interaction. If direct adjustments are not possible, the system offers the option to modify configurations based on a present arrangement.

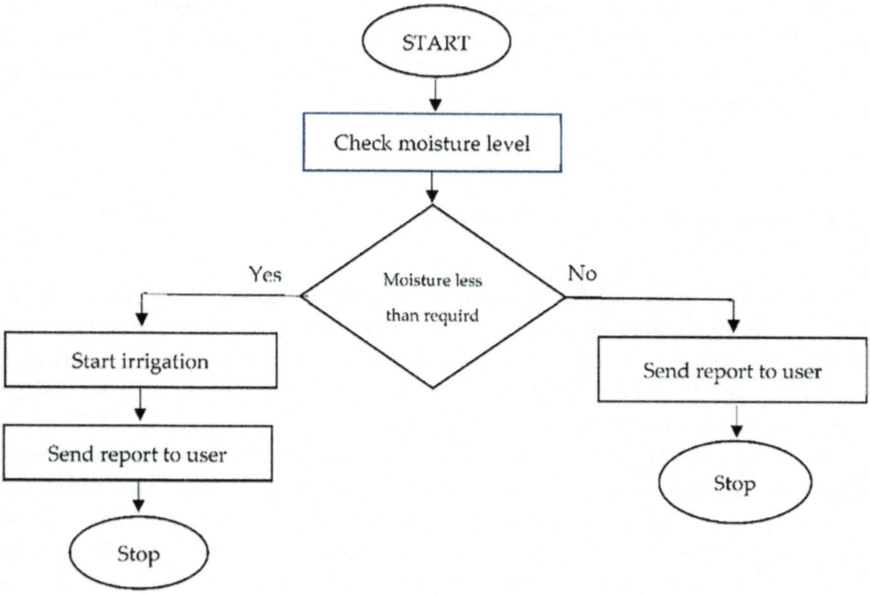

Fig. 4 Block schematic of the irrigation system powered by IoT

9.7 *Flow Diagram*

In order to acquire data regarding the present soil conditions and moisture levels, which are used to modify the state of the irrigation system, the user first interacts with the system through a mobile application or other connected interface. In this study, the sensor node, also known as the mote, measures the temperature, humidity, and soil moisture around each date palm tree (Habib ur Rehman et al., 2020).

The information can be accessed through an automated system, allowing users to make informed decisions about watering based on the data provided by the application or system. This approach enables precise agricultural practices by closely monitoring field conditions and using real-time data to choose the most effective irrigation strategy (Habib ur Rehman et al., 2020).

9.8 *Data Gathering and Connectivity*

Every sensor node has been deployed and linked to the master node of the network. When communication is active, the master node sends information packets to each sensor hub using its unique ID. After the sensors provide data packets, soil parameters, such as soil moisture levels (e.g., 30%), are analyzed. The acidity or alkalinity of the soil is indicated by the pH value. Sensors assess the functionality of the apparatus, and the Arduino microcontroller unit's memory stores these quality metrics

(MCU). Based on predetermined conditions, the MCU executes particular operations in response to the stored values. The MCU participates in the operations of the Internet of Things (IoT) ecosystem by wirelessly transmitting data packets that it receives from the sensor node to the master node.

9.9 Custom Monitoring Dashboard

The dashboard functions like a mobile device and is accessed by a server. Users have the ability to monitor temperature, humidity, and water flow through the dashboard. In addition, customers can use the dashboard to manage the water pump, which includes an On/Off control button to manage the pump and view installed applications (Habib ur Rehman et al., 2020).

Direct current (DC) electricity from solar panels powers the sensors and electronics; the amount of power depends on the rating of the panels. Through the use of a charging kit, the battery is charged by DC power generated by the solar system. An inverter circuit receives this DC power after that. Older water pumps require AC power, Therefore, in order to convert DC electricity to AC, an inverter is required. We have utilized several hardware elements to put this system into operation, including solar panels, charging kits, power sources, and other items transformers (d). Using the software on a smartphone is now possible. But the software deployment of the device is not supported by the big, bloated network model.

9.10 Instantaneous Execution

In the IoT Smart Irrigation, a schematic diagram of the automated irrigation system is displayed. The system has a management unit, a soil moisture detecting unit, an on-site processing unit, and devices for managing the delivery of water. Data on farming is constantly stored in a remote database. Every field sensing mote is utilized to collect data on climatic conditions, such as soil moisture content. By modifying the system in response to the current climate in the field, the user can operate the water valve using a personal internet interface. Several factors influence water usage, including plant type, evapotranspiration rate, field size, valve and pipe types, valve spacing, and the frequency of daily watering. By considering these factors, we can precisely calculate the required water for a field using Eq. (1):

$$\text{Time} = \frac{\text{Area} \times \left(K_{\text{c}} + ET_{\text{r}}\right) \times 60}{\text{Period} \times F \times N \times 100} \tag{1}$$

where

- Time represents the irrigation duration in minutes
- K_c is the crop coefficient (ranging from between 7.04 and 11.7 mm/day)

- F is the water flow per valve expressed in m^3/h
- N is the number of valves
- In m^2, area refers to the field area

We performed three analyses using Eq. (1) with varying data inputs and compared them to a standard watering system:

1. Using forecast data and Eq. (1) to calculate irrigation times.
2. Utilizing sensor data and Eq. (1) to regulate the irrigation system.
3. Calculating and regulating irrigation schedules with sensor data and Eq. (1).

The Internet of Things (IoT) is growing increasingly vital due to its can link almost any object or device to a network. Numerous communication protocols and technologies, including popular networks like WiFi, LTE, and Bluetooth, make this possible. We seek to validate the deployment of a Wireless Sensor Network (WSN) in medium-scale crops for particular crop kinds and designated locations using the suggested technology (Habib ur Rehman et al., 2020).

The main objective of this research was to lower sensor device costs while improving their technical capabilities. This strategy paves the way for large-scale production of monitored and sustainable crops in the future.

10 Conclusion

The three main causes of fouling formation in emitters are physical clogging from water with a high sand content, biological clogging from reclaimed water, and chemical clogging from saline water and fertigation. It is necessary to develop several anticlogging characteristics for emitter material, such reducing microorganisms' activity, reducing weed growth, and preventing chemical precipitation. The effective and targeted use of wastewater is ensured using techniques like variable rate irrigation (VRI) and precision irrigation that raise water use and fertilizer distribution while reducing environmental hazards like runoff and groundwater contamination. These methods improve water management and improve sustainability by the use in modern monitoring systems and automation. All things considered, the utilization of treated wastewater for modern irrigation helps preserve freshwater supplies, improve sustainable farming practices, and guarantee environmental stewardship over a long time.

References

Abbas, G., et al. (2020). Applications of crop modeling in cotton production. In S. Ahmad & M. Hasanuzzaman (Eds.), *Cotton production and uses*. Springer. Print ISBN: 978-981-15-1471-5, Online ISBN: 978-981-15-1472-2.

Ahmad, A., et al. (2015). *Impact of climate change on the rice–wheat cropping system of Pakistan* (pp. 191–221). World Scientific Publishing Co. Pt. Ltd. ISBN: 978-1-78326-563-3.

Ahmed, M., et al. (2020). Abiotic stress tolerance in wheat and the role of silicon: An experimental evidence. In M. Hasanuzzaman (Ed.), *Agronomic crops*. Springer.

Akram, R., et al. (2017a). Role of antibiotics in climate change. In M. Hashmi, V. Strezov, & A. Varma (Eds.), *Antibiotics and antibiotics resistance genes in soils* (Soil biology) (Vol. 51). Springer.

Akram, R., et al. (2017b). Fate of antibiotics in soil. In M. Hashmi, V. Strezov, & A. Varma (Eds.), *Antibiotics and antibiotics resistance genes in soils* (Soil biology) (Vol. 51). Springer.

Akram, R., et al. (2018a). Fate of organic and inorganic pollutants in paddy soils. In M. Hashmi, V. Strezov, & A. Varma (Eds.), *Environmental pollution of paddy soils* (Soil biology) (Vol. 53). Springer.

Akram, R., et al. (2018b). Paddy land pollutants and their role in climate change. In M. Hashmi, V. Strezov, & A. Varma (Eds.), *Environmental pollution of paddy soils* (Soil biology) (Vol. 53). Springer.

Akram, R., et al. (2019a). Plant growth and morphological changes in rice under abiotic stress. In H. Mirza, M. Fujita, & J. K. Biswas (Eds.), *Advances in rice research for abiotic stress tolerance*. Woodhead Publishing. ISBN: 978-0-12-814332-2.

Akram, R., et al. (2019b). Global trends of E-waste pollution and its impact on environment. In M. Zaffar Hashmi & A. Varma (Eds.), *Electronic waste pollution*. Springer. ISBN: 978-3-030-26614-1.

Ali, M., et al. (2019). Role of ICT in crop management. In H. Mirza (Ed.), *Agronomic crops*. Springer. ISBN: 978-981-32-9150-8.

Ali, A., et al. (2021). Temporal analysis for anomaly detection in precipitation pattern over the selected area in Indus Basin of Pakistan. *Pure and Applied Geophysics., 178*, 651–669.

Amanet, K., Mubeen, M., et al. (2019). *Cotton production in Africa* (pp. 359–369). John Wiley & Sons, Ltd.

Balal, R. M., et al. (2022). Morphological, physiological, and biochemical modulations in crops under salt stress. In W. N. Jatoi, M. Mubeen, A. Ahmad, M. A. Cheema, Z. Lin, & M. Z. Hashmi (Eds.), *Building climate resilience in agriculture*. Springer.

Din, M. S. U., et al. (2022). World nations priorities on climate change and food security. In W. N. Jatoi, M. Mubeen, A. Ahmad, M. A. Cheema, Z. Lin, & M. Z. Hashmi (Eds.), *Building climate resilience in agriculture*. Springer.

El Sabagh, A., et al. (2020a). Elevated CO2 concentration improves heat-tolerant ability in crops.

El Sabagh, A., et al. (2020b). Drought and heat stress in cotton (Gossypium hirsutum L.): Consequences and their possible mitigation strategies. In M. Hasanuzzaman (Ed.), *Agronomic crops*. Springer. Print ISBN: 978-981-15-0024-4, Online ISBN: 978-981-15-0025-1.

El Sabagh, A., et al. (2020c). Nitrogen fixation of legumes under the family Fabaceae: Adverse effect of abiotic stresses and mitigation strategies. In M. Hasanuzzaman, S. Araújo, & S. Gill (Eds.), *The plant family Fabaceae*. Springer.

El Sabagh, A., et al. (2020d). Prospective role of plant growth regulators for tolerance to abiotic stresses. In T. Aftab & K. R. Hakeem (Eds.), *Plant growth regulators*. Springer.

El Sabagh, A., et al. (2021). Potential role of plant growth regulators in administering crucial processes against abiotic stresses. *Frontiers in Agronomy, 3*, 648694.

El Sabagh, A., et al. (2022). Phytohormones as growth regulators during abiotic stress tolerance in plants. *Frontiers in Agronomy, 4*, 765068.

Fahad, S., et al. (2019). Abiotic stress and rice grain quality. In H. Mirza, M. Fujita, & J. K. Biswas (Eds.), *Advances in rice research for abiotic stress tolerance*. Woodhead Publishing. ISBN: 978-0-12-814332-2.

Habib ur Rehman, M., et al. (2020). Climate resilient cotton production system: A case study in Pakistan. In M. Hasanuzzaman (Ed.), *Cotton production and uses*. Springer. Print ISBN: 978-981-15-1471-5, Online ISBN: 978-981-15-1472-2.

Hammad, H. M., et al. (2022). Assessing the impact of varying temperatures on growth of maize and nitrogen use efficiency under semi-arid environment. *Frontiers in Plant Science., 13*, 885479.

Hashmi, M. Z., et al. (2017). Management and regulation of antibiotics and antibiotics resistance genes in soils. In M. Hashmi, V. Strezov, & A. Varma (Eds.), *Antibiotics and antibiotics resistance genes in soils* (Soil biology) (Vol. 51). Springer.

Hasnain, A., et al. (2017). Antibiotics resistance genes. In M. Hashmi, V. Strezov, & A. Varma (Eds.), *Antibiotics and antibiotics resistance genes in soils* (Soil biology) (Vol. 51). Springer.

Hussain, S., et al. (2020). Irrigation scheduling for cotton cultivation. In S. Ahmad & M. Hasanuzzaman (Eds.), *Cotton production and uses*. Springer. Print ISBN: 978-981-15-1471-5, Online ISBN: 978-981-15-1472-2.

Hussain, S., et al. (2021a). Using space–time scan statistic for studying the effects of COVID-19 in Punjab, Pakistan: A guideline for policy measures in regional agriculture. *Environmental Science and Pollution Research, 30*, 42495–42508.

Hussain, S., et al. (2021b). Satellite-based evaluation of temporal change in cultivated land in southern Punjab (Multan region) through dynamics of vegetation and land surface temperature. *Open Geosciences, 13*, 1561–1577.

Hussain, S., et al. (2022a). Land use and land cover (LULC) change analysis using TM, ETM+ and OLI Landsat images in district of Okara, Punjab, Pakistan. *Physics and Chemistry of the Earth, 126*, 103117.

Hussain, S., et al. (2022b). Monitoring the dynamic changes in vegetation cover using spatio-temporal remote sensing data from 1984 to 2020. *Atmosphere, 13*, 1609.

Hussain, S., et al. (2022c). Spatiotemporal variation in land use land cover in the response to local climate change using multispectral remote sensing data. *Land, 11*, 595.

Javeed, H. M. R., et al. (2021a). Potential utilization of diluted seawater for the cultivation of certain summer vegetable crops: Physiological and nutritional implications. *Agronomy, 2021*(11), 1826.

Javeed, H. M. R., et al. (2021b). Lipoic acid combined with melatonin mitigates oxidative stress and promotes root formation and growth in salt-stressed Canola seedlings (Brassica napus L.). *Molecules, 26*, 3147.

Khaliq, M. A. (2022). Assessment of heavy metal accumulation and health risks in okra (Abelmoschus Esculentus L.) and spinach (Spinacia Oleracea L.) fertigated with wastewater. *International Journal of Food Contamination, 9*, 11.

Mehmood, A., et al. (2021). Exploring the potential of moringa leaf extract as bio stimulant for improving yield and quality of black cumin oil. *Scientific Reports, 11*, 24217.

Mubeen, M., et al. (2021). Effect of plant growth promoting bacteria and drought on spring maize. *Pakistan Journal of Botany, 53*(2), 731–739. https://doi.org/10.30848/PJB2021-2(38)

Sami ul Din, M., et al. (2019). Agronomic cropping systems in relation to climatic variability. In H. Mirza (Ed.), *Agronomic crops*. Springer. ISBN: 978-981-32-9150-8.

Shahzad, K., et al. (2022). Importance of carbon sequestration in the context of climate change. In W. N. Jatoi, M. Mubeen, A. Ahmad, M. A. Cheema, Z. Lin, & M. Z. Hashmi (Eds.), *Building climate resilience in agriculture*. Springer.

Tahir, M., et al. (2022). Efficacy of organic-based carrier material for plant beneficial rhizobacteria application in okra under normal and salt-affected soil conditions. *Journal of Applied Microbiology, 133*(2), 943–959.

Tariq, M., et al. (2019). Fundamentals of crop rotation in agronomic management. In H. Mirza (Ed.), *Agronomic crops*. Springer. ISBN: 978-981-32-9150-8.

Zahoor, S. A., et al. (2019). Improving water use efficiency in agronomic crop production. In M. Hasanuzzaman (Ed.), *Agronomic crops*. Springer.

Knowledge Transfer to Stakeholders in Use of Wastewater for Irrigation

Asif Sardar, Sidra Javed, and Aneela Afzal

Abstract Reuse of treated wastewater has the potential to play a crucial role in the implementation of sustainable water resource management practices within the agricultural sector. Nevertheless, the effectiveness of this technique relies on the acceptance and active involvement of the public, which necessitates a comprehensive study. This chapter provides a comprehensive assessment from the existing literature and published relevant studies pertaining to the uses of waste water technologies and their public acceptance. Additionally, it explores the significance of knowledge in promoting the adoption of these technologies among key stakeholders. The study's findings illustrate the potential aspects that can enhance awareness and understanding regarding the adoption of wastewater treatment technology and practices in the agriculture sector, as well as how farmers perceive health risks. In addition, we discuss the gender role in this setting and examine the primary instruments used to assess the consequences of risks. This chapter presents policy recommendations to relevant stakeholders based on the research findings.

Keywords Water management practices · Awareness · Understanding · Assessment · Potential aspects · Irrigation steps

The original version of the chapter has been revised. A correction to this chapter can be found at https://doi.org/10.1007/978-3-031-91883-4_24

A. Sardar (✉) · S. Javed
National Center of Industrial Biotechnology, PMAS-Arid Agricultural University, Rawalpindi, Pakistan

A. Afzal
National Center of Industrial Biotechnology, PMAS-Arid Agricultural University, Rawalpindi, Pakistan

Department of Agricultural Extension, PMAS-Arid Agriculture University, Rawalpindi, Pakistan

M. Mubeen et al. (eds.), *Innovations in Agricultural Water Management*, https://doi.org/10.1007/978-3-031-91883-4_12

1 Wastewater Irrigation in the Agriculture Sector: An Overview

Globally, the usage of wastewater is increasing as a result of the depletion of freshwater supplies. It serves various purposes, including the irrigation of agricultural land, the support of aquaculture, the watering of landscapes, the fulfilment of municipal and industrial needs, and the replenishment of artificial groundwater. If wastewater is treated appropriately and operates reliably, it can be utilized for any purpose. The utilization of wastewater in agriculture is the most well-established and enduring application. The majority of cultivated land that receives irrigation is located in or near urban areas where wastewater is produced (Zhang & Shen, 2019). According to a study by Singh (2021), almost 20 million ha of farmland are being irrigated using wastewater, including both treated and untreated water. Wastewater irrigation is a widespread and rapidly expanding practice that involves using wastewater for agricultural purposes. This is particularly common in low- and high-income nations with arid climates. It entails either the direct application of untreated wastewater or the indirect use of polluted waters from rivers and streams. This is not only prevalent in low-income countries but also in many middle-income nations. Urban farmers sometimes rely on heavily polluted water sources because of the scarcity or expensive nature of freshwater, as well as the insufficient capacity of wastewater treatment systems to meet urban growth. A substantial proportion of the population residing in emerging nations primarily relies on the agricultural industry. Agriculture serves as a means of earning income, promoting job possibilities, and guaranteeing food accessibility in the most of the South Asian countries (Sardar et al., 2020, 2021; Hammad et al., 2020; Zamin et al., 2020; Shahzad et al., 2020; Awais et al., 2020; Abbas et al., 2020; Mubeen et al., 2020).

In 2006, the World Health Organization (WHO) issued updated guidelines for the usage of wastewater in agricultural practices, as a component of a complete risk management strategy. The 2006 guidelines include health-based targets that correspond to the permissible level of disease burden resulting from the utilization of agricultural wastewater. Ungureanu et al. (2020) describe that wastewater refers to the water that has been previously utilized in home or industrial purposes. Wastewater comprises numerous contaminants, both biological and chemical, necessitating treatment prior to its potential reuse for alternative use. Potential contaminants encompass a range of substances such as inorganic solids, dissolved solids, organic matter, heavy metals, floating substances (such as oil and grease), microbes, and nutrients. Treatment encompasses a range of biological, chemical, and physical techniques to diminish the levels of contaminants (Abd-Elwahed, 2019; Kampouris et al., 2021).

There are two main kinds of reuse that have been practiced worldwide:

1. Potable uses encompass the utilization of reclaimed water, which has undergone thorough treatment, to enhance the supply of drinking water. In addition, drinkable water can be obtained indirectly by allowing reclaimed water to pass through the natural environment.

Table 1 Advantage, disadvantage, and associated risk of wastewater irrigation (Chojnacka et al., 2020; Singh, 2021; Ungureanu et al., 2020)

Advantage	Disadvantages	Risk associated with wastewater usage
Enhancement of the investments in wastewater irrigation infrastructure	Wastewater is typically generated consistently throughout the entire year, whereas the practice of using wastewater for irrigation is primarily restricted to the period of crop growth	Possible contamination of groundwater resulting from the presence of heavy metals, nitrates, and organic substances
Preservation of freshwater resources		
Aquifer recharge via infiltration of water (natural treatment)		
The utilization of nutrients found in wastewater, such as nitrogen and phosphate, leads to a decrease in the reliance on synthetic fertilizers. This, in turn, enhances soil fertility and results in improved crop yields	Certain compounds found in wastewater can reach quantities that are harmful to plants and can cause environmental harm	Possible endangerment to human Well-being by the transmission of infectious microorganisms
Cost reduction in treatment: Treating the pre-treated wastewater by irrigating the soil (without the need for additional treatment stages; however, the effectiveness depends greatly on the origin of the wastewater)		Possible soil damage caused by the accumulation of heavy metals and acidity
Minimization of environmental effects, including as eutrophication and adherence to minimum discharge criteria		

2. Non-potable applications encompass many uses, like irrigated agriculture, irrigation of parks, and public forestry areas (including golf courses, which is the most rapidly adopted form of reuse in Europe), aquaculture, aquifer recharge (indirect reuse), as well as industrial and urban uses. Table 1 provides a comprehensive overview of the numerous advantages, disadvantages, and risk considerations associated with wastewater irrigation.

2 Methodology

To understand the use of wastewater in the agricultural sector, its consequences, and the obstacles encountered by farmers in mitigating the effects of water scarcity, policymakers and local farmers should know the importance of this application and possess the knowledge to effectively execute it at the farm level. This study examines the existing literature on web-based papers, journal research articles, projects, and publications that focus on specific regions and local contexts. The review encompassed locally-accessible reports and projects pertaining to the use of waste water irrigation. The implications of waste water irrigation, health risks, and potential benefits to farmers were identified by analyzing previous study papers and the collected literature.

3 Capitalizing Farmers' Knowledge and Perceptions of Wastewater-Irrigated Agriculture

Capitalizing on farmers' knowledge and perception of wastewater-irrigated agriculture is important to understand how different crops respond to wastewater irrigation and can adapt their farming practices accordingly. Insights from farmers regarding the impact of using wastewater for irrigation on soil fertility and crop yield can contribute to a more comprehensive comprehension of the advantages and disadvantages of this activity. According to the literature (Antwi-Agyei et al., 2016; Wells et al., 2016; Hussain et al., 2022a, 2023a, 2023b, 2023c, 2024a, 2024b, 2024c; Naz et al., 2023; Muhammad et al., 2023; Khurshid et al., 2023), farmers' experiences and traditional practices play a crucial role in enhancing the efficiency of wastewater reuse in agriculture. Their effective irrigation practices, enhanced crop productivity, and few adverse consequences may incentivize more farmers to adopt wastewater irrigation. The efficacy of this activity relies on the farmers' familiarity with the geographical context and irrigation techniques employed on their land, as well as their awareness of potential health hazards. Khan et al. (2013) studied that in Pakistan, post-harvest contamination of agricultural products is a prevalent issue that poses a significant risk to consumers. On the whole, the production of wastewater irrigation does not impact the demand for produce. The survey revealed that both merchants and buyers seldom took into account the origin of the commodities and the watering practices when making purchases. Their preferences were primarily shaped by other factors, specifically the quality and crispness of the veggies. Facilitating good communication between farmers, merchants, and customers to encourage the adoption of risk-reduction measures via the multi-barrier technique is a viable approach to address this problem (Perez-Espejo et al., 2011).

4 Stakeholders' Attitude Toward the Utilization of Treated Wastewater for Irrigation

According to a study (Khanpae et al., 2020), policymakers and researchers are now showing their concern regarding the implementation of sustainable water management practices and the utilization of treated wastewater for irrigation in response to climatic variations. Farmers, governments, and agricultural communities are the stakeholders who have shown a positive disposition toward utilizing treated wastewater for irrigation purposes. The potential benefits are emphasized, such as increased agricultural productivity, less reliance on freshwater resources, and cost savings. The choice to utilize wastewater for irrigation is determined by environmental, economic, and geographical factors, in addition to the objective of promoting sustainable practices. The utilization of treated wastewater for irrigation may provide challenges for certain stakeholders. To priorities the preservation of human and environmental health and ensure water quality, individuals may express concerns regarding potential health risks, the presence of toxins in the treated water utilized for agricultural cultivation, and the impact on surrounding ecosystems. The literature, such as Irfan et al. (2022), Ullah et al. (2022), Naz et al. (2022), Hussain et al. (2022b), Deh-Haghi et al. (2020), Khan et al. (2013), and Saliba et al. (2018), illustrate that there are different views of different stakeholders regarding the utilization of treated wastewater for irrigation, which are impacted by their values, interests, and concerns. To achieve a consensus and maximize the advantages of this sustainable water management plan, it is crucial to priorities cooperation, address pertinent challenges, and emphasize the benefits.

When evaluating support for wastewater treatment and reuse, it is important to include the public's environmental consciousness and their concern for future generations. Consequently, providing the public with information about the treatment process and the quality of the water produced acts as a catalyst for the success of any reuse scheme. The study by Deh-Haghi et al. (2020), Hussain et al. (2020a, b), Hashmi et al. (2020), Zaman et al. (2020), Farid et al. (2019), Danish et al. (2019), Hammad et al. (2019), and Imran et al. (2019) found that the level of acceptability is greatly impacted by factors such as age, education, and understanding of wastewater reuse. Gender and income have a somewhat less impact. Moreover, the potential uses of processed wastewater have a substantial influence on the level of acceptance for its reutilization. It was shown that 70% of Californians expressed a preference for utilizing recycled wastewater for various applications such as irrigating sports fields, firefighting, road washing, and cultivating non-edible crops. Conversely, a mere 40% of the individuals surveyed expressed their support for utilizing treated wastewater for purposes such as growing vegetables, feeding cattle, and replenishing groundwater. According to a study conducted by Michetti et al. (2019), the majority of sociological analysis on reuse emphasizes the ability to influence society attitudes through two key factors: trust and perception of risk. Hence, the initial phase in formulating any water management plan includes comprehending the perceived hazards and the level of public consciousness regarding treated wastewater. Waleed et al. (2022), Hussain et al. (2022c), Habib-ur-Rahman et al. (2022), Khaliq et al. (2022), Mubeen et al. (2021), Ahmad et al. (2021), and Baawain et al. (2020) assert that in order to achieve a change in perspective and implement the necessary processes for sustainable wastewater reuse, it is essential to involve a diverse and extensive group of stakeholders. This is because no individual alone possesses the capacity to carry out all the required tasks. These issues demonstrate that promoting social learning is a crucial objective for enhancing stakeholders' engagement.

5 Significance of the Knowledge for Social and Cultural Aspects of Wastewater Usage in Agriculture

Wastewater irrigation, which involves using treated or untreated wastewater for agricultural uses, requires a thorough understanding and thoughtful consideration of its social and cultural implications (Michetti et al., 2019). From a sociological perspective, this knowledge is essential for ensuring the well-being and physical health of populations that rely on wastewater irrigation. Acquiring information about the possible health risks and taking suitable measures aids in protecting the local community from the adverse impacts of germs and pollutants (Al-Hazmi et al., 2023). Moreover, understanding of wastewater irrigation can additionally facilitate the establishment of guidelines and protocols for the safe utilization and disposal of wastewater, and it diminishes the environmental and health risks linked to it (Dawar et al., 2021; Ali et al., 2021; Bukhari et al., 2021; Rehman et al., 2021; Baawain et al., 2020).

Furthermore, societies demonstrate a wide range of viewpoints regarding the recycling of wastewater, which are shaped by cultural beliefs, practices, and traditions. An in-depth understanding of the societal reception, conflicts, and challenges associated with wastewater irrigation in different communities can be achieved by examining these cultural aspects. This knowledge can be utilized to develop efficient communication tactics, educational initiatives, and community engagement activities aimed at promoting the acceptance and implementation of wastewater irrigation practices in culturally sensitive ways (Michetti et al., 2019). The significance of comprehending the social and cultural aspects of wastewater irrigation lies in its ability to guarantee public health, conserve the environment, and promote sustainable farming practices. It facilitates educated decision-making, policy creation, and implementation of programs that take into account the cultural diversity and complexities of various people. Ultimately, this information improves a thorough and inclusive strategy to using wastewater for irrigation, resulting in advantages for both society and the environment (Ahmmadi et al., 2021; Ricart & Rico, 2019).

6 Wastewater-Irrigated Agriculture and the Perception of Its Health Consequences

A study by Kesari et al. (2021) described that the utilization of wastewater in underdeveloped nations poses distinctive obstacles. Indirect wastewater consumption involves the direct discharge of untreated domestic, agricultural, and occasionally industrial wastewater into urban waterways. In less developed countries, farmers intentionally collect wastewater on their property or use the sewer systems to dispose of it. Consequently, the act of irrigating agricultural crops with wastewater has arisen as an unplanned or spontaneous action, primarily because of the scarcity of water and insufficient planning. The worldwide apprehension regarding the safety of using wastewater for agricultural purposes is substantial. Insufficient treatment of wastewater before its application in irrigation can result in substantial concerns regarding soil health. The factors that contribute to soil degradation are salinization, toxicity caused by sodium ions, chloride, and boron, structural decay, decreased air circulation, and clogging of pores owing to suspended solids in the wastewater, that results in the soil's ability to transmit water. Moreover, a study by Natasha et al. (2023), Rasool et al. (2021), Shah et al. (2021), Mehmood et al. (2021), Rasool et al. (2020), Hussain et al. (2020a), and Shah et al. (2020) found that issues for agricultural crops, including the buildup of toxic metals, contamination by microorganisms, and hindered or erratic plant growth are caused by high nutrient concentrations. Moreover, it might have an adverse effect on groundwater by inducing the leaching of excessive amounts of nitrates. Lastly, it presents hazards to the well-being of farmers who utilize these fluids and the individuals who consume the irrigated products. The probability of acquiring a bacterial, protozoan, or fungal illness by contact with wastewater is considerably less when compared to the danger of infection from pathogenic worm eggs. This is due to the shorter persistence of these pathogens in the environment and the heightened immunity of the host. The risk of viral infections is minimal due to the host's robust immune defense against such illnesses (Gashaye, 2020). It is vital to recognize that the presence of detrimental bacteria in both processed and unprocessed wastewater does not inherently indicate that they would pose a threat to human well-being when utilized for irrigating plants. The survival of these diseases is affected by different factors, such as ultraviolet (UV) radiation and high temperatures in dry areas. Various factors, including decreased moisture levels, the presence of soil microbial communities, and the specific crop type, have a substantial impact on the potential of diseases to survive and then spread to humans via the food chain. Adopting wastewater treatment is the most strongly advised method for mitigating health risks and eradicating pathogens to protect the well-being of farmers and consumers who rely on crops irrigated with wastewater (Natasha et al., 2023; Pratap et al., 2023).

7 Current Practices to Encourage the Application of Wastewater Treatment Technologies in the Agriculture Sector

Wastewater irrigation is a low-cost and efficient technique that not only facilitates irrigation but also enriches the soil with important nutrients. Despite its benefits, using wastewater in agricultural production poses various obstacles that need careful planning and innovative approaches. It is critical to resolve these obstacles to guarantee the long-term responsible adoption of wastewater irrigation methods (Khan et al., 2022). Some of the practices to encourage the application of wastewater treatment technologies in the agriculture sector are discussed in the following text.

7.1 Consider Wastewater as a Resource for Requiring Good Management

Wastewater can serve as both a sewage and a reusable resource when looking over the perspective of integrated natural resources management (Lal et al., 2021). Recycling water for useful purposes like irrigation, crops, and livelihoods could be more economical for farmers. In addition, irrigation may use nutrients and "treat" effluent that would otherwise be expensive to treat or dispose of. To reduce wastewater treatment expenses, and boost agricultural output, many countries are emphasizing reuse of wastewater for irrigational purposes (Qadir et al., 2010).

The biggest challenge of developing countries for wastewater irrigation is determining the scale at which treatment is feasible, with a focus on separating industrial and domestic wastewater for safe reuse. Developing countries need to focus on the institutional and economic constraints, wastewater constituents and sources, and purpose for reuse when determining the best treatment method, rather than rigorous standards. If treatment is not feasible, the WHO recommends a variety of without-treatment measures to lessen potential health hazards after reuse (Drechsel et al., 2022).

Assigning property rights also encourages the effective use of wastewater in areas where farmers have limited water supply (Drechsel et al., 2022).

7.2 Implementing Economic Incentives

Wastewater management can be cheaper by improving freshwater institutions and policies. Sanitation systems and water supply are often institutionally and economically not connected, and even in some countries where the wastewater treatment structure is adequate, governmental systems have overlapping authorities that hinder policy execution. Incentives such as tradable permits, effluent standards, and levies can be employed to encourage businesses and households that discharge wastewater from point sources to enhance their wastewater management practices (Saad et al., 2017).

Incentives on wastewater irrigation are beneficial for water resource management. A reduction in the cost of water and incentives for the purchase of new equipment for treating wastewater improve the adoption of wastewater for irrigation by farmers (Drechsel et al., 2008).

7.3 Improvement in Financial Management

The financial resources in developing countries are scarce, preventing them from investing in or even maintaining effluent treatment facilities. The allocation of necessary resources may be facilitated by institutional frameworks and policies. The proper use of wastewater disposal for potable water will motivate water conservation and the re-use of wastewater rather than its release into water bodies or facilities managed by a wastewater treatment institution. Programs that generate revenue through the imposition of a charge per unit of effluent produced by water users (the polluter pays principle) can be conceptually justified, especially when the funds are utilized to develop infrastructure for wastewater collection, treatment, and reuse (Ranjan, 2021).

Amoah et al. (2011) describe that poor farmers may be compensated by policy changes that provide them with access to alternate irrigation water sources, monetary aid, and/or vocational training. Policies should design in a way that permits the deprived farmers to progressively reduce wastewater utilization while pursuing alternative means of subsistence may be more prudent than those that force abrupt disruptions in wastewater supply or merely restrict or prohibit wastewater use.

7.4 Consultation with Organizations and Individuals

Public agencies must conduct extensive consultations with farmers, businesses, and organizations regarding effluent generation and use. The effectiveness of wastewater reuse campaigns and associated food safety initiatives can be improved by the engagement of several stakeholders. Moreover, public policies for wastewater management are further improved by better communication between environmental institutions and government agencies with expertise in wastewater management (WHO, 2006b).

Governmental institution coordination plays a pivotal role in the implementation of policies to guarantee the public goals for wastewater management. For instance, the agriculture ministries, public health, water resources, and economic development need to work together so that their respective missions and programs do not interfere with one another. Successful inter-ministerial collaboration will reduce the overall cost of attaining public objectives.

7.5 Support Research, Development, and Outreach

A more comprehensive understanding of nutrient concentrations in soils, water supply, and crop needs could enable farmers to optimize the utilization of effluent nutrient content. Improvement in research and development sectors is needed especially in developing countries to know the beneficial impacts of effluent utilization in their respective agricultural setting (Drechsel et al., 2015).

Researchers and government organizations can do a better job of responding to real-world opportunities and risks if they have access to more accurate information about the present state of wastewater reuse for irrigation (WHO, 2006b). More comprehensive risk assessments are also required. Designing effective strategies to better manage wastewater and safeguard public health could be possible through the exact information about the quantity and quality of wastewater utilized for irrigation and the geographic distribution of wastewater. Offering incentives to the farmers WHO report on yields, irrigation rates, and observed effects on persons, soils, plants, and groundwater could aid in the spread of awareness. Additionally, public agencies and farmers may collaborate to develop effluent monitoring programs (Brar et al., 2022).

8 Gender Role to Increase the Reuse of Wastewater in Agriculture

A study by Sohail et al. (2022) stated that recognizing the diversity of the people in the community is essential for successful community engagement. It should include factors like age, degree of education, social status, economic status, and gender. Men and women are distinct, and it is critical to recognize this distinction in terms of their respective roles and areas of interest (Drechsel et al., 2008). The degree to which both sexes participate in and benefit from water development initiatives is influenced by a variety of gender-related factors. Women in many developing nations face several barriers, including poorer social status, fewer opportunities for advancement, fewer resources and services, more burdens due to poor health, and a lack of participation in decision-making at both the home and community levels (Saad et al., 2017). Thus, gender differences in knowledge, action, and decision-making and their repercussions on individuals, families, and communities must be balanced (Segal & Giudice, 2022). More productive, efficient, and cost-effective results may be achieved by adopting a gender-sensitive strategy. There are fewer mistakes in technical planning, better resource and financial management, and more openness when women are included in these processes (WHO, 2006b).

Recognizing gender roles and distinctions provides opportunities for strategic planning that optimizes social and economic progress while mitigating competition and conflicts concerning water resources. Further research is required to comprehend the gender implications in the water sector to bring positive change (Drechsel et al., 2015).

9 Financing and Policy Barriers to Wastewater Reuse for Irrigation in the Context of Stakeholders

Wastewater management faces a lot of financial and policy barriers, especially in developing countries. Many factors are responsible for this, especially inadequate funding, less technical information, or understanding of policy consequences, inadequate investment, lack of political will, or less institutional capabilities (Lal et al., 2021). Wastewater irrigation due to water scarcity, inadequate water quality, and inefficient utilization could impose detrimental impacts on public health, economic development, and the environment. The public sector typically lacks the resources and expertise necessary to effectively address environmental and health issues associated with wastewater irrigation, and it is often harder for them to enforce the regulations for wastewater treatment. Furthermore, governments may be hesitant to accept the utilization of treated wastewater for irrigation for fear of economic and environmental impacts on agriculture, preventing them from enacting food safety measures. Policies for treated wastewater irrigation must be designed from the condition when wastewater is generated, while it is being utilized on crops, and after products have been prepared for sale and consumption (Malisa-van der Walt & Taigbenu, 2023). The most important aspect to be considered while designing the policies for wastewater irrigation is social acceptance. Social acceptance toward wastewater reuse for agriculture could be possible through knowing about the sociodemographic factors, trust and knowledge, matters of choice, expressed disgust, costs and benefits, and environmental attitudes of the farmer (Prasad, 2023). Moreover, the policy structure should also include factors such as contact, education, water scarcity, risk awareness, available alternative water sources, stakeholder involvement, economics, and prior experience with treated wastewater irrigation.

10 Tools for Assessing the Risk of Using Wastewater in the Agriculture Sector in Accordance with the WHO Guidelines

The WHO sets guidelines for the safe utilization of wastewater, excreta, and gray water in agriculture with a specific focus on pathogenic risks (WHO, 2006b). Moreover, the WHO provides guidelines to the regions where just treatment cannot reduce the pathogen cycle, and additional non-treatment or post-treatment practices are required to accomplish acceptable risk reduction. This does not mean that standards in various nations are variable. The WHO also sets the tolerable disease burden per person per year (WHO, 2006b). Each country follows the WHO standards for safe wastewater irrigation according to the present condition, context, and managerial and human resource capabilities. It is advisable to adopt a step-by-step approach to risk reduction. It is crucial to acknowledge that the chosen approach includes various treatment and post-treatment approaches with the goals of securing food security, public health, and environmental protection (Drechsel et al., 2022).

The WHO implemented health-based targets to enhance the efficacy of risk management strategies (WHO, 2006b). These guidelines focus on health-based targets, which are measured in disability-adjusted life years, evaluate the hazards, and diseases and quantify the effectiveness of various pathogen barriers in reducing risk (WHO, 2006a). The WHO provides the guidelines according to various conditions as described in Table 2.

Assessing the risk of using wastewater in the agriculture sector in accordance with the World Health Organization (WHO) guidelines involves a comprehensive evaluation of various factors related to water quality, treatment processes, and potential health risks. Several tools and approaches can be used to fulfill the guidelines mentioned in Fig. 1.

Table 2 Wastewater treatment guidelines according to the WHO (Drechsel et al., 2022; WHO, 2006b)

Condition	WHO guidelines
The decision to allow or prohibit reuse should be based on	Dependence on various measures to prevent pathogens from treatment to consumption, to reduce health risks measured in DALYs, shifting focus from treating irrigation water to ensuring food is not contaminated. This approach enables the use of multiple measures to lower the risk of infection
Stakeholder capacity assumptions	Required the capacities to monitor risks, treat wastewater, and raise risk consciousness and willingness to implement preventative action
Risk assessment	Microbiological risk assessment using quantitative methods (using dose-response functions). Estimation of human exposure to chemical pollutants
Risk mitigation	If treatment is not practicable, rely on food chain actors' risk reduction alternatives (multi-barrier methods, crop limitation) and self-protection
The efficacy of systems in the face of external stressors, such as disasters like floods, lightning, and interruptions in power	The implementation of a multi-barrier system is expected to be more effective in mitigating risks in comparison to the situation before the disaster
Implementation potential	Big potential in places where garbage treatment is not available or is not enough of a problem. Low potential in nations with sophisticated wastewater treatment systems and established processes for reusing water of high quality
Flexibility	Ability to adjust to new circumstances with ease and use whatever level of technology necessary.
Stakeholder involvement and ownership	Plays a crucial part in reducing risk as explained in detail in the manual of sanitation safety planning
Low and middle-income country adoption	Minimal uptake or modification occurred

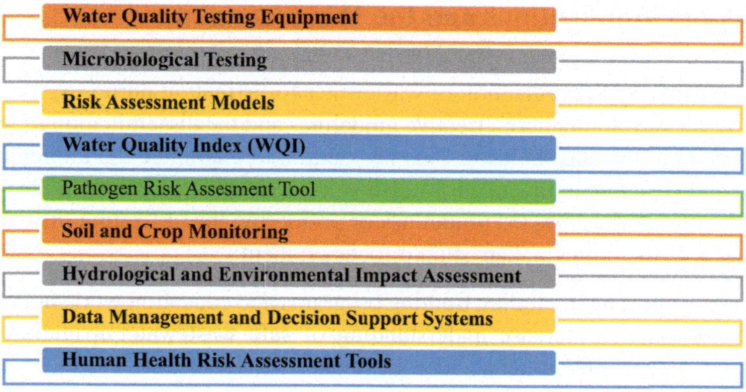

Fig. 1 Tools for assessing the risk of using wastewater in the agriculture sector in accordance with the WHO guidelines (Ahmed et al., 2022; Victor et al., 2008; WHO, 2006b)

11 Wastewater Lessons Learned and Its Potential Applications

Depletion of freshwater resources is caused by industrial growth, long-term urbanization, and unsustainable utilization of freshwater. Wastewater utilization for agriculture could help in wastewater management. Using wastewater in agriculture is an inexpensive method to compensate for the shortage of irrigation water while also supplying crops with vital macro- and micronutrients. To dispose of effluent and fulfill the water needs of agriculture in arid and semiarid areas, treated wastewater is the greatest option.

To meet water needs, many developing countries are forced to utilize raw or diluted wastewater for irrigation. As wastewater consumption rises, governmental organizations responsible for protecting human health and the environment will face new hurdles. Public agencies are responsible for implementing preventive measures, like isolating industrial effluents that eliminate the hazardous wastewater components, adopting farm-level efforts to prevent pathogen interaction with crops and farmers, and establishing post-harvest measures and campaigns to protect human health, in cases where funds for significant improvements in wastewater collection and treatment remains limited. Large sums will have to be spent over time on permanent sewage treatment plants (Ahmed et al., 2022). These should be implemented in stages, with consideration given more to local opportunities and limitations than to international criteria. Policies must preserve the welfare of poor farmers and merchants who rely on wastewater-irrigated agricultural production while investments are made (Victor et al., 2008).

12 Policy Implications and the Way Forward

When considering the utilization of wastewater for irrigation purposes, it is vital to prioritize the development of effective guidelines and the dissemination of information relevant to the stakeholders. This is essential to guarantee the implementation of safe and sustainable practices (Malisa-van der Walt & Taigbenu, 2023; Prasad, 2023). While establishing the guidelines or promoting the utilization of wastewater for irrigation, the policymakers should focus on them:

- Capacity building, including training programs and education workshops, should be prioritized to increase understanding of safe wastewater irrigation among farmers, extension workers, and policymakers.
- Research and innovation to improve treatment technologies and irrigation should be made.
- Cooperative partnerships between government agencies, research institutes, NGOs, and the private sector to facilitate knowledge sharing and collaboration.
- Strong monitoring systems and effective enforcement mechanisms are essential for routine water quality assessment and control.
- Incentives such as financial support are necessary to farmers, adopting safer practices.
- To promote to community involvement, infrastructure, flexibility, and internationalization determinants.
- Effective communication mechanisms are necessary to disseminate policies and guidelines that encourage open dialogue between policymakers, scientists, and farmers.
- Socioeconomic considerations to incorporate into policy formulation to protect the well-being of communities are dependent on wastewater-irrigated agriculture.
- Long-term planning, consistent with wastewater irrigation targets, ensures the resilience of wastewater irrigation if population growth, urbanization, and agricultural practices are anticipated it will continue.

These holistic approaches will be helpful to the policymakers to contribute to the responsible and beneficial use of wastewater for irrigation to protect public health and the environment.

13 Conclusions

This chapter delves into the primary determinants of acceptability, strategies, and tools for assessing the risk under the WHO guidelines for improving wastewater utilization in agriculture. This study found that public farmers opinion, lack of awareness, and knowledge about how to reuse treated wastewater in crop cultivation limit its use in agriculture. Potential uses, excessive water consumption demands, and household water scarcity significantly influence the acceptance of future applications of treated wastewater. The implementation of a well-rounded gender strategy also led to a rise in acceptance as women tend to be more attuned to environmental concerns and prefer sustainable solutions. The primary obstacle to willingness to adopt is the absence of interaction between farmers and decision-makers. Farmers' adoption to reuse treated wastewater depends on several factors, such as the crops' water requirements, familiarity with reuse legislation, evaluation of certain parameters' suitability, understanding of successful reuse on farms, and the benefits associated with reuse in agriculture. Factors such as daily water use, reuse benefits awareness, treatment plant knowledge, and gender roles influence consumers' views toward reuse. This chapter proposes customizing water resource policies through communication and awareness initiatives. This entails advocating for the advantages of utilizing treated wastewater for farmers, including the reduction in freshwater consumption, extraction and pumping expenses, and the recovery of nutrients for crops. Customers primarily seek information about wastewater treatment and water reuse for their daily use. The advantages include reduced freshwater resource utilization, lower water costs, and more stringent water quality regulations. The presence of quality certificates for agricultural items produced using treated wastewater may enhance the public acceptability of reclaimed water. The certifications procedure would enhance the promotion of these products as environmentally beneficial. All stakeholders should take into account to encourage the farmers and consumers to reuse wastewater. This aids in assessing their policy response to enhance adaptation to local requirements. It may be necessary to revise the regional water resources management plan and legislation to incorporate all of the steps described in this chapter to promote the use of treated wastewater in agriculture.

Acknowledgments The authors gratefully acknowledge PSDP funded project No. 321 "Establishment of National Center of Industrial Biotechnology for Pilot Manufacturing of Bio products Using Synthetic Biology and Metabolic Engineering Technologies at PMAS-Arid Agriculture University Rawalpindi," executed through the Higher Education Commission Islamabad, Pakistan. This study is a part of the above mentioned project.

References

Abbas, Y., et al. (2020). Valorization of solid waste biomass by inoculation for the enhanced yield of biogas. *Clean Technologies and Environmental Policy, 22*, 513–522.

Abd-Elwahed, M. S. (2019). Effect of long-term wastewater irrigation on the quality of alluvial soil for agricultural sustainability. *Annals of Agricultural Sciences, 64*(2), 151–160. https://doi.org/10.1016/J.AOAS.2019.10.003

Ahmad, M., et al. (2021). Hydrogen sulfide: A novel gaseous molecule for plant adaptation to stress. *Journal of Plant Growth Regulation, 40*, 2485–2501.

Ahmed, W., Safdar, U., Ali, A., Haider, K., Tahir, N., Sajid, S., Ahmad, M., Khalid, M. N., & Sattar, M. T. (2022). Sustainable water use in agriculture: A review of worldwide research. *International Journal of Agriculture and Biosciences, 11*(4). https://doi.org/10.47278/journal.ijab/2023.033

Ahmmadi, P., Rahimian, M., & Movahed, R. G. (2021). Theory of planned behavior to predict consumer behavior in using products irrigated with purified wastewater in Iran consumer. *Journal of Cleaner Production, 296*, 126359. https://doi.org/10.1016/J.JCLEPRO.2021.126359

Al-Hazmi, H. E., Mohammadi, A., Hejna, A., Majtacz, J., Esmaeili, A., Habibzadeh, S., Saeb, M. R., Badawi, M., Lima, E. C., & Mąkinia, J. (2023). Wastewater reuse in agriculture: Prospects and challenges. *Environmental Research, 236*, 116711. https://doi.org/10.1016/J.ENVRES.2023.116711

Ali, S., et al. (2021). Future climatic changes, extreme events, related uncertainties, and policy recommendations in the Hindu Kush sub-regions of Pakistan. *Theoretical and Applied Climatology, 143*, 193–209.

Amoah, P., Keraita, B., Akple, M., Drechsel, P., Abaidoo, R. C., & Konradsen, F. (2011). *Low-cost options for reducing consumer health risks from farm to fork where crops are irrigated with polluted water in West Africa*. International Water Management Institute. https://doi.org/10.5337/2011.201

Antwi-Agyei, P., Peasey, A., Biran, A., Bruce, J., & Ensink, J. (2016). Risk perceptions of wastewater use for urban agriculture in Accra, Ghana. *PLoS One, 11*(3), e0150603. https://doi.org/10.1371/JOURNAL.PONE.0150603

Awais, M., et al. (2020). Adaptation strategies for sunflower production under climate change for arid environment by using CROPGRO and APSIM-sunflower models. *Pakistan Journal of Agricultural Sciences., 57*(6), 1529–1538.

Baawain, M. S., Al-Mamun, A., Omidvarborna, H., Al-Sabti, A., & Choudri, B. S. (2020). Public perceptions of reusing treated wastewater for urban and industrial applications: Challenges and opportunities. *Environment, Development and Sustainability, 22*(3), 1859–1871. https://doi.org/10.1007/S10668-018-0266-0/FIGURES/10

Brar, B., Sonu, Naveen, & Rawat, J. (2022). Wastewater reuse for irrigation of a vegetable crops and its impacts. *The Pharma Innovation Journal, 11*(7), 111.

Bukhari, M. A., et al. (2021). Silicon mitigates drought stress in wheat (Triticum aestivum L.) through improving photosynthetic pigments, biochemical and yield characters. *Silicon, 13*, 4757–4772.

Chojnacka, K., Witek-Krowiak, A., Moustakas, K., Skrzypczak, D., Mikula, K., & Loizidou, M. (2020). A transition from conventional irrigation to fertigation with reclaimed wastewater: Prospects and challenges. *Renewable and Sustainable Energy Reviews, 130*, 109959. https://doi.org/10.1016/J.RSER.2020.109959

Danish, S., et al. (2019). Alleviation of chromium toxicity in maize by Fe fortification and chromium tolerant ACC deaminase producing plant growth promoting rhizobacteria. *Ecotoxicology and Environmental Safety, 185*, 109706.

Dawar, M., et al. (2021). The effect of biochar and nitrogen inhibitor on ammonia and nitrous oxide emissions and wheat productivity. *Journal of Plant Growth Regulation, 40*, 2465–2475.

Deh-Haghi, Z., Bagheri, A., Fotourehchi, Z., & Damalas, C. A. (2020). Farmers' acceptance and willingness to pay for using treated wastewater in crop irrigation: A survey in western Iran. *Agricultural Water Management, 239*, 106262. https://doi.org/10.1016/J.AGWAT.2020.106262

Drechsel, P., Keraita, B., Amoah, P., Abaidoo, R. C., Raschid-Sally, L., & Bahri, A. (2008). Reducing health risks from wastewater use in urban and peri-urban sub-Saharan Africa: Applying the 2006 WHO guidelines. *Water Science and Technology, 57*(9), 1461. https://doi.org/10.2166/wst.2008.245

Drechsel, P., Mahjoub, O., & Keraita, B. (2015). Social and cultural dimensions in wastewater use. In *Wastewater: Economic asset in an urbanizing world* (pp. 75–92). Springer. https://doi.org/10.1007/978-94-017-9545-6_5

Drechsel, P., Qadir, M., & Galibourg, D. (2022). The WHO guidelines for safe wastewater use in agriculture: A review of implementation challenges and possible solutions in the global south. *Water (Switzerland), 14*(6). https://doi.org/10.3390/w14060864

Farid, H. U., et al. (2019). Assessing seasonal and long-term changes in groundwater quality due to over-abstraction using geostatistical techniques. *Environmental Earth Sciences, 78*, 386.

Gashaye, D. (2020). Wastewater-irrigated urban vegetable farming in Ethiopia: A review on their potential contamination and health effects. *Cogent Food & Agriculture, 6*(1). https://doi.org/10.1080/23311932.2020.1772629

Habib-ur-Rahman, M., et al. (2022). Impact of climate change on agricultural production; Issues, challenges, and opportunities in Asia. *Frontiers in Plant Science, 13*, 925548.

Hammad, H. M., et al. (2019). Environmental factors affecting the frequency of road traffic accidents: A case study of sub-urban area of Pakistan. *Environmental Science and Pollution Research, 26*(12), 11674–11685.

Hammad, H. M., et al. (2020). Comparative effects of organic and inorganic fertilizers on soil organic carbon and wheat productivity under arid region. *Communications in Soil Science and Plant Analysis, 51*(10), 1406–1422.

Hashmi, M. Z., et al. (2020). Arsenic distribution and metabolism genes abundance in Paddy soils from Punjab and Sindh provinces, Pakistan. *Arabian Journal of Geosciences, 13*, 526.

Hussain, S., et al. (2020a). Study of land use/land cover changes using RS and GIS: A case study of Multan district, Pakistan. *Environmental Monitoring and Assessment, 192*, 2.

Hussain, S., et al. (2020b). Using GIS tools to detect the land use/land cover changes during forty years in Lodhran District of Pakistan. *Environmental Science and Pollution Research, 27*, 39676–39692.

Hussain, S., et al. (2022a). Spatiotemporal variation in land use land cover in the response to local climate change using multispectral remote sensing data. *Land, 11*, 595. https://doi.org/10.3390/land11050595

Hussain, S., et al. (2022b). Monitoring the dynamic changes in vegetation cover using spatio-temporal remote sensing data from 1984 to 2020. *Atmosphere, 2022*(13), 1609.

Hussain, S., et al. (2022c). Using space–time scan statistic for studying the effects of COVID-19 in Punjab, Pakistan: A guideline for policy measures in regional agriculture. *Environmental Science and Pollution Research, 30*, 42495–42508.

Hussain, S., et al. (2023a). Investigation of irrigation water requirement and evapotranspiration for water resource management in Southern Punjab, Pakistan. *Sustainability, 15*, 1768.

Hussain, S., et al. (2023b). Relation of land surface temperature with different vegetation indices using multi-temporal remote sensing data in Sahiwal region, Pakistan. *Geoscience Letters, 15*, 1768.

Hussain, S., et al. (2023c). Assessment of land use/land cover changes and its effect on land surface temperature using remote sensing techniques in southern Punjab, Pakistan. *Environmental Science and Pollution Research, 30*, 99202–99218.

Hussain, S., et al. (2024a). Assessment of future prediction of urban growth and climate change in district Multan, Pakistan using CA-Markov method. *Urban Climate, 53*, 101766.

Hussain, S., et al. (2024b). Agricultural land suitability analysis of southern Punjab, Pakistan using analytical hierarchy process (AHP) and multi-criteria decision analysis (MCDA) techniques. *Cogent Food & Agriculture, 10*(1), 2294540.

Hussain, S., et al. (2024c). Assessing the impact of land use land cover changes on soil moisture and vegetation cover in southern Punjab, Pakistan using multi-temporal satellite data. *Geology, Ecology, and Landscapes*, 1. https://doi.org/10.1080/24749508.2024.2338970

Imran, M., et al. (2019). Kinetic and equilibrium studies for cadmium biosorption from contaminated water using Cassia fistula biomass. *International journal of Environmental Science and Technology, 16*(7), 3099–3108.

Irfan, M., et al. (2022). Exploring the potential effect of Achnatherum splendens L. derived biochar treated with phosphoric acid on bioavailability of cadmium and wheat growth in contaminated soil. *Environmental Science and Pollution Research, 29*, 37676–37684.

Kampouris, I. D., Klümper, U., Agrawal, S., Orschler, L., Cacace, D., Kunze, S., & Berendonk, T. U. (2021). Treated wastewater irrigation promotes the spread of antibiotic resistance into subsoil pore-water. *Environment International, 146*, 106190. https://doi.org/10.1016/J.ENVINT.2020.106190

Kesari, K. K., Soni, R., Jamal, Q. M. S., Tripathi, P., Lal, J. A., Jha, N. K., Siddiqui, M. H., Kumar, P., Tripathi, V., & Ruokolainen, J. (2021). Wastewater treatment and reuse: A review of its applications and health implications. *Water, Air, and Soil Pollution, 232*(5), 1–28. https://doi.org/10.1007/S11270-021-05154-8/FIGURES/5

Khaliq, M. A., et al. (2022). Assessment of heavy metal accumulation and health risks in okra (Abelmoschus Esculentus L.) and spinach (Spinacia Oleracea L.) fertigated with wastewater. *International Journal of Food Contamination, 9*, 11.

Khan, M. U., Malik, R. N., & Muhammad, S. (2013). Human health risk from heavy metal via food crops consumption with wastewater irrigation practices in Pakistan. *Chemosphere, 93*(10), 2230–2238. https://doi.org/10.1016/J.CHEMOSPHERE.2013.07.067

Khan, M. M., Siddiqi, S. A., Farooque, A. A., Iqbal, Q., Shahid, S. A., Akram, M. T., Rahman, S., Al-Busaidi, W., & Khan, I. (2022). Towards sustainable application of wastewater in agriculture: A review on reusability and risk assessment. *Agronomy, 12*(6), 1397. https://doi.org/10.3390/AGRONOMY12061397

Khanpae, M., Karami, E., Maleksaeidi, H., & Keshavarz, M. (2020). Farmers' attitude towards using treated wastewater for irrigation: The question of sustainability. *Journal of Cleaner Production, 243*, 118541. https://doi.org/10.1016/J.JCLEPRO.2019.118541

Khurshid, N., et al. (2023). Exogenously applied nicotinic acid alleviates drought stress by enhancing morpho-physiological traits and antioxidant defense mechanisms in wheat. *Ecotoxicology and Environmental Safety, 263*, 115350.

Lal, K., Franco, D., Rosin, K. G., & Yadav, R. K. (2021). Urban and peri-urban agroforestry: Utilization of wastewater and degraded landscapes for environmental and livelihood security. In *Agroforestry for degraded landscapes: Recent advances and emerging challenges* (Vol. 2). Springer. https://doi.org/10.1007/978-981-15-6807-7_8

Malisa-van der Walt, R., & Taigbenu, A. (2023). Policy, laws, and guidelines of wastewater reuse for agricultural purposes in developing countries. In *Handbook of environmental chemistry* (Vol. 117). Springer. https://doi.org/10.1007/698_2022_864

Mehmood, A., et al. (2021). Exploring the potential of moringa leaf extract as bio stimulant for improving yield and quality of black cumin oil. *Scientific Reports, 11*, 24217.

Michetti, M., Raggi, M., Guerra, E., & Viaggi, D. (2019). Interpreting farmers' perceptions of risks and benefits concerning wastewater reuse for irrigation: A case study in Emilia-Romagna (Italy). *Water, 11*(1), 108. https://doi.org/10.3390/W11010108

Mubeen, M., et al. (2020). Evaluating the climate change impact on crop water requirement of cotton-wheat in semi-arid conditions using DSSAT model. *Journal of Water and Climate Change, 11*(4), 1661–1675.

Mubeen, M., et al. (2021). Effect of PGPR and drought on spring maize (Zia mays L.). *Pakistan Journal of Botany, 53*(2), 731–739.

Muhammad, H., et al. (2023). A paradigm shift towards beneficial microbes enhancing the efficiency of organic and inorganic nitrogen sources for a sustainable environment. *Land, 12*(3), 680.

Natasha, Shahid, M., Khalid, S., Murtaza, B., Anwar, H., Shah, A. H., Sardar, A., Shabbir, Z., & Niazi, N. K. (2023). A critical analysis of wastewater use in agriculture and associated health risks in Pakistan. *Environmental Geochemistry and Health, 45*(8), 5599–5618. https://doi.org/10.1007/S10653-020-00702-3/TABLES/6

Naz, S., Ahmad, S., et al. (2022). Modeling the impact of climate warming on potato phenology. *Eurpean Journal of Agronomy, 132*, 126404.

Naz, T., et al. (2023). Bio-fortification of two wheat cultivars with iron and zinc through their soil and foliar application in salt-factored soil: Growth, ionic, physiological, and biochemical modifications. *Journal of Plant Growth Regulation, 42*, 5727–5745.

Perez-Espejo, R., Aguilar Ibarra, A., Luis Escobedo-Sagaz, J., Perez-espejo, R., & Luis Escobedo-sagaz, J. (2011). Agriculture and water pollution: Farmers' perceptions in Central Mexico. *International Journal of Water Resources Development, 27*(1), 263–273. https://doi.org/10.1080/07900627.2010.537245

Prasad, M. N. V. (2023). Sustainable water use in agriculture—Circular economy approach. *Advances in Science, Technology and Innovation*. https://doi.org/10.1007/978-3-031-18165-8_5

Pratap, B., Kumar, S., Purchase, D., Bharagava, R. N., & Dutta, V. (2023). Practice of wastewater irrigation and its impacts on human health and environment: A state of the art. *International journal of Environmental Science and Technology, 20*(2), 2181–2196. https://doi.org/10.1007/S13762-021-03682-8/TABLES/3

Qadir, M., Wichelns, D., Raschid-Sally, L., McCornick, P. G., Drechsel, P., Bahri, A., & Minhas, P. S. (2010). The challenges of wastewater irrigation in developing countries. *Agricultural Water Management, 97*(4), 561–568. https://doi.org/10.1016/J.AGWAT.2008.11.004

Ranjan, R. (2021). Restoring natural wetlands through financial incentives based adoption of constructed wetlands on agricultural farms. *Journal of Cleaner Production, 317*, 128346. https://doi.org/10.1016/j.jclepro.2021.128346

Rasool, A., et al. (2020). Microbial diversity response in thallium polluted riverbank soils of the Lanmuchang. *Ecotoxicology and Environmental Safety, 187*, 109854.

Rasool, A., et al. (2021). Microbial diversity and community structure in alpine stream soil. *Geomicrobiology Journal, 38*(3), 210–219.

Rehman, M. H., et al. (2021). Influence of semi-arid environment on radiation use efficiency and other growth attributes of lentil crop. *Environmental Science and Pollution Research, 28*, 13697–13711.

Ricart, S., & Rico, A. M. (2019). Assessing technical and social driving factors of water reuse in agriculture: A review on risks, regulation and the yuck factor. *Agricultural Water Management, 217*, 426–439. https://doi.org/10.1016/J.AGWAT.2019.03.017

Saad, D., Byrne, D., & Drechsel, P. (2017). Social perspectives on the effective management of wastewater. In *Physico-chemical wastewater treatment and resource recovery*. InTech. https://doi.org/10.5772/67312

Saliba, R., Callieris, R., D'Agostino, D., Roma, R., & Scardigno, A. (2018). Stakeholders' attitude towards the reuse of treated wastewater for irrigation in Mediterranean agriculture. *Agricultural Water Management, 204*, 60–68. https://doi.org/10.1016/J.AGWAT.2018.03.036

Sardar, A., et al. (2020). Risk assessment of heavy metal(loid)s via Spinacia oleracea ingestion after sewage water irrigation practices in Vehari District. *Environmental Science and Pollution Research, 27*, 39841–39851.

Sardar, A., Kiani, A. K., & Kuslu, Y. (2021). Does adoption of climate-smart agriculture (CSA) practices improve farmers' crop income? Assessing the determinants and its impacts in Punjab province, Pakistan. *Environment, Development and Sustainability, 23*(7), 10119–10140. https://doi.org/10.1007/S10668-020-01049-6

Segal, T. R., & Giudice, L. C. (2022). Systematic review of climate change effects on reproductive health. *Fertility and Sterility, 118*(2). https://doi.org/10.1016/j.fertnstert.2022.06.005

Shah, A. H., et al. (2020). Assessment of arsenic exposure by drinking well water and associated carcinogenic risk in peri-urban areas of Vehari, Pakistan. *Environmental Geochemistry and Health, 42*, 121–133.

Shah, M. Z., et al. (2021). Synthesis of silver nanoparticles using Plantago lanceolata extract and assessing their antibacterial and antioxidant activities. *Scientific Reports, 11*, 20754.

Shahzad, H., et al. (2020). An integrated use of GIS, geostatistical and map overlay techniques for spatio-temporal variability analysis of groundwater quality and level in the Punjab Province of Pakistan, South Asia. *Water, 12*, 3555.

Singh, A. (2021). A review of wastewater irrigation: Environmental implications. *Resources, Conservation and Recycling, 168*, 105454. https://doi.org/10.1016/J.RESCONREC.2021.105454

Sohail, M. T., Mustafa, S., Ali, M. M., & Riaz, S. (2022). Agricultural communities' risk assessment and the effects of climate change: A pathway toward green productivity and sustainable development. *Frontiers in Environmental Science, 10*. https://doi.org/10.3389/fenvs.2022.948016

Ullah, I., et al. (2022). Comparative effects of biochar and NPK on wheat crops under different management systems. *Crop & Pasture Science*. Retrieved from https://www.publish.csiro.au/CP/CP21146

Ungureanu, N., Vlăduț, V., & Voicu, G. (2020). Water scarcity and wastewater reuse in crop irrigation. Sustainability 12(21), 9055. https://doi.org/10.3390/SU12219055.

Victor, R., Kotter, R., O'Brien, G., Mitropoulos, M., & Panayi, G. (2008). WHO guidelines for the safe use of wastewater, excreta and greywater, volumes 1–4. *International Journal of Environmental Studies, 65*(1), 157. https://doi.org/10.1080/00207230701846598

Waleed, M., et al. (2022). Evaluating the efficiency of coarser to finer resolution multispectral satellites in mapping paddy rice fields using GEE implementation. *Scientific Reports, 12*, 13210.

Wells, E. C., Zarger, R. K., Whiteford, L. M., Mihelcic, J. R., Koenig, E. S., & Cairns, M. R. (2016). The impacts of tourism development on perceptions and practices of sustainable wastewater management on the Placencia Peninsula, Belize. *Journal of Cleaner Production, 111*, 430–441. https://doi.org/10.1016/J.JCLEPRO.2014.08.050

WHO. (2006a). WHO guidelines for the safe use of wastewater, excreta and greywater. *Policy and regulatory aspects* (Vol. 1).

WHO. (2006b). WHO guidelines for the safe use of wastewater, excreta and greywater. *Wastewater use in agriculture* (Vol. 2).

Zaman, G., et al. (2020). Utilization of compost from bio-municipal solid waste improves saline-sodic soils and crop productivity in rice-wheat. *Compost Science & Utilization, 28*, 16.

Zamin, M., et al. (2020). Developing the first halophytic turfgrasses for the urban landscape from native Arabian desert grass. *Environmental Science and Pollution Research, 27*, 39702–39716.

Zhang, Y., & Shen, Y. (2019). Wastewater irrigation: Past, present, and future. *Wiley Interdisciplinary Reviews: Water, 6*(3), e1234. https://doi.org/10.1002/WAT2.1234

Ongoing Programs in On-Farm Water Management

Samra Tariq, Muhammad Mubeen, and Mazhar Ali

Abstract Effective management of on-farm water resources is critical for sustainable agriculture, particularly in regions facing water scarcity and climate variability. This chapter examines various initiatives and strategies aimed at optimizing water use efficiency in agricultural practices, with a focus on international organizations. Key interventions include the adoption of advanced irrigation technologies such as drip and sprinkler systems, the rehabilitation and enhancement of irrigation infrastructure, and the promotion of climate-smart agricultural practices. These efforts are supported by collaborative projects like the Punjab Resilient and Inclusive Agriculture Transformation (PRIAT) and the National Programme for Improvement of Watercourses in Pakistan (Phase-II). Such initiatives not only aim to increase agricultural productivity and profitability but also enhance resilience to climate change impacts. Moreover, The International Water Management Institute (IWMI) and the CGIAR Research Program on Water, Land and Ecosystems (WLE) are international organizations and research institutions that prioritize the advancement and advocacy of inventive approaches to water management. Their work is essential for attaining sustainable water utilization through the integration of scientific research, policy formulation, and practical implementation tactics. These organizations engage in study to comprehend water-related challenges, such as scarcity, pollution, and the effects of climate change. They also devise technology and strategies to tackle these problems. Their endeavors are directed on enhancing agricultural output, while simultaneously protecting ecosystems, improving livelihoods, and ensuring water security for communities worldwide.

S. Tariq
Department of Environmental Sciences, COMSATS University Islamabad, Vehari Campus, Vehari, Pakistan

M. Mubeen (✉) · M. Ali
Department of Biotechnology, COMSATS University Islamabad, Vehari Campus, Vehari, Pakistan
e-mail: muhammadmubeen@cuivehari.edu.pk

M. Mubeen et al. (eds.), *Innovations in Agricultural Water Management*, https://doi.org/10.1007/978-3-031-91883-4_13

Keywords Water utilization · Management · Water scarcity · Climate variability · Irrigation strategies

1 Introduction

Water is essential for agricultural production; successful and sustainable farming practices depend on the effective use of on-farm water supplies (ul Hasan et al., 2021). The term "on-farm water management" (OFWM) describes the methods and plans that farmers implement to maximize the utilization of water resources in their farming operations (Davies et al., 2020). Water conservation, irrigation system management, and incorporating water-efficient practices into agricultural production are all included in this (Levidow et al., 2014). Around 70% of all water withdrawals are used for agriculture, making it the world's largest user of freshwater (Chen et al., 2018). The need for water for agriculture is predicted to rise sharply as the global population grows and the consequences of climate change become more apparent (Piao et al., 2010). On the other hand, freshwater supplies are frequently scarce, and water shortage is a major problem in many areas (El Kharraz et al., 2012). Resolving these issues and guaranteeing the agricultural systems' long-term sustainability require efficient on-farm water management. These programs not only contribute to agricultural sustainability but also yield significant environmental and economic benefits. Efficient water management reduces pressure on freshwater sources, mitigates soil erosion, and minimizes nutrient leaching into groundwater (Schoumans et al., 2014). By conserving water and energy resources, farmers can lower production costs and improve profitability. Furthermore, sustainable water practices contribute to resilience against climate variability, ensuring stable agricultural yields and livelihoods for farming communities (Srivastav et al., 2021).

At its core, on-farm water management aims to optimize the application of water to crops, taking into account factors such as soil type, crop water requirements, weather conditions, and irrigation methods (Robert et al., 2018). Central to on-farm water management is the adoption of efficient irrigation techniques. Traditional methods like flood irrigation, which can result in significant water losses due to evaporation and runoff, are being replaced by more precise methods such as drip irrigation, sprinkler systems, and micro-irrigation. These technologies deliver water directly to the root zone of plants, reducing water wastage and improving water use efficiency (Feng et al., 2024; Hussain et al., 2023a, b, c, 2024a, b, c; Hammad et al., 2023; Abdullah et al., 2023; Kang et al., 2021). By matching irrigation schedules to crop needs and soil moisture levels, farmers can ensure that water application is timely and sufficient without overwatering. Moreover, soil management practices play a crucial role in on-farm water management. Techniques such as conservation tillage, mulching, and cover cropping help improve soil structure and water retention capacity. Healthy soils with enhanced organic matter content absorb and retain water more effectively, reducing the need for frequent irrigation and minimizing nutrient leaching into groundwater.

Another key aspect of on-farm water management is the integration of modern technologies and data-driven decision-making (Xin & Zazueta, 2016). Farmers are increasingly utilizing tools such as soil moisture sensors, weather forecasting, and remote-sensing technologies to monitor field conditions in real-time. This data allows for precise irrigation scheduling and adjustments, optimizing water use efficiency and crop performance. Additionally, advances in agricultural research and extension services provide farmers with the knowledge and support needed to implement best practices in water management (Farid et al., 2023; Haseeb et al., 2023; Alsuhaibani et al., 2023; Yang et al., 2023; Muhammad et al., 2023; Levidow et al., 2014). Enhancing agricultural productivity through on-farm water management contributes to environmental sustainability. Efficient water use reduces the pressure on freshwater sources, mitigates soil erosion, and minimizes the impact of agriculture on water quality (Schoumans et al., 2014). By conserving water resources and minimizing energy inputs associated with pumping and distribution, farmers can also reduce production costs and improve their economic resilience. The On Farm Water Management (OFWM) Wing of the Agriculture Department has a vital role in modernizing the global agriculture industry, making it more scientifically sophisticated, dynamic, and internationally interconnected. The OFWM has a crucial role in the strategic progress of agricultural and rural areas globally. The OFWM, established in 1976, has made significant contributions to socioeconomic development and rural economy by enhancing the transportation, utilization, and effectiveness of water resources through the implementation of efficient and productive technology.

The OFWM program employs an integrated development strategy that prioritizes the optimization of crop productivity while minimizing water consumption. This is achieved by implementing a range of climate-smart measures. These activities involve improving watercourses in canal-controlled areas, restoring irrigation systems in non-canal-controlled areas, implementing advanced irrigation systems like drip and sprinkler, providing precision land leveling services using LASER land levelling units in the private sector, supplying solar systems to power high-efficiency irrigation systems, building water storage ponds, conducting capacity development, raising awareness, and engaging in research and development. These programs are implemented through a cost-sharing agreement, involving the active involvement of the corporate sector and farming communities (Fig. 1).

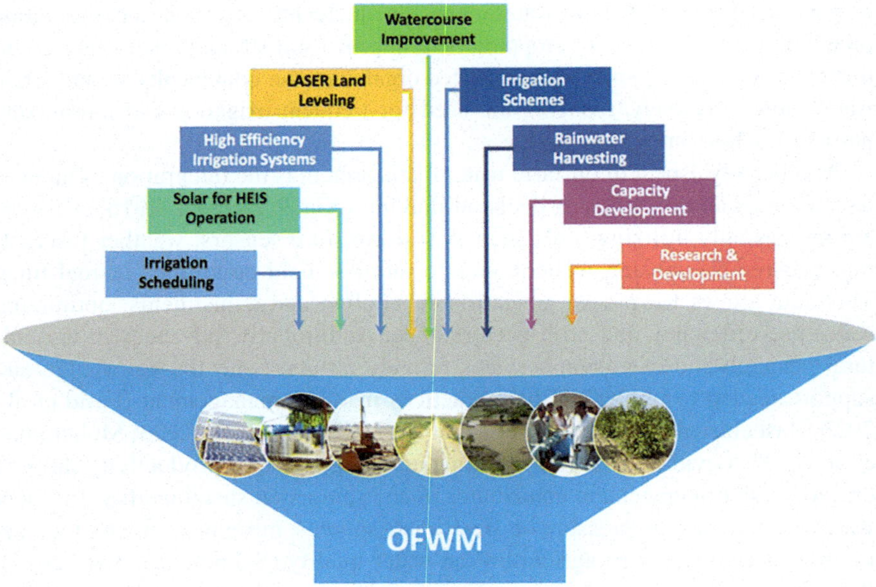

Fig. 1 Water management interventions through user participation

2 International Water Management Institute (IWMI)

The IWMI is a nonprofit organization focused on international water management research, operating under CGIAR (Sikka, 2018). Its main office is located in Colombo, Sri Lanka, with additional offices spread across Africa and Asia. The IWMI's commitment to farm water management begins with extensive research aimed at developing and disseminating innovative solutions tailored to diverse agricultural contexts (Ghosh et al., 2022). The IWMI also conducts comprehensive research and develops creative solutions that are best suited for various farming systems throughout the world in order to boost agricultural productivity while preserving water for future generations. The IWMI's work can be seen through its research initiatives in optimizing water use efficiency in farming (Reddy & Reddy, 2017). These include considering irrigation techniques that maximize yields while minimizing water waste, like drip irrigation, and precision farming.

In order to develop a thorough strategy for water use on farms, the IWMI also investigates integrated water management strategies, which take into account factors like crop water requirements, climate conditions, and soil health (Dirwai et al., 2021). In order to carry out these policies, which promote sustainable farming practices that result in water conservation efforts and a reduction in the ecological consequences on the environment, it works in collaboration with regional farmers, governments, and stakeholders. The IWMI promotes training and policy advocacy in addition to research. The IWMI plays a key role in influencing water management policies that support smallholder farmers and promote equitable distribution of water resources by collaborating with governments and international organizations (Dirwai et al., 2021). For example, the organization can increase farmers' ability to adopt effective irrigation systems for better farming by educating them about potential solutions and providing the essential instruments that would boost this purpose. To share results and start an international conversation about sustainable agriculture-based water control, it also publishes materials online and offline and organizes conferences.

3 World Bank's Agricultural Water Management Projects

The World Bank has launched Agricultural On-Farm Water Management Projects worldwide to address the need for improving water use efficiency and sustainability in agriculture, particularly in regions struggling with water scarcity or performing less sustainable irrigation practices (Gany et al., 2019). These projects collectively target improving crop productivity along with water conservation. Among other key components are the construction and rehabilitation irrigation infrastructure, including canals, pumps, and reservoirs to reduce water losses as well ensure equitability over all beneficiaries. These programs typically promote the use of highly effective irrigation techniques, such as sprinkler and drip systems that are implemented in fields with laser levelling to minimize water waste and maximize crop productivity. Moreover, their efforts are directed towards equipping farmers and communities with the necessary skills to manage their water resources. This includes training them in water-efficient farming techniques that require less moisture in the soil.

The World Bank also supports the institutional building and policy changes required to foster a climate that is conducive to more sustainable water use in agriculture (Shiferaw et al., 2024). In order to promote improved water governance and make investments in agricultural water infrastructure, this involved creating policy frameworks, promoting integrated water resource management (IWRM), and supporting public-private partnerships.

The Water Management Department is responsible for providing a range of services relating to the efficient and effective use of water resources (Kalogiannidis et al., 2023). One of the main services provided is the registration of Water Users Associations.

These organizations play a crucial role in overseeing the allocation of water resources and guaranteeing fair and equal access to water for all individuals (Mujtaba et al., 2024). Watercourse improvement is another significant service offered by the Water Management division. This entails the restoration and enhancement of irrigation systems to enhance water distribution and maximize the overall effectiveness of the system. Another service provided is precision land levelling, which enhances water efficiency by establishing a consistent field surface and minimizing water runoff (Meena et al., 2024).

In addition, the Water Management section has expertise in installing high efficiency irrigation systems (HEIS). These systems are engineered to optimize water efficiency and minimize loss, ultimately resulting in enhanced crop yields and more effective utilization of water resources. Training and capacity building are essential elements of the services provided by the Water Management wing (Trimmer et al., 2023). The organization offers training and assistance to farmers and water user associations to ensure they possess the requisite expertise and understanding to successfully and sustainably manage water resources. The Water Management Department is essential for effectively managing and maximizing water resources. Their wide range of services ensures the sustainable utilization of water for agricultural and other purposes.

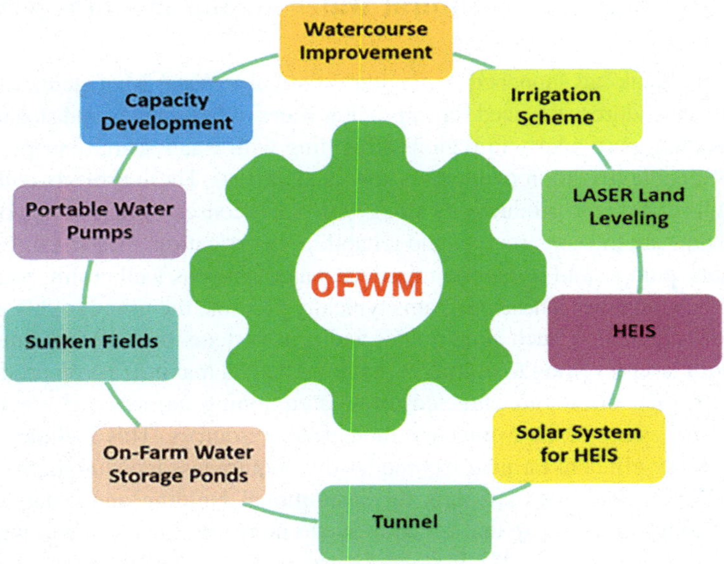

4 European Union's Water Framework Directive (WFD)

The primary objectives of the WFD are outlined in Article 4 of the Directive. In order to maintain good status and halt degradation, it requires member states to use their Programmes of Measures (PoMs) and River Basin Management Plans (RBMPs) to protect and, when necessary, restore water bodies (Kagalou & Latinopoulos, 2020). A state of good health includes both chemical and ecological well-being. The primary legislative framework is the Water Framework Directive (WFD) (De Smedt & Van Rijswick, 2024). Two subsidiary directives, referred to as daughter directives, provide assistance for this. One focuses on regulating the quality and quantity of groundwater, while the other focuses on regulating the quality of surface water. The WFD includes rules pertaining to the time limits for achieving the goals of the directive, as well as regulations concerning exceptions. The annexes to the Water Framework Directive (WFD) provide specific information on many aspects, such as monitoring obligations, the standards for evaluating the condition of water bodies, and the components of the River Basin Management Plans (RBMPs).

The Water Framework Directive requires river basin management plans to be submitted every 6 years. The last cycle of reporting took place in 2016, and the next one is slated for 2022. Each of the 180 river basins is covered by a river basin management plan, which includes an assessment of the water bodies within it, an analysis of the stressors they experience, and relevant methods to achieve good condition (Puharinen, 2024). Surface waters include a wide range of bodies of water, including lakes, rivers, coastal waterways, and transitional zones. The Water Framework Directive is now being implemented by the European Union (EU) and Norway. About 146,500 surface waterbodies and 15,000 groundwater bodies have been assessed in the EU and Norway in compliance with the Water Framework Directive.

The establishment of River Basin Districts (RBD) is a crucial element of the Water Framework Directive (Nadir & Ahmed, 2023). The establishment of these zones is determined by the river basin, which serves as the natural physical and hydrological unit, rather than being based on administrative or political boundaries. Due to the frequent occurrence of rivers crossing national borders, officials from many member states must collaborate and cooperate in order to effectively manage the basin, sometimes referred to as transboundary basins. The management of the river basins is governed by River Basin Management Plans, which outline the specific strategies and actions needed to achieve the objectives within the specified timeframe. They should be refreshed on a biennial basis.

The 2017 Water Environment (Water Framework Directive) Act in England and Wales updated the 2003 framework. Scotland and Northern Ireland have their own distinct legislation. The user's text is enclosed in tags. Prior to the UK's accession to the "Common Market," water body management was administered through the organization of river basins. This responsibility was initially held by the River Boards, followed by the River Authorities, then the Water Authorities, and ultimately by the Environment Agency in England, Natural Resources Wales in Wales, and SEPA in Scotland. The UK will maintain its long-standing heritage of river basin management in its approach to spatially managing river basins (Carter et al., 2024). Every River Basin District (RBD) will still be obligated to prepare a River Basin Management Plan. A River Basin District (RBD) is a geographical area consisting of one or more adjacent river basins, along with its corresponding groundwater and coastal waters.

5 CGIAR Research Program on Water, Land, and Ecosystems (WLE)

The primary objective of the CGIAR Research Program on Water, Land and Ecosystems (WLE) is to tackle the many difficulties associated with water management in agricultural systems (Jampani et al., 2023). The CGIAR Research Program on Water, Land and Ecosystems (WLE) is dedicated to addressing the intricate issues surrounding water management in agricultural systems, with the goal of ensuring food security, enhancing nutrition, supporting livelihoods, and fostering environmental sustainability. The initiative aims to create and execute inventive techniques that improve the efficiency of water use in agriculture while reducing negative impacts on the environment. This entails incorporating state-of-the-art studies on water management techniques, including efficient irrigation methods, rainwater collection, and the restoration of deteriorated landscapes. The WLE strives to enhance the resilience of agricultural communities to climate change and variability by supporting these practices, enabling them to flourish even in difficult situations.

The WLE program places great importance on efficient water management. The goal is to enhance water efficiency in agriculture by advancing and advocating for innovative irrigation methods and water-conserving technologies. The focus is on utilizing methods like drip and sprinkler irrigation to minimize water consumption. Furthermore, the initiative promotes the gathering and effective utilization of precipitation, employing methods that capture and store rainwater for agricultural purposes. These measures enhance water efficiency and contribute to the overall sustainability of agricultural systems. The WLE places significant emphasis on environmental sustainability. The program's objective is to guarantee that agricultural water management practices do not cause any harm to the environment. This entails the restoration of deteriorated landscapes, rejuvenating them for productive use by implementing soil conservation methods and other restoration approaches. The WLE enhances the health and productivity of these areas, so contributing to the conservation of natural ecosystems and biodiversity, ultimately fostering a more sustainable environment.

The program's influence extends beyond the advantages for the environment, as it also seeks to improve food security, nutrition, and livelihoods for agricultural communities. The Water, Land, and Ecosystems (WLE) program promotes consistent access to water for agricultural purposes, which in turn enhances food production stability and facilitates the cultivation of a wide range of crops that contribute to improved nutritional outcomes. Moreover, through the improvement in agricultural productivity, the initiative generates more lucrative income prospects for farmers. Additionally, it offers training and tools to promote sustainable farming methods, thus enhancing the local ability to effectively manage water.

In the following paragraphs, we would like to present some case study which is related to the OFWM work in Pakistan (a developing country).

6 A Case Study: The OFWM Activities in Pakistan

6.1 Punjab Resilient and Inclusive Agriculture Transformation (PRIAT) (2022–2023 to 2026–2027)

The aim of this project is to boost agriculture by increasing the efficiency of water usage on farms and improving the profitability of farmers. The project seeks achieve this by following several fundamental objectives, including:

- Improving the efficiency of water transportation by farm irrigation conveyance networks on farms in both canal and noncanal command areas.
- Promoting the use of innovative agricultural production methods, renewable energy sources, and on-farm water management technology to enhance the resilience of agriculture to climate change.
- Increasing the financial viability of farmers by promoting the cultivation of high-value crops and diversifying the agriculture value chain.
- Enhancing the capabilities and long-term viability of the private sector to provide services that assist climate-resilient irrigated agriculture.
- Enhancing the ability of stakeholders to adopt climate-smart and high-value agricultural methods to enhance the profitability of farms.
- Creating employment opportunities in agriculture for rural populations, which can enhance their quality of life and reduce poverty.

6.1.1 Components of the Project

1. The following techniques aim to improve the effectiveness of water transportation inside agricultural premises:
   - Enhancing the efficiency of 1000 watercourses to optimize water flow.
   - Expanding the lining of 2000 watercourses to reduce water loss caused by seepage.
   - Restoring and optimizing the function of 1000 watercourses that have become obsolete.
   - Implementing the construction of 3000 irrigation schemes in regions that are not covered by canal commands or riverine zones, with the aim of increasing the availability of irrigation.
   - Implementing on-farm irrigation advice services, including water budgeting and accounting, on 72 watercourses to aid farmers in efficiently managing water resources.
2. The following actions will be taken to encourage climate-smart production and equitable market access:
   - To improve product quality and market value, regenerative agriculture techniques, harvesting, processing, and agricultural value-adding machinery will be promoted. There will also be easier access to markets.

3. The following measures will be undertaken to promote climate-smart production:
   - Implementing high-efficiency irrigation systems across 40,000 acres to decrease water usage and enhance crop productivity.
   - The implementation of solar systems to provide energy for the high-efficiency irrigation systems covering 20,000 acres, with the aim of decreasing dependence on fossil fuels and reducing carbon emissions.
   - The supply of authorized orchard plants and vegetable seeds/seedlings on a 5000-acre area to enhance agricultural diversity and ability to withstand the effects of climate change.
   - Implementation of 1000 on-farm water storage ponds to bolster water security and mitigate the effects of droughts.

6.1.2 Project Cost

The project is estimated to have a total cost of Rs. 68,672.560 million, which would be financed by various sources. The World Bank will allocate a sum of Rs. 45,863.560 million, which is equivalent to USD 200 million or SDR 148.8 million. The Punjab Government would contribute Rs. 9072.000 million. Furthermore, farmers would contribute a total of Rs. 13,737.000 million towards the project. The awarded money will be used for many parts of the project, such as infrastructure development, capacity building, and other activities that align with the project objectives. Systematic financial monitoring and reporting will be carried out to ensure adherence to the project's budget and efficient utilization of funds.

- Total project cost: Rs. 68,672.560 million
- World Bank share: Rs. 45,863.560 million (USD 200 million/SDR 148.8 million)
- Punjab Government share: Rs. 9072.000 million
- Farmers' contribution: Rs. 13,737.000 million

The Environmental and Social Management Framework (ESMF) is a tool that is employed to evaluate the potential hazards and impacts of a program or series of subprojects when the specifics of each component are not yet known. It offers procedures, principles, rules, and guidelines for the identification and mitigation of environmental and social dangers and impacts.

6.2 National Program for Improvement of Watercourses in Pakistan (Phase-II): The Punjab Component (2019–2020 to 2024–2025)

The National Programme for Improvement of Watercourses in Pakistan (Phase-II)—The Punjab Component (2019–2020 to 2024–2025) is a project that receives funding from the Federal Public Sector Development Programme (PSDP) and the Provincial Annual Development Programme (ADP) in Pakistan. The objective is to enhance water efficiency and optimize agricultural techniques in the canal-irrigated regions of Punjab. The project encompasses the primary goals and elements:

6.2.1 Project Objectives

- Enhance the efficiency of water conveyance and optimize on-farm water usage by implementing water conservation strategies to increase irrigation water yield.
- Enhance the ability and long-term viability of the private sector in providing services to facilitate irrigated agriculture.
- Enhance the capabilities of stakeholders in effectively managing irrigation water to attain increased crop yields while minimizing production expenses.
- Contribute to enhancing water productivity, boosting agricultural production, increasing incomes for farmers, creating more employment possibilities in rural regions, improving living standards for farmers, and achieving positive environmental effects.

The primary activities addressed by the NPIW-II project consist of the following:

Watercourse Improvement

- Enhancing conveyance efficiency by improving 2500 unimproved watercourses, covering up to 50% of their whole length.
- Maximizing the potential of watercourse improvement by adding lining to 7500 partially improved watercourses, covering up to 50% of their length.

LASER Land Leveling
A total of 9500 LASER land levelers will be provided to farmers and service providers in order to enhance LASER land levelling services in the private sector.

On-Farm Water Storage
The objective is to construct 3000 on-farm water storage ponds in irrigated areas to store surplus canal and rainwater, which will be used to provide additional irrigation.

Project Area
The project would be executed in the canal-irrigated regions across the whole Punjab province.

Gestation Period
The project has a duration of 5 years, from 2019–2020 to 2023–2024, extended for 1 year till 2024–2025.

Financial Implications (Rs. Million)

- Total project cost: 46,255.356 million
- Government share: 28,589.822 million
- Farmers' contribution: 17,665.534 million

The objective of the National Programme for Improvement of Watercourses in Pakistan (Phase-II)—The Punjab Component is to increase water productivity, promote sustainable agriculture, and boost the livelihoods of farmers in Punjab. The project's objective is to enhance water conveyance efficiency, advocate for water conservation measures, and offer essential equipment and infrastructure. This will lead to higher crop yields, increased revenue, and more employment prospects, all while assuring improved water management and environmental outcomes.

7 Conclusion

Efficient on-farm water management is indispensable for ensuring agricultural sustainability, especially in developing regions like Punjab, Pakistan, where water resources are under pressure from increasing demand and climate variability. This study has highlighted various initiatives and programs aimed at optimizing water use efficiency, ranging from the adoption of modern irrigation technologies to infrastructure development and capacity building. Projects such as PRIAT and NPIW-II underscore the collaborative efforts between governments, international organizations, and local communities to enhance water productivity, promote climate-resilient agriculture, and improve livelihoods for farmers. International organizations like IWMI and initiatives such as CGIAR's WLE have played significant roles in researching and advocating for sustainable water management practices worldwide, influencing policies and fostering knowledge exchange. These efforts are crucial for addressing the challenges posed by water scarcity and climate change, ensuring equitable access to water resources, and enhancing agricultural resilience.

References

Abdullah, H., et al. (2023). Investigating on-site energy consumption patterns using vertical electrical sounding (VES) and GIS techniques. *Sustainability, 2023*(15), 16782. https://doi.org/10.3390/su152416782

Alsuhaibani, A. M., et al. (2023). Green buildings model: Impact of rigid polyurethane foam on indoor environment and sustainable development in energy sector. *Heliyon, 9*, 14451.

Carter, J. G., Karvonen, A., & Winter, A. (2024). Towards Catchment Scale Natural Flood Management: Developing evidence, funding and governance approaches. *Environmental Policy and Governance, 34*, 553.

Chen, B., Han, M., Peng, K., Zhou, S., Shao, L., Wu, X., et al. (2018). Global land-water nexus: Agricultural land and freshwater use embodied in worldwide supply chains. *Science of the Total Environment, 613*, 931–943.

Davies, S., Moeen, M. S., Shikoh, S. H., & Rizwan, N. (2020). Solarization of electric tube-wells for agriculture in Balochistan: Economic and environmental viability.

De Smedt, P., & Van Rijswick, H. F. (2024). The right to water and the right to a healthy environment: Does the water framework directive map out the way ahead? In *The right to a healthy environment in and beyond the anthropocene* (pp. 136–164). Edward Elgar Publishing.

Dirwai, T. L., Kanda, E. K., Senzanje, A., & Busari, T. I. (2021). Water resource management: IWRM strategies for improved water management. A systematic review of case studies of East, West and Southern Africa. *PLoS One, 16*(5), e0236903.

El Kharraz, J., El-Sadek, A., Ghaffour, N., & Mino, E. (2012). Water scarcity and drought in WANA countries. *Procedia Engineering, 33*, 14–29.

Farid, H. M., et al. (2023). Groundwater quality risk assessment using hydro-chemical and geospatial analysis. *Environment, Development and Sustainability, 25*, 8343–8365.

Feng, L., et al. (2024). Seasonal dynamics in land surface temperature in response to land use land cover changes using Google earth engine. *IEEE Journal of Selected Topics in Applied Earth Observations and Remote Sensing, 17*(2024), 17983.

Gany, A. H. A., Sharma, P., Singh, S., & drainage. (2019). Global review of institutional reforms in the irrigation sector for sustainable agricultural water management, including water users' associations. *Irrigation and Drainage, 68*(1), 84–97.

Ghosh, E., Kemp-Benedict, E., Huber-Lee, A., Nazareth, A., & Oudra, I. (2022). *Directions for sustainable agricultural water investments*. Food & Agriculture Organisation.

Hammad, H. M., et al. (2023). Impacts of COVID-19 pandemic on environment, society, and food security. *Environmental Science and Pollution Research, 30*, 99261–99272.

Haseeb, M., et al. (2023). Quantifying irrigation water demand and supply gap using remote sensing and GIS in Multan, Pakistan. *Environmental Monitoring and Assessment, 195*, 990.

Hussain, S., et al. (2023a). Assessment of land use/land cover changes and its effect on land surface temperature using remote sensing techniques in southern Punjab, Pakistan. *Environmental Science and Pollution Research, 30*, 99202–99218.

Hussain, S., et al. (2023b). Relation of land surface temperature with different vegetation indices using multi-temporal remote sensing data in Sahiwal region, Pakistan. *Geoscience Letters, 10*, 33.

Hussain, S., et al. (2023c). Investigation of irrigation water requirement and evapotranspiration for water resource management in southern Punjab, Pakistan. *Sustainability, 15*(15), 1768.

Hussain, S., et al. (2024a). Assessing the impact of land use land cover changes on soil moisture and vegetation cover in southern Punjab, Pakistan using multi-temporal satellite data. *Geology, Ecology, and Landscapes*. https://doi.org/10.1080/24749508.2024.2338970

Hussain, S., et al. (2024b). Agricultural land suitability analysis of southern Punjab, Pakistan using analytical hierarchy process (AHP) and multi-criteria decision analysis (MCDA) techniques. *Cogent Food & Agriculture, 10*(1), 2294540.

Hussain, S., et al. (2024c). Assessment of future prediction of urban growth and climate change in district Multan, Pakistan using CA-Markov method. *Urban Climate, 53*, 101766.

Jampani, M., Sena, D. R., & Matheswaran, K. (2023). Framework for advancing water resource sustainability and climate resilience through local-scale hydrological modeling in the Ganges Delta.

Kagalou, I., & Latinopoulos, D. (2020). Filling the gap between ecosystem services concept and river basin management plans: The case of Greece in WFD 20+. *Sustainability, 12*(18), 7710.

Kalogiannidis, S., Kalfas, D., Giannarakis, G., & Paschalidou, M. (2023). Integration of water resources management strategies in land use planning towards environmental conservation. *Sustainability, 15*(21), 15242.

Kang, J., Hao, X., Zhou, H., & Ding, R. (2021). An integrated strategy for improving water use efficiency by understanding physiological mechanisms of crops responding to water deficit: Present and prospect. *Agricultural Water Management, 255*, 107008.

Levidow, L., Zaccaria, D., Maia, R., Vivas, E., Todorovic, M., & Scardigno, A. (2014). Improving water-efficient irrigation: Prospects and difficulties of innovative practices. *Agricultural Water Management, 146*, 84–94.

Meena, R. P., Karnam, V., Sujatha, H. T., Tripathi, S. C., & Singh, G. (2024). Practical approaches to enhance water productivity at the farm level in Asia: A review. *Irrigation and Drainage, 73*(2), 770–793.

Muhammad, H., et al. (2023). A paradigm shift towards beneficial microbes enhancing the efficiency of organic and inorganic nitrogen sources for a sustainable environment. *Land, 12*, 680.

Mujtaba, G., Shah, M. U. H., Hai, A., Daud, M., & Hayat, M. (2024). A holistic approach to embracing the United Nation's Sustainable Development Goal (SDG-6) towards water security in Pakistan. *Journal of Water Process Engineering, 57*, 104691.

Nadir, H. M., & Ahmed, A. (2023). The critical review of the performance of the EU water framework directive in improving the rivers' water quality and the impediments in its implementation. *Novel Research in Science, 14*(4), 1–7.

Piao, S., Ciais, P., Huang, Y., Shen, Z., Peng, S., Li, J., et al. (2010). The impacts of climate change on water resources and agriculture in China. *Nature, 467*(7311), 43–51.

Puharinen, S.-T. (2024). *Normative environmental quality as a regulatory strategy in EU environmental law: Legal implications of water status objectives*. Itä-Suomen yliopisto.

Reddy, K. Y., & Reddy, L. N. (2017). Incentivizing water use efficiency-virtual water approach. *IWRA (India) Journal, 6*(1), 35–42.

Robert, M., Thomas, A., Sekhar, M., Raynal, H., Casellas, É., Casel, P., et al. (2018). A dynamic model for water management at the farm level integrating strategic, tactical and operational decisions. *Environmental Modelling & Software, 100*, 123–135.

Schoumans, O., Chardon, W., Bechmann, M., Gascuel-Odoux, C., Hofman, G., Kronvang, B., et al. (2014). Mitigation options to reduce phosphorus losses from the agricultural sector and improve surface water quality: A review. *Science of the Total Environment, 468*, 1255–1266.

Shiferaw, B. A., Reddy, V. R., & Sharma, B. (2024). Groundwater governance under climate change in India: Lessons based on evaluation of World Bank interventions. *International Journal of Water Resources Development, 40*(3), 401–424.

Sikka, A. K. (2018). Role of IWMI in water management for the betterment of agriculture in rural India. *Indian Farming, 67*(10), 16.

Srivastav, A. L., Dhyani, R., Ranjan, M., Madhav, S., & Sillanpää, M. (2021). Climate-resilient strategies for sustainable management of water resources and agriculture. *Environmental Science and Pollution Research, 28*(31), 41576–41595.

Trimmer, J. T., Qureshi, H., Otoo, M., & Delaire, C. (2023). The enabling environment for city-wide water service provision: Insights from six successful cities. *PLoS Water, 2*(6), e0000071.

ul Hasan, F., Fatima, B., & Heaney-Mustafa, S. (2021). A critique of successful elements of existing on-farm irrigation water management initiatives in Pakistan. *Agricultural Water Management, 244*, 106598.

Xin, J., & Zazueta, F. (2016). Technology trends in ICT–towards data-driven, farmer-centered and knowledge-based hybrid cloud architectures for smart farming. *Agricultural Engineering International: CIGR Journal, 18*(4), 275–279.

Yang, X., et al. (2023). Quantitative evaluation of soil water and wind erosion rates in Pakistan. *Remote Sensing, 15*, 2404.

Scope of Impact-Based Forecasting in On-Farm Management Programs

Dildar Hussain Kazmi, Zaheer Ahmed Babar, Asma Jawad Hashmi, Wajid Nasim, Beenish Nazakat Raja, Tariq Mahmood, Muhammad Ayaz, and Ghulam Mustafa Shah

Abstract Climate and its variability have been diagnosed as the major challenge for agricultural production over the globe. However, the recent events of weather extremes have shocked not only the planners but also the climate modelers. It has become more problematic for seasonal forecasters to precisely predict the outlook. The agricultural crops and orchards are merely exposed to bear the major damage. Importantly, the farmers of developing countries remained the most vulnerable due to their unawareness and lack of modern technologies. To deal and minimize the major losses imposed by the weather extremes, the crop scientists have put tremendous efforts and got success up to some extents. But the protection and sustainability of supporting material particularly the water availability has still been prone to erratic weather. Crop water requirement has already been ignited by global warming, especially over the tropical–extratropical regions. Chief amount of cereals and fruit has been cultivated through irrigated farming (like in south Asia), which involves a big cost. In the present era of scientific revolution, justified and smart utility of water may be achieved through innovative programs. Impact-based forecasting is an emerging field and capable to assist the farming community in smart cultivation practices. Now the international agencies have got aware of the issues and playing supportive roles to manage the farm-level practices. As a pilot, one of such projects has successfully been completed for a small region of Pakistan. However, it requires a good chemistry between the stakeholders including the

D. H. Kazmi (✉) · A. J. Hashmi · B. N. Raja · T. Mahmood · M. Ayaz
Pakistan Meteorological Department, National Agromet Centre, Islamabad, Pakistan

Z. A. Babar
Pakistan Meteorological Department, National Weather Forecasting Centre, Islamabad, Pakistan

W. Nasim
Department of Agronomy, Faculty of Agriculture and Environment (FA&E), Islamia University of Bahawalpur (IUB), Bahawalpur, Pakistan

G. M. Shah
College of Engineering (Key Laboratory for Clean Renewable Energy Utilization Technology, Ministry of Agriculture), China Agricultural University, Beijing, People's Republic of China

M. Mubeen et al. (eds.), *Innovations in Agricultural Water Management*,
https://doi.org/10.1007/978-3-031-91883-4_14

weather forecasters, agriculture organizations, etc., and the end-users. Above all, an accurate forecast, its timely dissemination, and active participation of agricultural community could help improve the prevailing situation.

Keywords Climate · Agriculture · IBF · Impact matrix · Pothohar region

1 Introduction

Climate change has now been a well-known fact and its impacts are more threatening for the underdeveloped regions like south Asia and Africa. This is mainly due to the lesser infrastructure and deficiency of modern technologies to curtail the losses and damages. In fact, the natural exposure or orientation remained a prime factor of vulnerability and is harder to cope with. Impacts of global warming and climate change are clearly visible through rapid recession of cryosphere in Hindukush Himalaya (HKH) where snowline has gone to higher elevation and most of the glaciers have been pushed into the negative mass balance. According to Mishra et al. (2019), with 1.5 °C increase in temperature, one-third ice mass of HKH glaciers will disappear and two-thirds of glaciers' ice mass will be depleted by the end of the twenty-first century. Summer monsoon has become highly variable over time and space, therefore occurring of storms. Storm water destroys the standing field crops, inflicts damage to infrastructure, and runs down to the Arabian Sea.

The climate hazards are happening more often in the recent years. Accordingly, the ecological cycle including the agriculture sector has become more vulnerable. In particular, the erratic rains have been making it harder to manage the water requirements for agricultural crops. In this way, the situation may become worse in the next few decades (Ali et al., 2017). In South Asia, Pakistan is a small agrarian country with varying features of topography as moving from north to south and from east to west. Monsoon rainfall (July–September) is the lifeline of Pakistan's water resources, which caters to the highest water demands of field crops during summers and gathers the reserves to meet the requirements of low flow periods in winter and fall (Naheed & Rasul, 2011). Being the agro-based economy, it is highly sensitive to any shock wave in the form of weak monsoon or mismanagement of available water resources. It is important to note that diversity of weather pattern is evident for the occurrence of strong El Nino during the recent summers and heavy downpour in a couple of succeeding years. Prolonged drought conditions and abnormal rains in a short span of summer monsoon have been threatening the whole ecology in the region.

Besides other sectors, agriculture is significantly affected by climate change. The water availability throughout the cropping season has a key role but it is likely to be disrupted in future by increasing temperatures and variable hydrological cycle (Kazmi & Rasul, 2012). Besides water, the recent episodes of climate variability have been affecting the soil features as well. The deficiency of natural resources has been posing serious threats to the food security of today's overpopulated world and needs special attention to cope with (FAO, 2019). Consequently, changing climate has been disrupting the crop production as well as the food quality and availability to the end-users. Future scenarios of climate are also showing more increase in temperatures with abnormal precipitation patterns, which would surely be a threat to agriculture sector. Nature has provided a giant tank of freezing waters in the north over the Himalaya, Karakoram, and Hindukush and the environment has been operating this tank in terms of temperature after the strong build-up of greenhouse gases.

It is evident from data analyses that the day temperatures are continuously increasing in the lower half of Pakistan while night temperatures are increasing particularly in the upper half. Rainfalls have become more uncertain, and seasonal rains are being shifted to alternate areas along with variation in the time of occurrence. Accordingly, moisture conditions in the air as well as in the soil along with other agro-meteorological parameters are modified. The population is on increase and so is the demand of resources like food, water, and energy. There are countries with sufficient food of their own or having the potential to import. But the situation is rather challenging for the developing countries like Pakistan.

As a matter of fact, climate change cannot be reversed and we have to live with it. Therefore, for sustainable agricultural production, adaptation to the changing climate is the only option left.

1.1 Challenges for Underdeveloped Regions

The impacts of climate change are being felt throughout the world and the people have started considering the problem. However, the Third World countries are more vulnerable with less infrastructure and capacity to modernize their agricultural practices. Kurukulasuriya et al. (2006) stated that the underdeveloped regions with lesser infrastructure are more vulnerable to the impacts of climate change. The countries of Africa and South Asia with arid to semi-arid conditions face difficulties in coping with the financial and technical deficiencies. In these regions, agriculture is the major occupation, which plays the most prominent role in earning national GDP. Most people are poor and not able to adopt modern agriculture technology for sustainable production (The World Bank, 2012; Vermeulen et al., 2012).

1.2 Local Climatology and Hydrological Factors

At present, Pakistan is facing serious threats in the context of worsening environment and climate change. Based on these conditions, the German watch Global Climate Risk Index (GGCRI) has ranked it as the eighth most vulnerable country to the climate change consequences (UNDP, Climate Promise-II, https://www.undp.

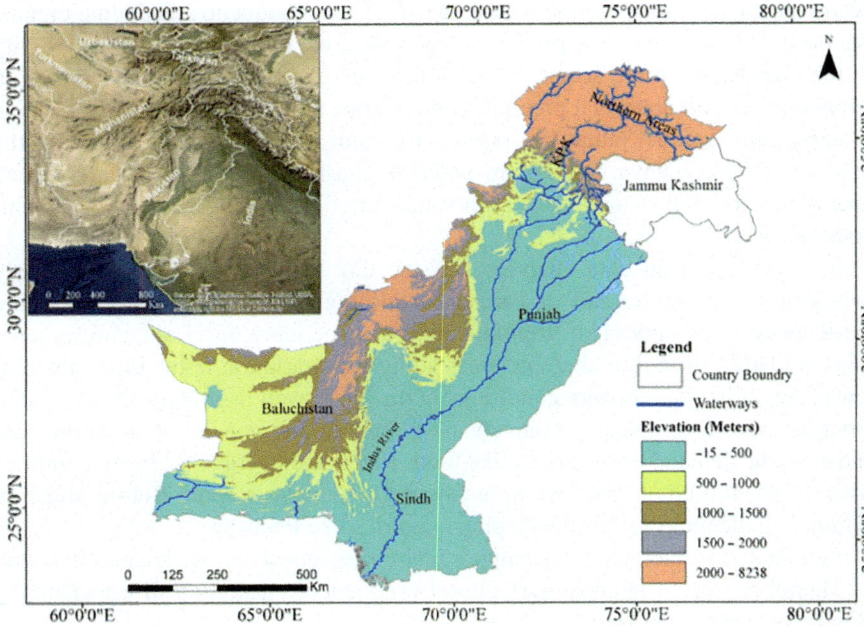

Fig. 1 Hydrological map of Pakistan. Topography borders with its neighboring countries are presented in the inset. (Source: Khan et al., 2021)

org/pakistan/projects/climate-promise-ii), while it is contributing below 1% to the global greenhouse gases with over 200 million population (GoP, 2022_i).

Central agricultural plains of the country are irrigated by the Indus water system (Fig. 1). The flows in this water system have been managed through the rains and glacier melting in the upper Indus basin (Chaudhry & Rasul, 2007). The summer monsoon during July to September has been the major source of water for irrigation and other purposes. It feeds the standing crops in Kharif and contribute a major role in maintaining the soil moisture for upcoming Rabi season (Naheed & Rasul, 2011; Rasul et al., 2004). Although summer monsoon over South Asia is extended over the period from June to September but its peak covers only the months of July and August (Ding & Ke, 2013), whereas the period between May and June has been the hottest and driest time of the year (Rasul et al., 2004). But the summer monsoon became unreliable based on its frequent fluctuations in the recent years. Consequently, occasional floods and droughts have been occurring in the downstream areas of lower half of the country (Rasul & Kazmi, 2013). Agriculture sector is directly affected by the abnormality of climate conditions. The global warming along with the inadequate water supply for the entire cropping season has been a major challenge in Pakistan (ECP, 2019).

Geographically, Punjab and Sindh provinces hold the major plains for agricultural practices based on their suitable climate and the water availability through Indus tributaries. There are mainly two cropping seasons, Rabi (October/November to March/April) and Kharif (April to September/October), in Pakistan. Wheat is the most important crop of Rabi season (winters) and stands for major cereal food for local population followed by rice and maize. On the other hand, cotton is one of the most important crops of Kharif season (summers) and remained a major source for foreign exchange.

In a recent study (Kazmi et al., 2023), it is revealed that the summer monsoon (June–September) contributes the principal share in annual rainfall followed by the spring season (March–April). The winter (December–February), the most important period for wheat, etc., has been pushed to the third place with lesser rains in comparison. On the other hand, extensive foggy conditions in the plains of Punjab and Sindh during the winters have been creating more troubles for wheat growth and development. Cloudy skies with no rain and sunshine impose negative impact on the crop in terms of poor photosynthesis. Besides, unusual rising temperatures and occasional showers/hails/storms in the late winters/spring in the recent years have been disrupting the harvesting and post-harvesting activities. Moreover, heatwaves in the early summers and excessive humid conditions during the monsoon period, etc., have emerged as stunning challenges for the healthier growth of cotton crop.

1.3 Significance of Climate Factor

In addition to quantitative strength, a periodic spatial distribution of rainfall is important for agricultural stability. It is more pronounced for the weaker soils with lesser water-holding capacity (Pratley, 2003). The meteorological conditions have major role in the growth of agriculture products and their significance become more vital at the particular stages of the crops, orchids, vegetables, etc. Rasul and Kazmi (2011) stated that out of five major crops, four are cultivated during Kharif (April/May to October/November) including rice, sugarcane, cotton, and Maize. Whereas wheat, a major cereal along with sesame, grams, etc., is grown in Rabi (October/November to April/May). So, the summer monsoon has a direct link with these prominent cash crops, either in standing position (Kharif crops) or being cultivated in the following periods (Rabi crops). Considering these facts, the rainfall and day/night temperatures for all the important periods have been focused in the following discussion.

2 Scope and Motivation

The situation demands a comprehensive study for exploring the possible ways to safeguard the important crops in the presence of extreme climate variability. The climate has continuously modified and may not be reversed to its norms, but we may find alternate options to minimize the expected damages through revising the ongoing practices and adopting effective techniques. It requires the collective efforts by the major stakeholders including the meteorologists, agri-scientists, policy makers, as well as the farming community.

In the present era of innovation and adoptability, why do not try alternate options to protect our agriculture in the presence of unavoidable climate extremes. In the recent projects related to Impact-Based Forecasting and Decision Support System for local farmers by UK Met Office, Regional Integrated Multi-Hazard Early Warning System for Africa and Asia (RIMES), and International Centre for Integrated Mountain Development (ICIMOD), it was revealed that the local farmers have become vigilant about the weather patterns. They knew the importance of agrometeorological conditions and were interested to obtain some effective roadmap to sustain their productivity. Also, not only the local farmers but also the relevant research organizations as well as the agriculture extension departments are looking for some reliable climate projections to devise their planning and practices.

3 Data and Methods

Besides regional data related to climate, hydrology, and agriculture, country-level information for Pakistan has been utilized. Monthly data of precipitation and air temperatures from observational network of Pakistan Meteorological Department (PMD) have been utilized for the period January 1961 till the end of 2023. For seasonal analysis, the whole data period has been divided into winter (December–February; DJF), spring (March–April; MA), pre-summer (May–June; MJ), summer monsoon (June–September; JJAS), and autumn (October–November; ON). For agriculture-related discussion, information related to cropping seasons Rabi (October/November to March/April) and Kharif (April to September/October) has been added.

Departure of weather data for the last 3 years has been calculated with reference to the recent climate normal 1991–2020. However, climate normal data has also been utilized for anomaly calculation based on last 60-year period starting from 1961.

4 Data Analysis and Important Findings

4.1 Climate Change During the Recent Periods

Climate variations have been evident from the changes in the atmospheric conditions and ultimately the meteorological features like rainfall and air temperatures. Rainfall and air temperatures have vital role in the growth of agricultural crops, vegetables, as well as orchards. Both elements have been changed due to the consequent effect of climate variability, especially the rainfall. In this context, seasonal rainfalls for two prominent rainy seasons, that is, summer (JJAS) and spring (MA), have been calculated following the pattern of climatic normal for the period 1961–2020 and shown here as a reference (Fig. 2a, b).

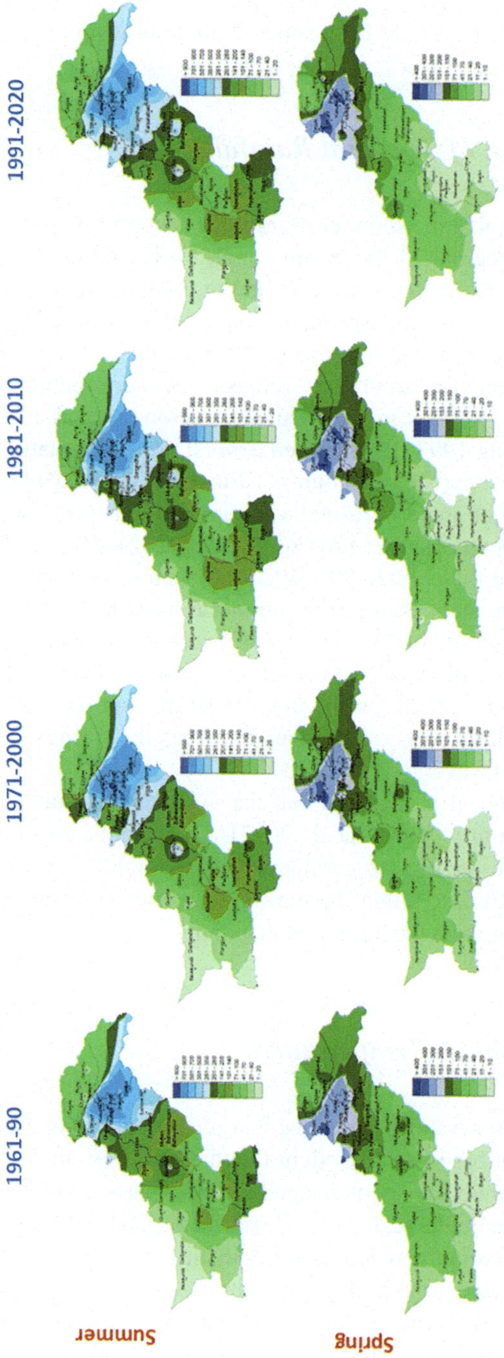

Fig. 2 Seasonal rainfall (Summer "JJAS" and Spring "MA") for 30 years based on climatic normal; 1961–2020

It is evident from these maps that rainfall for both summer and spring seasons has continuously been changing for most parts of the country. Therefore, detailed discussions in this context have been added in the following text.

4.2 Distribution of Seasonal Rainfall

Data analysis shows that the summer monsoon has not been producing the rainfalls for all agricultural regions of the country (Fig. 3). Kazmi et al. (2023) stated that particular zones of the country are now getting comparative rainfalls during the summers and springs. Also, the spring has replaced the winter rains and taken second place after the summer monsoon. El Niño-Southern Oscillation (ENSO) has largely been modifying the summer as well as the winter rainfalls and consequently affecting the standing crops in the particular regions (Rasul & Kazmi, 2013). A strong El-Nino during 1997–1998 caused a worst drought situation in the proceeding 4–5 years in most parts of the country (Rasul & Kazmi, 2011).

Based on the major share of rainfalls in summer and spring seasons, the departures of both have been calculated for last 3 years (2021–2023). The recent climate normal compiled for the period 1991–2020 has been taken as reference (Fig. 4). It is evident from these maps that during summer monsoon (JJAS), anomalous heavy downpours were observed in 2022 but it was merely confined to the particular parts of lower half or the northern belt. However, most parts of the country received additional rains in summer 2023. On the other hand, the spring 2023 got more water over central to southern parts of Pakistan. Whereas the springs (MA) for rest of 2 years received just the average rains.

On the other hand, it is revealed that the night temperatures are increasing in most parts of the country (Kazmi et al., 2023). Accordingly, the crops would be taking comparatively lesser time to get matured in the light of rapid accumulation of degree days. Therefore, to attain the normal pace of plants, more irrigation water would be required to meet the amplified demand.

4.3 Variation in Air Temperatures

Importance of air temperatures cannot be ignored for agriculture sector, particularly the diurnal variation based on day–night fluctuations. Both the day and night temperatures have continuously been changed for most parts of the country and consequently affecting the ecological cycle (Kazmi et al., 2015). The climate variability has been witnessed from the recent changes in air temperatures over Pakistan.

Both maximum and minimum temperatures have been plotted for two very important periods of the year, "spring (MA)" and "pre-summer (MJ)." Pre-summer has been focused here in this case based on its significance as the hottest period of the year and its prominent role in driving the summer monsoon in the following

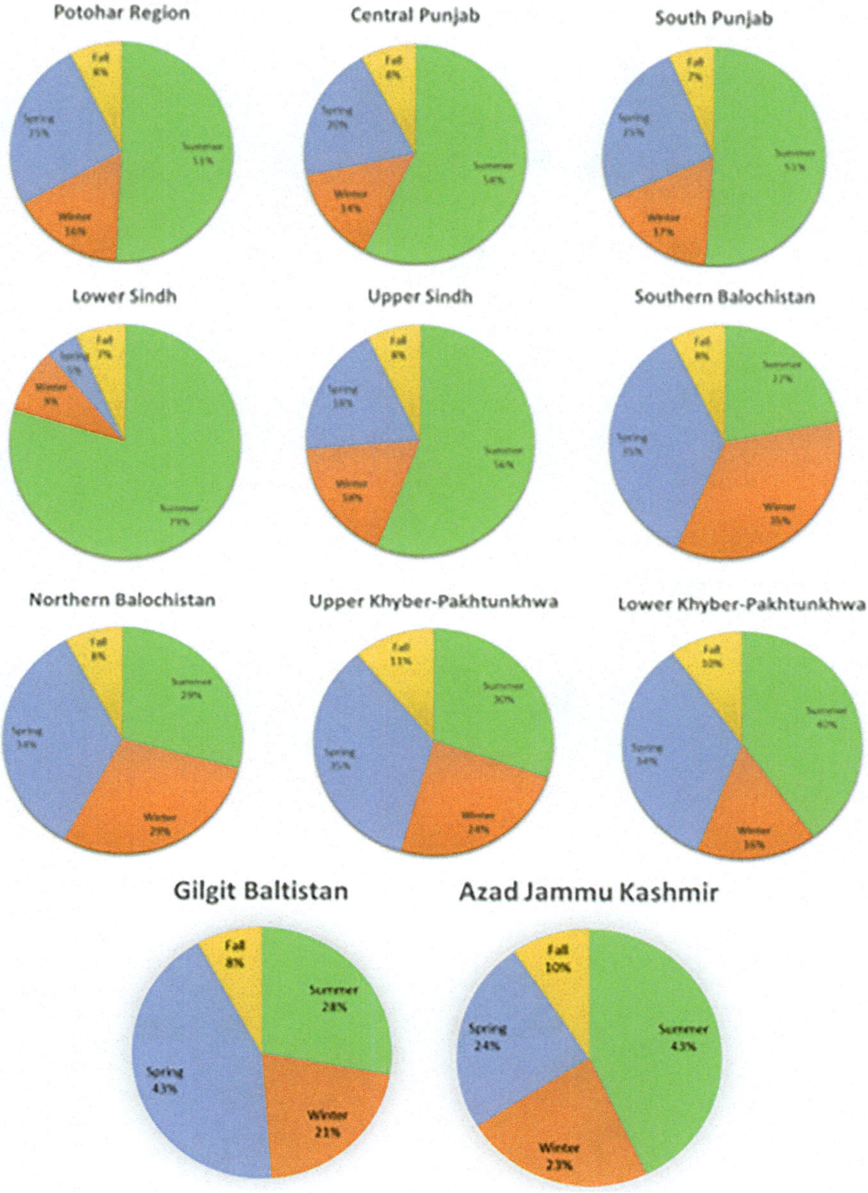

Fig. 3 Seasonal proportion (%) of rainfall for all the agroclimatic regions of Pakistan, based on data for 1991–2020. (Source: Kazmi et al., 2023)

months. The summer monsoon is very important for the region and majorly affected by the changes in the temperatures during May and June (Kazmi et al., 2015). It can be seen in Fig. 5, that the maximum or day temperatures for the spring remained above normal during 2021 and 2022, particularly in the latter year. As discussed

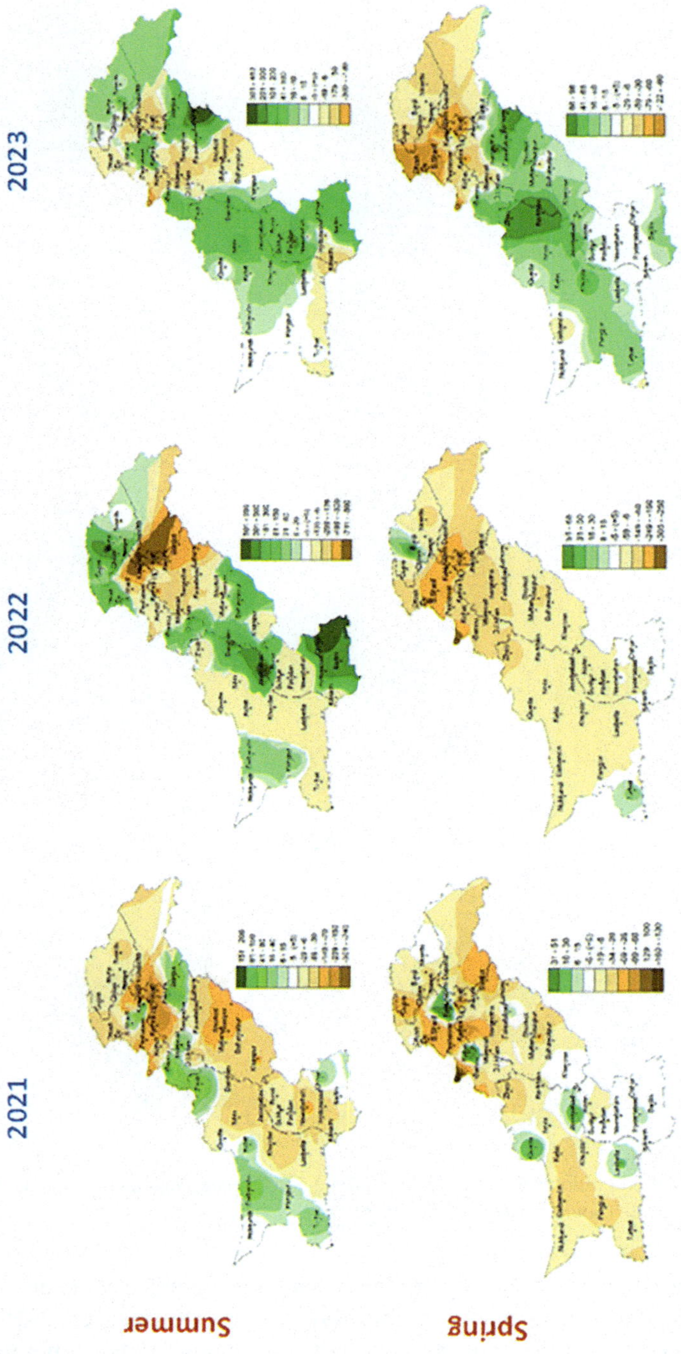

Fig. 4 Departure of seasonal rainfall (Summer "JJAS" and Spring "MA"; 2021–2023) from normal values (1991–2020)

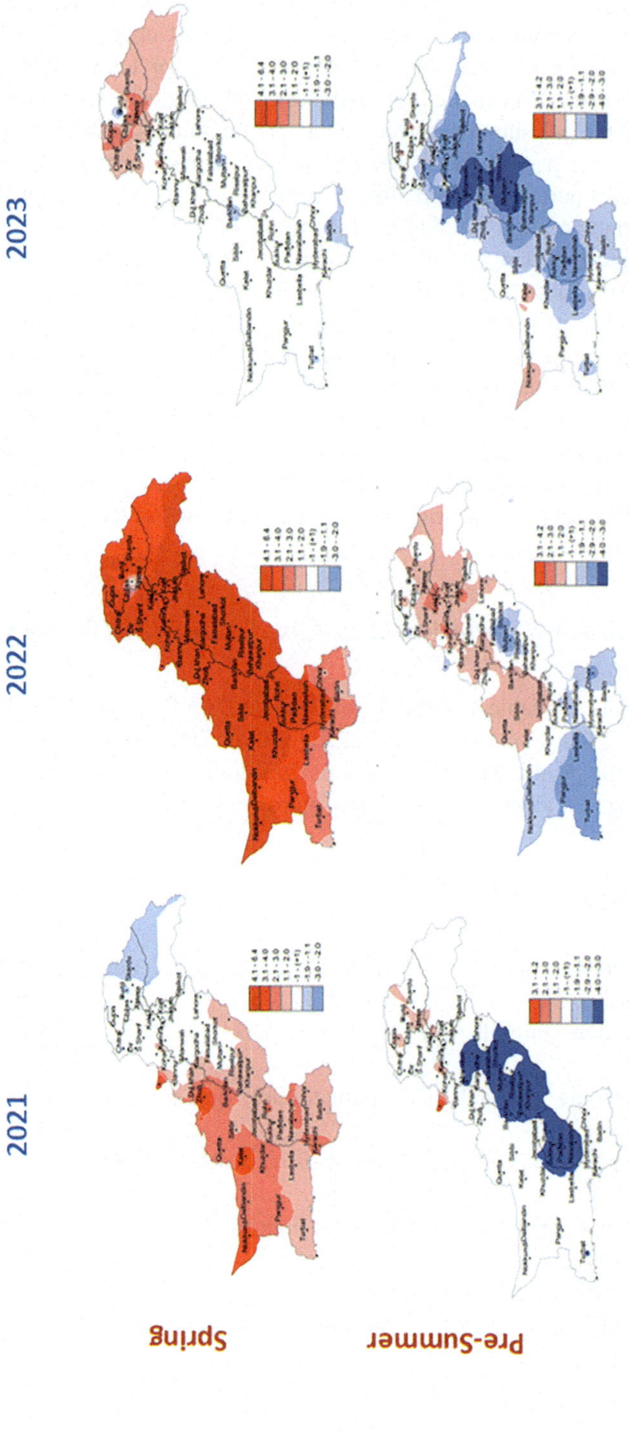

Fig. 5 Departure of maximum temperature (Spring "MA" and Pre-summer "MJ"; 2021–2023) from climatic normal "1991–2020"

earlier, this season is very important for the harvesting/post-harvesting activities of Rabi crops, especially wheat, which is grown throughout and is a major cereal food.

Also, April is generally the sowing month of Kharif crops, especially cotton, which is the major cash crop. On the other hand, higher temperatures in spring would push the rivers flow in the Indus tributaries due to glacier melting in the HKH region. During the pre-summer period, the day temperatures in 2023 remained well below normal over past plains followed by 2021. Consequently, the major rainy season in the post period has been affected in these 2 years. The same is depicted in the yearly climatic reports listed in the following text.

Besides maximum temperature, minimum or night temperature has very important role and been plotted against the former one. Minimum temperature is considered more significant for the winters but due to its usual pattern during the last 3 years, the relevant maps have not been included. The night temperatures for the spring during 2022 followed by 2021 were observed as above normal over most parts, particularly the central and western regions (Fig. 6). The higher values of night temperatures for a longer period had surely contributed in enhancing the mean temperature of the spring season. Accordingly, the heat units and other important factors related to Rabi and Kharif crops had been affected in the particular years.

4.4 Important Climatic Event in Last 5 Years (2019–2023)

Based on the data available at webpage of Climate Data Processing Centre (CDPC) of PMD (GoP, 2020; GoP, 2021; GoP, 2022_ii; GoP, 2023; GoP, 2024 https://cdpc.pmd.gov.pk/), the prominent weather events during the last 5 years have been listed in the following text along with the yearly maps for 4 years (Fig. 7).

4.4.1 Year 2019

- Pakistan's annual rainfall 21% above normal; above normal for Sindh and Balochistan.
- Above average rainfall during post-monsoon (OND) period, especially November.
- Close to average national mean temperature, first non-warmest year in past 5 years.
- Significant heatwaves in May and June; Jacobabad 51 °C on June 1–2, hottest days.
- The frequency of cyclones in Arabian Sea abnormally high; four cyclones formed including "Vayu."
- One of the strongest positive Indian Ocean Dipole (IOD) event.
- El Niño–Southern Oscillation remained neutral throughout the year.

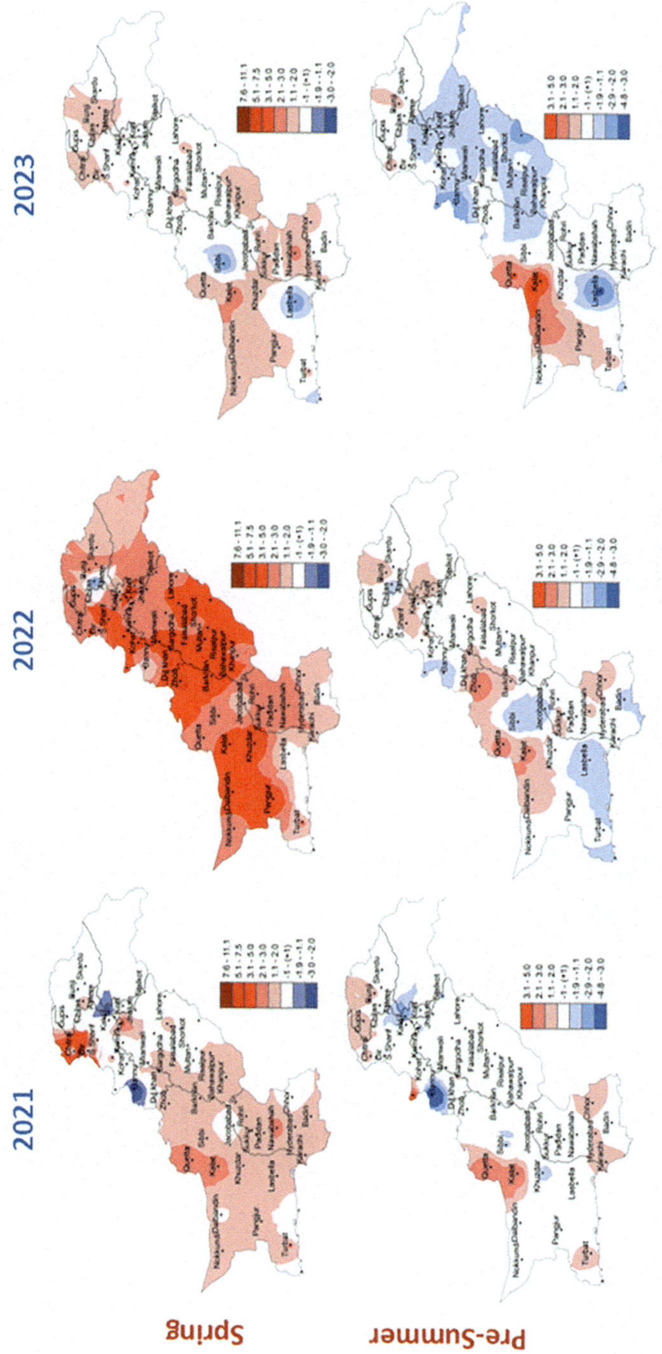

Fig. 6 Departure of minimum temperature (Spring "MA" and Pre-summer "MJ"; 2021–2023) from climatic normal "1991–2020"

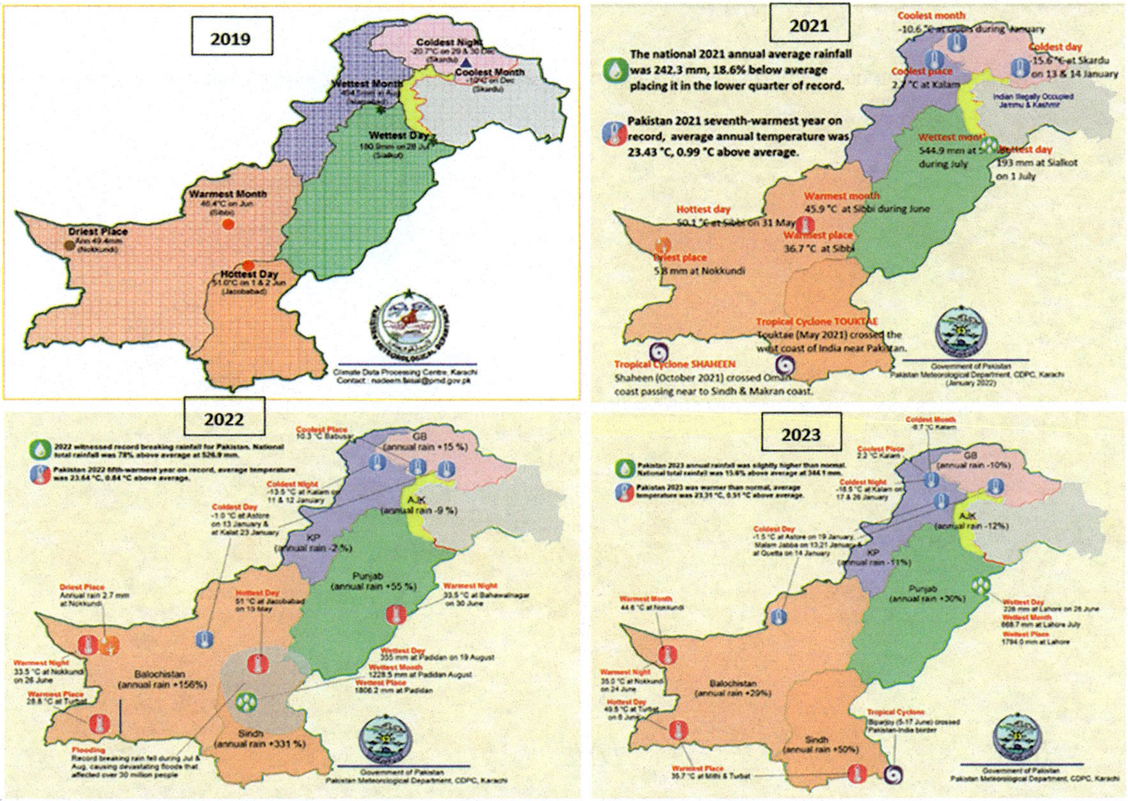

Fig. 7 Significant climate events recorded during 2019, 2021, 2022, and 2023. (Source: State of Pakistan Climate, available at: https://cdpc.pmd.gov.pk/)

4.4.2 Year 2020

- The national total rainfall for the year 2020 was 38% above average with 2020 being the fourth wettest year since 1961.
- Overall, regional rainfalls were in the ten highest of historical observations, especially for Sindh (+105.5%) and Balochistan (+46.3%).
- The year commenced with heavy rainfall with January (+133%) as second wettest month in last 60 years.
- Winter rainfall (JFM) over Punjab (+166%) was the ever highest seasonal on record.
- Sindh and Balochistan recorded August with highest ever rainfall in last 60 years.
- In Karachi, the three sites—Faisal-Base (588.0 mm), Masroor (436.0 mm), and Airport (367.0 mm)—recorded the wettest August in 48 years, 63 years, and 89 years, respectively.
- On August 28, Karachi-Faisal recorded 231.0 mm, the 24-h (1-day) heaviest rainfall, and a new record.
- Flooding affected most parts during August and September, particularly in Sindh and Punjab.
- Pakistan's annual national mean temperature was 0.22 °C above average, the ninth warmest year in the decade.
- Indian Ocean Dipole event generally remained neutral throughout the year.
- La Niña declared in September, reaching moderate strength by the end of the year.

4.4.3 Year 2021

- The most significant features of the year were the formation of cyclones Tauktae and Shaheen in the north Arabian Sea, which crossed the Indian and Oman coasts respectively affecting coastal areas of Pakistan.
- The national annual average rainfall was 242.3 mm, 18.6% below average, placing it in the lower quarter of record.
- The year 2021 remained the seventh warmest year on record across Pakistan with national average annual temperature of 23.43 °C, being 0.99 °C above average.
- National rainfall for the winter season (JFM) was largely below average, spring (AMJ) close to average, and monsoon (JAS) and post season (OND) slightly lower than average.
- The year commenced with much of country being affected by drought; especially, provinces of Sindh and Balochistan saw their driest month on record with no rainfall in the entire month.
- August 2021 saw one of its severe deficient rainfall (−55.5%) and ranked as the third driest month on record.
- During September and October 2021, national rainfall was largely above average.
- Indian Ocean Dipole event was mostly neutral throughout the year.
- La Niña weakened through the first half of 2021, reaching an ENSO-neutral state in May.

4.4.4 Year 2022

- A very heavy snowstorm occurred (16.5 in.) on January 7–8, 2022, at the hilly station, Murree and suburbs, which caused several deaths of tourists, due to freezing and carbon monoxide poisoning owing to being stranded in their vehicles whole night.
- The country witnessed ever highest and record-breaking rainfall during the year 2022 with national total rainfall being +77% of the average, since 1961.
- The country experienced the worst-ever devastating floods during the monsoon season, peaking in late August. July (181% above normal) and August (243% above normal) were the wettest on record. Sindh province was particularly severely affected, with Balochistan also hard-hit.
- Record-breaking rain triggered massive flood that caused over 1700 deaths, with 33 million people affected and 7.9 million displaced.
- The unprecedented intense rainfall in Balochistan, Sindh, and Southern Punjab in monsoon 2022 set many new daily and monthly total rainfall records.
- The annual national mean temperature for 2022, for Pakistan as a whole, was 0.84 °C above average, placing it as fifth warmest year on record.
- The pre-monsoon period was exceptionally hot in the country with March and April being hottest ever months on record. Daily maximum temperatures set new records at several locations in Sindh, Balochistan, and Southern Punjab.
- Six heatwaves occurred across the country during March–May 2022 with day/maximum temperatures being 5–12 °C above average.
- In maximum winds there were 81 events of 30 knots (and more) at different plain stations with 60 knots being the highest wind gust recorded at Lahore Airport on May 22.
- The La Niña conditions persisted for third year in a row, one of the rare phenomena, billed as a Triple-dip La Niña.
- Negative Indian Ocean Dipole (IOD) developed during June–September and then returned to neutral phase at the end of 2022.

4.4.5 Year 2023

- The extremely severe cyclonic storm "Biparjoy" was the first cyclonic storm that originated in the Arabian Sea in the year 2023.
- Biparjoy has been the longest duration cyclone over the Arabian Sea with 13 days and 3 h lifetime and covered about 2500 km translational distance.
- Annual rainfall during the year 2023 was slightly above normal with national total rainfall being +16% of the average.
- Annual rainfall was above average over Balochistan (+29%), Sindh (+50%), and Punjab (+30) while below normal over Khyber Pakhtunkhwa 'KP' (−11%), Azad Jammu Kashmir 'AJK' (−12%), and Gilgit Baltistan 'GB' (−10%).
- The annual rainfall of 2023 over Sindh and Punjab was significantly high and it ranked as the 16th and 12th wettest year, respectively, since 1961.

- The country witnessed one of the record-breaking seasonal (AMJ) 2023 rainfall (+92%), which stood as third heaviest seasonal rainfall since 1961.
- National monsoon (JAS) and post-monsoon (OND) seasons were near average while winter (JFM) was below average.
- Annual 2023 national mean temperature for Pakistan, as a whole, was 0.51 °C above average.
- Positive phases of Indian Ocean Dipole (IOD) and El Niño developed during June–September 2023.

5 Piloting Impact-Based Forecasting for Local Farmers of Pakistan

Based on the climate variations recorded in the recent period, there was a dire need to help protect the local agriculture for a sustainable future. In this regard, Meteorological Office of United Kingdom (UKMO) has initiated the first ever Impact-Based Forecasting project in the region for farmers of northern Punjab through funding from its 4-year "Asia Regional Resilience to a Changing Climate (ARRCC)" project. In this project, two districts of Pothohar region were selected to monitor peanut crop (March/April to August/September). The basic aim was to provide impact-based weather forecast to the local farmers to protect field crops from the weather extremes and to boost their final production. The field survey was processed in 2021 and after several meetings and trainings for the representatives of main stakeholders (regional agriculture extension department, research organizations, and the farming community), it had been officially launched in March 2022. Under the technical supervision of UKMO, PMD played a major role on technical grounds and ICIMOD was hired for coordinating local liaison.

In this project, two main trainings were arranged during February and March 2022 to discuss the technical proceedings of the final product. Impact tables were prepared after consultation with the primary stakeholders and terms and conditions were finalized to proceed (Figs. 8 and 9; Table 1). Also, in the final stages a stakeholders meeting was arranged at the end of July 2022.

5.1 Major Objectives (Copied from PMD's IBF Webpage)

- Enhancing regional collaboration and capability for provision of weather and climate services.
- Supporting development of regional and sub-regional forecasting and early warning systems.
- Improving capacity in focused countries to develop and disseminate impact-based forecast (across multiple timescales) to climate-sensitive sectors and vulnerable communities.

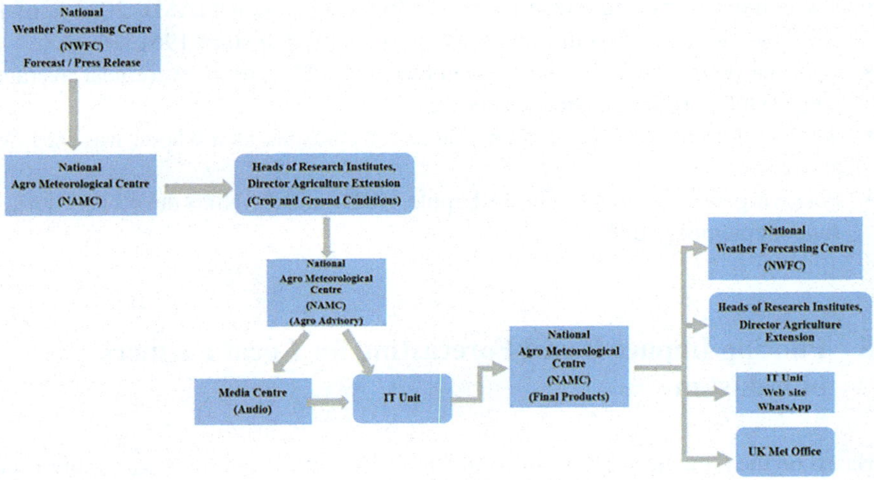

Fig. 8 Agreed SOPs to process the whole mechanism of Impact based Forecasting (IBF)

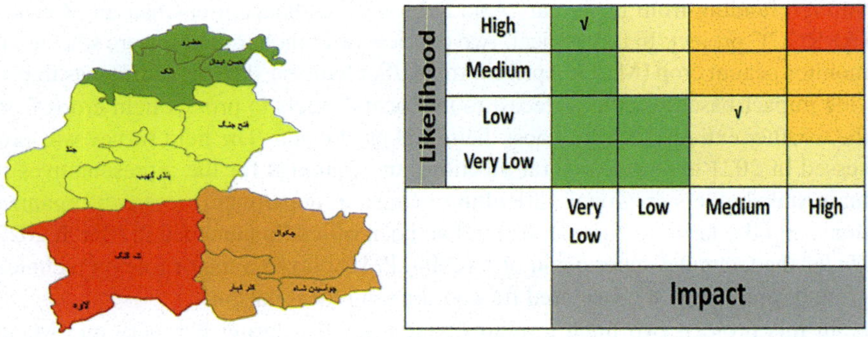

Fig. 9 Impact map for the region (subdistrict level, left) and impact matrix (right)

5.2 *Overview of the Pilot IBF*

- After completing the pre-requisites including the initial surveys, the project was officially launched in March 2022.
- Primarily two crops (peanut and maize) but later only a single crop "peanut" was considered.
- Two districts (Chakwal and Attock) were focused.
- Regular online meetings had been conducted between the technical teams of UKMO and PMD to monitor the proceedings.
- Training workshops were organized for all the stakeholders in February and March 2022.
- Final Stakeholders Meeting was arranged at the end of July 2022 to discuss the proceedings.

Table 1 Impact table for peanut crop at maturity stage during August/September

Rainfall Thresholds			
• 24 hours rainfall 2-10 mm with/without wind 3-10 knots • 5 days accumulated rainfall 10-50 mm with/without wind 3-10 knots			
	• 24 hours rainfall 10-30 mm with/without isolated thunderstorm/hail and/or wind 11 to 25 knots • 5 days accumulated rainfall 50-80 mm and/or isolated thunderstorm/hail with/without maximum wind gust 11 to 25 knots		
		• 24 hours rainfall >30mm and/or scattered Thunderstorm/Hail with/without wind>25 Knots • 5 days accumulated rainfall > 80 mm and/or thunderstorm/hail with/without maximum wind gust > 25 knots	
Impacts			
Minimal	**Minor**	**Significant**	**Severe**
• Little puddles in the fields • Muddy paths	• Puddles in the fields • Muddy paths and fields	• Water Inundation • Muddy paths and fields • Fungal diseases may start	• May produce localized flooding • Fields may become inaccessible • Weeds may appear rapidly • Pod damage may occur underground • Pods quality may be affected
Response Matrix			
Very Low	**Low**	**Medium**	**High**
• No action required • Remain updated with the weather forecast	• Be Vigilant • Monitoring of crops	• Be cautious • Be prepared for water drainage • Spraying of suitable fungicide or removal of plants	• Immediate drainage • Spray on weeds • Spraying of suitable fungicide or removal of damaged plants

- Timely dissemination of the IBF products was managed by utilizing PMD's webpage, WhatsApp groups, regional Frequency Modulation 'FM' radio station, as well as the agricultural field workers–farmers forums.
- This project was successfully completed in August/September 2022.

5.3 *Outcome of the Project (Pilot IBF)*

In the present age of climate variability and food security challenges, projects like IBF are more pronounced for a rainfed region. Pothohar region of Pakistan is known for its limited water resources and small land holdings (Rasul & Kazmi, 2013). Therefore, impact-based forecasting initiative is more meaningful for local farmers of this area. During whole period of consultation and practical implementation, the weather experts, agricultural research organizations, agri field staff, and the farmers remained in contact with each other through WhatsApp groups, telephone calls, and emails. Accordingly, a good coordination got established between all the stakeholders and it improved the confidence of the farmers on the scientific departments related to weather as well as the agricultural advisories.

Up to subdistrict level, the farmers had regularly been informed about the expected weather in the coming days along with the possible impacts and the experts advisories. Being small land owners, the farmers of this rainfed belt are very much concerned about any weather/climate phenomenon that may affect their crops. Through this project, they have been guided throughout the crop period, which improved their production. The project was successfully completed and had provided motivation to go ahead.

The findings of this pilot encouraged PMD to extend these services to rest of the parts as well. Therefore, the same services have been extended to the whole Pothohar region (four districts) for both cropping seasons, including Rabi and Kharif. Also, the coordination has been in progress for some other parts of the country as well, including Punjab and Sindh provinces. These efforts were made in the light of Food and Agriculture Organization of the United Nations (FAO)'s climate smart agriculture program, which would surely lead to farm-level management services. But above all, the cooperation and dedication of all the stakeholders in this regard would remain the key figure.

6 Summary and Way Forward

The climate has been changing over the globe and may continue with almost the same pace. But man must find its survival and for that reason the sustainability of food is the primary requirement. In today's world of technology and awareness, in addition to other sectors, agriculture also needs to be modified. Especially for the climatically vulnerable regions with lesser infrastructure to cope with, it is very important to adopt the climate variability for sustainable agriculture production.

In South Asia, Pakistan has minute contribution in the global greenhouse gases but is among the most vulnerable countries in the context of climate extremes. The evidence has been recorded in the recent years with anomalous behavior of air temperatures and their consequent effects on the seasonal rainfall. Therefore, the water availability for the major crops both in Rabi (wheat being the prominent crop) and Kharif (cotton being the prominent crop) seasons has become riskier.

Recently, impact-based forecasting has been tested for farming community of rainfed Pothohar region located in the north of Punjab province of Pakistan. Through a technical forum, all the regional stakeholders including the weather/agriculture experts and the farmers have been gathered to produce impact-based weather forecasts along with advisories. The pilot project has been completed successfully and opened a new era of hope and trust among the professionals and the local community. It is revealed that the agricultural products may be protected from the climate hazards through an organized way of IBF. Accordingly, on-farm management in climate context would also be achieved through similar projects in near future.

Acknowledgments The authors are highly grateful to Mr. Muhammad Bilal, Mr. Aftikhar Ahmed, and rest of technical staff working at NAMC (PMD), for the assistance in data compilation and graphics. Besides, they are thankful to the administrative heads of CDPC and NAMC for provision of required data to proceed for this study.

References

Ali, S., Liu, Y., Ishaq, M., Shah, T., Abdullah, I. A., & Izhar, U. D. (2017). Climate change and its impact on the yield of major food crops: Evidence from Pakistan. *Food, 2017*(6), 39. https://doi.org/10.3390/foods6060039. Retrieved from www.mdpi.com/journal/foods

Chaudhry, Q. Z., & Rasul, G. (2007). Global warming and expected snowline shift along northern mountains of Pakistan. In *Proceedings of 1st Asiaclic Symposium, Yokohama, Japan.*

Ding, T., & Ke, Z. (2013). A comparison of statistical approaches for seasonal precipitation prediction in Pakistan. *Weather and Forecasting, 28*, 1116. https://doi.org/10.1175/WAF-D-12-00112.1

ECP. (2019). Economic survey of Pakistan for the period 2018–19.

FAO. (2019). *Agriculture and climate change—Challenges and opportunities at the global and local level—Collaboration on climate-smart agriculture*. FAO (52 pp.).

GoP. (2020). State of Pakistan Climate in 2019. Retrieved from https://cdpc.pmd.gov.pk/Pakistan_Climate_2019.pdf

GoP. (2021). State of Pakistan Climate in 2020. Retrieved from https://cdpc.pmd.gov.pk/Pakistan_Climate_2020.pdf

GoP. (2022_i). Pakistan: Updated Nationally Determined Contributions 2021. Available at https://unfccc.int/sites/default/files/NDC/2022-06/Pakistan%20Updated%20NDC%202021.pdf

GoP. (2022_ii). State of Pakistan Climate in 2021. Retrieved from https://cdpc.pmd.gov.pk/Pakistan_Climate_2021.pdf

GoP. (2023). State of Pakistan Climate in 2022. Retrieved from https://cdpc.pmd.gov.pk/Pakistan_Climate_2021.pdf

GoP. (2024). State of Pakistan Climate in 2023. Retrieved from https://cdpc.pmd.gov.pk/Pakistan_Climate_2023.pdf

Kazmi, D. H., & Rasul, G. (2012). Agrometeorological wheat yield prediction in rainfed Potohar region of Pakistan. *Agriculture Sciences, 3*(2), 170–177.

Kazmi, D. H., Li, J., Rasul, G., et al. (2015). Statistical downscaling and future scenario generation of temperatures for Pakistan region. *Theoretical and Applied Climatology, 120*, 341–350. https://doi.org/10.1007/s00704-014-1176-1

Kazmi, D. H., Afzaal, M., Mubeen, M., Hussain, S., & Jatoi, W. N. (2023). Unpredictable weather and agriculture-based economy of developing countries. In W. N. Jatoi, M. Mubeen, M. Z. Hashmi, S. Ali, S. Fahad, & K. Mahmood (Eds.), *Climate change impacts on agriculture*. Springer. https://doi.org/10.1007/978-3-031-26692-8_4

Khan, U., Janjuhah, H. T., Kontakiotis, G., Rehman, A., & Zarkogiannis, S. D. (2021). Natural processes and anthropogenic activity in the Indus River sedimentary environment in Pakistan: A critical review. *Journal of Marine Science and Engineering, 9*(10), 1109. https://doi.org/10.3390/jmse9101109

Kurukulasuriya, P., Mendelsohn, R., Hassan, R., Benin, J., Deressa, T., Diop, M., Eid, H. M., Fosu, K. Y., Gbetibuo, G., Jain, S., et al. (2006). Will African agriculture survive climate change? *World Bank Economic Review, 20*, 367–388.

Mishra, A. E., Mukherji, A. E., Shrestha, A. B. E., & Wester, P. E. (2019). *The Hindu Kush Himalaya assessment: Mountains, climate change, sustainability and people*. Springer International Publishing.

Naheed, G., & Rasul, G. (2011). Investigation of rainfall variability for Pakistan. *Pakistan Journal of Meteorology, 7*(14), 25.

Pratley, J. (2003). *Principles of field crop production*. Oxford University Press.

Rasul, G., & Kazmi, D. H. (2011). Climate change and challenges to crop production. In *Proceedings of International Workshop on Plant Conservation & Reversing Desertification: A Way Forward*, October 12–13, 2009. PMAS Arid Agriculture University (pp. 81–99).

Rasul, G., & Kazmi, D. H. (2013). *Wheat yield & climate change in potohar region of Pakistan*. LAP LAMBERT Academic Publishing.

Rasul, G., Chaudhry, Q. Z., Zhao, S. X., & Zeng, Q. C. (2004). A diagnostic study of record heavy rainfall in twin cities Islamabad-Rawalpindi. *Advances in Atmospheric Sciences, 21*(6), 976–988.

The World Bank (WB). (2012). World Bank Health Nutrition and Population Statistics. Retrieved May 20, 2017, from http://databank.worldbank.org/data/home.aspx

Vermeulen, S. J., Aggrawal, P. K., Ainslie, A., Angelone, C., Campbell, B. M., Challinor, A. A., Hansen, J. W., Ingram, J. S. I., Jarvis, A., Kristjanson, P., et al. (2012). Options for support to agriculture and food security under climate change. *Environmental Science & Policy, 15*, 136–144.

Groundwater Sustainability and the Imperative of Effective Management

Muhammad Mohsin Waqas, Tanveer Ahmad, Usman Khalid Awan, Sikandar Ali, and Rao Husnain Arshad

Abstract Groundwater is a critical resource to ensure the food security of the swiftly increasing population in the world. The flexible availability of the groundwater has made it vulnerable. Especially the core challenges like overextraction, declining water tables, and water quality degradation threaten its long-term sustainability. This chapter explores the significance of effective management strategies in preserving groundwater resources and achieving sustainability in agricultural water use. The chapter begins with an overview of groundwater's role in agriculture and its ecological and economic importance. It addresses the challenges and risks faced by groundwater, emphasizing the need for proactive and comprehensive management solutions. The nexus between effective management practices and groundwater conservation is explored, outlining key principles for sustainability. The chapter delves into the significance of monitoring and data collection, highlighting advanced technologies for real-time groundwater assessment. Groundwater modeling is discussed as a valuable tool for predicting availability and understanding aquifer dynamics. Mitigating groundwater losses is central to this chapter, focusing on the critical role of advisory services in promoting sustainable groundwater use. Case studies illustrate the positive impact of advisory services on groundwater conservation. The integration of traditional knowledge with modern practices is discussed to foster sustainable water use at the local level. The economic aspect of groundwater sustainability is examined, including market-based approaches and incentives for

M. M. Waqas (✉) · T. Ahmad
Department of Agricultural Engineering, Khwaja Fareed University of Engineering and Information Technology, Rahim Yar Khan, Pakistan
e-mail: mohsin.waqas@kfueit.edu.pk

U. K. Awan
Center for Development Research (ZEF), University of Bonn, Bonn, Germany
e-mail: ukawan@uni-bonn.de

S. Ali · R. H. Arshad
Department of Irrigation and Drainage, University of Agriculture Faisalabad, Faisalabad, Pakistan
e-mail: sikandar_ali@uaf.edu.pk; husnain.arshad@uaf.edu.pk

M. Mubeen et al. (eds.), *Innovations in Agricultural Water Management*,
https://doi.org/10.1007/978-3-031-91883-4_15

farmers to adopt sustainable practices. Capacity building and knowledge transfer are highlighted as critical components of achieving sustainability. The chapter explores training programs for farmers, water managers, and advisory personnel, along with innovative approaches for efficient knowledge dissemination. Finally, the chapter presents a vision for the future, stressing the need for collective action and global cooperation in tackling challenges like climate change and population growth. Emerging trends and innovations in groundwater management encourage continuous improvement in sustainability efforts.

Keywords Groundwater · Governance · Management · Sustainability

1 Introduction

1.1 The Critical Role of Groundwater in Agricultural Water Management

Around the world, groundwater still forms the backbone of all freshwater supplies. Water is rapidly evolving into a scarce commodity due to its restricted availability. Awais et al. (2023) indicated that groundwater is being excessively used primarily in agriculture, threatening its future availability. Groundwater is important due to its global use in agriculture as a reliable source of water, especially when surface water resources are limited. It can be an alternative or additional resource for irrigation hence economically advantageous during a drought and necessary to maintain high efficiency for food production. Groundwater storage secures the dry season ensuring the availability of drinking water for animal-rearing village, thus helping promote agricultural development.

Managing aquifer recharge can help to sustain the groundwater, and not only does it stabilize levels, it also increase availability of water. Yet it is also true that human activities have perturbed the global water cycle to such an extent that 35% of all freshwater taken by humans globally, now comes from groundwater. It also highlights where human activity impacts groundwater and surface water storage in different ways. For instance, in Southern China, groundwater is recharged by surface water irrigation while in regions like the High Plains aquifer is known to recharge some of its reserved compartments for over 1000 years through return flows of withdrawn groundwater. Human water use also affects seasonal changes in stored terrestrial water except for some exceptions. Measuring human-induced changes in total terrestrial water storage using the Gravity Recovery and Climate Experiment (GRACE) satellite mission is difficult, especially over areas such as along the Mississippi Basin where trends among regions are highly complex (Frappart & Ramillien, 2018).

1.2 The Ecological and Economic Importance of Sustainable Groundwater Use

The sustainable use of groundwater is critically important both because it impacts the environment and also for economic benefits (Liu et al., 2020; Qasim et al., 2020; Mubeen et al., 2020; Rasool et al., 2020; Zainab et al., 2020). Environmentally sustainable groundwater management is a cornerstone for the maintenance of many sorts of ecosystems (e.g., wetlands and riparian zones) constituting important habitats in which biodiversity can be conserved. Recharging groundwater adequately is essential as it serves to maintain the good health of larger surface water bodies, and hence conserves diverse fauna characteristic of regions in arid belts. In adverse to water-scarcity contexts, sustainable use of groundwater resources helps build climate resilience by shaping healthy ecosystems with the capacity for carbon sequestration, an important tool in efforts to combat and adapt the impacts of climate change.

From an economic point of view, agriculture for both crops and livestock will collapse alongside many industries. Groundwater is fundamental to agricultural irrigation, which sustains food security and the rural economy. Groundwater supports key economic activities that rely on regular water supply to function effectively like in sustaining livestock and rural livelihoods, while industries and energy sectors rely heavily on groundwater for their operations. Besides, natural features sustained by below-ground sources attract tourism and recreational related businesses (Huneau, 2019).

1.3 Challenges and Threats to Groundwater Sustainability

Long-term availability and quality of groundwater are hindered by multiple challenges due to anthropogenic activities compounded with natural factors. The high rate of overextraction coupled with depletion caused by excessive pumping for different uses leads to declining groundwater levels and ultimately reduces water availability (Hoogesteger, 2022; Hussain et al., 2021a, b, 2022a, b, c; Waleed et al., 2022; Khaliq, 2022; Hammad et al., 2022; Tahir et al., 2022; Sabagh et al., 2022). Africa and Pakistan are the regions where groundwater is polluted by urbanization, industrialization, agriculture, mining, and poor sanitation. Overexploitation adds to this growing consumption and yet in many cases these resources are underused as well. We need to develop a reliable monitoring system, appreciate the impacts of groundwater on freshwater ecosystems, in particular aquatic ecologies, and manage cross-boundary basins that contain aquifers across national borders (Cook et al., 2022; Hammad et al., 2020; Hashmi et al., 2020; Shahid et al., 2020; Zamin et al., 2020; Hussain et al., 2020; Ali et al., 2019). Climate change effects should be addressed as well, which directly reflects results from climate changes. We also need to assess their overpumping consequences, particularly concerning vegetation cover, but also linkages between surface water resources and river salinization properties or its consequence where it goes through estuarine landscapes, not forgetting building capacity for managing this critical reserve underground. It can be contaminated by agricultural runoff; besides, industrial discharges and the improper disposal of waste are some sources that pollute the water supply contributing to health problems and environmental damage. Changes in precipitation patterns and temperatures associated with climate change alter aquifer recharge rates, that is, the amount of new water available to enter an aquifer. Urbanization and deforestation are the land use changes that have disturbed natural groundwater recharge process. These practices in conjunction with poor governance make the situation worse and put even further pressure on groundwater resources. While groundwater is an important resource, the absence of scientific and adequate data on its quantity in space and time questions consistent sound management approaches. In addition, due to the interconnectedness of surface water and groundwater, collective management is required in developing new ways for managing both resources effctively. Overcoming these challenges is essential to achieve sustainable groundwater management.

2 The Nexus of Effective Management and Groundwater Sustainability

2.1 Exploring the Relationship Between Effective Management and Groundwater Conservation

The importance of managing groundwater effectively cannot be overstated; it is a critical natural resource that must be sustainably managed. This will require setting safe-yield thresholds that avoid overextraction, and monitoring water levels and quality to be able to make good choices about groundwater use, as well as how it connects with surface waters. Recharge augmentation practices help recharge depleting aquifers, and pollution control measures seek to preserve the quality of groundwater (Echeverry et al., 2024; Mehmood et al., 2021; Sabagh et al., 2021; Javeed et al., 2021a, b; Ali et al., 2021; Mubeen et al., 2021). Educating and working with communities to teach the public about conserving groundwater help promote responsible use of water. While strong legal and regulatory frameworks are fundamental, however, it is important for more inclusive and sustainable approaches for equitable management of groundwater.

2.2 Key Principles of Sustainable Groundwater Management

Sustainable groundwater management is based on a set of principles to efficiently and effectively use available groundwater resources while ensuring long-term conservation (Wang et al., 2019; Fahad et al., 2019; Shah et al., 2019; Shahzad et al.,

2018; Conti & Gupta, 2016). The principles entail achieving a balance between groundwater withdrawals and natural recharge, monitoring groundwater levels and quality regularly, integrating surface water and groundwater management, promoting conservation of waters at both yield rate and use efficiency levels, encouraging non-conventional sources of supply, preventing pollution, equally involving stakeholders in the decision-making processes, practicing adaptive management, ensuring strong regulatory frameworks, etc. Through heeding these principles, we can expect to foster better sustainable use and protection of this invaluable natural resource.

2.3 Integrating Agricultural Water Demands with Long-Term Groundwater Preservation

Managing this resource to be able to both meet agricultural water demands and ensure long-term groundwater sustainability is key for sustainable agriculture. Water depletion in turn exacerbates climate change while leaving too little water for drinking, as agriculture is the largest consumer of water (Rahman et al., 2019; Ahmad et al., 2019; Sabagh et al., 2019; Danish et al., 2019; Iqbal et al., 2019; Singh, 2022). Demand in this sector will grow as populations increase and their diets improve, but also because weather patterns are being disrupted by global warming. It may be done with the use of water-efficient irrigation techniques, crop selection based on local water availability, or sustainable land management practices. Establish sustainable well-pumping restrictions and regulate aquifer withdrawals to avoid a drawdown from groundwater overextraction. Farmers can be encouraged to practice sustainable water practices with the help of public awareness and education. To evaluate the long-term effects of agricultural water use on aquifer health, proper measures should be taken to monitor and record groundwater usage at regular intervals so that appropriate adaptive management strategies can be followed. For effective water management schemes, the beneficiaries of water along with farmers and managers must work together.

3 Assessing Groundwater Resources: Monitoring and Data Collection

3.1 The Importance of Accurate and Reliable Groundwater Data

Groundwater represents a cornerstone of many water management and sustainability challenges, however responsible decision-making requires accurate groundwater data (Adelana, 2009). Overextraction is occurring then this can be prevented and it will ensure the supply of water when drought conditions prevail in the any area. Characterizing groundwater availability is critical for environmentally sustainable development, which requires that over exploition of resources must be avoided for future generations. Understanding changes to recharge rates and availability is critical in order to develop water management adaptation strategies under climate change, based on the wide range of groundwater data. They also support the health of ecosystems by maintaining wetlands and rivers throughout the droughts years. Groundwater quality monitoring is used to trace contamination sources and devise effective remediation strategies to automatically protect potable water resources. In addition, groundwater data inform infrastructure planning and can be used to reduce risks in project siting and design. In adhering to legal requirements, groundwater data aid in the equitable distribution of water and settling disputes between users. In conclusion, accurate and reliable data of the groundwater level are very important for sustainable water management and this valuable natural resource preservation.

3.2 Advanced Technologies for Real-Time Groundwater Monitoring

Real-time monitoring of the groundwater has enhanced significantly with advanced technologies, specifically wireless sensor networks and Internet of Things (IoT) devices. Accurate and timely land use information in the canal command area is a prerequisite for judicious water management. The application of remote sensing techniques and geographic information system (GIS) for the assessment of canal water deficit at the canal command level was estimated (Waqas et al., 2019).

3.3 Establishing Groundwater Monitoring Networks and Data-Sharing Platforms

Establishment of groundwater monitoring networks and data-sharing platform is essential for effective groundwater management and protection. They enable continuous monitoring and data acquisition and analysis to yield insights on changes in the groundwater. The steps involved are: to create monitoring networks, deploy wells and sensors, collect real-time data, validate and analyze the information, and visualize and report. Data-sharing or collaboration platforms enable transparency and public availability of the citizen-oriented stakeholder's involvement. These systems are important in resource management, early warning systems, research to policy translation, climate change resilience, adaptation planning transparency, public engagement, interagency coordination, etc.

4 Groundwater Modeling for Sustainable Resource Planning

4.1 *Understanding Groundwater Flow and Aquifer Dynamics*

Knowledge about groundwater flow and physical properties of aquifers helps in planning effective water resource management, and sustainable use of groundwater resources with their conservation. Groundwater flow is defined as the movement of water through subterranean aquifers and an examination of these groundwater compartments is known as aquifer dynamics, which encompass all processes that control its motion, storage, and distribution. Major components consist of porosity and permeability, aquifer categories, groundwater recharge and discharge, hydraulic gradient and flow direction, Darcy's Law, etc. (Khan et al., 2019; Amin et al., 2018a, b; Rahman et al., 2018; Nasim et al., 2018a, b; Hammad et al., 2018; Ali et al., 2018a, b; Tariq et al., 2018; Eshtawi et al., 2016).

4.2 *Modeling Tools for Predicting Groundwater Availability and Recharge*

Modeling tools (e.g., MODFLOW, Soil Water Assessment Tool, and Artificial Neural Networks) are used to predict groundwater availability and recharge by simulating numerous hydrological processes. They account for the exchange processes between precipitation, surface water, and soil/aquifers thus offering a holistic view of water availability models that can help interpret groundwater recharge patterns and support an integrated natural resource management planning.

4.3 *Applications of Groundwater Models in Decision-Making*

Groundwater models are frequently used in decision-making and serve as important tools for the analysis and management of groundwater resources. These models are based on mathematical views of hydrogeological processes and they simulate groundwater levels, flows, and quality, which can help in the decision-making process (Jakeman et al., 2016).

Management and allocation of groundwater: Water models are very important components in the development of ground-administration plans. This fact allows the models to be used as decision support tools for local water management practitioners to infer information regarding available groundwater, groundwater volume change predictions, and the effects of management decisions. Water managers can obtain estimates of sustainable groundwater yields for different pumping rates and scenarios of groundwater abstraction, enabling them to issue competitive allocations among potential users.

Aquifer recharge and vulnerability: Knowledge of aquifer recharge and vulnerability is indispensable for safeguarding groundwater quality. They provide clear spatial information that helps decision-makers identify areas with high recharge potential and also on the other hand evaluate which aquifer is susceptible to contamination. Also, this knowledge enables policymakers to regulate proper land use and implementation of protective measures for sensitive recharge areas, which would otherwise create a potential contamination threat.

Groundwater modeling for water resources planning and climate change mitigation: In water resources planning and climate change mitigation, the use of groundwater models prevails. These models are useful for policymakers to measure the variables associated with change that influence groundwater and recharge water availability from climate impact. Such information is necessary to formulate flexible strategies for sustainable management of groundwater resources under changed climatic conditions aiming at coping with variability to ensure water security.

Groundwater models are effective tools for infrastructure impact assessments: Large infrastructure projects like the transport sector often come with apprehension about their possible adverse impact on existing groundwater resources. The models project altered groundwater levels and flow due to channeling and spillover of flood plains as well as surface–groundwater exchanges. This is important information when trying to minimize impacts on surrounding aquifers and ecosystems and, more generally, how we can develop in a way that keeps groundwater-dependent environments healthy.

Groundwater remediation planning: Groundwater models are useful tools for planning and optimizing groundwater remediation efforts. These models can be used by decision-makers to predict the plume movement and lifespan in porous media, which could steer toward developing codified approaches for efficient as well as cost-effective remediation strategies. The effect will allow the scientists to determine which groundwater remediation methods work best for contaminated sites. Groundwater models are crucial for emergency response and drought planning. In times of water shortage or drought, the models can be employed by policymakers to forecast where wells may go dry and what that will mean for the drinking water supply. This information is useful to be prepared for the worst-case scenario, and in managing water stocks and quality effectively in times of scarcity.

Public awareness and stakeholder participation: Groundwater models can help in creating awareness among the public about groundwater dynamics, which further also helps to engage stakeholders for better water management decisions. Through both the simple thousands-of-lines simulation model and the millions-of-degrees scenario analyses, visual communication shall be introduced from these models to enhance such complicated hydrological information communicated to the public and stakeholders. Involving stakeholders in framing the dispositions facilitates better comprehension, buy-in, and possession of groundwater management projects.

5 Groundwater Extraction and Recharge Management

5.1 Balancing Groundwater Extraction with Recharge Rates

The balance between the critical practice of groundwater extraction and the necessary recharge to keep water availability is one of the greatest concerns in the current-day management of water resources. Groundwater is one of our most important sources of fresh water used in agriculture and municipality, hence its proper management also plays an essential role in ensuring drinking security, especially in areas susceptible to water scarcity. This chapter of the book unravels these complex issues, outlining challenges, solutions, and sustainable measures that have been taken in managing groundwater abstraction vis-à-vis recharge based on intricate water demand versus supply dynamics.

Groundwater is a component of the hydrological system and it is coming under increasing pressure due to overextraction for different purposes. On the other hand, vulnerability to saltwater intrusion occurs due to proximity of the coastline. This is followed by extremely regulated governance structures requiring strict rules and monitoring to determine the amount of water extracted and pumped into the ground as opposed to whether it is naturally or artificially recharged. Further, it discusses the recharge management practices—the man-made and other recharge methods like infiltration basins and trenches, injection wells, bore holes, and percolation ponds. Succinct yet wide-ranging in scope, the book covers issues as far-reaching as climate change and fossil fuel extraction alongside a focused exploration of local water policy implementation at varied scales.

5.2 Sustainable Pumping Practices and Water Use Efficiency

Sustainable water management is linked to sustainable pumping methods and has supplied intelligent few breakthrough means of managing this vital commodity more efficiently, especially from populations of dry lands challenged with their existing resource depletion versus demand hike. These methods are designed so as to maximize the efficiency of groundwater and surface water use, preserve water resources for future generations, and further protect the environment. To regulate the permissible drawl of water for various uses and the state of health and trends in groundwater levels, it is essential to carry out monitoring of groundwater so that authorities can set pumping limits with regard to assessments made from sustainable yield. At the other end of the spectrum, too much surface water is available causing excessive drawdown, which can be limited by surface water and groundwater conjunctive use. Methods of managed aquifer recharge (MAR), such as artificial recharge, encourage the sustainable storage and filling regime in groundwater.

Through the adoption of energy efficient, well-managed pumping systems that are time efficient, emissions are lowered across the life cycle. This makes for all-round better operations. A pumping schedule that satisfies the demand and the yield from the aquifer would distance customers from unnecessary water intake even during deficit periods or save it at all. Water use efficiency is to save water in agriculture as well as in homes and industries. Eco-friendly irrigation methods assist in minimizing water pollution and wastage of this precious resource by recycling used water or utilizing water saving technologies. Waste areas and areas that can be improved are identified by water audit. Also, water environment campaigns encourage responsible conduct of people on water. Periodic assessment of water losses through leakage in distribution systems helps to contain the quantity of water that is wasted as water that is lost due to degrading of lands is untreated; however, efficient water flow in a system enhances the distribution of water and decreases losses of it while conserving the beneficial resource.

5.3 Enhancing Natural Groundwater Recharge Through Artificial Means

Water reuse is the practice of recycling water for various uses. The natural processes of rain and snow melt and evaporation would have generally replenished the aquifers and saturated soil as the surface runoff water from precipitation events does this nearby and does not take catchment areas into account. Similarly, in urban areas though most of the water physically contains pollutants such as bacteria and sewage, there is enough significance in most of the water, physically or technically, to be considered amenable to some form of use and hence further treatment or technological innovations. Desalination although expensive in practice is some lever for balancing the imbalance between rates of abstracting salt water and the rates of effluent releasing into the open water bodies. In milder climates and at low relative angles, slopes often oriented toward the transferring torrents might imply the spillover more similar to flooding in a controlled configuration.

Engineering structures like recharge ponds or basins catch and store rainwater, which otherwise would have only flowed downstream in the surface runoff. These structures make sure that water seeps into the aquifer floor thus replenishing the water table. Rainwater harvesting is characterized by the inclusion of processes where rainwater is routinely collected and stored such as collection of rain water from roofs and other non-porous surfaces. The obtained water is then used to carry out agricultural activities or used in homes and the excess water is pumped back into recharge wells or infiltration basins. Artificial means of recharge has brought up an acceptable concept of managing water levels where oversupply is maintained reasonably, even though water is of certain quality and can be utilized either for drinking or industry production. It assists in combating the problem of reducing groundwater and promotes appropriate distribution and effective application of water in various fields. Figure 1 depicts the mechanism of artificial groundwater recharge.

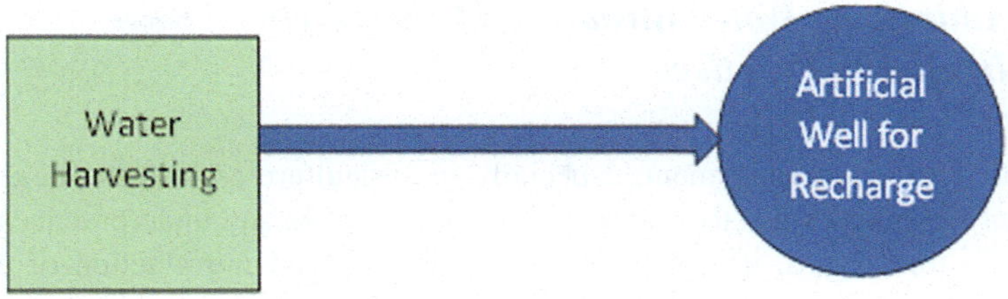

Fig. 1 Mechanism for artificial groundwater management

6 Mitigating Groundwater Losses: Challenges and Strategies

6.1 Identifying Factors Contributing to Groundwater Losses in Agriculture

To evaluate the role of various factors in the groundwater losses in agriculture, one must understand the factors that cause depletion of groundwater in farming. This includes the problem of farming wastes; they include poor irrigation practices, controversial excessive groundwater use, high water use in growing of crops, and absence of water management.

- **Inefficient irrigation techniques**: It is a common practice in many agricultural regions to use flood irrigation, which leads to wastage of huge amount of water from evaporation and runoff. Failure to apply proper irrigation methods also leads to high demand of water making it worse as there is more dependence on groundwater abstraction.
- **Overextraction of groundwater**: Excessive pumping of groundwater at a rate faster than the aquifer recharge leads to creeping depletion of the aquifer causing long-standing reduction of groundwater levels and diminished water supply.
- **Excessive water demand for crop irrigation**: Certain crop species like cover crops draw too much water, while agriculture in one part seems to have over-relied on such plants lacking consideration of groundwater's threat to depletion.
- **Lack of water conservation practices**: This leads to waste of water that is not needed, and hence one of the reasons for high groundwater extraction more than required is the absence of water saving devices and practices. There are no water conservation measures, which lead to overexploitation of water sources, especially underground water. This is particularly relevant in studies of climate and development so as to address the issues of sustainable water management for agriculture land, but also in order that groundwater resource use is directed in a sustainable way.

6.2 *Addressing Water Wastage and Inefficient Irrigation Practices*

Sustainable water management, especially in agriculture as it is the largest consumer of fresh water, has to address factors like wastage and underpractices of irrigation. In this regard, use of water-saving devices and introduction of efficient techniques including drip and sprinkler irrigation, accompanied with water-saving practices, may be employed to counter such a trend. Effective irrigation methods also minimize water losses through evaporation, drainage, and deep infiltration, making it possible to apply the required amount of water for crop production. Practices like soil moisture sensors and automated irrigation save time and effort for farmers and ensure that water can be applied to where it is needed most without applying too much water. They are also useful in minimizing water losses through repair of leaks and increasing the efficiency of water usage as well as reducing the need to repair and maintain the distribution networks. Consequently, water users adopt matching strategies for crop's water requirements and scheduling to improve productivity with respect to water availability levels. These measures not only lead to water conservation but also mitigate extensive pressure on the precious ground and surface water sources.

6.3 *Implementing Water Saving Techniques Through Advisory Services*

Advisory services have become an essential instrument for improving water use among the stakeholders in agriculture. These services provide farmers with concrete recommendations on how best to conserve water and these recommendations are tailored for each individual farmer. On a farm, they come in handy in addressing problems in a boarded manner and getting solutions specific to each situation. The farmers are informed of how to use the new technologies within the farm aimed at conserving water. Advisories promote the use of water-conserving devices like drip irrigation and educate the farmers on their installation and operational skills. They also aid in efficient irrigation scheduling and promote strategies to reduce wastage of water. Real-time advisory services depend on the use of information obtained through remote sensing and forecasts of weather conditions and help to rationalize water consumption. RDTs help farmers finance the adoption of water-saving devices and implement Irrigation Water Management Plans. And then, advisory services expand the horizon of insight among farmers and relative interactions develop among them. It even raises the consciousness of the residents concerning the pollution and erosion caused by usage of the water resources and instils the idea of responsible irrigation.

7 The Role of Groundwater Irrigation Advisory Services

Groundwater irrigation advisory (GIA) is a specialized service specifically designed for farmers to assist them in the optimal utilization of groundwater for irrigation purposes. They assist farmers in the efficient management of water resources and groundwater through the application of measures aimed at improving irrigation efficiency in a more applicable manner. This service assists farmers in better water practices and improves irrigation by incorporating the types of water the plants need. Following these recommendations yields crop enhancement while promoting water conservation for the farmers and integrating them into sustainable agriculture.

7.1 Customized Guidance for Sustainable Groundwater Usage

Policy- and legal-measures aimed at sustainable use of groundwater will however seek to understand in depth the motivators and barriers affecting farmers. There are processes of analysis that include determining the type of crops and their growing conditions, soil quality, and even the environmental conditions in order to offer water-conservative farming techniques. Hence less water is lost by the application of techniques such as drip irrigation as well as using predictive analytics regarding the immediate weather conditions to distribute water in a desired manner. These technologies are then used in the field to increase water efficiency in irrigation. Planning for irrigation, safeguarding of wetland areas, selection of plants' and animals' products, compliance with law concerning water saving, promotion of groundwater management, etc. are all advised in action not only covering irrigation.

8 Participatory Approaches to Groundwater Management

8.1 Engaging Local Communities and Stakeholders in Decision-Making

It is crucial to involve the local inhabitants and other stakeholders into making decisions regarding groundwater management if the results of such decisions are to be sustainable. Hearing their voices makes space for stakeholders and strategies that will be beneficial to the community. The key approaches include the following:

Community meetings and workshops: Such community meetings offer a forum every now and then to deal with farmers, leaders, and organization officers on groundwater-related issues.

Participant mapping and data collection: Residents are involved in the mapping of water sources and collection of information to identify water placement strategy that can be developed by the people.

Local Advisory Committees: Different committees can be used purposefully in order to enhance the process of priority setting, planning, and monitoring/naming control in an opened and less controlled format.

Capacity building: Training stakeholders on the basics of groundwater enables them to provide relevant and appropriate input in the decision-making processes and the lobby processes.

Community-based research and monitoring: When members of the community engage themselves in research, the credibility of the inputs given increases, promoting the implementation of evidence-based strategies.

Incentives and rewards: Motivating practices to maintain sustainability of some methods invites the ownership and the involvement in the practice.

Collaborative partnerships: Thanks to partnerships we can pool resources and tackle the water-related problems of this region.

Public awareness campaigns: General public campaigns and activity of local promoters help to create an understanding that is supportive.

Water Users' Associations: Associations help in sharing and making decisions about resources.

Inclusive policy formulation: Policies formulated this way are likely to be embraced at the grassroots level since they respond to the needs of the people.

8.2 Incorporating Traditional Knowledge with Modern Groundwater Management Practices

The combination of ancient knowledge and modern methods empowers the development of sustainable groundwater management approaches. Such strategies include local knowledge passed on from generation to generation on the behavioral patterns of aquifers, hydrological cycles, or the need for bunds in the catchment area. Doing this together with modern methods makes it better.

- **Local ecosystem insights**: It gives a community-based perspective regarding how aquifers and surface waters interact with a view of enhancing our understanding of ecological groundwater resources management toward making it more effective. It results in all-encompassing and sustainable solutions.
- **Conservation synergy**: Practices, such as rainwater harvesting and terracing, which are common practices combined with modern high-level technology, assist in mitigating water shortage and adverse climate changes.
- **Ownership and community involvement**: Including native knowledge ensures proper marketability and stakeholder mobilization. Chambers said the Lazar study and other studies have shown that gaining local knowledge is one of the

ways to get the members of the community to adhere to sustainable water use policies.

- **Culturally sensitive**: It fosters interactions between the communities, practitioners, and policymakers while articulating indigenous and scientific knowledge in a respectful manner.
- **Inclusive co-creation**: Management of groundwater resources becomes fair, effective, and culturally appropriate by combining old traditional and present-day new methods. And this practice also provides further benefits from the social and natural resource perspectives.

8.3 Building Consensus for Sustainable Groundwater Policies and Practices

- Formulating a sustainable groundwater policy is vital to achieving equitable and efficient management of water resources. This means a "collaborative partnership," which involves multiple "actors," like communities, farmers—preferably in cooperative formats—thinking of governments, and thinking in terms of responsibilities rather than opportunities. It promotes co-managed values, and it ensures that solutions do not move away from the agreed-upon agenda.
- **Open communication and listening**: Spending time talking about and ensuring shared responsibility. One of the most fundamental approaches to deepening respect, inclusion, and equity in any recruitment center is to begin at the top and promote open communication through engagement. Open discussion allows all interested parties to express themselves and gives them responsibility; if they are given a voice then it will bring about more commitment in implementing agreements (that take less time to do).
- **Transparency and data**: Drought management will require coordination and information sharing; this necessitates the need to improve on the transparency of groundwater data. Hence information should be able to promote different people in quality and meaningful inputs to the formation of policies.
- **Dialogues and workshops**: Might offer valuable opportunities for exchanging and seeking mutual understandings, thereby fostering trust, and where possible, acquiring some common denominators.
- **Balancing interests**: Achievable are solutions that do not infringe on the interests of each stakeholders proving to be sustainable for public principles.
- **Valuing traditional knowledge**: This primary data in conjunction with westernized science enhances decision-making capabilities and creates better understanding of the region.
- **Guided leadership**: Discussions are facilitated by impartial moderators and policy formulation is carried out with fairness, proportionality, and equality, and thus we come to full circle.

- **Concrete action**: Targets and assessment criteria are formulated so that the mere conclusion of agreements is converted into action. Progress of each item is the most essential and thus must be measured correctly.
- **Continuous evolution**: Practices to forge agreement develop and are never completed.
- **Empowering stakeholders**: Collaboration fosters joint ownership, leading to sustainable and fair results.

9 Policy and Governance for Groundwater Sustainability

9.1 Regulatory Frameworks for Groundwater Management

Regulatory frameworks are established by governments and other authorities to make sure that underground water can be tapped in a sustainable manner. The mechanism for groundwater management is depicted in Fig. 2. Such regulatory measures include laws, guidelines and rules, extraction (correct method of evacuating some earth), allocation (approach to make sure that everybody gets some scanty stock), and conservation (doing more with less). Some common aims may include the desired results of minimizing the overuse, preventing the degradation, and guaranteeing equity.

- **Defining extraction rights**: The frameworks that can determine who can extract which resource. These structures address and even the rate of extraction so that overextraction does not occur.
- **Vigilant monitoring**: Monitoring groundwater level and quality, and recharge aids good management to avoid depleting effects.
- **Environmental protection**: Minimum environmental flows in rivers and control of contamination by sea water intrusion mitigate destructive effects to environment.
- **Promoting conservation**: In terms of accuracy and mapping capacity, rewards acquisition (money for effective irrigation system) motivates to donate water.
- **Inclusive decision-making**: Participation of communities, owners of the stakeholders, and experts in relation to policy formulation and active implementation to the different needs.
- **Balancing enforcement**: Management wears a different face that advocates for and ensures compliance, which is yet another critical aspect of governance; penally enforcing mechanisms are the most effective means of halting extraction procedures when they are not sanctioned.
- **Balancing development**: The ground is that it is possible to accommodate regional differences as regards groundwater availably and water use due to regulatory flexibility.

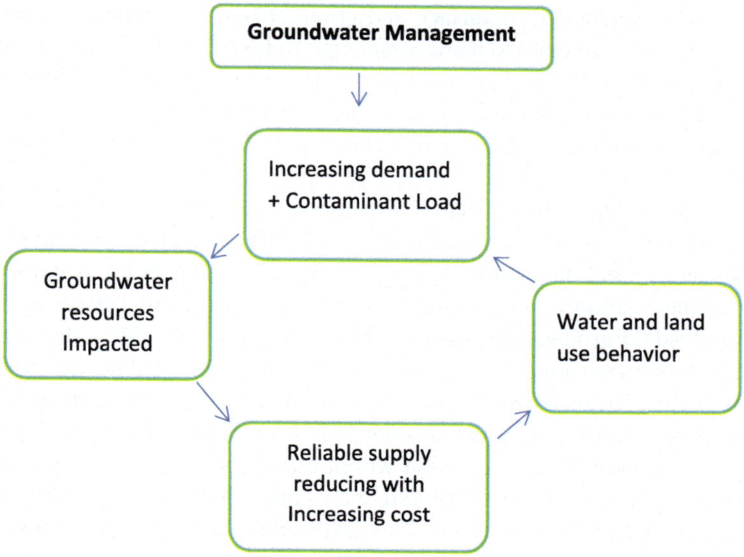

Fig. 2 Groundwater management

9.2 *Integrating Groundwater Conservation into Agricultural and Environmental Policies*

Integrating groundwater conservation into agriculture and environmental management is an adequate measure in the management of water resources since it seeks to preserve the dwindling groundwater resources. This is especially important, for as much water is embedded in agriculture; agriculture, in turn, uses water in excess and inappropriately. Technically, the availability of farmed-water return flows and its usage and reuse in agricultural landscaping design is a direct approach to solving the problem of sustainable groundwater use. Conservation measures could be added to agriculture-oriented technologies to improve irrigation efficiency, water-saving technologies, and responsible groundwater exploitation. For instance, the agricultural policy can induce the upgradation and use of the latest methods such as sprayed drip irrigation, soil moisture sensors, and other high-technology interventions by issuing monetary rewards or grants subsidies. The adoption of such measures encourages farmers to incorporate water-efficient technologies that minimize water wastage and enhance the sustainability of groundwater resources.

Water is probably most useful in relation to the practices and policies that see it as being held in groundwater. The implementation of such provisions of environmental law as groundwater protection measures, can help to stop pollution of the aquifers and control their sustainable use. Riparian, watersheds, and different reservoirs protection can enhance the capacity of aquifers through natural replenishment. Incorporating groundwater conservation measures into an effective overarching water resource management policy fosters a better relationship between the management of surface and groundwater. For instance, some protocols may consist of integrated management measures acknowledging the interconnection of surface water and groundwater systems. Such measures would assist to ensure that any decisions that are taken concerning water allocation do not unduly compromise the quantity of the sustainable yield of groundwater or environmental flows.

Furthermore, the policies that tackle other water use in the cases of non-agricultural activities, such as the industrial and urban, sometimes help the situation. Reducing water demand in these sectors reduces the heat on the groundwater resources and so allows padding against the extraction pressure. Educating the public on the need for groundwater sustainability is crucial in promoting the integration of groundwater management into policy. Use of government-led public information campaigns can educate citizens, farmers, and water users in general on the need and ways of conserving groundwater resources. The most basic of this is embedding policies on groundwater conservation within the agriculture and the environment, and enabling policy decisions to capture the nexus. Policy development will hence lead to more sustainable societies and less exposure of government water resources for risk.

9.3 Governance Challenges and Opportunities in Promoting Sustainable Groundwater Use

Promoting sustainable groundwater use involves navigating various governance challenges and capitalizing on opportunities to ensure effective and equitable water management. Here is a detailed overview of the governance challenges and opportunities in promoting sustainable groundwater use.

9.3.1 Challenges

1. **Fragmented governance**: Different government bodies or departments bear some responsibilities toward the management of groundwater thereby leading to fragmented governance. There are risks in this decentralized construct including policy conflicts and even poor water use management without extensive cooperation and coordination among these bodies both locally between different sections of a river basin and inter-country within shared basins.
2. **Limited data and monitoring**: The data available on all the aspects of groundwater, viz., availability, rate of utilization, and recharge, is quite limited and it turns out difficult to use it to evidence-based decision-making within the organizations. Poor monitoring tools and grid density factors among others have proved to be a major bottleneck to the proper assessment and management of groundwater resources.

3. **Overextraction and depletion**: There is overextraction and therefore depletion; for example, the groundwater withdrawal is progressing at a level that exceeds the natural recharge and is motivated simply by rising needs. The depletion of the aquifers of the resource that supports life underground, the loss of the water table, and the impacts of overextraction will lead to environmental degradation.
4. **Institutional capacity**: Many parts of the world are likely to have inadequate institutional capacity with respect to the formation and execution of even basic groundwater management policies.
5. **Equitable access**: Attaining equitable distribution of groundwater resources among different categories of users in particular regions of water scarcity is a governance issue.
6. **Regulatory enforcement**: There are two ways in which groundwater is subject to enforcement and each one presents a different set of rulemaking challenges that in turn determine whether the prohibitive rules should be employed.
7. **Climate change and variability**: Changes in precipitation patterns caused as a result of climatic change, together with higher temperatures, can further worsen the problem of groundwater resources calling for innovative strategies of governance.
8. **Land-use changes**: Change of land use without planning, especially due to urbanization and agricultural encroachment, can impact on groundwater recharge/socio-hydrological systems and therefore needs land governance.

9.3.2 Opportunities

1. **Integrated water resource management (IWRM)**: Integrating the elements of IWRM will encourage cooperation between different sectors and different users of water and hence, a more integrated management strategy concerning groundwater will be adopted.
2. **Data and technology**: Technological breakthroughs, such as remote sensing and groundwater modeling as well as sensor networks, can help enhance the efficiency with which data on groundwater resources is collected, monitored, and analyzed.
3. **Policy and institutional reforms**: More effective regulation and improved governance can improve the management and the conservation of groundwater resources. Management strategies that allow for flexibility can also help in managing the uncertainties that develop out of the mobility of groundwater.
4. **Stakeholder engagement**: Involving people such as local communities and farmers, even industries and non-governmental organizations (NGOs), in decisions about the management of water resource increases the acceptance and knowledge of these concepts.
5. **Water pricing and economic incentives**: Reasonable charging for the supply of available water and economic motivation for the conservation of water generally will promote responsible use of groundwater sources and efficiency.

6. **Capacity building**: Emphasis should be put on the training and capacity improvement of water managers, the local population, and farmers in regard to sustainable use of groundwater.
7. **Trans boundary cooperation**: For aquifers or basins that are shared with other countries, encouragement of trans-boundary collaboration and agreements would ensure that there is cooperation on management of certain groundwater resources and resolve issues on groundwater resource management at the international borders.
8. **Research and innovation**: Encouraging research and innovation in the area of groundwater management impacts the creation of new technologies, best practices, and decision support systems.

10 Economic Incentives for Groundwater Conservation

10.1 Exploring Market-Based Approaches to Groundwater Sustainability

These approaches are perceived as an advanced method of addressing problems like scarcity, aggravation, and emergence of more resources through overexploitation, which have been experienced around the globe. Such efforts use financial benefits and market-based approaches to assist in the rational use and protection of deep aquifers. In Australia, the National Water Initiative (NWI) and its websites require an administrative approval for water rights trading, which is the legal acceptance of the tradeable water rights concept. This is nothing but freeing the water rights that can be allocated for sale and purchasing by other customers, so that water uses can be shifted to areas where it is needed instead of wasting the surplus that permits restrictions of allocation states. Consequently, the commerce in rights to utilize the water is more productive of that water and even increases its availability in the localities as observed in the Murray-Darling Basin.

Within the United States, groundwater banking is recognized as one of the market-based policies in groundwater management. This arrangement provides for the temporary diversion of groundwater rights between parties in periods of surplus allowing for the replenishment of usable water during drought periods. Groundwater banking has been a success in states such as California and Arizona where it alleviates water scarcity and enhances sustainable use of groundwater. Water management has also witnessed some market-based mechanisms in China. The country has tested systems for trading water rights where water users can purchase or sell their allocated water usage rights in a controlled environment such as a province. Within this context, water rights trading has been beneficial in improving water distribution efficiency and lessening the problem of groundwater overextraction. Chile, too, has espoused the market-oriented water management framework at the fore of innovations. Under the Water Code of the Country, it is possible to trade both surface and underground water through designated water markets. It has been observed that, in Chile, water trading improves water efficiency, decreases wasteful practices, and better meets a variety of water uses. These facts illustrate how market-based solutions have proved useful in enhancing the efforts of the legal mechanisms in the achievement of sustainable exploitation of the groundwater resources.

Through the use of market-based strategies to promote conservation and efficient use of water, there is the opportunity to significantly protect below-the-surface water reserves in the years to come, as well as develop the economy while protecting the environment.

10.2 Economic Instruments and Incentives for Farmers to Adopt Sustainable Practices

The adoption of economic instruments and incentives is critical in motivating farmers to implement relevant practices, and farmers have become more cost-effective in doing so due to the management of water as well as groundwater. These mechanisms are designed to motivate farmers in an economic sense, while trying to achieve the larger objectives of environmental sustainability and rational use of resources. Such economic instruments as governmental programs have been applied to farms in an attempt to assist farmers to apply sustainable practices more effectively, and numerous studies have focused on their utility. Subsidy is one of the common economic instruments. Governments, non-governmental organizations, or any cooperatives offer subsidies to farmers who adopt environmentally friendly practices such as water-efficient irrigation practices or manage soil erosion. Given that farmers have to make a large initial investment in the new technologies, subsidies are an international common practice that reduces such expenses and encourages farmers to adopting the new practices.

Another powerful economic instrument is water pricing. For example, revenue-based water pricing can create incentives for water use efficiency among the farmers. Farmers tend to carefully conserve water and use advanced water-saving technologies and practices when water prices are raised owing to scarcity. Payments for Ecosystem Services (PES) is an economic mechanism where payments are made for certain types of ecosystem services—for instance, farmers are paid for the environmental benefits achieved through sustainable farming activities. Some examples might include farmers of such crops as oilseeds meadow farming who will get money for improvement of water quality, and restoring natural ecosystems. Carbon markets provide an economic motivation for the farmers to adopt the practices that help in carbon sequestration and cutting back the greenhouse gases emissions. Geared toward these carbon market scenarios, farmers could improvise selling carbon credits gained from the adoption of such systems as agroforestry, cover cropping, or reduced tillage.

Farmers are also financially motivated to incorporate sustainability into their work, owing to the market and compliance certifications. As a result of organic or sustainable certifications, they may attract more customers and thereby receive better prices for their produce, which is good for the profitability of farmers better aimed at such markets. It has been shown in the literature that these economic mechanisms have been successful in persuading farmers to switch to some sustainable actions and conserve resources. The impact of these incentives is not always independent of other conditions like how much money support is available, how open the markets are, how clear the policies are, and how knowledgeable the farmers are.

10.3 The Role of Subsidies and Pricing Mechanisms in Sustainable Groundwater Use

Considerations of utilization and mechanisms of pricing in water resources management with applied examples for irreversible management of groundwater, are regarded as key economic instruments for invoking changes. They also facilitate making such key management decisions related to the users of groundwater and encourage those users to consider this resource in a wider time span. Each one of these aspects is different with respect to timing as the authors describe how they utilize these tools and cite the evidence (Toan, 2016).

10.3.1 Subsidies

Subsidies are monetary benefits that are either supplied by the government or various organizations to encourage adopting practices that result in sustainable utilization of groundwater resources. Furthermore, they are capable of significantly reducing the expenses related to the acquisition of technologies that save water, the deployment of adequate irrigation systems, or the observation of best practices in water conservation. By decreasing the economic constraints posed, subsidies help in mainstreaming sustainable practices to a diverse group of water users such as smallholder farmers and disadvantaged households. Evidence suggests that subsidies are useful in enforcing more responsible usage of groundwater. For example, in India, grower adoption of drip irrigation systems was driven by government subsidies without negatively affecting productive water use or crop yields.

10.3.2 Pricing Mechanisms

Water pricing mechanisms are strategies used to charge users according to the amount of groundwater that they waste in any way. Well-formulated pricing mechanisms serve to impart a certain economic aspect of overcoming the given water problem while promoting its launch. Increasing water prices in times of scarcity or in areas with very little groundwater also makes people use water judiciously and seek ways to reduce its usage. Studies support the use of water pricing to encourage sustainable extraction of groundwater resources. In California, water pricing mechanisms were associated with reduced groundwater production by farmers, as well as increased adoption of water inspiring irrigation systems, technologies, and devices. An equivalent case was observed in Pakistan that higher water prices reduce groundwater extraction and improve water use efficiency.

11 Capacity Building and Knowledge Transfer

11.1 Training Programs for Farmers, Water Managers, and Advisory Personnel

Enhancing the knowledge level of farmers, water managers, and other advisory personnel in training programs is one of the important elements for the improvement of the sustainable management of groundwater resources. The objective of these programs is to raise the awareness and uptake of knowledge practices, appropriate water saving techniques, and prudent management of groundwater (Kuschminder, 2014). When stakeholders receive proper training, they are able to make informed decisions, apply sustainable methods, and play an active role in the preservation and proper use of groundwater resources.

11.1.1 Training Programs for Farmers

By training farmers with knowledge and skills on best groundwater utilization and management practices, they are given the chances for increasing water efficiency further and accepting water-saving measures. Typically, such programs include efficient application of irrigation, soil moisture, crop selection and management, or growing water aware plants. They are trained on the best practices and application of water resources such that wastages are reduced and the yield in terms of crops is enhanced without compromising the ecological environment. Farmer training can happen in the form of workshops, demos, field visits, and other capacity-enhancing activities. Such training opportunities where farmers participate in brainstorming on the issues are very effective because they enhance appreciation of the local difficulties and the addressing of these problems in a manner that is tailor-made for the farmers.

11.1.2 Training Programs for Water Managers

Groundwater undertakings render a significant role of supervision and actualization of the sustainable management of groundwater resources. Training programs for water managers include learning of basic functions such as hydrology, groundwater monitoring, development of database, data analysis, and the use of various forms of decision science. The participants also appreciate the legal and political aspects of groundwater management and learn how to deal with the stakeholders. The capacity development for water managers may take the form of technical training, workshops, seminars, as well as the application of modeling to study groundwater behavior patterns and forthcoming strategies for solving sustainable water allocation issues. These processes assure that they will continuously upgrade their skills and knowledge; thus water managers will be in a position to tackle the groundwater-related issues and help in formulating viable groundwater management plans.

11.1.3 Training Programs for Advisory Personnel

Advisory staff, such as agricultural extension services and water practitioners, have the task of sharing information and encouraging the best practices with farmers and water users. Advisory training programs are designed to cultivate skills in communication, the newest trends in sustainable groundwater development, and new techniques. Such training programs enable advisory staff to be oriented to the modern practices and technologies so that they can offer relevant and sound guidance to farmers and other stakeholders. When advisory staff are trained, they participate in the dissemination of best practices in the use of groundwater resources among practitioners. Good training programs have to be dynamic and flexible to adapt to changes and new emerging issues in groundwater policies and practices. However, if all of them participate in building the capacity of farmers, water managers, and advisory staff, then there can be an integrated approach to the protection of groundwater resources for future and more sustainable water management.

11.2 Educating Stakeholders on the Value of Groundwater Conservation

Educating stakeholders about groundwater's importance is pivotal for sustainable management.

Public campaigns: Mass media, neighborhood, and local events focus on public education campaigns, which help to disseminate the groundwater resource in agriculture, drinking water, and the ecosystem.

Engaging farmers: Specific programs have been made available to show the benefits reaped through them such as increased crop yields or reduced expenses on water use.

Empowering managers: Capacity building provides water managers with the knowledge required for the development of effective country-level groundwater management strategies.

School curriculum: Developing groundwater awareness among the young generation ensures the conservation of water resources in the future.

Community participation: Making decisions with the involvement of local people strengthens their connection with groundwater.

Partnership amplification: Outreach will be further enhanced through partnerships with NGOs, academic institutions, and the private sector.

11.3 Technology for Efficient Knowledge Dissemination

The use of technology for effective knowledge transfer is one of the methods in changing the paradigm in the sustainable management of groundwater resources and stakeholders' awareness (Karthic, 2024). The information technology provides the means to overcome spatial and temporal distances, especially in the context of the widespread access to information and stimulating educational efforts. Special websites, mobile phones and schools, and self-serve kiosks filled with pages of different educational materials and interactive maps and graphs make it easy for each stakeholder to obtain current information on groundwater resources, the conservation measures, and the strategies for sustainable management. It is possible to conduct such interaction through webinar and online workshops with a heterogeneous audience from different geographical places making it easy for people to participate and discuss. Other sources of information that are equally important in environmental outreach such as radio and television broadcast and social media present a wide platform of integrating educational materials, advocacy, and groundwater conservation success stories using appealing pictures and videos. Geographic information system (GIS) and remote sensing technologies are assets in groundwater resource mapping, spatial assessment of groundwater levels, and evaluation of groundwater resources changes, thereby providing the stakeholders with important information to make choices. Besides, elements like online training modules, e-learning platforms, and data visualizing tools also enhance or simplify complex concepts making the concept doable even for the average person. Thus, where intellectual information is to be disseminated, the use of technology has proved very useful in creating opportunities for working globally, sharing knowledge, and enhancing skill development among people and places, which contributes to improving community water management assistance globally. Therefore, technology becomes a vital strategic weapon against the information abyss providing opportunity creation to the sensitization as well as stewardship of water resources to the future generations.

12 A Vision for the Future: Advancing Groundwater Sustainability

12.1 Emphasizing the Need for Collective Action and Global Cooperation

Because groundwater has boundaries that are held, affects relationships and connections that go beyond countries and borders also cannot be overemphasized (Mays, 2013). There is an overextraction and pollution of groundwater that transcends borders, hence calls for cooperation in water issues. Increased demand because of growing population meets the pressing need for the sustainable use of groundwater resources across all countries.

- **Collective action**: Governments, NGOs, academia, and the business sector collaborate on water issues. Treating water as a global resource promotes fair distribution and use of information.
- **Transboundary considerations**: Most aquifers are located on different territories, which requires international treaties that each nation must abide by. These benefits help to avert water resource depletion and pollution while allowing equitable distribution.
- **Knowledge exchange**: Cooperation at the international level allows sharing experiences on how to manage groundwater sustainably. Developed countries provide resources, technical assistance, and capacity building to developing countries.
- **Unified response**: Climate, increase in people, and reduction in water availability need a collaborative approach. Groundwater resources are at risk and require global effort.
- **Holistic well-being**: Global cooperation is not only for protecting the world's environment but also for achieving social equilibrium and developing the economy. By undertaking collaborative approaches, water security and environmental stability are preserved.
- **Shared vision**: More children's lives and education on water security issues bring the nations together for a sustainable and green world. This cooperation nurtures and develops a balanced methodology for decorative management.

12.2 Emerging Trends and Innovations in Groundwater Management

Groundwater management concepts are modernizing and aimed at ensuring sustainability and conservation. Among them, as they deal with water scarcity and other environmental problems, are technology, cooperation, and research (Shaikh & Birajdar, 2024).

- **Smart water management**: Groundwater or its aquifers can be monitored in real time and optimally used through the use of IoT and data analytics (adapted from Original Text).
- **Managed aquifer recharge (MAR)**: Droughts make increased water availability possible, and flooding basins and other injection wells can supply additional water to aquifers.
- **Artificial intelligence (AI) and machine learning**: Thanks to more sophisticated modeling and forecasting, groundwater appraisal has improved considerably.
- **Groundwater banking and trading**: Incentives are created to limit water use and trading to ensure sustainable distribution.
- **Nature-based solutions**: Rainwater harvesting, and wetland restoration help to recharge groundwater and preserve ecosystems.
- **Decentralized systems**: Stakeholder participation and community stewardship enhance sustainable groundwater management.
- **Promising future**: These innovations will complement and be feasible with suitable policies that will provide water and an ecologically balanced environment for posterity.

12.3 Strategies for Adapting to Climate Change and Population Growth

Adapting to climate change and population growth demands a holistic approach to sustainable groundwater management. Several key strategies enhance resilience and resource availability. Homer-Dixon and Watts (2003) described the groundwater fluctuation under climate change in the irrigated area. The key driving factors for adapting to climate change and population growth are given as:

- **Water-efficient practices**: Maximize water use in all sectors by employing these techniques, such as drip irrigation, soil moisture sensors, and rainwater harvesting.
- **Managed aquifer recharge (MAR)**: Assured infiltration basins and injection wells are in place so as not to waste aquifers that are useful for storing water ultimately.
- **Climate-resilient plans**: Plan to integrate these water supplies with climate information in anticipation of changes in the water supply.
- **Community involvement**: Environmental education and awareness should help people understand the importance of resources and climate changes and thus the need to shift attitudes.
- **Policy reforms**: Proper pricing and allocation of resources and water use monitoring programs encourage organizable utilization of water.
- **Research and monitoring**: Resources should be spent on understanding how climate and population affect groundwater systems and how adaptation will occur.

- **Resilient future**: Combining these strategies provides a buffer to the stresses and strains of climate change and population increase, which is needed for proper water management.

13 Conclusion

Groundwater management stands as a pivotal element in ensuring sustainable water resource utilization and resilience against the challenges posed by climate change, population growth, and environmental degradation. Accurate and reliable groundwater data form the foundation for informed decision-making, supporting strategies that balance extraction with recharge, protect water quality, and guide infrastructure planning. Advanced technologies like IoT-enabled monitoring systems and modeling tools empower resource managers to predict and adapt to changing groundwater dynamics, ensuring the preservation of this vital resource for future generations. These efforts are complemented by the establishment of collaborative data-sharing platforms, which foster transparency, public engagement, and interagency coordination.

Effective groundwater management also hinges on participatory approaches that actively involve local communities and stakeholders in decision-making. By integrating traditional knowledge with scientific insights, such inclusive strategies can promote sustainable practices, resolve conflicts, and enhance public awareness of groundwater dynamics. With the adoption of innovative technologies, sustainable pumping methods, and artificial recharge techniques, coupled with advisory services tailored to local needs, groundwater resources can be managed judiciously. Together, these measures underscore the importance of a holistic and collaborative approach to groundwater management, ensuring the availability and quality of this indispensable resource for both current and future needs.

References

Adelana, S. M. A. (2009). Monitoring groundwater resources in sub-Saharan Africa: Issues and challenges. In *Groundwater and Climate in Africa (Proceedings of the Kampala Conference, June 2008)* (Vol. 334, pp. 103–113). IAHS-AISH Publication.

Ahmad, A., Shahid, M., et al. (2019). Residues of endosulfan in cotton growing area of Vehari, Pakistan: An assessment of knowledge and awareness of pesticide use and health risks. *Environmental Science and Pollution Research, 26*, 20079–20091.

Ali, S., et al. (2018a). Effects of sucrose and growth regulators on the microtuberization of potato germplasm. *Pakistan Journal of Botany, 50*(2), 763–768.

Ali, S., et al. (2018b). In vitro effects of gibberellic acid on morphogenesis of CIP potato explants and acclimatization of plantlets in field. *In Vitro Cellular & Developmental Biology, 54*(1), 104–111.

Ali, S., Eum, H., et al. (2019). Assessment of climate extremes in future projections downscaled by multiple statistical downscaling methods over Pakistan. *Atmospheric Research, 222*, 114–133.

Ali, A., et al. (2021). Temporal analysis for anomaly detection in precipitation pattern over the selected area in Indus Basin of Pakistan. *Pure and Applied Geophysics, 178*, 651–669.

Amin, A., et al. (2018a). Evaluation and analysis of temperature for historical (1996–2015) and projected (2030–2060) climates in Pakistan using SimCLIM climate model: Ensemble application. *Atmospheric Research, 213*, 422–436.

Amin, A., et al. (2018b). Regional climate assessment of precipitation and temperature in southern Punjab (Pakistan) using SimCLIM climate model for different temporal scales. *Theoretical Applied Climatology, 131*, 121–131.

Awais, M., Arshad, M., Ahmad, S. R., Nazeer, A., Waqas, M. M., Aziz, R., Shakoor, A., Rizwan, M., Chauhdary, J. N., Mehmood, Q., & Ahmad, M. (2023). Simulation of groundwater flow dynamics under different stresses using MODFLOW in Rechna Doab, Pakistan. *Sustainability (Switzerland), 15*(1). https://doi.org/10.3390/su15010661

Conti, K. I., & Gupta, J. (2016). Global governance principles for the sustainable development of groundwater resources. *International Environmental Agreements: Politics, Law and Economics, 16*(6), 849–871. https://doi.org/10.1007/s10784-015-9316-3

Cook, P. G., Shanafield, M., Andersen, M. S., Bourke, S., Cartwright, I., Cleverly, J., Currell, M., Doody, T. M., Hofmann, H., Hugmann, R., Irvine, D. J., Jakeman, A., McKay, J., Nelson, R., & Werner, A. D. (2022). Sustainable management of groundwater extraction: An Australian perspective on current challenges. *Journal of Hydrology: Regional Studies, 44*, 101262. https://doi.org/10.1016/j.ejrh.2022.101262

Danish, S., et al. (2019). Alleviation of Cr toxicity in maize by Fe fortification and Cr tolerant ACC deaminase producing PGPR. *Ecotoxicology & Environmental Safety, 185*, 109706.

Echeverry, S. M. V., Marques, A., & Yoshinaga Pereira, S. (2024). Sustainability and groundwater in the context of water-energy-food nexus: A review of the literature. *Derbyana, 45*, 1–14. https://doi.org/10.69469/derb.v45.833

Eshtawi, T., Evers, M., Tischbein, B., & Diekkrüger, B. (2016). Integrated hydrologic modeling as a key for sustainable urban water resources planning. *Water Research, 101*, 411–428. https://doi.org/10.1016/j.watres.2016.05.061

Fahad, S., et al. (2019). Suppressing photorespiration for the improvement in photosynthesis and crop yields: A review on the role of S-allantoin as a nitrogen source. *Journal of Environmental Management, 237*, 644–651.

Frappart, F., & Ramillien, G. (2018). Monitoring groundwater storage changes using the Gravity Recovery and Climate Experiment (GRACE) satellite mission: A review. *Remote Sensing, 10*(6). https://doi.org/10.3390/rs10060829

Hammad, H. M., et al. (2018). Uptake and toxicological effects of pharmaceutical active compounds on maize. *Agriculture, Ecosystem and Environment, 258*, 143–148.

Hammad, M. H., et al. (2020). Comparative effects of organic and inorganic fertilizers on soil organic carbon and wheat productivity under arid region. *Communication in Soil Science and Plant Analysis, 51*(10), 1406–1422.

Hammad, H. M., et al. (2022). Assessing the impact of varying temperatures on growth of maize and nitrogen use efficiency under semi-arid environment. *Frontiers in Plant Science., 13*, 885479.

Hashmi, M. Z., et al. (2020). Arsenic distribution and metabolism genes abundance in Paddy soils from Punjab and Sindh provinces, Pakistan. *Arabian Journal of Geosciences, 13*, 526.

Homer-Dixon, T., & Watts, M. (2003). The future of megaconferences population and environment funding poverty and environmental insecurity in South Asia water and power in southern Africa, Issue 9. Retrieved from www.wilsoncenter.org/ecsp

Hoogesteger, J. (2022). Regulating agricultural groundwater use in arid and semi-arid regions of the Global South: Challenges and socio-environmental impacts. *Current Opinion in Environmental Science and Health, 27*, 100341. https://doi.org/10.1016/j.coesh.2022.100341

Huneau, F. (2019).

Hussain, S., Mubeen, M., Ahmad, A. et al. Using GIS tools to detect the land use/land cover changes during forty years in Lodhran District of Pakistan. (2020). *Environ Sci Pollut Res, 27*, 39676–39692. https://doi.org/10.1007/s11356-019-06072-3

Hussain, S., et al. (2021a). Using space–time scan statistic for studying the effects of COVID-19 in Punjab, Pakistan: A guideline for policy measures in regional agriculture. *Environmental Science and Pollution Research, 30*, 42495–42508.

Hussain, S., et al. (2021b). Satellite-based evaluation of temporal change in cultivated land in Southern Punjab (Multan region) through dynamics of vegetation and land surface temperature. *Open Geosciences, 13*, 1561–1577.

Hussain, S., et al. (2022a). Land use and land cover (LULC) change analysis using TM, ETM+ and OLI Landsat images in district of Okara, Punjab, Pakistan. *Physics and Chemistry of the Earth, 126*, 103117.

Hussain, S., et al. (2022b). Monitoring the dynamic changes in vegetation cover using spatio-temporal remote sensing data from 1984 to 2020. *Atmosphere, 13*, 1609.

Hussain, S., et al. (2022c). Spatiotemporal variation in land use land cover in the response to local climate change using multispectral remote sensing data. *Land, 11*, 595.

Iqbal, J., et al. (2019). Purple nutsedge (Cyperus rotundus) control through interference by summer crops. *International Journal of Agriculture Biology, 21*, 1083–1088.

Jakeman, A. J., Barreteau, O., Hunt, R. J., Rinaudo, J.-D., & Ross, A. (2016). Erratum to: Integrated groundwater management. In *Integrated groundwater management*. Springer. https://doi.org/10.1007/978-3-319-23576-9_29

Javeed, H. M. R., et al. (2021a). Lipoic acid combined with melatonin mitigates oxidative stress and promotes root formation and growth in salt-stressed Canola seedlings (Brassica napus L.). *Molecules, 26*, 3147.

Javeed, H. M. R., et al. (2021b). Potential utilization of diluted seawater for the cultivation of certain summer vegetable crops: Physiological and nutritional implications. *Agronomy, 11*, 1826.

Karthic, M. U. (2024, March). *Poverty eradication—SCMS Journal*, April–June 2011.

Khaliq, M. A. (2022). Assessment of heavy metal accumulation and health risks in okra (Abelmoschus Esculentus L.) and spinach (Spinacia Oleracea L.) Fertigated with wastewater. *International Journal of Food Contamination, 9*, 11.

Khan, A., et al. (2019). Managing tillage operation and manure to restore soil carbon stocks in wheat–maize cropping system. *Agronomy Journal, 111*(5), 2600–2609.

Kuschminder, K. (2014). Knowledge transfer and capacity building through the temporary return of qualified nationals to Afghanistan. *International Migration, 52*(5), 191–207. https://doi.org/10.1111/imig.12065

Liu, Y., Wang, P., Ruan, H., Wang, T., Yu, J., Cheng, Y., & Kulmatov, R. (2020). Sustainable use of groundwater resources in the transboundary aquifers of the five central Asian countries: Challenges and perspectives. *Water (Switzerland), 12*(8). https://doi.org/10.3390/W12082101

Mays, L. W. (2013). Groundwater resources sustainability: Past, present, and future. *Water Resources Management, 27*(13), 4409–4424. https://doi.org/10.1007/s11269-013-0436-7

Mehmood, A., et al. (2021). Exploring the potential of moringa leaf extract as bio stimulant for improving yield and quality of black cumin oil. *Scientific Reports, 11*, 24217.

Mubeen, M., et al. (2020). Evaluating the climate change impacts on water requirements of cotton-wheat in semi-arid conditions using DSSAT model. *Journal of Water and Climate Change, 11*(4), 1661–1675.

Mubeen, M., et al. (2021). Effect of plant growth promoting bacteria and drought on spring maize. *Pakistan Journal of Botany, 53*(2), 731–739. https://doi.org/10.30848/PJB2021-2(38)

Nasim, W., et al. (2018a). Future risk assessment by estimating historical heat wave trends with projected heat accumulation using SimCLIM climate model in Pakistan. *Atmospheric Research, 205*, 118–133.

Nasim, W., et al. (2018b). Radiation efficiency and nitrogen fertilizer impacts on sunflower crop in contrasting environments. *Environmental Science and Pollution Research, 25*(2), 1822–1836.

Qasim, M. Z., et al. (2020). The potential applications of picotechnology in biomedical and environmental sciences. *Environmental Science and Pollution Research, 27*, 133–142.

Rahman, M. H. R., Ashfaq, A., et al. (2018). Multi-model projections of future climate and climate change impacts uncertainty assessment for cotton production in Pakistan. *Agricultural and Forest Meteorology, 253–254*, 94–113.

Rahman, H. M., et al. (2019). Application of CSM-CROPGRO-cotton model for cultivars and optimum planting dates. *Field Crops Research, 238*, 139–152.

Rasool, A., et al. (2020). Quantification of Tl (I) and Tl (III) based on microcolumn separation through ICP-MS in river sediment pore water. *Environmental Science and Pollution Research, 27*, 9686–9696.

Sabagh, A. E. L., et al. (2019). Drought and salinity stresses in barley: Consequences and mitigation strategies. *Australian Journal of Crop Science, 13*(6), 810–820.

Sabagh, A. E. L., et al. (2021). Potential role of plant growth regulators in administering crucial processes against abiotic stresses. *Frontiers in Agronomy, 3*, 648694.

Sabagh, A. E. L., et al. (2022). Phytohormones as growth regulators during abiotic stress tolerance in plants. *Frontiers in Agronomy, 4*, 765068.

Shah, G. M., et al. (2019). Composting of municipal solid waste by different methods improved the growth of vegetables and reduced the health risks of cadmium and lead. *Environmental Science and Pollution Research, 26*(6), 5463–5474.

Shahid, M. A., et al. (2020). Insights into the physiological and biochemical impacts of salt stress on plant growth and development. *Agronomy (MDPI), 10*, 938.

Shahzad, K., et al. (2018). Tillage and biochar effects on wheat productivity under arid conditions. *Crop Science, 59*, 1–9.

Shaikh, M., & Birajdar, F. (2024). Advancements in remote sensing and GIS for sustainable groundwater monitoring: Applications, challenges, and future directions. *International Journal of Research in Engineering, Science and Management, 7*(3), 16–24. Retrieved from https://journal.ijresm.com/index.php/ijresm/article/view/2955

Singh, A. (2022). Judicious and optimal use of water and land resources for long-term agricultural sustainability. *Resources, Conservation and Recycling Advances, 13*, 200067. https://doi.org/10.1016/j.rcradv.2022.200067

Tahir, M., et al. (2022). Efficacy of organic-based carrier material for plant beneficial rhizobacteria application in okra under normal and salt-affected soil conditions. *Journal of Applied Microbiology, 133*(2), 943–959.

Tariq, M., et al. (2018). The impact of climate warming and crop management on phenology of sunflower-based cropping systems in Punjab, Pakistan. *Agricultural and Forest Meteorology, 256–257*, 270–282.

Toan, T. D. (2016). Water pricing policy and subsidies to irrigation: A review. *Environmental Processes, 3*(4), 1081–1098. https://doi.org/10.1007/s40710-016-0187-6

Waleed, M., et al. (2022). Evaluating the efficiency of coarser to finer resolution multispectral satellites in mapping paddy rice fields using GEE implementation. *Scientific Reports, 12*, 13210.

Wang, D., Fahad, S., et al. (2019). Morphological acclimation to agronomic manipulation in leaf dispersion and orientation to promote "Ideotype" breeding: Evidence from 3D visual modeling of "super" rice. *Plant Physiology and Biochemistry, 135*, 499–510.

Waqas, M.M., Awan, U.K., Cheema, M.J.M. et al. (2019). Estimation of Canal Water Deficit Using Satellite Remote Sensing and GIS: A Case Study in Lower Chenab Canal System. *J Indian Soc Remote Sens 47*, 1153–1162. https://doi.org/10.1007/s12524-019-00977-9

Zainab, N., et al. (2020). Deciphering metal toxicity responses of flax (Linum usitatissimum L.) with exopolysaccharide and ACC-deaminase producing bacteria in industrially contaminated soils. *Plant Physiology and Biochemistry, 152*, 90–99.

Zamin, M., et al. (2020). Developing the first halophytic turfgrasses for the urban landscape from native Arabian desert grass. *Environmental Science and Pollution Research, 27*, 39702–39716.

Future of Wastewater Reuse

Shehryar Liaqat, Samra Tariq, Muhammad Mubeen, Saeed Ahmad Qaisrani, Asad Amin, and Nazim Hussain

Abstract This chapter explores one of the most pressing issues in agricultural water management in the present era: "the sustainable and effective reuse of wastewater." Rapid population growth and climate change have immense pressure on freshwater resources used for sustainable agriculture, which encourage us to reuse the wastewater. Furthermore, wastewater treatment is a practical way to address the water shortage in agriculture; however, it could cause potential health risks and environmental pollution. This pressure urges the researcher to explore novel solutions to meet the rising water demands in agriculture other than relying only on conventional freshwater sources. The risks and advantages of functional practices of wastewater reuse in agriculture are thoroughly reviewed in this chapter. It also examines the possible effects of wastewater on soil health, crop productivity, and environmental sustainability. It examines the different levels of wastewater treatment required for diverse agricultural needs. Additionally, it addresses the challenges linked to the implementation of policies and the adoption of new technologies associated with wastewater reuse. This chapter highlights the innovations which prioritize the development of technologies for wastewater treatment. Digitalization, precision farming, and artificial intelligence improve efficiency, reduce risks, and maximize resource use in agriculture from wastewater management. Using wastewater reuse in agriculture has economic and environmental benefits as well.

S. Liaqat · S. Tariq · S. A. Qaisrani
Department of Environmental Sciences, COMSATS University Islamabad, Vehari Campus, Vehari, Pakistan

M. Mubeen (✉)
Department of Biotechnology, COMSATS University Islamabad, Vehari Campus, Vehari, Pakistan
e-mail: muhammadmubeen@cuivehari.edu.pk

A. Amin
Queensland Alliance for Agriculture and Food Innovation, St. Lucia, QLD, Australia

N. Hussain
Faculty of Agricultural Sciences and Technology, Bahauddin Zakariya University, Multan, Pakistan

M. Mubeen et al. (eds.), *Innovations in Agricultural Water Management*,
https://doi.org/10.1007/978-3-031-91883-4_16

Moreover, wastewater reuse systems can reduce greenhouse gas emissions and save energy. The significance of raising awareness and gaining stakeholder support is also emphasized to ensure that wastewater reuse practices are successfully adopted and put into practice.

Keywords Wastewater treatment · Soil health · Crop productivity · Environmental sustainability · Nutrient pollution

1 Introduction

As our global population continues to grow, so does the demand for food production. Agriculture remains the largest consumer of freshwater resources, accounting for approximately 70% of total water withdrawals worldwide (Velmurugan et al., 2020). Yet, water scarcity is becoming increasingly prevalent due to climate change, over-extraction of groundwater, and pollution of freshwater sources. As the world's population grows, so does the demand for water. We need water for drinking, cooking, bathing, sanitation, and agriculture. As more and more people come into the world, we will need to find ways to produce more food without using more water.

One of the biggest challenges we face is that water is not evenly distributed around the world. Some regions are naturally water-rich, while others are water-scarce (Deng et al., 2021). Population growth is putting additional strain on water resources in water-scarce regions. For example, in the Middle East and North Africa region, water scarcity is already a major problem (du Plessis, 2023). The region is home to just 5% of the world's population, but it has only 1% of the world's renewable freshwater resources. Climate change is making the water situation even worse in the region (Haarstrick & Bahadir, 2022).

Climate change is disrupting the global water cycle. This is leading to more extreme weather events, such as droughts and floods (Yildiz et al., 2022). It is also causing glaciers and snowpack to melt, which is reducing the amount of freshwater available. Droughts are becoming more frequent and severe in many parts of the world. This is reducing crop yields and making it difficult to produce enough food (Nguyen et al., 2023). Floods are also becoming more common, and they can damage infrastructure and contaminate water supplies. Water scarcity directly affects agriculture, which is heavily reliant on water for irrigation (Ingrao et al., 2023). By addressing water scarcity, we can promote sustainable farming practices and ensure an adequate water supply for growing crops, thus contributing to food security.

Many of the world's conventional freshwater sources, such as rivers and lakes, are polluted with agricultural runoff, industrial waste, and sewage (Kakade et al., 2021). Additionally, many aquifers are overused, which is leading to depletion. Sustainable freshwater solutions are those that meet the needs of the present without compromising the ability of future generations to meet their own needs. Some examples of sustainable freshwater solutions include rainwater harvesting, desalination, and wastewater recycling.

Wastewater reuse is the practice of treating wastewater from domestic, industrial, or agricultural sources and then using it for beneficial purposes, such as irrigation, aquaculture, and industrial processes. Wastewater reuse can help to reduce the demand for freshwater in agriculture, which is a major consumer of water globally. This is especially important in arid and semi-arid regions, where water resources are scarce. Wastewater reuse is a promising solution for addressing the challenges of water scarcity, nutrient pollution, and soil health in agricultural water management (Al-Hazmi et al., 2023). It is a sustainable and cost-effective way to meet the growing demand for water in agriculture and protect freshwater resources for future generations (Segovia-Hernández et al., 2023).

Domestic wastewater was used for irrigation by prehistoric civilizations (since the Bronze Age). Thereafter, wastewater was used for disposal, irrigation, and fertilization purposes by Hellenic civilizations and later by Romans in areas surrounding cities (e.g., Athens and Rome) (Tzanakakis et al., 2020). In more recent history, the "sewage farms" (i.e., wastewater application to the land for disposal and agricultural use) were operated in Bunzlau (Silesia) in 1531 and in Edinburgh (Scotland) in 1650, where wastewater was used for beneficial crop production (Angelakis & Snyder, 2015).

Today, planning of projects for the wastewater treatment and reuse of effluents is significantly increasing in several countries (Miller, 2006). The main (re)uses of treated wastewater are: irrigation (both agricultural and landscape), recharge of aquifers, seawater barriers, industrial applications, dual-distribution systems for toilet flushing, and other urban uses. International organizations, such as the World Bank, the Food and Agriculture Organization (FAO) of the United Nations, and the World Health Organization (WHO), estimate that the average annual increase in the reused volume of such water in the USA, China, Japan, Spain, Israel, and Australia ranges from up to 25 (Luthy et al., 2020).

2 The Need for Wastewater Reuse in Agriculture

Wastewater reuse in agriculture is the practice of using treated wastewater to irrigate crops. It is a sustainable and cost-effective way to manage water resources, especially in areas where freshwater is scarce (Hussain et al., 2019). There are several reasons why wastewater reuse in agriculture is necessary:

Water scarcity: Climate change and population growth are putting increasing pressure on freshwater resources. Wastewater reuse can help to reduce the demand for freshwater and ensure a reliable supply of water for agriculture.

Nutrient pollution: Wastewater contains nutrients such as nitrogen and phosphorus, which are essential for crop growth. However, when wastewater is discharged into rivers and lakes, it can cause nutrient pollution (Nie et al., 2018). Wastewater reuse in agriculture allows these nutrients to be recycled and used to fertilize crops, while also reducing nutrient pollution of water bodies.

Public health: Wastewater can contain harmful pathogens that can cause diseases such as cholera and typhoid. However, when wastewater is properly treated, it can be safely used to irrigate crops. Wastewater reuse in agriculture can therefore help to protect public health.

Wastewater reuse in agriculture can also offer a number of economic and environmental benefits (Arena et al., 2020). For example, wastewater reuse can help to reduce the cost of wastewater treatment and disposal, and it can also help to improve soil health and crop yields. Agriculture is the largest user of water globally, accounting for about 70% of all freshwater withdrawals (Ayyam et al., 2019). This is because crops need water to grow, and irrigation is used to supplement rainfall in many parts of the world. As the demand for water from other sectors, such as industry and households, also increases, competition for water resources is likely to intensify. This could lead to water shortages and conflicts, particularly in regions where water resources are already scarce (Zhang et al., 2020). Water scarcity can lead to reduced crop yields and livestock production. This could have a negative impact on food security and nutrition. Agriculture can have a significant impact on the environment, including through water pollution and the depletion of groundwater resources. The growing demand for water in agriculture is likely to exacerbate these environmental problems (Cassardo & Jones, 2011).

3 Risks and Benefits of Wastewater Reuse

3.1 Benefits of Wastewater Reuse

- Wastewater reuse can help to conserve water resources by reducing the demand for freshwater. This is especially important in regions where water resources are scarce.
- Wastewater reuse can help to protect the environment by reducing the amount of wastewater that is discharged into rivers and lakes. This can help to improve water quality and reduce pollution.
- Wastewater reuse can help to reduce costs for businesses and households by providing a source of water that is often less expensive than freshwater. This can be especially beneficial for industries that use large amounts of water.

3.2 Risks of Wastewater Reuse

- Wastewater can contain harmful pollutants, such as bacteria, parasites, and viruses. If wastewater is not treated properly before reuse, it can pose a risk to human health.

- Wastewater can also contain pollutants that can harm the environment. For example, wastewater that contains high levels of nutrients can contribute to algae blooms and eutrophication.
- Wastewater reuse can be controversial, and there may be public resistance to using wastewater for certain purposes, such as irrigation of food crops.

4 Innovations in Wastewater Treatment

Innovations in wastewater treatment are constantly emerging, as researchers and engineers work to develop new and more efficient ways to remove pollutants from wastewater (Saravanan et al., 2021). Membrane bioreactors (MBRs) combine the biological treatment process with membrane filtration to produce high-quality reclaimed water. MBRs can be used to treat a wide range of wastewater streams, including municipal wastewater, industrial wastewater, and agricultural wastewater.

Advanced oxidation processes (AOPs) use a combination of chemicals and energy to generate highly reactive oxidants that can degrade a wide range of pollutants, including organic matter, emerging contaminants, and microorganisms. AOPs can be used as a stand-alone treatment process or as a polishing step to further improve the quality of reclaimed water. Electrodialysis (ED) is a desalination technology that uses electrical energy to remove salts and other ions from wastewater (Kehrein et al., 2020). ED can be used to produce high-quality reclaimed water for a variety of applications, including drinking water augmentation.

Constructed wetlands are artificial wetlands that are designed to treat wastewater. Constructed wetlands are a low-cost and sustainable treatment option that is effective at removing organic matter, nutrients, and suspended solids from wastewater (Gupta et al., 2016). Algal wastewater treatment systems use algae to remove pollutants from wastewater. Algae are very effective at removing nutrients and other organic matter from wastewater. Additionally, algae can be harvested and used to produce biofuels or other valuable products (Wu et al., 2015).

Ozone is a powerful oxidant that can be used to remove a wide range of pollutants from wastewater. Ozone is more effective at removing some pollutants than traditional chemicals, such as chlorine (Rekhate & Srivastava, 2020). However, ozone is also more expensive to generate. Digitalization involves the use of digital technologies to collect, manage, and analyze data. This data can be used to improve decision-making at all levels of the agricultural supply chain, from the farm to the fork. Precision farming is a data-driven approach to agriculture that uses sensors and other technologies to monitor crop conditions and apply inputs, such as water, fertilizer, and pesticides, in a targeted and precise manner (Pathak et al., 2018).

Digital sensors can be used to monitor soil moisture, crop health, and other factors in real time. This data can be used to guide irrigation, fertilization, and pest control practices, ensuring that inputs are applied only when and where they are needed. Precision farming technologies such as variable rate irrigation and variable rate fertilization can be used to apply inputs in a more targeted and precise manner (Balafoutis et al., 2017). This can help to reduce waste and improve yields. Innovative treatments can help to reduce the need for pesticides and herbicides, which can have a number of negative consequences, including environmental damage and human health risks. For example, crop breeding programs can develop varieties that are more resistant to pests and diseases, and new pest control methods can be developed that are more targeted and effective.

Improving crop resilience to climate change can help to improve the resilience of crops to climate change (Dhankher & Foyer, 2018). For example, crop breeding programs can develop varieties that are more tolerant to drought, heat, and salinity. Additionally, new management practices can be developed to help crops cope with the challenges of climate change.

5 Implications of Wastewater on Soil Health and Crop Productivity

Wastewater is a complex mixture of water and dissolved solids, including organic matter, nutrients, and pollutants (Yildiz, 2012). The quality of wastewater varies depending on the source, such as domestic, industrial, or agricultural wastewater. Wastewater is a complex mixture of water and dissolved solids, including organic matter, nutrients, and pollutants (Crini & Lichtfouse, 2018). The quality of wastewater varies depending on the source, such as domestic, industrial, or agricultural wastewater. The application of wastewater to soil can have both positive and negative implications for soil health and crop productivity.

5.1 Positive Implications of Wastewater on Soil Health and Crop Productivity

- Wastewater contains essential nutrients such as nitrogen, phosphorus, and potassium, which can improve soil fertility and crop yields.
- Wastewater can help to improve soil structure by increasing organic matter content and aggregation.
- Wastewater can help to improve soil water retention capacity, which can be beneficial for crops in drought-prone areas.
- The nutrients in wastewater can reduce the need for chemical fertilizers, which can save farmers money and reduce environmental impact.

5.2 *Negative Implications of Wastewater on Soil Health and Crop Productivity*

- Wastewater can contain harmful pollutants such as heavy metals, pathogens, and pharmaceuticals. These pollutants can accumulate in soil and crops, and they can also contaminate groundwater and surface water.
- Wastewater can contain high levels of salts. The application of wastewater can lead to salinization of the soil, which can reduce crop yields.
- Some of the pollutants in wastewater can be phytotoxic, meaning that they can damage plants.
- The high organic matter content in wastewater can lead to reduced oxygen levels in the soil, which can damage plant roots.

The study of the effects of wastewater application on soil health and crop growth is a complex and ongoing area of research. However, a number of studies have been conducted to investigate the potential benefits and risks of this practice. One of the most comprehensive studies on the effects of wastewater application on soil health and crop growth was conducted by the Food and Agriculture Organization of the United Nations (FAO).

Wastewater is a complex mixture of water and dissolved solids, including organic matter, nutrients, and pollutants. The quality of wastewater varies depending on the source, such as domestic, industrial, or agricultural wastewater.

6 Environmental Sustainability in Wastewater Reuse in Agriculture

In light of the growing global challenges, wastewater reuse in agriculture is a crucial strategy for improving environmental sustainability. The various aspects of environmental health could be significantly impacted by this practice. Wastewater can be a valuable source of nutrients for crops when properly treated (Obaideen et al., 2022). It typically contains nitrogen, phosphorus, and potassium, which are essential nutrients for plant growth (Shrestha et al., 2020). These nutrients can be used in fields in place of or in addition to chemical fertilizers. The production and application of synthetic fertilizers are decreased by using nutrient-rich wastewater in place of chemical fertilizers. As a result, the production of fertilizer uses less energy and emits fewer greenhouse gases. Nutrient pollution in water bodies is one of the most important environmental problems in the world today (Chen et al., 2019). Excess nutrients from agricultural runoff can lead to eutrophication, algal blooms, and oxygen-depleted dead zones in aquatic ecosystems. Properly treated wastewater application can reduce nutrient runoff by keeping nutrients on agricultural lands and out of waterways (Rosemarin et al., 2020). Wastewater that is rich in nutrients can improve soil fertility and structure. Organic material in wastewater encourages soil aggregation, which boosts soil's ability to retain water and lowers soil erosion. Agribusiness that is more resilient and sustainable benefits from improved soil health. Reusing wastewater also helps to conserve freshwater resources in addition to recycling nutrients. Using treated wastewater for irrigation in areas where water scarcity is a serious issue can ease the burden on freshwater sources, protecting these important ecosystems (Ungureanu et al., 2020). Moreover, it minimizes the energy required to pump and transport freshwater, further reducing the carbon footprint of agriculture.

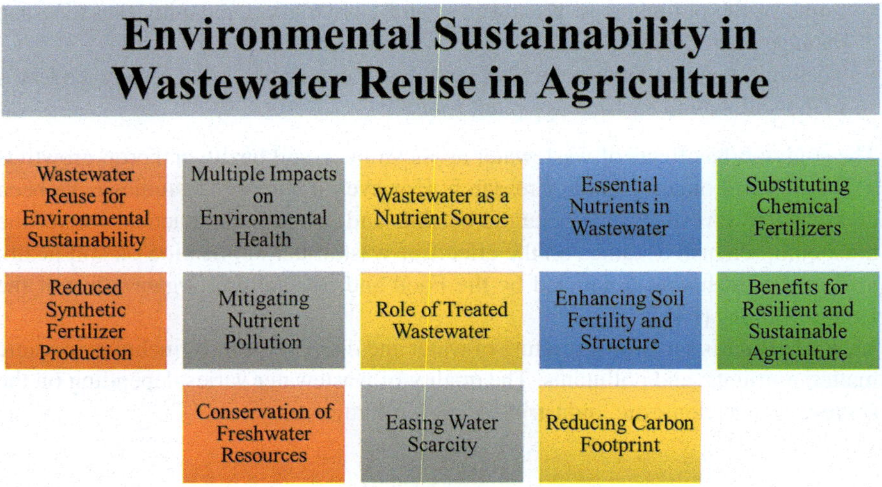

7 Tailoring Wastewater Treatment for Agricultural Needs

There is no one-size-fits-all method for treating wastewater for agricultural use. Regarding water quality, different agricultural applications have specific needs. For wastewater treatment processes to be optimized, it is essential to comprehend these particular requirements. Compared to other agricultural uses, wastewater used for irrigation typically requires less rigorous treatment. The removal of pathogens and contaminants that might affect crop health is the main concern. Filtration, sedimentation, and disinfection are frequently used irrigation treatment techniques. By using these procedures, the water is made safe for use with crops. Water quality is crucial in aquaculture and livestock farming. Pathogens and pollutants can have a direct impact on both the safety of the finished food products and the health of the animals (Obaideen et al., 2022). As a result, sophisticated treatment techniques like biological treatment, membrane filtration, and ultraviolet (UV) disinfection are frequently required. It's crucial to strike a balance between the intensity of the care and the costs involved. Economically, over-treating wastewater for agricultural use may not be feasible (Sun et al., 2019). Consequently, treatment plans should be customized to meet the unique requirements of each application while taking cost-effectiveness into account.

Additionally, considerations like the availability of resources, regional environmental conditions, and regulatory standards should be made when choosing treatment technologies. The ability to adapt wastewater reuse to different agricultural needs is made possible by flexible treatment methods (Rizzo et al., 2020).

8 Challenges in Policy Implementation and Technology Adoption

It can be difficult to create and put into effect regulations that effectively control wastewater reuse in agriculture. In order to protect public health and the environment and promote sustainable practices, policymakers must strike a balance. Because of initial investment costs, farmers and agricultural businesses may be reluctant to adopt wastewater reuse technologies. To promote adoption, policymakers must look into financial incentives and subsidies. The public might be worried about the security of agricultural goods produced using recycled wastewater. To address these issues and foster public trust, effective communication and education campaigns are crucial. To support agricultural reuse, wastewater treatment infrastructure needs to be created and maintained. This can be very difficult, especially in areas with few resources. To create comprehensive policies, policymakers frequently need to coordinate efforts across several government agencies, including agriculture, environment, and health.

For farmers, especially smallholders, investing in cutting-edge wastewater treatment technologies can be financially difficult (Abideen et al., 2023). It might be necessary to use government subsidies and incentives to promote adoption. In order to effectively operate and maintain wastewater treatment systems, farmers and agricultural workers may need training. Construction of the infrastructure required for wastewater treatment and distribution can be labor-intensive, especially in rural areas. Systems for treating wastewater must be compatible with agricultural equipment and procedures already in use (Fito & Van Hulle, 2021). To reduce disruptions, integration should be seamless. Continuous monitoring and compliance measures, which can be resource-intensive, are needed to ensure that treated wastewater complies with regulatory standards. Policymakers and stakeholders must collaborate to develop strategies that encourage the adoption of wastewater reuse technologies and the implementation of efficient regulatory frameworks in order to address these challenges.

9 Raising Awareness and Obtaining Support from Stakeholders

Implementing wastewater reuse practices in agriculture successfully depends on increasing stakeholders' understanding and support (Morris et al., 2021). Informing communities about the advantages and safety of wastewater reuse can be accomplished in large part through public awareness campaigns and educational initiatives. The target audiences for these campaigns should include policymakers, consumers, and farmers. By putting demonstration projects into action, successful wastewater reuse in agriculture can be demonstrated with real-world examples. These initiatives can serve as role models and allay worries about effectiveness and safety. Building trust and addressing issues can be facilitated by interacting with local communities and involving them in decision-making processes. When creating wastewater reuse programs, community feedback should be taken into account. Collaboration between governmental organizations, agricultural groups, environmental organizations, and research institutions can aid in combining resources, subject matter knowledge, and outreach initiatives. These alliances can help with advocacy for responsible wastewater management and knowledge exchange. Case studies of effective wastewater reuse initiatives can be highlighted to illustrate the advantages and dispel common myths. Positive outcomes in the real world can be used as persuasive tools to win support. Communication about the advantages of wastewater reuse and its safety must be open and transparent. Avoid using technical jargon, and present information in a way that is clear and understandable to a wide audience. Building a solid foundation of awareness and support within the communities it serves is essential for the success of wastewater reuse in agriculture (Sohail et al., 2021). In order to do this, effective communication and engagement are essential.

10 Making the Necessary Preparations for a Sustainable Agricultural Future

As a sustainable and environmentally conscious solution to the problems of water scarcity, improved soil fertility, and reduced environmental impact of farming practices, wastewater reuse in agriculture has a bright future (Ungureanu et al., 2020). However, to fully realize this potential, policymakers, agricultural stakeholders, researchers, and the general public must work together. Reusing wastewater in agriculture plays a crucial role because it can address several issues at once. We can lessen the environmental effects of chemical fertilizers and mitigate nutrient pollution in water bodies by recycling wastewater nutrients (Lam et al., 2020). Additionally, wastewater reuse contributes to sustainable water management strategies, particularly in areas where freshwater resources are at risk of scarcity. Water quality standards are met while cutting back on unnecessary expenses when wastewater treatment is specifically adapted to the needs of the agricultural sector. It is a test of resource management and effective use of the tools at hand. It's crucial to overcome the difficulties posed by adopting new technologies and policies. For wastewater reuse in agriculture to be possible, policymakers must navigate the complexities of regulation, economic factors, and public perception. Farmers also require assistance in implementing wastewater treatment technologies into their operations. The first steps are raising awareness and gaining support from stakeholders. These initiatives help to inform decision-making while also fostering a sense of acceptance and trust. Partnerships, effective demonstration projects, and effective communication are essential in this regard. Reusing wastewater in agriculture can help address water scarcity while also paving the way for a more sustainable agricultural future. It supports international initiatives to lessen pollution, protect water resources, and improve food security. We can help the agricultural industry become resilient and environmentally conscious by embracing responsible wastewater management practices. Although there are difficulties, there is also great potential for progress. It is a call to action, a commitment to eco-friendly procedures, and a financial commitment to the industry's futures.

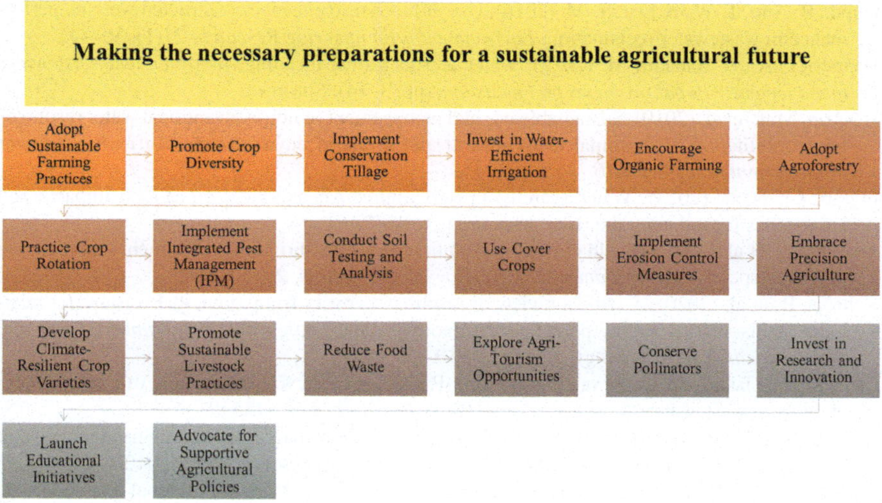

References

Abideen, A. Z., Sundram, V. P. K., & Sorooshian, S. (2023). Scope for sustainable development of small holder farmers in the palm oil supply chain—A systematic literature review and thematic scientific mapping. *Logistics, 7*(1), 6.

Al-Hazmi, H. E., et al. (2023). Wastewater treatment for reuse in agriculture: Prospects and challenges. *Environmental Research, 236*, 116711.

Angelakis, A. N., & Snyder, S. A. (2015). Wastewater treatment and reuse: Past, present, and future. *Water, 7*, 4887–4895.

Arena, C., Genco, M., & Mazzola, M. R. (2020). Environmental benefits and economical sustainability of urban wastewater reuse for irrigation—A cost-benefit analysis of an existing reuse project in Puglia, Italy. *Water, 12*(10), 2926.

Ayyam, V., et al. (2019). Water resources and the changing needs. In *Coastal ecosystems of the tropics—Adaptive management* (pp. 153–173). Springer.

Balafoutis, A., et al. (2017). Precision agriculture technologies positively contributing to GHG emissions mitigation, farm productivity and economics. *Sustainability, 9*(8), 1339.

Cassardo, C., & Jones, J. A. A. (2011). Managing water in a changing world. *Water, 3*, 618–628.

Chen, X., et al. (2019). Multi-scale modeling of nutrient pollution in the rivers of China. *Environmental Science & Technology, 53*(16), 9614–9625.

Crini, G., & Lichtfouse, E. (2018). Wastewater treatment: An overview. In *Green adsorbents for pollutant removal: Fundamentals and design* (pp. 1–21). Springer.

Deng, J., et al. (2021). The impact of water scarcity on Chinese inter-provincial virtual water trade. *Sustainable Production and Consumption, 28*, 1699–1707.

Dhankher, O. P., & Foyer, C. H. (2018). Climate resilient crops for improving global food security and safety. *Plant, Cell & Environment, 41*, 877–884.

du Plessis, A. (2023). Water resources from a global perspective. In *South Africa's water predicament: Freshwater's unceasing decline* (pp. 1–25). Springer.

Fito, J., & Van Hulle, S. W. (2021). Wastewater reclamation and reuse potentials in agriculture: Towards environmental sustainability. *Environment, Development and Sustainability, 23*, 2949–2972.

Gupta, P., Ann, T.-W., & Lee, S.-M. (2016). Use of biochar to enhance constructed wetland performance in wastewater reclamation. *Environmental Engineering Research, 21*(1), 36–44.

Haarstrick, A., & Bahadir, M. (2022). Water and its global meaning. In *Water and wastewater management: Global problems and measures* (pp. 3–14). Springer.

Hussain, M. I., et al. (2019). Sustainable use and management of non-conventional water resources for rehabilitation of marginal lands in arid and semiarid environments. *Agricultural Water Management, 221*, 462–476.

Ingrao, C., et al. (2023). Water scarcity in agriculture: An overview of causes, impacts and approaches for reducing the risks. *Heliyon, 9*, e18507.

Kakade, A., et al. (2021). World eutrophic pollution of lake and river: Biotreatment potential and future perspectives. *Environmental Technology & Innovation, 23*, 101604.

Kehrein, P., et al. (2020). A critical review of resource recovery from municipal wastewater treatment plants–market supply potentials, technologies and bottlenecks. *Environmental Science: Water Research & Technology, 6*(4), 877–910.

Lam, K. L., Zlatanović, L., & van der Hoek, J. P. (2020). Life cycle assessment of nutrient recycling from wastewater: A critical review. *Water Research, 173*, 115519.

Luthy, R. G., Wolfand, J. M., & Bradshaw, J. L. (2020). Urban water revolution: Sustainable water futures for California cities. *Journal of Environmental Engineering, 146*(7), 04020065.

Miller, G. W. (2006). Integrated concepts in water reuse: Managing global water needs. *Desalination, 187*(1–3), 65–75.

Morris, J., et al. (2021). Barriers in implementation of wastewater reuse: Identifying the way forward in closing the loop. *Circular Economy and Sustainability, 1*, 413–433.

Nguyen, T. T., et al. (2023). Security risks from climate change and environmental degradation: Implications for sustainable land use transformation in the Global South. *Current Opinion in Environmental Sustainability, 63*, 101322.

Nie, J., et al. (2018). Causes, assessment, and treatment of nutrient (N and P) pollution in rivers, estuaries, and coastal waters. *Current Pollution Reports, 4*, 154–161.

Obaideen, K., et al. (2022). The role of wastewater treatment in achieving sustainable development goals (SDGs) and sustainability guideline. *Energy Nexus, 7*, 100112.

Pathak, R., Barzin, R., & Bora, G. C. (2018). Data-driven precision agricultural applications using field sensors and unmanned aerial vehicle. *International Journal of Precision Agricultural Aviation, 1*(1), 19.

Rekhate, C. V., & Srivastava, J. (2020). Recent advances in ozone-based advanced oxidation processes for treatment of wastewater—A review. *Chemical Engineering Journal Advances, 3*, 100031.

Rizzo, L., et al. (2020). Best available technologies and treatment trains to address current challenges in urban wastewater reuse for irrigation of crops in EU countries. *Science of the Total Environment, 710*, 136312.

Rosemarin, A., et al. (2020). Circular nutrient solutions for agriculture and wastewater—A review of technologies and practices. *Current Opinion in Environmental Sustainability, 45*, 78–91.

Saravanan, A., et al. (2021). Effective water/wastewater treatment methodologies for toxic pollutants removal: Processes and applications towards sustainable development. *Chemosphere, 280*, 130595.
Segovia-Hernández, J. G., Contreras-Zarazúa, G., & Ramírez-Márquez, C. (2023). Sustainable design of water-energy-food nexus: A literature review. *RSC Sustainability, 1*, 1332.
Shrestha, J., et al. (2020). Role of nutrients in rice (Oryza sativa L.): A review. *Agrica, 9*(1), 53–62.
Sohail, M. T., et al. (2021). Farmers' awareness about impacts of reusing wastewater, risk perception and adaptation to climate change in Faisalabad District, Pakistan. *Polish Journal of Environmental Studies, 30*, 4663–4675.
Sun, H., et al. (2019). Water reclamation and reuse. *Water Environment Research, 91*(10), 1080–1090.
Tzanakakis, V. A., Paranychianakis, N. V., & Angelakis, A. N. (2020). Water supply and water scarcity. *Water, 12*, 2347.
Ungureanu, N., Vlăduț, V., & Voicu, G. (2020). Water scarcity and wastewater reuse in crop irrigation. *Sustainability, 12*(21), 9055.
Velmurugan, A., et al. (2020). Water demand and salinity. In *Desalination-challenges and opportunities* (pp. 1–11). IntechOpen.
Wu, H., et al. (2015). A review on the sustainability of constructed wetlands for wastewater treatment: Design and operation. *Bioresource Technology, 175*, 594–601.
Yildiz, B. S. (2012). Water and wastewater treatment: Biological processes. In *Metropolitan sustainability* (pp. 406–428). Elsevier.
Yildiz, V., Hatipoglu, M. A., & Kumcu, S. Y. (2022). Climate change impacts on water resources. In *Water and wastewater management: Global problems and measures* (pp. 17–25). Springer.
Zhang, D., et al. (2020). Water scarcity and sustainability in an emerging economy: A management perspective for future. *Sustainability, 13*(1), 144.

Water Shortage and Role of Biotechnology

Muhammad Tahir, Aqdus Ali, Tanzeela Bashir, Sania Fareed, Muhammad Mubeen, Muhammad Suffian, Abu Bakr Umer Farooq, Shehzad Mehmood, Mazhar Saleem, Urooj Khan, Sobia Parveen, and Ayman El Sabagh

Abstract Water shortage is a critical and widespread problem all over the world. Water scarcity is commonly classified into two types: economic scarcity, which occurs when there is insufficient water infrastructure, and physical scarcity, which occurs when there is a paucity of water due to regional ecological circumstances. In Sub-Saharan Africa, millions lack access to clean water, with regions like the Sahel and the Horn of Africa facing severe droughts exacerbated by climate change. Similarly, in North Africa, countries such as Egypt and Morocco are heavily dependent on overused water resources like the Nile River. The Middle East is among the most water-stressed regions, with homeland such as Jordan, Saudi Arabia, and Yemen grappling with severe shortages due to arid climates, growing populations, and regional water disputes. Pakistan wastes one-third of the planet's water supply, placing it at number 14 out of 17 "extremely high-water risk" countries. Over 80% of the nation's population suffers from "severe water scarcity." In 1962, there were 5229 m^3 of water per person in Pakistan; today, there are only 1187. Furthermore, Pakistan has one of the lowest wastewater treatment rates in the world, treating just 1% of wastewater. Due to the availability of healthy and plentiful soil, Pakistan's

M. Tahir (✉) · T. Bashir · M. Mubeen (✉) · A. B. U. Farooq · S. Mehmood · M. Saleem · U. Khan · S. Parveen
Department of Biotechnology, COMSATS University Islamabad, Vehari Campus, Vehari, Pakistan
e-mail: muhammad_tahir@cuivehari.edu.pk; muhammadmubeen@cuivehari.edu.pk

A. Ali · S. Fareed
Department of Environmental Sciences, COMSATS University Islamabad, Vehari Campus, Vehari, Pakistan

M. Suffian
Department of Biochemistry, The Islamia University of Bahawalpur, Bahawalpur, Pakistan

A. El Sabagh
Department of Agronomy, Faculty of Agriculture, Kafrelsheikh University, Kafrelsheikh, Egypt

Department of Field Crops, Faculty of Agriculture, Siirt University, Siirt, Turkey

M. Mubeen et al. (eds.), *Innovations in Agricultural Water Management*,
https://doi.org/10.1007/978-3-031-91883-4_17

arid region—which includes the majority of the districts in the interior provinces of Sindh and Southern Punjab—has enormous potential for agricultural development. The major drawback, though, is that these regions frequently experience extremely high temperatures, little rainfall, and a lack of irrigation water. Plants that are under water stress experience osmotic stress, which lowers their rate of vegetative growth and leaf expansion, speeds up senescence and abscission, and sometimes causes them to spread their roots farther in an attempt to find water. Prolonged drought stress results in a decrease in antioxidant enzymes, such as ascorbate, peroxidase, catalase, and glutathione peroxidases, which lower crop performance.

Crop water scarcity increases the synthesis of reactive oxygen species (ROS) and ethylene, a hormone that inhibits growth, which lowers plant productivity and growth.

Numerous strategies particularly development and cultivation of less water requiring cultivars have been developed to cope with water shortage. A new sustainable method that helps plants survive water scarcity is the isolation of plant growth-promoting rhizobacteria (PGPR) and their use as bioinoculants. The stated PGPR synthesize growth hormones, such as gibberellins, 1-aminocyclopropane-1-carboxylate deaminase (ACC) deaminase, exopolysaccharides, cytokinin, and indole-3-acetic acid (IAA), which boost crop yield, root proliferation, water and nutrient uptake, and cell protection against water stress.

Keywords Economic scarcity · Physical scarcity · Water-stressed regions · High temperatures · Little rainfall · Affected plants

1 Introduction

Water scarcity is a growing global concern, affecting millions of people and ecosystems, especially in areas with limited freshwater supplies. As climate change worsens and number of people rise, effectively managing water supplies has become an urgent concern. Thus, management of water supple combined with creative water conservation measures provides feasible solutions to water scarcity by promoting sustainable agricultural practices, increasing plant drought tolerance, and improving wastewater treatment procedures. It is investigated that there are several biotechnological techniques and microbiological interventions aimed at reducing water scarcity and promoting environmental sustainability.

2 Strategies to Cope with the Water Shortage

Managing water storage effectively is crucial for various purpose like enhancing agricultural sustainability and resilience against drought. Several strategies can be implemented to cope with water shortages. One significant approach is rainwater harvesting, where innovations can optimize the collection and use of rainwater for irrigation and nonpotable household purposes. This reduces reliance on treated water, making it possible to utilize collected rainwater in agricultural settings, thereby conserving resources. Research indicates that the quality of harvested rainwater can meet safety standards, making it suitable for irrigation, which is essential for maintaining crop health (Razali et al., 2022; Sardar et al., 2020; Hammad et al., 2020; Zamin et al., 2020; Shahzad et al., 2020; Awais et al., 2020; Abbas et al., 2020; Mubeen et al., 2020; Hussain et al., 2020a, b).

Using adequate containers for water storage is also crucial applications, notably for keeping treated wastewater or fertilizer solutions for hydroponics. Containers need to be hygienic, impervious to leaks, and constructed from secure materials like stainless steel or food-grade plastic. The prevention of leaks and contamination, which can negatively impact biotechnological activities like microbial cultures or plant growth, depends on routine examination and maintenance of these storage systems (Slavik et al., 2020). Furthermore, as variations in water quality can have a substantial effect on the performance of microorganisms or plants, routine testing of the water is necessary to guarantee that it satisfies the safety requirements needed for biotechnological applications. The quality of water determines whether it is suitable for human consumption. Water quality is determined by the composition of the water, which is affected by both natural and human processes. When water characteristics (microbiological, physical and chemical) exceed permissible limits, human health is at risk. Many agencies, such as the Centers for Disease Control and the World Health Organization, set exposure levels or safe limits for chemical pollutants in drinking water (Hammad et al., 2018; Awais et al., 2018; Rasool et al., 2018; Fahad et al., 2018).

By diversifying water sources, for example, through boreholes, wells, or desalination systems, one can increase water resilience, lessen the strain on individual sources, and reduce the risks associated with changes in rainfall patterns brought on by climate change (Hashmi et al., 2020; Zaman et al., 2020; Farid et al., 2019, Danish et al., 2019; Hammad et al., 2019; Imran et al., 2019). Involving the community is also essential since it promotes the creation of group water storage options like water-sharing plans or community reservoirs. Responsible use of existing resources is ensured by educating people about sustainable water management techniques. Complying with safety and environmental requirements requires an understanding of local water legislation and licenses pertaining to water storage. Furthermore, it is critical to take into account how climate change may affect water supply and storage systems, modifying sources or capacity as needed to accommodate changing situations (Rahman et al., 2022; Afzal et al., 2018; Amin et al., 2018; Khan et al., 2018; Hashmi et al., 2018; Adnan et al., 2018; Awais et al., 2017a). Communities can improve their water management procedures and deal with shortages of water more skillfully by putting these solutions into practice (Rahman et al., 2022).

3 Role of Biotechnology

By using genetic and molecular techniques to improve plants' capacity to flourish in water-scarce environments, biotechnology is essential to the development of drought-tolerant and lower water-requiring cultivars.

3.1 Creation of a Cultivar That Requires Less Water

Biotechnology significantly contributes to the development of cultivars that necessitate reduced water usage by employing genetic and molecular techniques to improve water-use efficiency and drought resilience in crops. In agriculture, biotechnology offers a wide range of products, multiple technological approaches, and various applications. Three main areas—transgenic plant production, somatic cell genetics, and micropropagation—are where plant in vitro technologies hold great promise. Rapid progress is being made in hybrid seed production, and new biotechnologies such as somatic hybridization, mutagenesis, and gene editing have revealed the roles of particular nuclear, mitochondrial, and chloroplast DNA components in the division of cells, reproduction, and flowering time (Haque et al., 2022; Abbas et al., 2017; Fahad et al., 2017; Saud et al., 2017; Zia et al., 2017; Amin et al., 2017a). Here's an overview of its role in this context:

3.1.1 Genetic Modification

Biotechnology enables the direct modification of a plant's genetic makeup by introducing genes or gene variants that enhance water-use efficiency and drought tolerance. Scientists can identify and insert specific genes responsible for traits such as reduced transpiration rates and improved root systems (Zhang et al., 2023). Our knowledge of the processes underlying drought tolerance in plants is expanded by the discovery of quantitative characteristics and genes linked to water and nutrient uptake. Genes those control root system architecture are important for availability of water and nutrient, making them the primary targets of breeding plans meant to increase drought resistance (Xiong et al., 2021).

3.1.2 Marker-Assisted Breeding

Biotechnology promotes marker-assisted breeding programs by identifying molecular markers associated with drought tolerance features, enabling breeders to effectively select and cross plants with desirable qualities. This speeds the creation of drought-resistant cultivars. The 1980s saw the development of molecular markers, which were essential for describing quantitative features that are influenced by several genes. Conventional markers and DNA markers are the two kinds of molecular markers utilized in plant breeding in genetics. Numerous systems based on polymorphism-detection methods have been developed, including DNA sequencing, PCR (polymerase chain reaction), and Southern blotting (nucleic acid hybridization). Traditional markers include morphological and biochemical markers (Xu et al., 2022).

3.1.3 Understanding Drought Tolerance Mechanisms

Targeted breeding efforts are informed by biotechnology techniques like transcriptomics and genomics, which give researchers a better understanding of the molecular mechanisms and genetics driving drought tolerance. Several plant metabolic pathways are adversely affected by drought stress, which lower production. However, in order to evade, withstand, or prevent the impacts of drought, plants can modify their morphological and physiological characteristics by activating or deactivating particular genes. Numerous genes, transcription factors, and signal transduction pathways are triggered by this stress. The ability to modify agricultural features more precisely because of advancements in genome editing technologies, especially Cas9 (Clustered Regularly Interspaced Short Palindromic Repeats/CRISPR-associated protein 9), has completely changed agriculture. A dependable and effective technique for developing varieties of plants resistant to various climatic difficulties, such as drought, is CRISPR technology. The creation of plant species with improved drought tolerance is greatly aided by the use of genome editing techniques like CRISPR (Raza et al., 2022).

Transgenic cultivars are made possible by biotechnology, which introduces genes from other organisms or plant species that can withstand drought. These genes can provide characteristics including increased stress response mechanisms, decreased stomatal conductance, and better water retention. Furthermore, biotechnology aids in the identification of stress-responsive genes in plants that can be improved or activated to increase the plants' resistance to water scarcity. This comprises genes related to stress signaling pathways, antioxidant defense mechanisms, and osmotic control (Munaweera et al., 2022). Cultivars that have been modified to have deeper and more effective root systems. Plants with improved root architecture are more drought-resistant because they can draw water from deeper soil layers (Varshney et al., 2021; Shakeel et al., 2017; Jabran et al., 2017a; Amin et al., 2017b; Mirza et al., 2017; Hammad et al., 2017). Precision farming helps to optimize water management techniques with the help of biotechnology instruments like remote sensing. These tools ensure more effective use of water by tracking soil moisture levels and modifying irrigation schedules in real-time depending on data (Singh et al., 2022). By creating biopesticides and biofungicides that are resistant to drought, biotechnology also helps protect crops from water stress and reduce production losses. Biotechnology plays an important role in ensuring global food security by promoting the development of drought-resistant cultivars that sustain crop yields even in areas with limited water resources, thereby reducing the impact of drought-induced shortages in food supplies (Wang, 2022).

Furthermore, by lowering the demand for irrigation, biotechnology-driven cultivars with reduced water demands support sustainable agriculture. It reduces soil deterioration, lessens the environmental effect of agriculture, and aids in water resource conservation. In conclusion, by utilizing genetic alteration, a molecular understanding of drought tolerance, and modern breeding methods, biotechnology plays a critical role in creating cultivars with lower water requirements. These developments address the issues of water shortage and climate change by promoting more resilient and sustainable farming techniques.

3.2 Plants Drought-Tolerance

By applying cutting-edge genetic and molecular techniques, it plays a very important role in creating plants that can withstand drought. In order to preserve cellular homeostasis, drought-tolerant plants adapt to water stress by storing water through processes like stomatal closure, changing root growth to reach deeper reservoirs, and building up osmolytes and solutes like proline and sugars. Through genetic modification, researchers can introduce particular genes that enable plants to adapt to water scarcity and flourish in drought-prone environments. New cultivars are developed more quickly through marker-assisted breeding, which speeds up the plants selection with drought-tolerant features (Naqvi et al., 2022). To increase a crop's capacity to endure in water-scarce conditions, transgenic plants are produced by transferring genes from species that are resistant to drought, such as desert plants. Additionally, studies have identified stress tolerance pathways that can be altered to improve drought resilience, including as osmotic regulation and defenses against antioxidants. By tracking soil moisture and modifying irrigation techniques to minimize water waste, certain technologies, such precision agriculture and remote sensing, maximize water management (Gillani et al., 2017; Jabran et al., 2017b; Ahmad et al., 2017; Nasim et al., 2017; Mahmood et al., 2017). It also makes it possible to analyze large databases, find important genes for drought resistance, and enhance plant breeding techniques. In order to preserve crops under stress, biotechnology also helps create biopesticides and biofungicides from microorganisms that can withstand drought (Perveen et al., 2023). Finally, by sustaining crop yields in areas with erratic rainfall, drought-tolerant plants contribute to the world's food supply and environmental sustainability by lowering the need for excessive irrigation (Munaweera et al., 2022; Tiwari et al., 2023).

3.3 Plant Growth-Promoting Rhizobacteria (PGPR)

Since Anton von Leeuwenhoek's discovery of microscopic creatures in 1683, these bacteria have been utilized to stimulate the growth and development of plants. The Hiltner first identified the "rhizosphere or microbe storehouse"—the area of soil surrounding plant roots—more than a century ago in 1904; it is a region where plant roots strongly interact with microfauna, plants, and soil to influence the chemical and biological characteristics of the earth. Numerous microbial species, especially prokaryotes like viruses, bacteria, and archaea as well as eukaryotes like fungi, algae, protozoa, oomycetes, and nematodes, can be found in the rhizosphere zone (Basu et al., 2021; Awais et al., 2017b; Nasim et al., 2016a, b, c; Mubeen et al., 2016). In order to support plant growth and development, bacteria called plant growth-promoting rhizobacteria (PGPR) are present in the rhizosphere, leaves, and close to roots. They release regulatory compounds. There are two types of plant growth-promoting rhizobacteria (PGPR): (1) intercellular (iPGPR), which are found in root cell structures such as Allorhizobium, Bradyrhizobium, Mesorhizobium, and Rhizobium; and (2) extracellular (ePGPR), which are found in the rhizosphere zone or root cortex cells such as Arthrobacter, Agrobacterium, Azotobacter, Azospirillum, Burkholderia, Bacillus, Pseudomonas, Erwinia, Flavobacterium, Micrococcus, Chromobacterium, and Caulobacter (Gupta et al., 2015). Through direct or indirect methods, plant growth-promoting microorganisms (PGPR) improve the development, growth, and productivity of plants. Direct mechanisms take place inside the targeted plant, enabling them to immediately affect and interact with plant activities, whereas indirect mechanisms take place outside the plant (Goswami et al., 2016). By facilitating nitrogen fixation, nutrient delivery, enzyme synthesis, and plant hormone synthesis, PGPR directly supports plant development. According to studies, PGPR, which can be found inside or outside of plant tissues, promotes growth and development by boosting the microbial population, water supply, and nutrient availability, allowing plants to withstand environmental challenges (Jeyanthi & Kanimozhi, 2018). In indirect mechanisms, PGPR are present outside the plant thus protecting plants from negative effects of various soil microbes and increase nitrogen and nutrients availability in root zone that can be used by the plants. The plant growth-promoting microorganisms (PGPR) are eco-friendly, environmental stresses tolerant like drought, heat, and salinity, and can easily colonize rhizospheres. These properties enhance plant growth, productivity, and development, providing protection to plants (Basu et al., 2021; Rasool et al., 2016; Alghabari et al., 2016; Fahad et al., 2016a, b, c, d; Shahid et al., 2014; Masood et al., 2013; Nasim et al., 2012). PGPRs have gained recognition for their long-term benefits in rural harvests, enhancing plant growth and productivity by 20–30%, promoting plant development, promoting organic soil, reestablishing fruitfulness, and protecting plants from diseases and environmental stresses (Nazir et al., 2018).

3.4 Drought-Tolerant PGPR

Throughout the year, plants are subjected to a variety of biotic and abiotic stressors, which are the primary cause of plant mortality and productivity loss. However, drought is the most significant environmental stressor that has a significant negative impact on plants (Khan et al., 2019). Climate change is causing longer periods of drier conditions and more dry years, affecting agricultural processes and food production in some countries, leading to low crop yield in dry regions worldwide (Niu et al., 2018). Drought conditions cause the production of oxygen species within plants, which are harmless in low levels and help plants respond to environmental stresses through molecular signaling and transduction processes. However, prolonged drought conditions can lead to protein oxidation, lipid peroxidation, damage to cell nucleic acid, and programmed cell death (PCD), ultimately causing plant cell death (Asghari et al., 2020). Plants with symbiotic relationships with PGPRs produce nonenzymatic antioxidants like phenols, ascorbic acid, and flavonoids during drought conditions. These nonenzymatic antioxidants mitigate the harmful effects of reactive oxygen species released during drought, enhancing plant survival and highlighting the importance of plant resistance mechanisms (Asghari et al., 2020). During drought stress, PGPR produce various phytohormones that act as abiotic resistance elicitors, for example, 1-aminocyclopropane-1-carboxylate deaminase (ACC deaminase) phytohormones such as indole acetic acid (IAA) signal molecules such as LCOs (lipo-chitooligosaccharides) and AHLs (*N*-acyl homoserine lactone) and BVCs (bacterial volatile compounds) that help plant to survive in harsh condition (Rosier et al., 2018). We can say that PGPR is capable of promoting plant growth and development by secreting various chemicals during the prevailing environmental stresses (Cavite et al., 2021).

3.4.1 Biochemical Characteristics of Drought-Tolerant PGPR

The rhizosphere contains plant growth-promoting rhizobacteria (PGPR), which support plants in a variety of ways to ensure appropriate growth and development such as supply nitrogen by free atmospheric nitrogen fixation, by providing protection from soil borne diseases and by producing number of phytohormones (gibberellins, auxins, and cytokinin's), enzymes ACC deaminase (1-aminocyclopropane-1-carboxylate deaminase), and antioxidants (phenols or ascorbic acids) that assist plants to cope stresses of environment (Leontidou et al., 2020). Plant growth-promoting rhizobacteria (PGPR) produce signaling molecules and hormones similar to those produced by plants, including IAA (indole acetic acid), cytokinin, abscisic acid, and gibberellin. Indole acetic acid is the most common chemicals produced by most PGPR (Mellidou & Karamanoli, 2022). But indole acetic acid produced by PGPR depends on various factors such as precursors availability or ability of plant to uptake IAA and other growth promoting hormones produced by soil bacteria. According to research, all PGPR produce vital growth hormones that support plant development, growth, and productivity like IAA, fixed nitrogen, ammonia, siderophore, and antioxidant enzymes during drought conditions, such as peroxides and catalase, which remove reactive oxygen species and prevent cell damage or death under abiotic environmental stresses (Kumari et al., 2018).

3.4.2 Production of Indole-3-Acetic Acid

IAA (indole-3-acetic acid) is a plant hormone produced by PGPR that aids plant cell division, seed germination, and development of root, facilitating the transportation of water and nutrients from the ground, accelerating plant growth. Because of its indole ring and L-tryptophan product, PGPR and rhizosphere fungi create the auxin phytohormone IAA (indole-3-acetic acid), which host plants required for growth and development (Myo et al., 2019). Because it is produced by plant growth-promoting rhizobacteria (PGPR) and has no effect on bacterial cells, IAA improves the connection between plants and soil microbes. IAA helps plants withstand a variety of environmental challenges, including extreme drought, by stimulating their roots, which increases their ability to absorb water from deeper soil and draw in more rhizosphere bacteria. Recent research indicates that the plant growth-promoting rhizobacteria (PGPR) produce a tiny amount of indole-3-acetic acid (IAA) through 1-Trp-independent activities. These processes produce IAA via distinct pathways from those that produce 1-tryptophan (1-Trp). Numerous environmental stressors, such as an acidic soil pH, carbon levels, or osmotic stress, might affect the generation of IAA by bacteria that promote plant growth.

3.4.3 Production of ACC Deaminase

The most prevalent enzyme that is crucial in controlling plant ethylene, growth, development, and productivity is 1-aminocyclopropane-1-carboxylate (ACC) deaminase, which is produced by bacteria that promote plant growth and alters the plant. Under drought stress condition, PGPR produce ACC that helps plant to survive by bringing down the ethylene level in the symbiont plants and increasing root surface area. IAA invigorate the enzyme known as ACC synthetase that is involved in converting SAM into ACC which is then used by the symbiont plant by absorbing ACC on the roots or seeds, while sometimes, microbes in the soil absorb a tiny quantity of ACC that is released by plant roots or seeds and hydrolyze to produce ammonia and ketobutyrate. Its reported that PGPR that consist of ACC deaminase regulate ethylene in symbiont plant that helps the plants to survive and grow in dry soil during drought condition.

3.4.4 Exopolysaccharide Production

The most important part of the extracellular matrix is exopolysaccharides (EPS), which typically account for 40–95% of the weight of bacteria (Naseem & Bano, 2014). The heavier molecular weight polymers in the bacteria's EPS mixture adhere to surfaces and aid in soil aggregation (Latif et al., 2022). Bacteria produce slime EPS and capsular EPS, which are long chains composed of lipids, proteins, nucleic acids, and polysaccharides with molecular weights ranging from 10 to 30 kDa (Naseem et al., 2018). Bacteria can create polyionic EPS, which contain functional groups like carboxyl group, phosphoric acid, and hydroxyl group, because of uronic acids or ketal-connected pyruvate groups (Ghosh et al., 2019). By transfer of sugar nucleotide diphospho precursors to undecaprenyl phosphate and isoprenoid lipid carriers, PGPR build the basic repeating unit at the cytoplasmic membrane. Outside the cell, the lipid-linked intermediates undergo polymerization after being carried over the membrane. After assembly, the EPS can be adhered to the cell surface to create a capsule or released into medium as slime. The rate and composition of EPS are influenced by the growth phase of bacteria, the composition of the soil, and external conditions (Camaille et al., 2021). Exopolysaccharides (EPS) are essential for sustaining plant development and moisture levels under challenging circumstances. They protect roots from dehydration by enveloping them in a rhizosheath. Additionally, EPS are involved in microbial aggregation, surface adhesion, bioremediation, and long-term plant-microbe interactions (Khan & Bano, 2019).

3.4.5 Antioxidant Enzyme Production

ROS (reactive oxygen species) comprising hydrogen peroxide, oxide ions, and hydroxide ions OH– are produced when environmental stressors affect the electron transport pathway, which causes immediate cell damage (Ghorbanpour et al., 2013). In order to fight oxidative stress, plants with PGPR produce a range of antioxidants, both enzymatic and nonenzymatic, including catalase, also known as Catalase (CAT), SOD (superoxide), GPx (glutathione peroxidase), APX (ascorbate peroxidase), and GR (glutathione reductase). These enzymes are essential in the breakdown of ROS, which protects the vital functions of symbiont plants (Camaille et al., 2021). According to the current research, enzymes of antioxidant like superoxide dismutase, reduced glutathione oxidase, and catalase guard the plants from harm brought on by oxidative stress (Sagar et al., 2020). In contrast to phenolic compounds, nonenzymatic antioxidants include flavanones, tocopherols, alkaloids, carotenoids, cysteine, reduced glutathione, and ascorbic acid (Chieb & Gachomo, 2023). Plant growth-promoting rhizobacteria (PGPR) can decline the oxidative degradation in plants those are under stress by disrupting antioxidant defense. Plants inoculated with Bacillus and Enterobacter species exhibited increased activity of antioxidant enzymes such as APX, SOD, and CAT under drought stress in comparison to plants that were not inoculated (Etesami & Maheshwari, 2018).

3.4.6 Application of the Drought-Tolerant PGPR as Bioinoculant

An efficient way to stop environmental degradation, assure high agricultural output, and enhance soil health is to promote sustainable agriculture, minimize the use of synthetic agrochemicals, and use materials obtained from biowaste as well as the biological and genetic potential of crop plants (Basu et al., 2021). By colonizing the rhizosphere or interior of a plant, microbial inoculants or microbiological base substances are put on seeds, surfaces, or soil to support growth and development by providing vital nutrients. Since they are vital components of soil, PGPR are necessary for controlling crops by addressing fertility issues in the soil, degradation, stress, diseases of plants, and growth (Verma et al., 2019). By colonizing plant roots, microbial organisms that function as bioinoculants foster symbiotic or free-living partnerships that advance plant development. These PGPR carry out their functions in a number of ways including helping plants assimilate nutrients and shielding them against soil-borne illnesses (Chaudhary & Shukla, 2019). Although PGPR bioformulations can be implemented to plants in a variety of ways, the most popular bio-inoculation methods for promoting vegetable growth are seed coating and soil drenching. Foliar sprays are another option for preventing disease, and phosphate-solubilizing bacteria (PSB) hydrolyze organic and inorganic insoluble P compounds as PGPR to produce soluble P forms that plants can readily use. Plant roots can obtain essential nutrients (such N, P, and K) via bio-inoculating with PGPR, which will accelerate growth and increase biomass (Kumar et al., 2021). So, as an environmentally benign substitute for chemical fertilizers, government and private companies ought to promote the use of biofertilizers. Additionally, it is important to inform farmers about the advantages of PGPR-based biofertilizers for sustainable agriculture in future.

3.5 Treatment of Wastewater

The process of purging water of impurities, contaminants, and dangerous materials so that it can be reused or released into the environment is known as the treatment of wastewater (Ahmed et al., 2022). It involves physical, chemical, and biological processes such as sedimentation, and filtration and microbial degradation to remove suspended solids and break down organic materials. Advanced techniques like membrane filtration, activated carbon adsorption, and advanced oxidation are used to further purify the water. And the treatment addresses pathogens through disinfection method like chlorination or UV radiation and manage sludge by-products for safe disposal or reuse (Al-Hazmi et al., 2023).

Types of Wastewater Treatment

Wastewater treatment involves a variety of methods and processes to effectively remove contaminants and pollutants from water. There are several types of wastewater treatment processes, each with its own specific objectives and methods (Kamaruddin et al., 2023). Here are some of the main types of wastewater treatment:

1. **Preliminary treatment**

Preliminary treatment in wastewater management involves removing large debris and small particles that could harm equipment. Screening removes large objects and debris from the wastewater using screens or grates. Screening is used to remove coarse materials from wastewater that could block mechanical equipment and pipework. Common types of screening machines include bar racks and screens (Donald et al., 2022).

Grit removal targets smaller, abrasive particles such as sand, ash, and gravel that wear down equipment. Grit is typically removed in a specially designed grit channels, either manually or mechanically, to prevent damage and inefficiency in later treatment stages (Kolosovska & Bauer, 2022).

2. **Primary treatment**

In primary treatment, sedimentation is used to allow larger suspended solids to settle out by gravity, forming sludge at the bottom of tanks. About 25–50% of the incoming biochemical oxygen demand (BOD), 50–70% of the total suspended solids (SS), and 65% of the grease and oil are eliminated following basic treatment. The wastewater produced by primary sedimentation units is referred to as primary effluent. While primary sedimentation eliminates some organic phosphorus, organic nitrogen, and heavy metals from solids, it does not influence colloidal or dissolved components.

3. **Secondary treatment**

Secondary treatment eliminates soluble organic materials and suspended particles that were not removed by main treatment. Organic pollutants are normally eliminated by biological processes in which microbes take them as nutrition and convert them into water, energy, and carbon dioxide for their own growth. The three primary biological treatment processes are the trickling filter, activated sludge process, and oxidation pond. These processes help maintain the dissolved oxygen balanced in receiving waterbodies.

- **Biological treatment (activated sludge process)**
  The activated sludge process breaks down organic matter in wastewater using aerobic bacteria, which are often found in aeration tanks. Wastewater enters the aeration tanks where bacteria consume the pollutants, converting them into carbon dioxide, water, and energy. Air is continuously pumped into the aeration tank to supply the necessary oxygen for the aerobic microorganisms, while mixing to keep the microorganisms in contact with the wastewater. The resulting mixture called activated sludge is then transferred to a secondary clarifier. Some of the deposited activated sludge is returned to the aeration tank. This is called "return sludge" and is done to keep a healthy population of microorganisms for continuing treatment (Waqas et al., 2023).
- **Trickling filters**
  Trickling filters use a medium such as stone and plastic over which wastewater flows. As the wastewater trickles down, microorganisms on the medium

break down organic waste. The biofilm of bacteria on the medium grows as it comes into contact with wastewater. Sewage can be recycled and several trickling filters connected in succession to improve treatment. Natural air convection provides the required aeration as organic matter undergoes biological oxidation within the biofilms (Arous et al., 2023).

- **Rotating biological contactors (RBCs)**
  Rotating biological contactors (RBCs) are cylindrical drums or discs that are partially immersed in wastewater and rotate slowly. A biofilms of bacteria and fungi are exposed to wastewater by the spinning motion, where they decompose organic matter (Ruiz-Munoz et al., 2023). This biological treatment process reduces the levels of organic pollutants, including chemical oxygen demand (COD) and biochemical oxygen demand (BOD), and some nutrition making them suitable for small to medium-sized treatments plants. With regular maintenance and monitoring to assure peak performance, they provide an affordable treatment option with little energy needs and resistance to shock loading (Waqas et al., 2023).
- **Sequential batch reactors (SBRs)**
  Sequential batch reactors (SBRs) use a sequence of steps, including filling, aeration, settling, the decanting process and idle phase, to treat wastewater in distinct, successive batches. Wastewater enters the reactor during the filling phase, and then microorganisms use the oxygen they receive to break down organic materials. The clarified effluent is decanted after settling, which permits particles to sink to the bottom after aeration (Noor et al., 2023). The reactor may go into an idle period after decanting to allow for additional particle settling. Flexible operation, effective nutrient removal, adaptability to changing influent conditions, and customization for particular treatment objectives are just a few benefits that SBRs provide. SBRs must be closely watched and managed, nevertheless, to ensure that every stage of the treatment operation goes according to plan. The sequencing and timing of the different phases are critical to achieving optimal treatment performance. Advanced automation and control systems are often used to manage SBR operations effectively (He et al., 2023).

4. **Tertiary treatment**

Tertiary treatment in wastewater management provides an additional level of purification beyond secondary treatment enhancing the quality of effluent before it discharges into the environment or its use for nonpotable purposes. It improves effluent quality by removing residuals pathogens, organic matter, and various pollutants that could negatively effects receiving water bodies. The selection of appropriate filtration method for tertiary treatment is influenced by factors such as the target treatment standard influence wastewater quality (Zagklis & Bampos, 2022).

- **Filtration**
  Granular media filters, such as sand and other materials, are commonly consumed in the treatment of tertiary wastewater to withdraw solid and fine sus-

pended flecks (Ahmed et al., 2021). Both organic and inorganic materials are successfully removed by these filters, which include reverse osmosis, multimedia filtration, micro-, and ultra-filtration, and sand filtration. The two and three media filter kinds are the most widely used. A bed of sand would be positioned beneath an anthracite layer in a typical two-media filter configuration. A garnet layer beneath the sand layer is a typical three-media filter configuration (Aragaw & Bogale, 2023).

- **Biological nutrient removal (BNR)**
  Biological nutrient removal (BNR) is a wastewater treatment procedure aimed at removal excess nutrients (phosphorous and nitrogen) to avoid environmental issues like algal blooms, oxygen depletion, and ecological imbalances. BNR utilizes microbial processes to transform and remove these nutrients, utilizing two primary methods, nitrification–denitrification and enhanced biological phosphorus removal (EBPR) (Wang et al., 2023). Nitrification involves converting ammonia (NH_3) to nitrite (NO_2^-) through ammonia-oxidizing and nitrite-oxidizing bacteria, while denitrification reduces nitrate (NO_3^-) to nitrogen gas (N_2) or nitrogen oxides in oxygen limited conditions. EBPR focuses on the uptake and storage of phosphorus by polyphosphate accumulating organisms which release phosphorus during anaerobic conditions for removal. BNR processes are often implemented in advanced treatment of wastewater system, for example, activated sludge or sequencing batch reactor, and require precise control of operational conditions to ensure effective nutrient removal (Carvalho et al., 2023).
- **Disinfection**
  A critical stage in wastewater treatment is disinfection, which uses techniques like ozone treatment, UV irradiation, or chlorination to get rid of or render inactive any remaining microorganisms. Chlorine is commonly used as a final disinfectant with residuals chlorine levels regulated to meet World Health Organization standards up to 5 mg/L in water and a minimum of 0.5 mg/L after 30 min of contact time for safe consumption. Fluoride is a frequent component to drinking water that helps with oral health and prevents dental decay (Shekhawat et al., 2023).

5. **Advanced treatment**

A semi-permeable membrane is used in the membrane filtering process to extract particles, impurities, and microbes from liquids. Membrane filtration, which is powered by pressure, concentration gradients, or electrical potential, is essential for treating wastewater and purifying water because it lets only certain substances through while obstructing larger particles (Chadha et al., 2022). A wide range of organic and inorganic contaminants can be effectively eliminated from liquids and gasses using activated carbon. It is a common option for treating water and air as well as in a variety of industrial processes due to its huge surface area and high absorption capability (Srivastava et al., 2021). Strong oxidants such as O_3, H_2O_2, or UV radiation are used in advanced oxidation processes (AOPs) to degrade persistent organic and inorganic contaminants in water. The highly reactive oxidative species (ROS) produced by these techniques can break down complex contaminants that are ineffectively addressed by conventional treatments. AOPs are renowned for their capacity to break down complicated chemicals and function in moderate conditions (Agabo-García et al., 2023).

3.5.1 On-Farm Wastewater Treatment

The management and treatment of wastewater produced on agricultural lands are referred to as "on-farm wastewater treatment." This covers the treatment of wastewater from a range of farming operations, including crop irrigation, animal husbandry, and industrialization. In order to prevent water contamination, safeguard public health, and advance sustainable farming practices, on-farm wastewater treatment is crucial. These are some typical techniques and factors to take into account while treating wastewater on farms.:

3.5.1.1 Sedimentation and Separation

To eliminate suspended solids from wastewater, sedimentation and separation methods such as sediment basins and settling ponds are employed. While sludge builds up at the bottom of settling ponds and needs to be periodically removed, particles can settle by slowing down the water flow, resulting in a discharge of clarified water. Sediment basins, commonly utilized in construction or erosion-prone locations, catch runoff and enable sediment to sink out before releasing treated water at regulated rates. Both systems depend on gravity and upkeep to assure effective sediment removal and improved water quality (El Jery et al., 2023).

3.5.1.2 Biological Treatment

Biological treatment methods like constructed wetlands and lagoon systems utilize natural processes to treat wastewater. Constructed wetlands, either surface flow and subsurface flow, use vegetation and microbial communities to remove nitrogen, phosphorus, and suspended solids from water (Zagklis & Bampos, 2022). Lagoon system, which can be aerobic or anaerobic, uses ponds to promote the natural breakdown of organic materials and sedimentation. These systems offer ecological benefits, supporting biodiversity while keeping in mind water quality, but require careful management to mitigate environmental threats loke pollution and habitat destruction (Thompson et al., 2022).

3.5.1.3 Nutrient Management

Nutrient management focuses on recovering and reusing essential nutrients like nitrogen and phosphorus from wastewater. Nutrient recovery capturing, recycling, and reusing nutrients, such as nitrogen and phosphorus, from various waste streams for beneficial purposes help reduce pollution and support agriculture by reusing them as fertilizers. Controlled irrigation involves the precise application of treated wastewater to crops, ensuring that nutrients are applied efficiently without overloading the soil. This method is commonly used for high-value crops like fruits and vegetables. However, the complexity of managing water and nutrient variability in larger farms requires advanced sensor systems and control algorithms. The high costs associated with these technologies can limit their widespread adoption (Gamal et al., 2023).

3.5.1.4 Composting and Solid Waste Management

Composting is a sustainable method that uses microorganisms to decompose organic materials, like scraps of kitchen, agricultural waste, and yard waste into nutritious soil amendments called compost. This process decreases landfill waste, amends structure of soil, enhances water preservation, and provides valuable nutrients for plants. It needs the right amount of moisture, aeration, and materials that are both carbon- and nitrogen-rich (Manga et al., 2023). Vermicomposting is a type of composting in which organic matter is broken down by earthworms, especially red worms, creating worm casting, a nutrient-rich compost. This mechanism accelerates decomposition and produces nutrient-rich material that benefits the health of both plants and soil (Kauser & Khwairakpam, 2022).

3.5.1.5 Recycling and Reuse

Once the wastewater has been treated to a certain degree, it can be safely reused for flushing, irrigation or nonpotable purpose and in this way water is conserved. A sustainable method of managing water resources is recycling and reusing wastewater, which entails the treatment and reuse of wastewater for a variety of nonpotable uses. It lessens pollution and preserves freshwater supplies and promotes environmental sustainability. There are several contexts in which wastewater recycling and reuse can be applied, such as cities, factories, farms, and more (Kumari et al., 2023).

3.5.2 Biological Wastewater Treatment

Biological wastewater treatment utilizes microbes to decompose organic matter and remove contaminants from wastewater, commonly used in municipal and industrial facilities. Activated sludge and wastewater are combined in an aeration tank, where aerobic microorganisms break down organic contaminants. Following aeration, the water–sludge combination is separated in a clarifier, which separates the sludge from the treated water. Biofilm reactors and trickling filters use media, coated with microorganisms to break down contamination as wastewater passes over them. Anaerobic digestion treats sludge in the absence of O_2, making biogas and stabilized sludge. Constructed wetlands use plants and microorganisms to naturally filter and treat wastewater (Abdelrahman et al., 2023).

3.5.3 Aerobic Digestion of Wastewater

Aerobic digestion is a biological process that uses oxygen-loving microbes to collapse organic matter and pollutants in the waste water. In wastewater treatment infrastructure, it is frequently employed as a subsequent treatment step. This procedure involves introducing wastewater into aeration tanks, where mechanical aerators give oxygen to promote the growth of aerobic bacteria that break down organic matter into CO_2, H_2O, and microbial biomass (Yao et al., 2023). The activated sludge is separated in a secondary clarifier. The sludge is often recirculated, while excess is removed and further treated. Aerobic digestion removes a considerable amount of organic matter, pathogens, and nutrients, from wastewater, leading to improved water quality before discharge into the environment. It also reduces odors associated with anaerobic processes and produces a more stable and less voluminous sludge compared to anaerobic digestion.

3.5.4 Anaerobic Digestion of Wastewater

Anaerobic digestion of wastewater is a biological treatment technique that degrades organic materials in the absence of oxygen. Anaerobic bacteria transform organic materials into biogas, which is mostly composed of CH_4 and CO_2, as well as digestate, a partially stabilized sludge (Abdelrahman et al., 2023). The procedure takes place in digesters and involves four stages: hydrolysis (breaking down complex organic molecules), acidogenesis (producing volatile fatty acids and alcohols), acetogenesis (generating acetate, CO_2, and hydrogen), and methanogenesis occurs when methanogenic archaea generate methane. Methane-rich biogas can be collected and utilized as an alternative source of energy. Anaerobic digestion reduces organic waste, lower pathogens levels, and produces valuable by-products like nutritious residue, which can be utilized as fertilizer. It is widely used in wastewater treatment plants and industrial operations for energy recovery and waste management.

References

Abbas, G., et al. (2017). Quantification the impacts of climate change and crop management on phenology of maize-based cropping system in Punjab, Pakistan. *Agricultural Forest Meteorology, 247*, 42–55.

Abbas, Y., et al. (2020). Valorization of solid waste biomass by inoculation for the enhanced yield of biogas. *Clean Technologies and Environmental Policy, 22*, 513–522.

Abdelrahman, A. M., Kosar, S., Gulhan, H., Cicekalan, B., Ucas, G., Atli, E., et al. (2023). Impact of primary treatment methods on sludge characteristics and digestibility, and wastewater treatment plant-wide economics. *Water Research, 235*, 119920.

Adnan, M., et al. (2018). Phosphate-solubilizing bacteria nullify the antagonistic effect of soil calcification on bioavailability of phosphorus. *Scientific Reports, 7*, 22653.

Afzal, M., et al. (2018). Current status and future possibilities of molecular genetics techniques in Brassica napis. *Biotechnology Letters, 40*(3), 479–492.

Agabo-García, C., Repetto, G., Albqmi, M., & Hodaifa, G. (2023). Evaluation of the olive mill wastewater treatment based on advanced oxidation processes (AOPs), flocculation, and filtration. *Journal of Environmental Chemical Engineering, 11*, 109789.

Ahmad, S., et al. (2017). Quantification of climate warming and crop management impacts on cotton phenology. *Plants, 6*(7), 1–16.

Ahmed, S., Mofijur, M., Nuzhat, S., Chowdhury, A. T., Rafa, N., Uddin, M. A., et al. (2021). Recent developments in physical, biological, chemical, and hybrid treatment techniques for removing emerging contaminants from wastewater. *Journal of Hazardous Materials, 416*, 125912.

Ahmed, S. F., Mehejabin, F., Momtahin, A., Tasannum, N., Faria, N. T., Mofijur, M., et al. (2022). Strategies to improve membrane performance in wastewater treatment. *Chemosphere, 306*, 135527.

Alghabari, F., et al. (2016). Gibberellin-sensitive Rht alleles confer tolerance to heat and drought stresses in wheat at booting stage. *Journal of Cereal Science, 70*, 72–78.

Al-Hazmi, H. E., Mohammadi, A., Hejna, A., Majtacz, J., Esmaeili, A., Habibzadeh, S., et al. (2023). Wastewater treatment for reuse in agriculture: Prospects and challenges. *Environmental Research, 236*, 116711.

Amin, A., et al. (2017a). Comparison of future and base precipitation anomalies by SimCLIM statistical projection through ensemble approach in Pakistan. *Atmospheric Research, 194*, 214–225.

Amin, A., et al. (2017b). Optimizing the phosphorus use in cotton by using CSM-CROPGRO-cotton model for semi-arid climate of Vehari-Punjab, Pakistan. *Environmental Science and Pollution Research, 24*(6), 5811–5823.

Amin, A., et al. (2018). Simulated CSM-CROPGRO-cotton yield under projected future climate by SimCLIM for southern Punjab, Pakistan. *Agricultural Systems, 167*, 213–222.

Aragaw, T. A., & Bogale, F. M. (2023). Role of coagulation/flocculation as a pretreatment option to reduce colloidal/bio-colloidal fouling in tertiary filtration of textile wastewater: A review and future outlooks. *Frontiers in Environmental Science, 11*, 1142227.

Arous, F., Kyriacou, S., Ennouri, H., Bessadok, S., & Jaouani, A. (2023). An energy-efficient two-stage passively aerated trickling filter for high-strength wastewater treatment and reuse. *Water and Environment Journal, 37*, 218–231.

Asghari, B., Khademian, R., & Sedaghati, B. (2020). Plant growth promoting rhizobacteria (PGPR) confer drought resistance and stimulate biosynthesis of secondary metabolites in pennyroyal (*Mentha pulegium* L.) under water shortage condition. *Scientia Horticulturae, 263*, 109132.

Awais, M., et al. (2017a). Modeling the water and nitrogen productivity of sunflower using OILCROP-SUN model in Pakistan. *Field Crops Research, 205*, 67–77.

Awais, M., et al. (2017b). Nitrogen and plant population change radiation capture and utilization capacity of sunflower arid environment. *Environmental Science and Pollution Research, 24*(21), 17511–17525.

Awais, M., et al. (2018). Potential impacts of climate change climate change impact assessment and adaptation strategies for sunflower in Pakistan. *Environmental Science and Pollution Research, 25*(14), 13719–13730.

Awais, M., et al. (2020). Adaptation strategies for sunflower production under climate change for arid environment by using CROPGRO and APSIM-Sunflower models. *Pakistan Journal of Agricultural Sciences, 57*(6), 1529–1538.

Basu, A., Prasad, P., Das, S. N., Kalam, S., Sayyed, R., Reddy, M., et al. (2021). Plant growth promoting rhizobacteria (PGPR) as green bioinoculants: Recent developments, constraints, and prospects. *Sustainability, 13*, 1140.

Camaille, M., Fabre, N., Clément, C., & Ait, B. E. (2021). Advances in wheat physiology in response to drought and the role of plant growth promoting rhizobacteria to trigger drought tolerance. *Microorganisms, 9*, 687.

Carvalho, V., Fradinho, J., Oehmen, A., & Reis, M. A. M. (2023). Long term operation of a phototrophic biological nutrient removal system: Impact of CO2 concentration and light exposure on process performance. *Journal of Environmental Management, 334*, 117490.

Cavite, H. J. M., Mactal, A. G., Evangelista, E. V., & Cruz, J. A. (2021). Biochemical characteristics and inoculation effects of multi-trait plant growth-promoting rhizobacteria on upland rice (*Oryza sativa* L. cv PSB Rc23) seedling growth. *Archives of Microbiology, 203*, 3533–3540.

Chadha, U., Selvaraj, S. K., Thanu, S. V., Cholapadath, V., Abraham, A. M., Manoharan, M., et al. (2022). A review of the function of using carbon nanomaterials in membrane filtration for contaminant removal from wastewater. *Materials Research Express, 9*, 012003.

Chaudhary, T., & Shukla, P. (2019). Bioinoculants for bioremediation applications and disease resistance: Innovative perspectives. *Indian Journal of Microbiology, 59*, 129–136.

Chieb, M., & Gachomo, E. W. (2023). The role of plant growth promoting rhizobacteria in plant drought stress responses. *BMC Plant Biology, 23*, 407.

Danish, S., et al. (2019). Alleviation of chromium toxicity in maize by Fe fortification and chromium tolerant ACC deaminase producing plant growth promoting rhizobacteria. *Ecotoxicology and Environmental Safety, 185*, 109706.

Donald, A., Asif, M., & Felicien, S. (2022). A review on the centralised municipal sewage and wastewater treatment unit processes. *MOJ Ecology & Environmental Sciences, 7*, 31–38.

El Jery, A., Kosarirad, H., Taheri, N., Bagheri, M., Aldrdery, M., Elkhaleefa, A., et al. (2023). An application of ultrasonic waves in the pretreatment of biological sludge in urban sewage and proposing an artificial neural network predictive model of concentration. *Sustainability, 15*, 12875.

Etesami, H., & Maheshwari, D. K. (2018). Use of plant growth promoting rhizobacteria (PGPRs) with multiple plant growth promoting traits in stress agriculture: Action mechanisms and future prospects. *Ecotoxicology and Environmental Safety, 156*, 225–246.

Fahad, S., et al. (2016a). Exogenously applied plant growth regulators enhance the morpho-physiological growth and yield of rice under high temperature. *Frontier in Plant Sciences, 7*(1250), 1–13.

Fahad, S., et al. (2016b). Responses of rapid viscoanalyzer profile and other rice grain qualities to exogenously applied plant growth regulators under high day and high night temperatures. *PLoS One, 11*(7), 1–13.

Fahad, S., et al. (2016c). Exogenously applied plant growth regulators affect heat-stressed rice pollens. *Journal of Agronomy and Crop Science, 202*, 139–150.

Fahad, S., et al. (2016d). A combined application of biochar and phosphorus alleviates heat-induced adversities on physiological, agronomical and quality attributes of rice. *Plant Physiology and Biochemistry, 103*, 191–198.

Fahad, S., et al. (2017). Crop production under drought and heat stress: Plant responses and management options. *Frontier in Plant Sciences, 8*(1147), 1–16.

Fahad, S., et al. (2018). Consequences of high temperature under changing climate optima for rice pollen characteristics concept and perspectives. *Archives of Agronomy and Soil Science, 64*(11), 1473–1488.

Farid, H. U., et al. (2019). Assessing seasonal and long-term changes in groundwater quality due to over-abstraction using geostatistical techniques. *Environmental Earth Sciences, 78*, 386.

Gamal, Y., Soltan, A., Said, L. A., Madian, A. H., & Radwan, A. G. (2023). Smart irrigation systems: Overview. *IEEE Access*. https://doi.org/10.1109/ACCESS.2023.3251655

Ghorbanpour, M., Hatami, M., & Khavazi, K. (2013). Role of plant growth promoting rhizobacteria on antioxidant enzyme activities and tropane alkaloid production of *Hyoscyamus niger* under water deficit stress. *Turkish Journal of Biology, 37*, 350–360.

Ghosh, D., Gupta, A., & Mohapatra, S. (2019). A comparative analysis of exopolysaccharide and phytohormone secretions by four drought-tolerant rhizobacterial strains and their impact on osmotic-stress mitigation in *Arabidopsis thaliana. World Journal of Microbiology and Biotechnology, 35*, 1–15.

Gillani, R. A., et al. (2017). Biosorption analysis of Agrobacterium tumefaciens 12b3 against Cr (1 III) and Pb(II) by 2 kinetic modeling and equilibrium studies. *Desalination and Water Treatment, 67*, 206–214.

Goswami, D., Thakker, J. N., & Dhandhukia, P. C. (2016). Portraying mechanics of plant growth promoting rhizobacteria (PGPR): A review. *Cogent Food & Agriculture, 2*, 1127500.

Gupta, G., Parihar, S. S., Ahirwar, N. K., Snehi, S. K., & Singh, V. (2015). Plant growth promoting rhizobacteria (PGPR): Current and future prospects for development of sustainable agriculture. *Journal of Microbial and Biochemical Technology, 7*, 96–102.

Hammad, H. M., et al. (2017). Maize plant nitrogen uptake dynamics at limited irrigation water and nitrogen. *Environmental Science & Pollution Research, 24*(3), 2549–2557.

Hammad, H. M., et al. (2018). Offsetting land degradation through nitrogen and water management during maize cultivation under arid conditions. *Land Degradation and Development, 29*(5), 1366–1375.

Hammad, H. M., et al. (2019). Environmental factors affecting the frequency of road traffic accidents: A case study of sub-urban area of Pakistan. *Environmental Science and Pollution Research, 26*(12), 11674–11685.

Hammad, H. M., et al. (2020). Comparative effects of organic and inorganic fertilizers on soil organic carbon and wheat productivity under arid region. *Communications in Soil Science and Plant Analysis, 51*(10), 1406–1422.

Haque, M. I., Singh, P. K., Ghuge, S., Kumar, A., Rai, A. C., Kumar, A., & Modi, A. (2022). A general introduction to and background of plant tissue culture: Past, current, and future aspects. In *Advances in plant tissue culture* (pp. 1–30). Elsevier.

Hashmi, M. Z., et al. (2018). PCB118-induced cell proliferation mediated by oxidative stress and MAPK signaling pathway in HELF cells. *Dose-Response, 16*, 1–8.

Hashmi, M. Z., et al. (2020). Arsenic distribution and metabolism genes abundance in Paddy soils from Punjab and Sindh provinces, Pakistan. *Arabian Journal of Geosciences, 13*, 526.

He, J., Zhang, Q., Li, M., Ming, T., Feng, J., Peng, H., et al. (2023). Evaluating the influence of gradient applied voltages on electro-enhanced sequence batch reactor treating aniline wastewater: System performance, microbial community and functional genes. *Journal of Cleaner Production, 389*, 136077.

Hussain, S., et al. (2020a). Study of land use/land cover changes using RS and GIS: A case study of Multan district, Pakistan. *Environmental Monitoring and Assessment, 192*, 2.

Hussain, S., et al. (2020b). Using GIS tools to detect the land use/land cover changes during forty years in Lodhran District of Pakistan. *Environmental Science and Pollution Research, 27*, 39676–39692.

Imran, M., et al. (2019). Kinetic and equilibrium studies for cadmium biosorption from contaminated water using Cassia fistula biomass. *International journal of Environmental Science and Technology, 16*(7), 3099–3108.

Jabran, K., et al. (2017a). Growth and physiology of basmati rice under conventional and water-saving production systems. *Archives of Agronomy and Soil Science, 63*(10), 1465–1476.

Jabran, K., et al. (2017b). Water-saving technologies affect the grain characteristics and recovery of fine-grain rice cultivars. *Environmental Science and Pollution Research, 24*, 12971–12981.

Jeyanthi, V., & Kanimozhi, S. J. (2018). Plant growth promoting rhizobacteria (PGPR)-prospective and mechanisms: A review. *Journal of Pure and Applied Microbiology, 12*, 733–749.

Kamaruddin, S. A., Liong, C.-Y., Abdul-Rahman, A., Nasir, N. A. N. M., & Zakarya, I. A. (2023). Artificial neural network models for wastewater treatment plant. *AIP Conference Proceedings, 2500*, 020040.

Kauser, H., & Khwairakpam, M. (2022). Organic waste management by two-stage composting process to decrease the time required for vermicomposting. *Environmental Technology & Innovation, 25*, 102193.

Khan, N., & Bano, A. (2019). Exopolysaccharide producing rhizobacteria and their impact on growth and drought tolerance of wheat grown under rainfed conditions. *PLoS One, 14*, e0222302.

Khan, N., et al. (2018). Interaction between PGPR and PGR for water conservation and plant growth under drought condition. *Biologia, 73*(11), 1083–1098.

Khan, N., Bano, A., Rahman, M. A., Guo, J., Kang, Z., & Babar, M. A. (2019). Comparative physiological and metabolic analysis reveals a complex mechanism involved in drought tolerance in chickpea (*Cicer arietinum* L.) induced by PGPR and PGRs. *Scientific Reports, 9*, 2097.

Kolosovska, T., & Bauer, S. K. (2022). Evaluation of grit properties at a medium-capacity wastewater treatment plant: A case study. *Resources, Conservation & Recycling Advances, 14*, 200080.

Kumar, M., Giri, V. P., Pandey, S., Gupta, A., Patel, M. K., Bajpai, A. B., et al. (2021). Plant-growth-promoting rhizobacteria emerging as an effective bioinoculant to improve the growth, production, and stress tolerance of vegetable crops. *International Journal of Molecular Sciences, 22*, 12245.

Kumari, P., Meena, M., & Upadhyay, R. (2018). Characterization of plant growth promoting rhizobacteria (PGPR) isolated from the rhizosphere of *Vigna radiata* (mung bean). *Biocatalysis and Agricultural Biotechnology, 16*, 155–162.

Kumari, A., Maurya, N. S., Kumar, A., Yadav, R. K., & Kumar, A. (2023). Options for the disposal and reuse of wastewater sludge, associated benefit, and environmental risk. In *Sewage management*. IntechOpen.

Latif, M., Bukhari, S. A. H., Alrajhi, A. A., Alotaibi, F. S., Ahmad, M., Shahzad, A. N., et al. (2022). Inducing drought tolerance in wheat through exopolysaccharide-producing rhizobacteria. *Agronomy, 12*, 1140.

Leontidou, K., Genitsaris, S., Papadopoulou, A., Kamou, N., Bosmali, I., Matsi, T., et al. (2020). Plant growth promoting rhizobacteria isolated from halophytes and drought-tolerant plants: Genomic characterisation and exploration of phyto-beneficial traits. *Scientific Reports, 10*, 14857.

Mahmood, F., et al. (2017). Economic and environmental impacts of introducing grain legumes in farming systems of Midi-Pyrénées region (France): A simulation approach. *International Journal of Plant Production, 1*(1), 65–87.

Manga, M., Muoghalu, C., Camargo-Valero, M. A., & Evans, B. E. (2023). Effect of turning frequency on the survival of fecal indicator microorganisms during aerobic composting of fecal sludge with sawdust. *International Journal of Environmental Research and Public Health, 20*, 2668.

Masood, N., et al. (2013). Whitefly (Aleurolobus barodensis Mask) spatio-temporal trends in semi-arid agro-ecological zones of Pakistan. *Pakistan Journal of Agricultural Sciences, 50*(1), 69–74.

Mellidou, I., & Karamanoli, K. (2022). Unlocking PGPR-mediated abiotic stress tolerance: What lies beneath. *Frontiers in Sustainable Food Systems*, 6. https://doi.org/10.3389/fsufs.2022.832896

Mirza, N., et al. (2017). Constitutional tolerance and chlorophyll fluorescence of Boehmeria nivea L in response to the antimony (Sb) and arsenic (As) co-contamination. *Toxicological and Environmental Chemistry, 99*(2), 265–272.

Mubeen, M., et al. (2016). Application of CSM-CERES-maize model in optimizing irrigated conditions. *Outlook on Agriculture, 45*(3), 01–12.

Mubeen, M., et al. (2020). Evaluating the climate change impact on crop water requirement of cotton-wheat in semi-arid conditions using DSSAT model. *Journal of Water and Climate Change, 11*(4), 1661–1675.

Munaweera, T., Jayawardana, N., Rajaratnam, R., Dissanayake, N. J. A., & Security, F. (2022). Modern plant biotechnology as a strategy in addressing climate change and attaining food security. *Agriculture & Food Security, 11*(1), 1–28.

Myo, E. M., Ge, B., Ma, J., Cui, H., Liu, B., Shi, L., et al. (2019). Indole-3-acetic acid production by *Streptomyces fradiae* NKZ-259 and its formulation to enhance plant growth. *BMC Microbiology, 19*, 1–14.

Naqvi, R. Z., Siddiqui, H. A., Mahmood, M. A., Najeebullah, S., Ehsan, A., Azhar, M., et al. (2022). Smart breeding approaches in post-genomics era for developing climate-resilient food crops. *Frontiers in Plant Science, 13*, 972164.

Naseem, H., & Bano, A. (2014). Role of plant growth-promoting rhizobacteria and their exopolysaccharide in drought tolerance of maize. *Journal of Plant Interactions, 9*, 689–701.

Naseem, H., Ahsan, M., Shahid, M. A., & Khan, N. (2018). Exopolysaccharides producing rhizobacteria and their role in plant growth and drought tolerance. *Journal of Basic Microbiology, 58*, 1009–1022.

Nasim, W., et al. (2012). Effect of nitrogen on growth and yield of sunflower under semiarid conditions of Pakistan. *Pakistan Journal of Botany, 44*(2), 639–648.

Nasim, W., et al. (2016a). Correlation studies on nitrogen for sunflower crop across the agroclimatic variability. *Environmental Science and Pollution Research, 23*(4), 3658–3670.

Nasim, W., et al. (2016b). Evaluation of the OILCROP-SUN model for sunflower hybrids under different agro-meteorological conditions of Punjab-Pakistan. *Field Crops Research, 188*, 17–30.

Nasim, W., et al. (2016c). Modelling climate change impacts and adaptation strategies for sunflower in Punjab-Pakistan. *Outlook on Agriculture, 45*(1), 39–45.

Nasim, W., et al. (2017). Response of sunflower hybrids to nitrogen application grown under different agro-environments. *Journal of Plant Nutrition, 40*(1), 82–92.

Nazir, N., Kamili, A. N., & Shah, D. (2018). Mechanism of plant growth promoting rhizobacteria (PGPR) in enhancing plant growth: A review. *International Journal of Management, Technology And Engineering, 8*, 709–721.

Niu, X., Song, L., Xiao, Y., & Ge, W. (2018). Drought-tolerant plant growth-promoting rhizobacteria associated with foxtail millet in a semi-arid agroecosystem and their potential in alleviating drought stress. *Frontiers in Microbiology, 8*, 2580.

Noor, A., Kutty, S. R. M., Baloo, L., Affam, A. C., Jagaba, A. H., Ghaleb, A. A. S., et al. (2023). Parametric optimization of additive manufactured biocarrier submerged in sequencing batch reactor for domestic wastewater treatment. *Heliyon, 9*, e14840.

Perveen, K., Alfagham, A. T., Debnath, S., Bukhari, N. A., Wei, D.-Q., Alshaikh, N. A., & Alwadai, A. S. (2023). Enriching drought resistance in *Solanum lycopersicum* using Abscisic acid as drought enhancer derived from *Lygodium japonicum*: A new-fangled computational approach. *Frontiers in Plant Science, 14*, 1106857.

Rahman, M. M., Nguyen, R., & Lu, L. (2022). Multi-level impacts of climate change and supply disruption events on a potato supply chain: An agent-based modeling approach. *Agricultural Systems, 201*, 103469.

Rasool, A., et al. (2016). Arsanic and heavy metal contaminations in the tube well water of Punjab, Pakistan and risk assessment: A case study. *Ecological Engineering, 95*, 90–100.

Rasool, A., et al. (2018). A review of global outlook on fluoride contamination in groundwater with prominence on the Pakistan current situation. *Environmental Geochemistry and Health, 40*(4), 1265–1281.

Raza, A., Tabassum, J., Fakhar, A. Z., Sharif, R., Chen, H., Zhang, C., et al. (2022). Smart reprograming of plants against salinity stress using modern biotechnological tools. *Critical Reviews in Biotechnology, 43*, 1–28.

Razali, M. M. A. M., Hamzah, N., Daud, N. M., & Bakar, A. A. A. (2022). A preliminary study on the utilization of rainwater harvesting system for non-potable usage. *AIP Conference Proceedings, 2532*, 060002.

Rosier, A., Medeiros, F. H., & Bais, H. P. (2018). Defining plant growth promoting rhizobacteria molecular and biochemical networks in beneficial plant-microbe interactions. *Plant and Soil, 428*, 35–55.

Ruiz-Munoz, A., Siles, J., Márquez, P., Toledo, M., Gutiérrez, M., & Martín, M. (2023). Odor emission assessment of different WWTPs with extended aeration activated sludge and rotating biological contactor technologies in the province of Cordoba (Spain). *Journal of Environmental Management, 326*, 116741.

Sagar, A., Sayyed, R., Ramteke, P., Sharma, S., Marraiki, N., Elgorban, A. M., et al. (2020). ACC deaminase and antioxidant enzymes producing halophilic Enterobacter sp. PR14 promotes the growth of rice and millets under salinity stress. *Physiology and Molecular Biology of Plants, 26*, 1847–1854.

Sardar, A., et al. (2020). Risk assessment of heavy metal(loid)s via Spinacia oleracea ingestion after sewage water irrigation practices in Vehari District. *Environmental Science and Pollution Research, 27*, 39841–39851.

Saud, S., et al. (2017). Effects of nitrogen supply on water stress and recovery mechanisms in Kentucky bluegrass plants. *Frontier in Plant Sciences, 8*(983), 01–18.

Shahid, M., et al. (2014). EDTA-enhanced phytoremediation of heavy metals: A review. *Soil and Sediments Contamination, 23*, 389–416.

Shahzad, H., et al. (2020). An integrated use of GIS, geostatistical and map overlay techniques for spatio-temporal variability analysis of groundwater quality and level in the Punjab Province of Pakistan, South Asia. *Water, 12*, 3555.

Shakeel, M., et al. (2017). Environment polluting conventional chemical control compared to an environmentally friendly IPM approach for control of diamondback moth, Plutella xylostella (L), in China: A review. *Environmental Science and Pollution Research, 24*, 14537–14550.

Shekhawat, S. S., Saini, P., Upadhyay, A., Pareek, N., Arora, S., Gupta, A. B., et al. (2023). Treatment of clinical laboratory sewage using a decentralized treatment unit and risk reduction for its reuse in irrigation using hybrid disinfection. *Journal of Environmental Management, 344*, 118684.

Singh, A., Mehrotra, R., Rajput, V. D., Dmitriev, P., Singh, A. K., Kumar, P., et al. (2022). Geoinformatics, artificial intelligence, sensor technology, big data: Emerging modern tools for sustainable agriculture (pp. 295–313).

Slavik, I., Oliveira, K. R., Cheung, P. B., & Uhl, W. (2020). Water quality aspects related to domestic drinking water storage tanks and consideration in current standards and guidelines throughout the world—A review. *Journal of Water and Health, 18*(4), 439–463.

Srivastava, A., Gupta, B., Majumder, A., Gupta, A. K., & Nimbhorkar, S. K. (2021). A comprehensive review on the synthesis, performance, modifications, and regeneration of activated carbon for the adsorptive removal of various water pollutants. *Journal of Environmental Chemical Engineering, 9*, 106177.

Thompson, M., Moussavi, S., Li, S., Barutha, P., & Dvorak, B. (2022). Environmental Life Cycle Assessment of small water resource recovery facilities: Comparison of mechanical and lagoon systems. *Water Research, 215*, 118234.

Tiwari, A. K., Mishra, H., Nishad, D. C., & Pandey, A. (2023). Sustainable water management in agriculture: Irrigation techniques and water conservation (Vol. 53).

Varshney, R. K., Barmukh, R., Roorkiwal, M., Qi, Y., Kholova, J., Tuberosa, R., et al. (2021). Breeding custom-designed crops for improved drought adaptation. *Advanced Genetics (Hoboken, NJ), 2*(3), e202100017.

Verma, M., Mishra, J., & Arora, N. K. (2019). Plant growth-promoting rhizobacteria: Diversity and applications. In *Environmental biotechnology: For sustainable future* (pp. 129–173). Springer.

Wang, X. (2022). Managing land carrying capacity: Key to achieving sustainable production systems for food security. *Land, 11*(4), 484.

Wang, K., Zhou, C., Zhou, H., Jiang, M., Chen, G., Wang, C., et al. (2023). Comparison on biological nutrient removal and microbial community between full-scale anaerobic/anoxic/aerobic process and its upgrading processes. *Bioresource Technology, 374*, 128757.

Waqas, S., Harun, N. Y., Sambudi, N. S., Bilad, M. R., Abioye, K. J., Ali, A., et al. (2023). A review of rotating biological contactors for wastewater treatment. *Water, 15*, 1913.

Xiong, R., Liu, S., Considine, M. J., Siddique, K. H., Lam, H. M., & Chen, Y. J. P. P. (2021). Root system architecture, physiological and transcriptional traits of soybean (*Glycine max* L.) in response to water deficit: A review. *Physiologia Plantarum, 172*(2), 405–418.

Xu, C., Feng, Y., Li, H., Yang, Y., & Wu, R. (2022). Research progress of phosphorus adsorption by attapulgite and its prospect as a filler of constructed wetlands to enhance phosphorus removal from mariculture wastewater. *Journal of Environmental Chemical Engineering, 10*, 108748.

Yao, H.-Y., Guo, H., Shen, F., Li, T., Show, D.-Y., Ling, M., et al. (2023). Anaerobic-aerobic treatment of high-strength and recalcitrant textile dyeing effluents. *Bioresource Technology, 379*, 129060.

Zagklis, D. P., & Bampos, G. (2022). Tertiary wastewater treatment technologies: A review of technical, economic, and life cycle aspects. *Processes, 10*, 2304.

Zaman, G., et al. (2020). Utilization of compost from bio-municipal solid waste improves saline-sodic soils and crop productivity in rice-wheat. *Compost Science & Utilization, 28*, 16.

Zamin, M., et al. (2020). Developing the first halophytic turfgrasses for the urban landscape from native Arabian desert grass. *Environmental Science and Pollution Research, 27*, 39702–39716.

Zhang, K., Xue, M., Qin, F., He, Y., & Zhou, Y. (2023). Natural polymorphisms in ZmIRX15A affect water-use efficiency by modulating stomatal density in maize. *Plant Biotechnology Journal, 21*, 2560–2573.

Zia, Z., et al. (2017). Effect of water management and silicon on germination, growth, phosphorus and arsenic uptake in rice. *Ecotoxicology and Environmental Safety, 144*, 11–18.

Opportunities of Big Data in Agricultural Water Management

Khadija Shabbir, Mazhar Ali, Muhammad Mubeen, Qurat-ul-ain Maken, Saeed Ahmad Qaisrani, Fahd Rasul, and Ashfaq Ahmad

Abstract This study delves into the transformative potential of big data analytics (BDA) and artificial intelligence (AI) in revolutionizing water resource management (WRM). By harnessing the power of machine learning, BDA supports sustainability goals, mitigates environmental damage from industrialization, and optimizes resource efficiency exemplified by water utilities using analytics to enhance operational resilience and reduce strain. This transformative tool not only addresses current global challenges but also paves the way for future advancements in ecological and industrial systems. Traditional statistical methods, data mining, and machine learning techniques are employed in groundwater research. Statistical methods, though useful for structured data and small samples, are limited in handling large, heterogeneous, and noisy data. Data mining combines machine learning and statistical techniques to discover patterns and insights in large datasets. Machine learning, a subset of artificial intelligence, encompasses supervised, unsupervised, and rein-

K. Shabbir · Qurat-ul-ain Maken
Department of Environmental Sciences, COMSATS University Islamabad, Vehari Campus, Vehari, Pakistan
e-mail: saeed.qaisrani@cuivehari.edu.pk

S. A. Qaisrani (✉)
Department of Environmental Sciences, COMSATS University Islamabad, Vehari Campus, Vehari, Pakistan

Department of Biotechnology, COMSATS University Islamabad, Vehari Campus, Vehari, Pakistan
e-mail: saeed.qaisrani@cuivehari.edu.pk

M. Ali · M. Mubeen (✉)
Department of Biotechnology, COMSATS University Islamabad, Vehari Campus, Vehari, Pakistan
e-mail: muhammadmubeen@cuivehari.edu.pk

F. Rasul
Agro-Climatology Lab, Department of Agronomy, University of Agriculture, Faisalabad, Pakistan

A. Ahmad
Ghazi University, Dera Ghazi Khan, Pakistan

M. Mubeen et al. (eds.), *Innovations in Agricultural Water Management*,
https://doi.org/10.1007/978-3-031-91883-4_18

forcement learning, and is applied in groundwater research for simulating groundwater flow, predicting water quality, and estimating recharge rates, offering a promising approach for analyzing complex, nonlinear relationships and making predictions. The future of water resource management (WRM) will involve integrating higher-resolution data, remote sensing, and Internet of Things (IoT) devices to enable more accurate and proactive management. The integration of AI, BDA, and IoT devices is proposed as a future direction for WRM, enabling more accurate and proactive management, improved forecasting, and enhanced decision-support systems. This research aims to inform the development of innovative, data-driven solutions for addressing global water challenges, promoting sustainable development, and ensuring water security for future generations.

Keywords Water management · Data analysis · Statistical methods · Data mining · Integration of AI · BDA · IoT devices

1 Introduction

"Big Data" refers to the collection of information gathered from daily interactions with digital goods and services, from smart devices (such as computers and mobile phones), multiple banking establishments, online transactions, multiple social media platforms, and multiple ground-and/or water-connected sensors (Malgaonkar et al., 2016).

2 Significance

Big data has been increasingly significant in our social, political, and economic life in the last few decades (Sun & Scanlon, 2019). Machines are now capable of managing millions, or even billions, of real-world synchronous occurrences each second and therefore demonstrate superior skills in automatic data analysis (Javed et al., 2017).

While the application of analysis of big data in other sectors is relatively new (Srinivasan, 2018), it has already had a substantial impact on how humans relate to the world via new technology in every business.

This includes a number of opportunities for job creation and fabrication, the provision of efficient energy and transportation services, and the preparation of disaster and climate change management plans. Businesses and international governments may now analyze massive datasets to make better-informed strategic decisions thanks to big data. Additionally, it is employed to fulfil human sustainability objectives (Sabagh et al., 2019; Farid et al., 2019a, 2019b; Tariq et al., 2008; Awais et al., 2018; Amin et al., 2018a, 2018b, 2018c; Hammad et al., 2018; Malgaonkar et al., 2016). Big data has the potential to be incredibly useful in the future, as well as a revolutionary improvement for the investigation of environmental issues, that encompasses whatever surrounding us. Living things in this ecosystem communicate with one another and with animate objects in a specific way to adapt to the circumstances in which they live, but many environmental disasters have occurred as a result of human industrialization, affecting life on Earth directly or indirectly. Big data can be utilized to mitigate this.

Digitalization and the current wave of new technology also appear to be helping the hydrocarbons, healthcare, and water sectors more than they did in the past. Water utilities may be able to function more efficiently by reducing load and stress by examining the most recent innovations in technology, software, and information analytics. These technologies allow utilities to obtain a greater grasp of their operations, interactions, and system behavior.

Data science and big data are steadily taking over the ability to get fresh data that supports improved supply chains, customer connections, dependability, and optimization. Water utilities receive data from a variety of industries via computerized maintenance management systems (CMMS), laboratory information management systems (LIMS), and flow data, electronic surveillance, dissolved oxygen (DO) evaluations, and air circulation, among other sources from Supervisory Control and Data Acquisition (SCADA) systems. A large portion of this substantial data collection—which comes from SCADA systems—has been in use for some years.

Nevertheless, both the quantity and quality of the data are diminished by the data collection process. Computer systems contain data silos that don't always communicate with one another. Operators of water and wastewater treatment plants may now comprehend and utilize this hitherto unutilized data to maximize plant performance and dependability thanks to the Internet Age's ability to integrate it in a unified and meaningful fashion. We can transform all that data into clear, actionable information that makes us more proactive and capable of making decisions thanks to big data projects and new data management solutions. According to a study by Zare and Koch (2018), groundwater level forecast is more affected by aquifer water extraction when climate change is taken into account. Neural networks and the adaptive neuro-fuzzy inference system (ANFIS) were used to monitor these phenomena (Banadkooki et al., 2020; Nasim et al., 2018; Awais et al., 2017; Amin et al., 2017a, 2017b; Farid et al., 2017; Nasim et al., 2016). In a further study, a model that predicted when the next water pipeline breakdown would occur was created using an extreme learning machine (ELM) neural network that was fed data from over 9500 pipe failures in Canada. This prevented additional water loss across the pipeline network. One of the key issues raised was the water pipeline networks; specifically, it was noted that maintaining and fixing these networks had become quite expensive in various cities across the globe (Sattar et al., 2019).

3 Big Data in Water Quality Measurement

For living things, water is the most essential natural resource and, depending on the quality of the water, can cause either health or illness. Modern water quality monitoring systems currently require more sophisticated techniques due to the rising levels of water contamination in reservoirs. There are two stages of using wireless monitoring systems for water: gathering and storing data, and distributing and displaying that data. Initially, the procedure gathers data from multiple sensors (temperature, pH, DO, etc.) and transmits it to the base station.

Furthermore, because it provides the administration with real-time data, they are able to detect changes in the water quality parameter early on and provide sufficient time to put risk- and damage-reduction measures into place. Following this gearbox, the system automatically switches off and waits for a predetermined amount of time in order to conserve energy. This prolongs the life of the sensor or sensors and lowers battery maintenance expenses. When this process is used effectively, it may monitor water quality parameters precisely and cover a vast region with reduced labor and costs for physical and chemical features surveillance (Demetillo et al., 2019).

4 Data Mining

Data mining is the process of continuously and methodically gathering large amounts of data, storing it, and then using computer-based techniques to analyze it. According to Che et al. (2013), data mining "uncovers interesting patterns and relationships hidden in a large volume of raw data, and the results tapped out may help make valuable predictions or future observations in the real world." We refer to this strategy as predictive analytics. It seeks to direct or even regulate behavior (Gandomi & Haider, 2015).

These analytics, however, call for additional algorithms that can classify and forecast using regularities, patterns, and linkages in addition to networks that are not immediately apparent. Thus, it is essential to comprehend the types of data that have been mined (such as sources and their constraints) as well as the methods used by the algorithms to analyze them. For this reason, diagnostic analytics (Why it occurs?) and descriptive analytics (What occurred?) are important (Banerjee et al., 2013).

5 Challenges Arising to Water Resource Management

5.1 Urbanization and Rapid Growth

The world's population is expected to reach nine billion people by 2050, with most of the expansion taking place in urban areas. In wealthy nations, population growth has stabilized, and there is already a sizable urban population. In developing nations, however, the urban population is growing. According to Tucci, 2017, urbanization intensifies competition for limited natural resources (land, water, and air) that are necessary for human habitation, production, and leisure.

5.2 Using AI and Big Data Analytics to Boost the Management of Water Resources

The advancement of technology has presented water resource management (WRM) with more and more challenging challenges. For the purpose of maximizing water allocation, utilization, and conservation, Chen et al. (2020b, 2020a), has demonstrated that the integration of artificial intelligence (AI) with BDA is a revolutionary strategy. The following subsection describes how proactive resource management, informed decision-making, and long-term results are made possible by the combination of AI and big data analytics (BDA), which can enhance WRM.

5.2.1 Data Insights Help Inform Decision-Making

WRM professionals now have access to a multitude of data-driven insights thanks to AI and BDA. These tools offer a comprehensive picture of water mechanics by merging data from several sources, including weather stations, sensors, and remote sensing satellites. Decision-makers can gain real-time access to information on availability of water, quality, consumption patterns, and ecological condition by utilizing modern data analytics. This type of informed decision-making is required for the development of policies that effectively address pollution, dearth of water, and climate-driven uncertainty (Du et al., 2022).

5.2.2 The Utilization of Predictive Algorithms for Effective Planning

AI-enabled predictive modelling is the foundation of contemporary WRM. This type of modelling estimates the amount of water accessible under various climatic conditions and actions by humans by simulating a variety of scenarios with historical and real-time data. The creation of adaptable flood control, drought mitigation, and ideal reservoir operation plans is facilitated by predictive models. Planners can proactively adjust to changing situations by quantifying the possible effects of different options, which minimizes risks and maximizes resources (Yereseme et al., 2022).

5.2.3 The Efficacy of Water Distribution and Conservation

In order to meet the demands of industry, metropolitan areas, agriculture, and habitats, water allocation must be optimized. AI-powered algorithms fed real-time information may modify water distribution in response to modifying circumstances. Furthermore, by detecting breaches in distribution networks and accurately applying irrigation in agriculture, such methods help identify and address water-use inadequacies. This improves the overall performance of water systems while also conserving water (Bajaj et al., 2022).

5.2.4 Early Warning Systems for Extreme Events

Drastic weather events, such as droughts and floods, can have disastrous consequences for water supply. Early warning systems driven by artificial intelligence (AI) analyze atmospheric and hydraulic information to identify potential disasters (Chen et al., 2022). These strategies provide adequate warning time for evacuating vulnerable areas, implementing mitigation measures, and allocating resources for disaster response. Using real-time data, those in charge may improve public security while also reducing the impact of catastrophic occurrences on infrastructure and water supplies.

5.2.5 Ecology Oversight and Rebuilding

Artificial intelligence and BDA enhance ecosystem-based WRM approaches. These technologies use remote sensing and data integration approaches to evaluate both terrestrial and aquatic environments. These insights guide restoration operations by identifying trouble spots, evaluating the impact of restoration efforts on water flow and quality, and tracking the projects' long-term results. This integrated strategy promotes ecosystem welfare and efficient handling of water (Jinyue et al., 2022).

5.2.6 Collaboration with Stakeholders and Transparency

Engagement of stakeholders is essential to the success of WRM. By providing readily accessible data platforms, AI and BDA contribute to inclusive and transparent decision-making processes. These platforms allow stakeholders, such as governments, corporations, and communities, to collaborate on resource management and get real-time data. This increases self-worth, reduces conflict, and encourages participation in conservation endeavors.

5.2.7 Strategies for Adaptation in the Context of the Changing Climate

Climate change makes WRM more complicated. Climate models driven by artificial intelligence (AI) forecast changes in temperature, precipitation, and water availability. By integrating these estimations within management tactics, decision-makers can develop flexible and resilient plans to changing circumstances. The iterative nature of AI allows methods to be continuously improved as fresh data becomes available (Himeur et al., 2023).

5.2.8 Regulation Observance and Enforcement

Artificial intelligence (AI) and BDA improve regulatory control by automation of compliance inspection. These devices have the ability to quickly identify breaches such as unlawful water use and environmental accidents. The algorithms used in machine learning examine large datasets to find patterns of noncompliance, enabling regulators to take swift enforcement action and discourage similar infractions in the future.

5.2.9 Controlling Drought

AI and BDA have the potential to improve drought management by offering immediate information on availability of water and usage. The best way to distribute water resources during droughts, for example, can be determined by using AI algorithms to analyze data on water availability and usage (Kikon & Deka, 2022).

5.2.10 Flood Management

AI and BDA can help manage floods by providing real-time data on the levels of water and velocity of water. For example, AI systems can be used to analyze data from water level sensors and satellite photos in order to predict and respond to potential floods (Yereseme et al., 2022).

These are just a few examples of how AI and BDA are being used to manage water resources. There are many potential uses for AI and BDA in this area, and new and creative methods to WRM are becoming more and more necessary to address the issues in this field and promote sustainable and efficient water management techniques (Sohn et al., 2023). Water resource management will be made more sustainable and effective with the usage of AI and BDA in WRM (Matin Saddiqi et al., 2023).

6 Methods in Big Data Analytics

The true value of large data is realized when it is transformed into useful information. Large-scale data mining uses a broad range of advanced computational, statistically significant, numerical, and graphical approaches to transform data into meaningful information (Malgaonkar et al., 2016). We go into greater detail on large-scale data analytics in this chapter, emphasizing the details that are critical to turning big data about groundwater into practical knowledge.

6.1 Big Data Analytics: What Is It?

Chen and Zhang (2014) defines big data analytics as sophisticated analytics applied to large amounts of information. Machine learning is one of many instruments and methods used in big data analytics that have been around for a long time (Watson, 2014). The importance of these sophisticated analytical methods has just now become apparent with the rise in big data. Advanced analytical techniques work effectively with very large, diverse datasets and require less data for preliminary processing than standard analytics procedures (Tsai et al., 2015).

Machine learning, for instance, can work on data that is either structured or unstructured, whereas conventional statistics is only effective on organized information. One of the main distinctions among conventional analytics and big data analytics involves the sites needed for processing; big data typically requires parallel processing techniques to analyze these large datasets, while traditional analytics techniques are typically meant to run on just one device (Tsai et al., 2015).

Only when data size is decreased using approaches like principal component analysis or genetic algorithms do traditional analytical methods work well. Furthermore, these techniques are not the best for processing in simultaneous manner. However, we can process and analyze huge and heterogeneous groundwater information quickly by combining big data analytics via conventional methods. This combination helps groundwater managers make better decisions by transforming large amounts of data into useful knowledge.

6.1.1 Statistical Methods

Statistical methods refer to the conventional data analysis techniques that have long been central to traditional empirical research (Chen et al., 2012). These approaches are deeply grounded in the disciplines of statistical and mathematical sciences (Sun & Huo, 2019; Sun & Scanlon, 2019). These methods are intended to carry out tasks such as identifying associations between data points, segmenting and grouping data through clustering, classifying information, detecting anomalies, and performing regression and predictive analyses on structured datasets (Chen et al., 2012).

Multivariate regression analysis, for instance, makes it possible to forecast the results for factors that are dependent by quantifying the causal links between different variables. These methods, like geostatistics, are still often used today to simulate groundwater processes and harvest information from groundwater samples. For instance, statistical methods are applied in the design of groundwater monitoring systems, assessing groundwater quality, and simulating groundwater flow. However, they do have limitations. Statistical methods are not fully optimized to manage large volumes of heterogeneous, high-dimensional, and noisy data (Gandomi & Haider, 2015). Traditional statistical techniques are better suited for working with samples of population data, which are then used to make inferences about the entire population based on statistical significance (Gandomi & Haider, 2015). Big data analytics, on the other hand, analyzes most, if not all, of the information found in a population, making the idea of relevance in statistics less applicable. Furthermore, huge data frequently requires parallel processing environments, which make it difficult to apply classic statistical approaches (Chen & Zhang, 2014). Nevertheless, these techniques can still be useful for handling conventionally organized groundwater data when included into big data analytics systems.

6.2 Data Mining

According to Sun and Huo (2019), data mining is the process of using big data statistical instruments to massive datasets in order to find new insights, such as patterns, correlations between variables, sets of related data points, or outcome predictions. Numerous methods related to machine learning and statistical analysis are applied. Data mining was utilized prior to big data became popular, despite being frequently linked to enormous datasets (Hariri et al., 2019). Some conventional analytical techniques are currently being modified to deal with processing large amounts of data. For example, conventional clustering methods such as K-means are now modified to handle big datasets by dividing them into smaller samples that may be handled concurrently by several processors. The complete dataset is then represented by combining the outcomes of individual samples (Chen & Zhang, 2014). The Clustering Large Applications Based on Randomized Search (CLARANS) method and the Clustering Large Applications (CLARA) technique are typical algorithms for this strategy (Chen & Zhang, 2014). According to Lin et al. (2020), data mining provides not just for data that is structured; it can also be used for written words, pictures, audio, and video files, and other analytics. Data mining is an essential stage within a big data analytics project since it is a process that turns unprocessed data into useful information.

6.3 AI and Machine Learning Method

Among the most commonly utilized big data analysis approaches in the scientific field is machine learning, which is an aspect of artificial intelligence (AI). Machine learning is built on autonomous learning methods, which are central for numerous AI applications. These approaches are effective for analyzing massive, multidimensional, and complicated nonlinear networks (Srivastava et al., 2013). For example, the interactions between various components of natural hydrological systems are frequently nonlinear. When modeled using typical statistical analysis, these systems can provide oversimplified and erroneous results. In such circumstances, machine learning techniques are more applicable. Machine learning is often classified into three distinct groups: reinforcement learning, supervised learning, and unsupervised learning (Jordan & Mitchell, 2015). Supervised machine learning is based on classified information that is utilized for assessment throughout the training phase (Schrider & Kern, 2018). The data with labels is made up of data points that have been associated with predefined properties, allowing the algorithm to learn from them. The model determines predicted outputs via a reiterative back-propagation learning procedure. During this procedure, the method or guidelines are refined to most effectively anticipate the labeled outcome using the input data. Using unsupervised machine learning, the model operates on unlabeled data, which means there are no predetermined outputs to guide the training process. Instead, the algorithm clusters data by identifying hidden patterns and groups. Learning through reinforcement techniques, on the other hand, use labeled data which falls somewhere among supervised and unsupervised methods. Rather than providing explicit correct answers, this data provides feedback on whether a particular action is suitable. The system gets signals that reward right behaviors and punish wrong ones, which aid in model training (Jordan & Mitchell, 2015).

Machine learning is typically utilized for four main tasks: regression, categorizing, aggregation, and interaction (Ali et al., 2016). The supervised machine learning techniques are typically used for regression tasks (e.g., linear regression) as well as classification tasks (e.g., k-nearest neighbor). Unsupervised machine learning is concerned with clustering (k-means clustering) or association (Apriori method) (Alashwal et al., 2019). Reinforcement training can be very useful for determining the appropriate actions within an environment by maximizing rewards. These machine learning algorithms are information-based, requiring a significant quantity of evidence to get reliable results. Their advantage is that they rely on information from the real world rather than generating predetermined conclusions about the framework. When combined with an analytical comprehension of procedures, these methods can reveal previously unknown relationships.

Historically, groundwater dynamics have been simulated primarily using physics-based numerical simulations and classical statistics. To produce reliable simulations, these models require considerable prior knowledge of the aquifer as well as detailed data on a variety of aquifer parameters. This intricacy makes it difficult to create numerical models. A study contrasting machine learning approaches and numerical models for simulating groundwater phenomena in the Heihe River Basin in Northwestern China found that machine learning outperformed numerical models (Chen et al., 2020b, 2020a). As a result, AI-based groundwater models represent a feasible replacement to classic physics-based processes.

While the use using machine learning for groundwater investigations is still in its early stages, some case studies show that it has great potential. For example, multiple machine learning algorithms have been applied to model and predict groundwater levels (Gong et al., 2018). This program is very useful for estimating future groundwater levels, which can aid in handling groundwater. Likewise, quality of groundwater monitoring has been accomplished by multivariate cluster analysis (Alahmadi, 2019). Ransom et al. (2017) demonstrated how to use an enhanced regression tree structure to simulate and foresee levels of nitrate in the Central Valley aquifer in California, USA. Huang et al. (2019) studied the use of different machine learning methods to estimate replenishment of groundwater. Furthermore, data mining approaches have been used for locating water bodies at the surface through Landsat snaps, demonstrating the utility of visual analytics in hydraulic studies (Isikdogan et al., 2017). Machine learning and artificial intelligence have the potential for precisely analyzing and foreseeing situations in the real world, while also recommending best actions to achieve desired results. This expertise could be critical in generating data based on information solutions for better handling of groundwater.

6.4 Analysis of Uncertainty

Managing uncertainty ranks as one of the most difficult aspects of collecting and evaluating large amounts of data. Big data's massive, highly varied, and multifaceted nature frequently leads to uncertainty. Such qualities present problems with informal, incoherent, incomplete, and turbulent information into the data analytics process. The extraction of data from numerous sources in different ways complicates the task of maintaining data quality. Stringent data quality standards are necessary because, for example, data derived from online posting cannot be generated using robust scientific approaches (Conrado et al., 2016; Bendler et al., 2014). The accuracy of the results can be reduced if information integrity and ambiguity are not addressed early in the analysis process. This can have compounding consequences throughout the enormous data chain of value (Hariri et al., 2019).

Furthermore, improper applications may arise from insufficient instruction and comprehension of the subtleties in different big data analytical algorithms (Chung et al., 2018). This emphasizes how crucial it is to use the right methods for large data analysis, which calls for a solid foundation in statistics. Tasks like outlier detection, duplication reduction, addressing missing data, and dataset unification are traditional approaches to managing uncertainty (Chung et al., 2018). It can still be difficult to identify intrinsic flaws in massive data, even with careful data preparation. Wang and He (2016) investigated a number of methods for reducing statistical learning errors in big data analytics. These tactics entail incorporating methods into the analytics process, including theory of probability, Bayesian concept, Shannon's entropy, rough set theory, and fuzzy set theory. It is important to proceed cautiously when integrating remote sensing along with other sensor-based evaluations via the standpoint of groundwater statistics. According to other studies, these sources frequently have uncertainties associated with different types of remote-sensing mistakes (Loew et al., 2017). Furthermore, monitoring mistakes, insufficient data collection, and insufficient information can all contribute to the unpredictability of local-scale factors like the levels of groundwater. Implementing strategies to lower data uncertainty and make sure the techniques are suitable for the kind of data being examined are essential.

7 Frameworks and Tools for Big Data Analytics for Geospatial Data

Groundwater scientists who want to use enormous amounts of data for groundwater management may quickly feel overloaded by the wide range of technologies, approaches, and strategies available. By combining different hardware and software modules according to a specific big data processing architecture, big data servers are enterprise-scale alternatives created to meet unique industry needs. Through a series of procedures, including gathering data, saving it, preparation, analytics-driven transformation, and information sharing, these platforms seek to fully utilize the potential of big data (Cavanillas et al., 2016).

7.1 Challenges in Using Big Data for Groundwater Management Applications

Sivarajah et al. (2017) states that while professionals encounter several difficulties when attempting to apply big data analytics, these may be generally classified into three groups:

1. The attributes of massive amounts of data alone (quantity, velocity, diversity, etc.) are related to data challenges.
2. Deal with obstacles are related to data capture, cooperation, and modification, as well as how to choose the most appropriate framework for analysis and deliver outcomes.
3. Difficulties with administration include matters like confidentiality, privacy, regulation, confinement, and so forth.

8 Conclusions

A route to more efficient, flexible, and sustainable water management techniques is illuminated by the innovative convergence of AI and BDA. The most current and anticipated applications of these innovations were covered in this article, along with their revolutionary potential in tackling the complex and numerous issues surrounding water supplies. From the previously unheard-of-precision of flood predictions to the continuous tracking of water quality, AI and BDA have opened up an entirely novel arena of data-driven analysis.

The computing power of AI algorithms combined with the capacity to integrate large and diverse datasets from multiple sources has enabled those who manage water resources to go beyond conventional limits and explore the intricacies of water systems. The investigation illustrated the wide-ranging and complex uses of AI and BDA in predicting water supply, distributing resources optimally, and enhancing infrastructure upkeep.

Furthermore, these technologies bridge surveillance gaps, educate communities, and deliver earlier unobtainable data to decision-makers. The research into the most modern new routes demonstrated the importance of collaboration among different collaborators. Researchers, legislators, technologists, and the public must harness the power of new technologies while remaining vigilant about moral issues, security of data, and equal opportunity for benefits. On the contrary, the approach of combining AI and BDA into WRM is hampered by various hurdles. Experts must work together to tackle technical difficulties, data boundaries, and the need for cross-disciplinary skills. However, the rewards are significant: a durable water future that traverses the uncertain prospects of climate change, promotes thriving ecosystems, and provides equal accessibility to this valuable resource. To summarize, focusing on these creative avenues marks an overhaul in our approach to WRM. With AI and BDA as our consultants, we start on a journey of transformation that reimagines our awareness, plans, and behaviors for protecting one of the most vital assets for existence on Earth.

As we forge forth on this new path, we intend to leave a legacy of sustainable water management for people to come.

9 Future Directions

Because technology is always evolving, the combination of AI and BDA presents an array of chances to enhance WRM even more. These technologies' potential effects on solving intricate WRM problems increase as they advance. Several exciting potential directions for integrating AI and BDA into the WRM space are described in this part of the article. Higher-resolution temporal and spatial data will probably be used for subsequent AI and BDA applications in WRM. Reliable data on the availability of water, alterations to land use, and ecological condition can be gathered using techniques for remote sensing like satellites and drones. These datasets can be combined with state-of-the-art AI algorithms to provide those in charge with more accurate perspectives, enabling more rapid and focused responses. The spread of Internet of Things devices that can collect and send information in real time will be essential to WRM in coming years. These devices, which are integrated into urban areas, habitats, and water distribution systems, can consistently offer data on a variety of characteristics, including weather, outflow rates, and the quality of water.

AI systems can process these continuous data streams to produce instant observations, enabling proactive management and quick reactions to changing conditions. Modeling for forecasting extreme occurrences like floods, droughts, and water quality emergencies is bound to be improved by AI and BDA. These models can produce forecasts that are more accurate and dependable by combining advanced machine learning algorithms with improved data intake. This will decrease the impact on communities and water systems by making it possible for regulators to take defensive action. The incorporation of AI and BDA into systems that assist with decisions is probably going to be the main focus for upcoming technologies. A user-friendly interface that combines intricate data and modeled results into practical suggestions will be made available to decision-makers by these systems. Such tools will make insights based on data more accessible and enable even nontechnical people to make accurate WRM decisions.

The primary objective of future WRM will be on cooperative strategies that involve various communities and sectors. By offering dynamic visualization of data and outcome modeling tools, AI and BDA can support collaborative decision-making. With the use of these tools, stakeholders can explore many scenarios, analyze negotiations, and create flexible plans that take into account a range of objectives. WRM models will gradually integrate socioeconomic information from BDA and AI. Through the analysis of financial patterns, shifting demographics, and behaviors, these technologies can offer a thorough picture regarding water consumer demand, apply patterns, and societal implications. More fair and multifaceted management techniques will be promoted by this cohesive strategy. Future studies might focus on methods to make AI models easier to grasp. Making sure that decisions are simple to comprehend becomes crucial as AI algorithms get more complicated. Administrators will be able to trust and understand the logic behind AI-driven suggestions if AI developments become explicable. AI and BDA adaptation efforts for small and resource-constrained areas are gaining traction. In order to efficiently employ these technologies in areas with inadequate infrastructure, future paths will involve the creation of models that demand fewer computational resources and data inputs. Problems concerning data privacy, algorithmic prejudice, and equal availability of incentives need to be considered as AI and BDA become more incorporated into WRM. The subsequent efforts will include the creation of ethical guidelines and rules that guarantee the appropriate and inclusive use of such innovations.

References

Alahmadi, F. S. Groundwater quality categorization by unsupervised machine learning in groundwater quality categorization by unsupervised machine learning in Madinah, Western Kingdom of Saudi Arabia. In *Proceedings of the International Geoinformatics Conference*, Riyadh, Saudi, 10–14 February 2019.

Alashwal, H., El Halaby, M., Crouse, J. J., Abdalla, A., & Moustafa, A. A. (2019). The application of unsupervised clustering methods to Alzheimer's disease. *Frontiers in Computational Neuroscience, 13*, 1–9.

Ali, A., Qadir, J., ur Rasool, R., Sathiaseelan, A., Zwitter, A., & Crowcroft, J. (2016). Big data for development: Applications and techniques. *Big Data Analytics, 1*, 2.

Amin, A., et al. (2017a). Optimizing the phosphorus use in cotton by using CSM-CROPGRO-cotton model for semi-arid climate of Vehari-Punjab, Pakistan. *Environmental Science and Pollution Research, 24*, 5811–5823.

Amin, A., et al. (2017b). Comparison of future and base precipitation anomalies by SimCLIM statistical projection through ensemble approach in Pakistan. *Atmospheric Research, 294*, 214–225.

Amin, A., et al. (2018a). Regional climate assessment of precipitation and temperature in southern Punjab (Pakistan) using SimCLIM climate model for different temporal scales. *Theoretical and Applied Climatology, 131*(1–2), 121–131.

Amin, A., et al. (2018b). Evaluation and analysis of temperature for historical (1996-2015) and projected (2030-2060) climates in Pakistan using SimCLIM climate model: Ensemble application. *Atmospheric Research, 213*, 422–436.

Amin, A., et al. (2018c). Simulated CSM-CROPGRO-cotton yield under projected future climate by SimCLIM for southern Punjab, Pakistan. *Agricultural Systems, 167*, 213–222.

Awais, M., et al. (2017). Nitrogen and plant population change radiation capture and utilization capacity of sunflower in semi-arid environment. *Environmental Science and Pollution Research, 24*(21), 17511–17525.

Awais, M., et al. (2018). Potential impacts of climate change and adaptation strategies for sunflower in Pakistan. *Environmental Science and Pollution Research, 25*(14), 13719–13730.

Bajaj, A., Singh, S. P., & Nayak, D. (2022). Impact of water markets on equity and efficiency in irrigation water use: A systematic review and meta-analysis. *Agricultural Water Management, 259*, 107182.

Banadkooki, F. B., Ehteram, M., Ahmed, A. N., Teo, F. B., Fai, C. M., Afan, H. A., Sapitang, A., & Shafie, A. E. (2020). Enhancement of groundwater-level prediction using an integrated machine learning model optimized by whale algorithm. *Natural Resources Research, 29*, 3233–3252.

Banerjee, A., Bandyopadhyay, T., & Acharya, P. (2013). Data analytics: Hyped up aspirations or true potential? *Vikalpa, 38*(4), 1–11.

Bendler, J., Wagner, S., Brandt, T., & Neumann, D. (2014). Taming uncertainty in big data: Evidence from social media in urban areas. *Business and Information Systems Engineering, 6*, 279–288.

Cavanillas, J. M., Curry, E., & Wahlster, W. (2016). New horizons for a data-driven economy: A roadmap for usage and exploitation of big data. In J. M. Cavanillas, E. Curry, & W. Wahlster (Eds.), *Europe*. Springer. isbn 9781420037050.

Che, D., Safran, M., & Peng, Z. (2013). From big data to big data mining: Challenges, issues, and opportunities. In B. Hong, X. Meng, L. Chen, W. Winiwarter, & W. Song (Eds.), Database systems for advanced applications: 18th international conference, DASFAA 2013, international workshops: BDMA, SNSM, SeCoP, Wuhan, China, April 22–25, 2013. Proceedings (pp. 1–15). Springer. https://doi.org/10.1007/978-3-642-40270-8_1.

Chen, K., Chen, H., Zhou, C., Huang, Y., Qi, X., Shen, R., et al. (2020b). Comparative analysis of surface water quality prediction performance and identification of key water parameters using different machine learning models based on big data. *Water Research, 171*, 115454.

Chen, H., Chiang, R. H. L., & Storey, V. C. (2012). Business intelligence and analytics: From big data to big impact. *MIS Quarterly, 36*, 1165–1188.

Chen, C., He, W., Zhou, H., Xue, Y., & Zhu, M. (2020a). A comparative study among machine learning and numerical models for simulating groundwater dynamics in the Heihe River basin, northwestern China. *Scientific Reports, 10*, 1–13.

Chen, L., Yan, H., Yan, J., Wang, J., Tao, T., Xin, K., Li, S., Pu, Z., & Qiu, J. (2022). Short-term water demand forecast based on automatic feature extraction by one-dimensional convolution. *Journal of Hydrology, 606*, 127440.

Chen, P. C. L., & Zhang, C. Y. (2014). Data-intensive applications, challenges, techniques and technologies: A survey on big data. *Information Sciences, 275*, 314–347.

Chung, Y., Servan-Schreiber, S., Zgraggen, E., & Kraska, T. (2018). Towards quantifying uncertainty in data analysis & exploration. *Bulletin of the Technical Committee on Data Engineering (IEEE-CS) 41*, 15–27.

Conrado, S. P., Neville, K., Woodworth, S., & O'Riordan, S. (2016). Managing social media uncertainty to support the decision making process during emergencies. *Journal of Decision Systems, 25*, 171–181.

Demetillo, A. T., Japitana, M. V., & Taboada, E. B. (2019). A system for monitoring water quality in a large aquatic area using wireless sensor network technology. *Sustainable Environment Research, 29*(12), 1–8.

Du, B., Huang, S., Guo, J., Tang, H., Wang, L., & Zhou, S. (2022). Interval forecasting for urban water demand using PSO optimized KDE distribution and LSTM neural networks. *Applied Soft Computing, 122*, 108875.

Farid, H. U., et al. (2017). Site-specific aquifer characterization and identification of groundwater potential areas in district Rahim Yar Khan, Pakistan. *Polish Journal of Environmental Studies, 26*(1), 17–27.

Farid, H. U., et al. (2019a). Estimation of infiltration models parameters and their comparison to simulate the site-specific soil infiltration characteristics. *International Journal of Agricultural and Biological Engineering, 12*(3), 84–91.

Farid, H. U., et al. (2019b). Identification of influencing factors for optimal adoptability of high efficiency irrigation system (HEIS) in Punjab, Pakistan. *Sarhad Journal of Agriculture, 35*(2), 539–549.

Gandomi, A., & Haider, M. (2015). Beyond the hype: Big data concepts, methods, and analytics. *International Journal of Information Management, 35*(2), 137–144. https://doi.org/10.1016/j.ijinfomgt.2014.10.007

Gong, Y., Wang, Z., Xu, G., & Zhang, Z. (2018). A comparative study of groundwater level forecasting using data-driven models based on ensemble empirical mode decomposition. *Water, 10*, 730.

Hammad, H. M., et al. (2018). Offsetting land degradation through nitrogen and water management during maize cultivation under arid conditions. *Land Degradation and Development, 29*, 1366–1375.

Hariri, R. H., Fredericks, E. M., & Bowers, K. M. (2019). Uncertainty in big data analytics: Survey, opportunities, and challenges. *Journal of Big Data, 6*, 44.

Himeur, Y., Elnour, M., Fadli, F., et al. (2023). AI-big data analytics for building automation and management systems: A survey, actual challenges and future perspectives. *Artificial Intelligence Review, 56*, 4929–5021. https://doi.org/10.1007/s10462-022-10286-2

Huang, X., Gao, L., Crosbie, R. S., Zhang, N., Fu, G., & Doble, R. (2019). Groundwater recharge prediction using linear regression, multi-layer perception network, and deep learning. *Water, 11*, 1879.

Isikdogan, F., Bovik, A. C., & Passalacqua, P. (2017). Surface water mapping by deep learning. *IEEE Journal of Selected Topics in Applied Earth Observations and Remote Sensing, 10*, 4909–4918.

Javed, M., Lu, X., & Kumar, D. (2017). *Characterization of big data stream processing pipeline: A case study using Flink and Kafka* (pp. 1–10). Big Data Computing, Applications and Technologies.

Jinyue, C., Chen, S., Rao, F., Li, D., Jiang, H., Wang, C., Peng, Y., Jia, K., & Hicks, B. J. (2022). Remote sensing big data for water environment monitoring: Current status, challenges, and future prospects. *Earth's Futures, 10*(2), e2021EF002289.

Jordan, M. I., & Mitchell, T. M. (2015). Machine learning: Trends, perspectives, and prospects. *Science, 349*, 255–260.

Kikon, A., & Deka, P. C. (2022). Artificial intelligence application in drought assessment, monitoring and forecasting: A review. *Stochastic Environmental Research and Risk Assessment, 36*(5), 1197–1214.

Lin, Y. T., Der Yang, M., Han, J. Y., Su, Y. F., & Jang, J. H. (2020). Quantifying flood water levels using image-based volunteered geographic information. *Remote Sensing, 12*, 706.

Loew, A., Bell, W., Brocca, L., Bulgin, C. E., Burdanowitz, J., Calbet, X., Donner, R. V., Ghent, D., Gruber, A., Kaminski, T., et al. (2017). Validation practices for satellite-based earth observation data across communities. *Reviews of Geophysics, 55*, 779–817.

Malgaonkar, S., Soral, S., Sumeet, S., & Parekhji, T. (2016). Study on big data analytics research domains. In *5th international conference on reliability, Infocom technologies and optimization (trends and future directions) (ICRITO)* (pp. 200–206).

Matin Saddiqi, M., Zhao, W., Cotterill, S., & Dereli, R. K. (2023). Smart management of combined sewer overflows: From an ancient technology to artificial intelligence. *Wiley Interdisciplinary Reviews: Water, 10*(3), e1635.

Nasim, W., et al. (2016). Correlation studies on nitrogen for sunflower crop across the agroclimatic variability. *Environmental Science and Pollution Research, 23*, 3658–3670.

Nasim, W., et al. (2018). Future risk assessment by estimating historical heat wave trends with projected heat accumulation using SimCLIM climate model in Pakistan. *Atmospheric Research, 205*(2018), 118–133.

Ransom, K. M., Nolan, B. T., Traum, J. A., Faunt, C. C., Bell, A. M., Ann, J., Gronberg, M., Wheeler, D. C., Rosecrans, C. Z., Jurgens, B., et al. (2017). A hybrid machine learning model to predict and visualize nitrate concentration throughout the Central Valley aquifer, California, USA. *Science of the Total Environment, 601–602*, 1160–1172.

Sabagh, A. E. L., et al. (2019). Drought and salinity stresses in barley: Consequences and mitigation strategies. *Australian Journal of Crop Science, 13*(6), 810–820.

Sattar, A. M. A., Ertugrul, O. F., Gharabaghi, B., Mcbean, E. A., & Cao, J. (2019). Extereme learning machine model for water network management. *Neural Computing and Applications, 31*, 157–169.

Schrider, D. R., & Kern, A. D. (2018). Supervised machine learning for population genetics: A new paradigm. *Trends in Genetics, 34*, 301–312.

Sivarajah, U., Kamal, M. M., Irani, Z., & Weerakkody, V. (2017). Critical analysis of big data challenges and analytical methods. *Journal of Business Research, 70*, 263–286.

Sohn, S., Huong, V. T., Nguyen, P.-D., Ly, N. H., Jang, S., Lee, H., Lee, C., et al. (2023). Equilibria of semi-volatile isothiazolinones between air and glass surfaces measured by gas chromatography and Raman spectroscopy. *Environmental Research, 218*, 114908.

Srinivasan, S. (2018). Guide to big data applications. *Studies in Big Data, 26*, 1–4.

Srivastava, P. K., Han, D., Ramirez, M. R., & Islam, T. (2013). Machine learning techniques for downscaling SMOS satellite soil moisture using MODIS land surface temperature for hydrological application. *Water Resources Management, 27*, 3127–3144.

Sun, Z., & Huo, Y. (2019). The spectrum of big data analytics. *Journal of Computer Information Systems, 61*, 154.

Sun, A. Y., & Scanlon, B. R. (2019). How can big data and machine learning benefit environment and water management: A survey of methods, applications, and future directions. *Environmental Research Letters, 14*(7), 1–3.

Tariq, M., et al. (2008). The impact of climate warming and crop management on phenology of sunflower-based cropping systems in Punjab, Pakistan. *Agricultural and Forest Meteorology, 256–257*(2018), 270–282.

Tsai, C. W., Lai, C. F., Chao, H. C., & Vasilakos, A. V. (2015). Big data analytics: A survey. *Journal of Big Data, 2*, 1–32.

Tucci, C. E. M. (2017). Urbanization and water resources. In C. de Mattos Bicudo, J. Galizia Tundisi, & M. Cortesão Barnsley Scheuenstuhl (Eds.), *Waters of Brazil*. Springer. https://doi.org/10.1007/978-3-319-41372-3_7

Wang, X., & He, Y. (2016). Learning from uncertainty for big data: Future analytical challenges and strategies. *IEEE Systems, Man, and Cybernetics Magazine, 2*, 26–31.

Watson, H. J. (2014). Tutorial: Big data analytics: Concepts, technologies, and applications. *Communications of the Association for Information Systems, 34*, 1247–1268.

Yereseme, A. K., Surendra, H. J., & Kuntoji, G. (2022). Sustainable integrated urban flood management strategies for planning of smart cities: A review. *Sustainable Water Resources Management, 8*(3), 85.

Zare, M., & Koch, M. (2018). Groundwater level fluctuations simulation and prediction by ANFIS- and hybrid wavelet-ANFIS/fuzzy C-means (FCM) clustering models: Application to the Miandarband plain. *Journal of Hydro-environment Research, 18*, 63–76. https://doi.org/10.1016/j.jher.2017.11.004

Remote Sensing for Agricultural Water Management

Sajjad Hussain, Alishba Sajjad, Muhammad Mubeen, and Shankar Karuppannan

Abstract Remote sensing (RS) plays a significant role in agricultural water management by providing valuable information for enhancing crop productivity, optimizing water use, and ensuring sustainable water resource management. The RS explores the topics such as detecting water stress in crops, estimating crop water requirements through evapotranspiration modeling, assessing water availability and mapping, optimizing irrigation scheduling, and monitoring and mitigating water-related issues like floods, salinity, and waterlogging. The combination of RS with geographic information systems (GIS) is also discussed as a means to enhance decision-making in water management. It then introduces RS as a powerful tool for monitoring and assessing water resources in agriculture. RS technologies, including satellite, aerial, and drone RS, as well as ground-based RS, along with their respective applications and properties are also dealt with. Also discussed are future challenges and trends for agricultural water management, including advancements in technology, data accessibility, and integration with artificial intelligence and machine learning. It highlights the need for continued research and development to harness the full potential of RS in addressing water scarcity and ensuring food security in agricultural systems. It emphasizes the potential of RS in improving water use efficiency (WUE), enhancing crop yields, and supporting sustainable water management practices. Overall, RS offers important opportunities for improving

S. Hussain (✉) · A. Sajjad
Department of Environmental Sciences, COMSATS University Islamabad, Vehari Campus, Vehari, Pakistan

M. Mubeen
Department of Biotechnology, COMSATS University Islamabad, Vehari Campus, Vehari, Pakistan

S. Karuppannan
Department of Applied Geology, College of Applied Natural Science, Adama Science and Technology University, Adama, Ethiopia

Department of Research Analytics, Saveetha Institute of Medical and Technical Sciences (SIMATS), Saveetha University, Chennai, Tamil Nadu, India

M. Mubeen et al. (eds.), *Innovations in Agricultural Water Management*,
https://doi.org/10.1007/978-3-031-91883-4_19

agricultural water management practices, promoting sustainable water use and enabling informed decision-making in agriculture.

Keywords Water management · Remote sensing · Crop

1 Introduction

Remote sensing (RS) has revolutionized agriculture water management by providing valuable insights into water resources, crop health, and irrigation practices. This powerful technology utilizes sensors mounted on ground-based platforms or satellites, to collect data on different land and vegetation parameters without direct physical contact. The information obtained through remote sensing plays a pivotal role in optimizing water management strategies, enhancing crop productivity, and promoting sustainable agricultural practices.

One of the key applications of RS in agriculture water management is the estimation of evapotranspiration (ET). Evapotranspiration is the joint process of water loss from plants through transpiration and soil surface through evaporation. Precise estimation of ET is vital for determining crop water requirements and scheduling irrigation practices. Remote sensing-based models, like the Surface Energy Balance Algorithm for Land (SEBAL) and the Simplified Surface Energy Balance Index (S-SEBI), utilize thermal infrared sensors to the assessment of ET at regional scales. These models provide valuable insights into the water consumption patterns of crops and enable efficient irrigation scheduling, thereby reducing water waste and improving water use efficiency (WUE).

Another important aspect of RS in agriculture water management is the assessment of soil moisture content. Soil moisture also plays a serious role in plant growth and water availability. Microwave sensors operating in the microwave portion of the electromagnetic spectrum can penetrate vegetation as well as measure soil moisture levels. Satellite-based missions, such as the Soil Moisture Active Passive (SMAP) mission, provide high-resolution soil moisture data that enable the observing of soil moisture dynamics at large scales. This information aids in optimizing irrigation practices, managing drought conditions, and reducing water stress on crops.

Vegetation health monitoring is another key application of RS in agriculture water management. Remote sensing data, particularly in the optical and near-infrared areas of the electromagnetic spectrum, provide valuable information on vegetation indices, for example Normalized Difference Vegetation Index (NDVI) and Enhanced Vegetation Index (EVI). These indices are indicative of vegetation vigor, biomass production, and water stress levels. By monitoring changes in vegetation indices over time, remote sensing enables the identification of areas experiencing nutrient deficiencies, water stress, or pest infestations. This allows for targeted interventions, precise irrigation management, and improved crop health.

Overall, remote sensing has transformed agriculture water management by providing valuable information on evapotranspiration, soil moisture, vegetation health, and water resources. The combination of RS data into decision support systems (DSS) allows for more efficient water management, reduced environmental impact, and improved agricultural productivity.

2 Basics of Agricultural Water Management

Agricultural water management is a serious component of sustainable crop production, ensuring efficient water use and maximizing yield potential. This section provides an overview of the basics of agricultural water management, focusing on key principles and practices employed in the field.

Irrigation methods are a central aspect of agricultural water management. Smith et al. (2023) discuss various irrigation techniques and their application in agricultural settings. Their research highlights the significance of selecting the suitable irrigation method based on factors like soil characteristics, crop type, and water availability. Understanding the advantages as well as limitations of different irrigation methods is necessary for optimizing water use efficiency (WUE) and minimizing water losses.

Water conservation is another crucial aspect of agricultural water management. Martinez et al. (2023) explore water conservation practices for sustainable agricultural water management. Their study emphasizes the importance of implementing measures such as mulching, soil moisture monitoring, and precision irrigation to reduce water wastage and promote efficient water utilization. By adopting these practices, farmers can mitigate water scarcity risks, conserve water resources, as well as enhance the overall sustainability of agricultural systems.

3 Remote Sensing Technologies for Water Management

Remote sensing involves the acquisition of information about an object or phenomenon without making physical contact. It utilizes various sensors mounted on satellites, aircraft, or drones to collect data about the Earth's surface. The RS techniques provide valuable information on vegetation dynamics, evapotranspiration, soil moisture, and water bodies, which are vital for effective agricultural water management (Thenkabail et al., 2019).

- **Monitoring Vegetation Dynamics**: Remote sensing allows the observing of vegetation growth and health parameters, like NDVI, vegetation water content, as well as leaf area index (LAI). These parameters indicate the vigor and water stress levels of crops. By analyzing RS data, farmers can detect the areas of low vegetation vigor and adjust irrigation scheduling accordingly, ensuring optimal WUE.

- **Estimating Soil Moisture**: Soil moisture is a key parameter for irrigation management. RS techniques, such as microwave sensors, can estimate soil moisture content across large areas. By assimilating soil moisture information with crop water necessities, farmers can determine the optimal irrigation timing and amount to avoid over-irrigation or under-irrigation, leading to improved WUE.
- **Evapotranspiration Estimation**: Evapotranspiration (ET) represents the combined water loss from the soil surface and transpiration from plants. Accurate estimation of ET is crucial for determining crop water requirements and scheduling irrigation. The RS-based models, like SEBAL and S-SEBI, utilize satellite data to estimate ET at different spatial scales. These models give the important information for efficient irrigation scheduling and water allocation.
- **Monitoring Water Bodies**: Remote sensing plays a vital role in monitoring water bodies, such as reservoirs, lakes, and rivers, which are essential sources of irrigation water. Satellite-based sensors can also give the information about water storage levels, water quality, and the extent of water bodies. These data help in managing water allocation, identifying water availability, and detecting water stress in agricultural areas.
- **Integration with Geographic Information Systems (GIS)**: RS data can be combined with GIS to create spatially explicit maps and decision support tools for agricultural water management. GIS enables the analysis and visualization of RS data, allowing farmers and water managers to identify areas of high or low water availability, track changes over time, and make informed decisions about water allocation and irrigation planning.

3.1 Aerial and Drone Remote Sensing

Water management is a critical component of sustainable resource utilization and conservation. Traditional methods of data collection and monitoring for water bodies often face limitations in terms of cost, time, as well as spatial coverage. On the other hand, the emergence of aerial and drone RS technologies have reformed the field, offering cost-effective and effective solutions for water management. This section aims to explore the applications of aerial and drone remote sensing in water management and highlight their significance.

Water Quality Monitoring: Aerial and drone remote sensing techniques have proven valuable for water quality monitoring. Multispectral sensors mounted on aerial platforms enable the detection and analysis of water constituents. For instance, chlorophyll-a, suspended sediments, and dissolved organic matter can be quantified using spectral reflectance measurements. Hyperspectral sensors further enhance the identification and quantification of specific water quality parameters by capturing data at narrower spectral bands.

Bathymetry and Depth Mapping: Accurate knowledge of water depth is critical for various water management applications. LiDAR (light detection and ranging) sensors mounted on aerial platforms or drones enable the creation of high-resolution depth maps. By emitting laser pulses and measuring the time taken for the reflected signals to return, LiDAR technology provides detailed bathymetric data. This information aids in the characterization of underwater topography and sediment distribution.

Flood Monitoring and Management: Aerial and drone RS technologies play an essential role in flood monitoring and management. These techniques capture imagery and elevation data, providing real-time information on flood extent, water levels, and floodplain mapping. The combination of RS data with hydrological models assists in effective emergency response planning, flood risk assessment, and the development of flood inundation models.

Water Resource Assessment: Assessing water resources is important for maintainable water management. Aerial and drone remote sensing techniques facilitate the estimation of surface water resources, including lakes, reservoirs, and rivers. By combining multispectral and thermal sensors, evapotranspiration rates and water stress indices can be derived, assisting in water resource planning and irrigation management.

3.2 *Wetland Mapping and Conservation*

Wetlands are ecologically significant ecosystems that require careful management and conservation. Aerial and drone remote sensing techniques enable wetland mapping and monitoring, providing valuable information on wetland extent, vegetation cover, and hydrological dynamics. These data aid in assessing wetland health, identifying stressors, and formulating conservation strategies (Liu et al., 2019).

One study conducted by Smith et al. (2018) demonstrated the application of ground-based remote sensing for monitoring water quality parameters in a reservoir. The authors utilized a combination of spectroradiometers and hyperspectral sensors to measure the concentrations of various water quality indicators, such as chlorophyll-a and turbidity. The outcomes presented a strong relationship between the remote sensing data and traditional water quality measurements, indicating the potential of ground-based RS as a reliable tool for water quality monitoring (Smith et al., 2018). In another study, Johnson et al. (2019) observed ground-based LiDAR RS to assess the water levels and floodplain dynamics of a river system. The researchers used a terrestrial laser scanner to capture high-resolution elevation data along the river corridor. The obtained data enabled accurate mapping of flood extents and identification of potential flood hazard zones. The study demonstrated the effectiveness of ground-based LiDAR RS in floodplain management and flood risk assessment (Johnson et al., 2019).

Satellite RS plays an important role in water management by providing valuable information about the Earth's water resources. It enables the monitoring and assessment of various water-related parameters, including water availability, quality, and usage. The following section highlights the importance of satellite RS in water management with corresponding citations.

3.3 Satellite Remote Sensing

Satellite RS allows for the monitoring of changes in water bodies, like lakes, rivers, and reservoirs, over large spatial extents as well as long time periods (Liu et al., 2019).

The RS data can also be used to estimate key hydrological variables like evapotranspiration, soil moisture, and precipitation, providing insights into the water cycle and hydrological processes (Senay et al., 2013).

The use of satellite-derived data in water management facilitates the identification of areas prone to drought, flood, or water scarcity, enabling proactive measures and effective allocation of water resources (Wu et al., 2017).

Satellite-based sensors can detect water quality parameters like turbidity, temperature, and chlorophyll-a concentration, aiding in the assessment and monitoring of water pollution and ecological health (Zhang et al., 2018).

4 Applications of Remote Sensing in Agricultural Water Management

Irrigation Scheduling: RS data, like satellite imagery and thermal sensors, can be used to assess crop water requirements and determine optimal irrigation scheduling (Lopez-Urrea et al., 2018).

Crop Water Stress Detection: RS techniques, including multispectral and thermal imaging, enable the detection of water stress in crops by assessing their physiological characteristics, like leaf temperature and vegetation indices (Thenkabail et al., 2019).

Evapotranspiration Estimation: Remote sensing-based algorithms, like the SEBAL and the Simplified Surface Energy Balance Index (S-SEBI) approach, allow for accurate assessment of evapotranspiration rates at regional scales (Allen et al., 2011).

Water Quality Monitoring: RS provides valuable information on water quality parameters, like turbidity, chlorophyll-a concentration, and harmful algal blooms, aiding in the management of irrigation water sources and reservoirs (Gitelson et al., 2016).

Soil Moisture Mapping: Microwave RS techniques, like synthetic aperture radar (SAR), can be used to map soil moisture content, enabling farmers to optimize irrigation practices and improve WUE (Entekhabi et al., 2010).

4.1 *Crop Water Requirements Estimation*

Crop water requirements refer to the amount of water required by crops to meet their evapotranspiration (ET) demands throughout their growth cycle. Accurate estimation of crop water requirements is crucial for efficient irrigation management and water resource planning in agricultural systems.

Reference Evapotranspiration (ET0): Reference evapotranspiration is the water loss from a well-watered, uniform grass cover under standardized environmental conditions. It serves as a baseline for assessing crop water necessities. The most widely used method for estimating ET0 is the Penman-Monteith equation, suggested by the Food and Agriculture Organization (FAO).

Crop Coefficient (Kc): The crop coefficient is a dimensionless factor that represents the ratio of crop evapotranspiration (ETc) to reference evapotranspiration (ET0) for a specific crop and growth stage. Kc values vary throughout the crop's life cycle, reflecting changes in water demand due to various factors such as leaf area development and canopy cover. Kc values can be obtained from literature or empirical models.

Crop Evapotranspiration (ETc) Calculation: ETc is the actual water requirement of a specific crop during its growth stages. ETc is designed by multiplying the ET0 by the crop coefficient (Kc) for a given time period. This provides an estimate of the amount of water required to meet crop water necessities (Mebrie et al., 2023).

Adjustments for Environmental Factors: In addition to ET0 and Kc, adjustments are often made to account for specific environmental factors that affect crop water requirements. Various correction factors and adjustment methods are available to refine the estimation process and account for local conditions (Mebrie et al., 2023).

Accurate estimation of crop water requirements is essential for optimizing irrigation scheduling, preventing over- or under-irrigation, and promoting water use efficiency in agricultural systems. By considering the specific water needs of crops at different growth stages, farmers and irrigation managers can make up-to-date decisions regarding irrigation timing, duration, and volume to ensure optimal crop health and productivity.

4.2 *Irrigation Scheduling and Optimization*

Irrigation Scheduling and Optimization play an important role in maximizing WUE and crop productivity in agricultural systems. By determining the optimal timing and amount of irrigation water, farmers can minimize water waste, reduce production costs, and conserve water resources. Here is an overview of irrigation planning and optimization:

Soil Moisture-Based Scheduling: Soil moisture-based scheduling involves observing the moisture content in the root zone to determine when irrigation is needed. Different methods, like soil moisture sensors or tensiometers, can be used to measure soil moisture. Irrigation is scheduled when the soil moisture drops below a certain threshold, ensuring that crops do not experience water stress.

Evapotranspiration-Based Scheduling: Evapotranspiration (ET)-based scheduling estimates the water needs of crops by considering the water lost through evaporation and transpiration from the plant. ET0 is calculated using weather data, and ETc is determined by multiplying ET0 with a Kc specific to the crop and growth stage. Irrigation is scheduled when the cumulative ETc reaches a certain level.

Weather Forecasting: Incorporating weather forecasts into irrigation scheduling allows farmers to anticipate rainfall events and adjust irrigation accordingly. If significant rainfall is predicted, irrigation can be reduced or skipped to avoid overwatering. Conversely, in the absence of rainfall, irrigation can be scheduled to supplement soil moisture.

Irrigation Methods and Efficiency: The choice of irrigation method has an important impact on scheduling and optimization. Various methods, like sprinkler irrigation, surface irrigation, or drip irrigation, have varying levels of efficiency and uniformity in water distribution. High-efficiency irrigation systems, such as drip or micro-irrigation, minimize water losses and allow for precise application, enhancing scheduling accuracy.

Crop Water Stress Observing: Monitoring crop water stress indicators, like leaf water potential or canopy temperature, provides valuable insights into crop water requirements. By regularly assessing these indicators, farmers can detect the water stress and adjust irrigation schedules accordingly, ensuring optimal plant health and minimizing water waste.

Irrigation Scheduling Tools and Technologies: Various tools and technologies are available to assist in irrigation scheduling and optimization. These include DSS that integrate weather data, soil moisture measurements, and crop information to provide real-time recommendations for irrigation scheduling.

Water Management Strategies: Implementing water management strategies, such as deficit irrigation or regulated deficit irrigation, can optimize water use while maintaining crop productivity. These strategies involve intentionally applying less water than the crop's full water requirements during specific growth stages, based on the crop's tolerance to water stress. This approach promotes water use efficiency and can be particularly useful in water-limited regions.

Monitoring and Evaluation: Continuous monitoring and evaluation of irrigation practices are essential for fine-tuning schedules and optimizing water use. Regular assessment of soil moisture, crop conditions, and irrigation system performance helps identify any inefficiencies or issues that may require adjustment, improving overall irrigation effectiveness.

4.3 Monitoring and Mitigation of Water-Related Issues

Monitoring and mitigation of water-related issues are critical for ensuring the sustainable management of water resources.

Water Quantity Monitoring: Monitoring the quantity of water resources, like rivers, lakes, and groundwater, helps track changes in water availability over time. This information is vital for understanding water scarcity issues and managing water allocation.

Watershed Management: Implementing integrated watershed management approaches, including land use planning, soil conservation practices, and reforestation, helps protect water quality and reduce erosion and sedimentation.

Water Use Efficiency: Promoting water use efficiency through practices like water-saving irrigation techniques, water recycling, and leak detection and repair in water distribution systems helps optimize water allocation as well as reduce wastage.

Early Warning Systems and Flood Monitoring: Monitoring river levels and rainfall patterns, and using RS technologies can contribute to effective flood monitoring and early warning systems, allowing timely mitigation and evacuation measures.

Drought Monitoring and Preparedness: Monitoring indicators such as precipitation, soil moisture, and vegetation health assists in early detection of drought situations. Drought preparedness measures include water conservation campaigns, drought-resistant crop varieties, and improved water storage and distribution systems.

Water Pollution Control: Implementing pollution control measures, such as wastewater treatment, storm water management, and industrial effluent regulations, helps reduce the contamination of water bodies and protect human and ecological health.

Ecosystem Restoration: Restoring degraded aquatic ecosystems through measures like wetland rehabilitation, riparian zone protection, and habitat restoration initiatives helps enhance water quality and biodiversity.

Public Awareness and Education: Promoting public awareness and education on water-related issues fosters responsible water use practices and encourages community participation in water monitoring and conservation efforts.

4.4 Water Resource Assessment and Planning

Water resource planning and assessment are essential for sustainable water management and ensuring the availability of water for various sectors.

Water Availability Assessment: Assessing water availability involves evaluating the quality and quantity of water resources within a given region. This includes analyzing factors such as surface water and groundwater availability, rainfall patterns, and hydrological modeling.

Water Demand Projection: Projecting water demand involves estimating future water requirements for numerous sectors, like agriculture, industry, and domestic use. This assessment considers factors such as economic development, population growth, and changes in water use efficiency. It emphasizes stakeholder participation, water governance, and the integration of land and water planning (Rangwala et al., 2023).

Water Allocation and Prioritization: Allocating water resources among competing uses requires establishing priorities and considering the needs of different sectors, ecosystems, and water-dependent communities. This process involves policy development, legal frameworks, and stakeholder consultations.

Climate Change Adaptation: Assessing the effects of climate change on water resources and developing adaptation strategies are crucial for long-term water resource planning. This includes incorporating climate change scenarios, considering changing precipitation patterns, and evaluating the resilience of water infrastructure.

Water–Energy–Food Nexus: The water–energy–food nexus method recognizes the interdependencies as well as trade-offs between energy, water, and food production. Assessing these interconnections helps identify synergies and optimize resource use across sectors.

Environmental Flow Assessment: Environmental flow assessments determine the minimum amount of water required to maintain the health of aquatic ecosystems. These assessments consider ecological needs, habitat requirements, and the impacts of water abstractions.

Water Pricing and Economic Instruments: Economic instruments, such as water pricing and market-based mechanisms, can play a role in water resource planning by promoting efficient water use and generating revenue for infrastructure development and maintenance.

Water Infrastructure Development: Planning for water infrastructure, such as dams, reservoirs, and water treatment plants, requires assessing water availability, demand projections, and environmental considerations. This involves engineering studies, environmental impact assessments, and cost–benefit analyses.

5 Integration of RS with GIS and DSS

The integration of RS with GIS and DSS has revolutionized the field of environmental monitoring and resource management.

Remote Sensing Data Acquisition: RS involves capturing information about the Earth's surface using sensors attached on aircraft, satellites, or drones. Different types of sensors, such as optical, thermal, or radar, collect data in various spectral ranges and resolutions (Abdullah et al., 2023).

GIS Data Management: GIS is a system for capturing, analyzing, storing, and displaying spatial data. It provides a framework for managing and integrating diverse data layers, including remote sensing imagery, topographic data, land use maps, and socio-economic data.

Image Classification and Analysis: Remote sensing imagery can be classified into different land cover or land use classes using image processing techniques. These classifications enable the extraction of valuable information, such as urban areas, vegetation cover, or water bodies (Abdullah et al., 2023).

Spatial Analysis and Modeling: GIS provides tools for spatial analysis, allowing the integration of RS data with other geospatial information. This integration enables spatial modeling, such as predicting crop yields, assessing flood vulnerability, or identifying suitable locations for infrastructure development.

Decision Support Systems: DSS integrate remote sensing and GIS data to provide tools and models for decision-making. These systems assist in tasks such as natural resource management, land use planning, disaster response, and environmental impact assessment (Sowmiya Narayanan & Manimaran, 2023).

Environmental Monitoring: The integration of RS and GIS in DSS facilitates real-time or near-real-time monitoring of environmental parameters. This includes monitoring vegetation health, water quality, air pollution, and ecosystem dynamics.

Precision Agriculture: Remote sensing, GIS, and DSS have been widely used in precision agriculture. They enable site-specific management of crops by assessing plant health, nutrient deficiencies, irrigation needs, and yield prediction.

6 Future Trends and Challenges

Future trends and challenges of RS for agricultural water management are:

Advanced Sensors and Platforms: Future trends in RS for agricultural water management will likely involve the development of more advanced sensors and platforms. This includes the use of hyperspectral sensors that can provide more complete information about soil moisture, crop health, and water stress. Emerging technologies such as unmanned aerial vehicles (UAVs) and drones offer the potential for high-resolution and on-demand data acquisition.

Integration of Multiple Data Sources: Incorporating data from multiple RS platforms, like satellites, UAVs, and ground-based sensors, will enhance the accuracy and reliability of agricultural water management assessments. Combining optical, thermal, and radar data can provide comprehensive information about crop water requirements, irrigation performance, and hydrological dynamics.

Artificial Intelligence and Machine Learning: Future developments in RS for agricultural water management will heavily rely on the use of artificial intelligence (AI) and machine learning techniques. These technologies can improve data processing, image classification, and data-driven decision support systems. They can also help identify patterns, forecast crop water needs, and optimize irrigation scheduling.

Data Fusion and Assimilation: Future trends will focus on the fusion and assimilation of RS data with other sources, like soil information, weather data, as well as crop models. A more thorough understanding of the agricultural water system may result from this integration, improving resource allocation and decision-making.

Real-Time Monitoring and Decision Support: Advancements in RS technologies will enable near-real-time monitoring of crop water status, soil moisture, and water availability. This real-time information, combined with decision support systems, can provide timely and accurate recommendations for irrigation scheduling, water allocation, and drought mitigation.

Water–Energy–Food Nexus: Future remote sensing applications will increasingly consider the water–energy–food nexus, which recognizes the interdependencies between water resources, energy production, and food security. Remote sensing can provide valuable insights into optimizing WUE, enhancing agricultural productivity, and promoting sustainable resource management across these interconnected sectors.

7 Challenges

Data Availability and Accessibility: Despite the advancements in remote sensing technology, challenges remain in terms of data accessibility and availability. Access to high-quality and up-to-date RS data, especially at a finer spatial resolution, can be limited, particularly in developing regions. Efforts to improve data sharing, open-access policies, and data integration are essential.

Data Interpretation and Validation: Interpreting and validating remote sensing data for accurate agricultural water management applications can be challenging. Ground truth data collection, calibration, and validation are necessary to ensure the reliability and accuracy of RS products. Integration with field observations and ground-based measurements is crucial for validation purposes.

Scalability and Adaptability: Implementing remote sensing technologies for agricultural water management at larger scales and across diverse agricultural systems can be complex. It requires considering the variability in crop types, soil conditions, and irrigation practices. Developing scalable and adaptable remote sensing approaches that can be tailored to different regions and cropping systems is essential.

Cost and Infrastructure: The cost of RS data acquisition and analysis can be a barrier to widespread adoption, especially for small-scale farmers and resource-constrained regions. Additionally, the availability of supporting infrastructure, such as computing resources and Internet connectivity, is essential for effective utilization of remote sensing technologies.

User Engagement and Capacity Building: To fully leverage the potential of RS for agricultural water management, user engagement and capacity building are crucial. Training programs, knowledge-sharing platforms, and stakeholder involvement are needed to enhance understanding, adoption, and effective utilization of remote sensing technologies among farmers, water managers, and decision-makers.

8 Conclusion

RS plays a vital role in agricultural water management by providing valuable information on crop water requirements, irrigation efficiency, and water resource monitoring. It offers a range of benefits, including improved water allocation, enhanced irrigation scheduling, and increased productivity. Looking ahead, future trends in RS for agricultural water management include the use of advanced sensors and platforms, integration of multiple data sources, utilization of artificial intelligence and machine learning, data fusion and assimilation, real-time monitoring, and consideration of the water–energy–food nexus. These advancements have the potential to revolutionize the field and enable more precise and sustainable water management practices. On the other hand, challenges like data availability and accessibility, data interpretation and validation, scalability and adaptability, cost and infrastructure, and user engagement as well as capacity building need to be addressed to fully harness the potential of RS in agricultural water management. Overcoming these challenges will require collaborative efforts, technological innovations, and capacity building initiatives. Overall, remote sensing holds great promise for optimizing water use in agriculture, improving water resource management, and fostering sustainable agricultural practices. By leveraging the power of remote sensing technology, we can contribute to global food security, water conservation, and environmental sustainability in the face of growing water challenges and a changing climate.

References

Abdullah, H. M., Mohana, N. T., Khan, B. M., Ahmed, S. M., Hossain, M., Islam, K. S., et al. (2023). Present and future scopes and challenges of plant pest and disease (P&D) monitoring: Remote sensing, image processing, and artificial intelligence perspectives. *Remote Sensing Applications: Society and Environment, 32*, 100996.

Allen, R. G., Tasumi, M., Morse, A., Trezza, R., Wright, J. L., & Bastiaanssen, W. (2011). Satellite-based energy balance for mapping evapotranspiration with internalized calibration (METRIC)—Model. *Journal of Irrigation and Drainage Engineering, 133*, 380–394. https://doi.org/10.1061/(ASCE)IR.1943-4774.0000361

Entekhabi, D., Njoku, E. G., O'Neill, P. E., Kellogg, K. H., Crow, W. T., Edelstein, W. N., Entin, J. K., Goodman, S. D., Jackson, T. J., Johnson, J., Kimball, J., Piepmeier, J. R., Koster, R. D., Martin, N., McDonald, K. C., Moghaddam, M., Moran, S., Reichle, R., Shi, J. C., Spencer, M. W., Thurman, S. W., Tsang, L., & Van Zyl, J. (2010). The soil moisture active passive (SMAP) mission. *Proceedings of the IEEE, 98*, 704–716. https://doi.org/10.1109/JPROC.2010.2043918

Gitelson, A. A., Rundquist, D. C., Kaufman, Y. J., & Stark, R. (2016). Remote sensing of chlorophyll concentration in crops: Advances and limitations. In *Photogrammetric engineering and remote sensing* (pp. 1003–1013).

Johnson, E. F., Williams, C. J., & Anderson, S. P. (2019). Ground-based lidar remote sensing for floodplain management: A case study of the XYZ river. *Water Resources Research, 55*(7), 5612–5628.

Liu, W., Wang, L., Li, Z., & Li, Y. (2019). Remote sensing of water body changes: Detection, monitoring, and analysis. *Remote Sensing, 11*(8), 927.

Lopez-Urrea, R., Ballesteros, R., García-Sánchez, F., & Cuartero, A. (2018). Estimating crop water requirements using satellite imagery: A review. *Water, 10*, 164. https://doi.org/10.3390/w10020164

Martinez, R., Garcia, E., & Rodriguez, D. (2023). Water conservation practices for sustainable agricultural water management. *Water Resources Research, 29*(1), 45–56.

Mebrie, D. W., Assefa, T. T., Yimam, A. Y., & Belay, S. A. (2023). A remote sensing approach to estimate variable crop coefficient and evapotranspiration for improved water productivity in the Ethiopian highlands. *Applied Water Science, 13*(8), 168.

Rangwala, T., Mukheibir, P., & Fane, S. (2023). A review and comparative analysis of IWCM concepts in Australia and similar jurisdictions. *Water, 15*(7), 1369.

Senay, G. B., Budde, M., Verdin, J. P., & Melesse, A. M. (2013). A coupled remote sensing and simplified surface energy balance approach to estimate actual evapotranspiration from irrigated fields. *Sensors, 13*(12), 16344–16367.

Smith, A. B., Johnson, C. D., & Brown, T. J. (2018). Application of ground-based remote sensing for water quality monitoring in a reservoir. *Journal of Environmental Monitoring, 20*(3), 456–464.

Smith, J., Johnson, A., & Thompson, C. (2023). Irrigation methods and their application in agricultural water management. *Journal of Agricultural Engineering, 57*(2), 78–91.

Sowmiya Narayanan, K. J., & Manimaran, A. (2023). Recent developments in geographic information systems across different application domains: A review. *Knowledge and Information Systems, 66*, 1523–1547.

Thenkabail, P. S., Dheeravath, V., Biradar, C. M., Noojipady, P., Li, Y., Velpuri, M., Gumma, M., & Pandey, S. (2019). Remote sensing of crop water productivity of cotton and wheat in western India. *International Journal of Applied Earth Observation and Geoinformation, 74*, 83–107. https://doi.org/10.1016/j.jag.2018.08.005

Wu, B., Duan, Z., & Wang, H. (2017). Remote sensing-based identification of drought-induced vegetation decline: A case study in the North China Plain. *Remote Sensing, 9*(7), 640.

Zhang, Y., Liu, X., Liu, X., & Xu, H. (2018). Retrieval of water quality parameters in turbid inland waters using satellite remote sensing: Current status, challenges, and future outlook. *Remote Sensing, 10*(12), 1986.

UAVs' Scope in Agricultural Water Management

Qurat-ul-ain Maken, Sajjad Hussain, Muhammad Mubeen, Samra Tariq, and Saeed Ahmad Qaisrani

Abstract Unmanned aerial vehicles (UAVs), sometimes known as drones, have become highly influential technologies in agricultural water management. They provide innovative solutions to improve water production, precision irrigation, and resource monitoring. This research examines the various functions of UAVs in transforming water management methods, namely in the agriculture industry. Unmanned aerial vehicles (UAVs), equipped with sophisticated sensors and imaging technologies, allow for the immediate gathering and analysis of data. This enables accurate monitoring of the condition of crops, soil moisture levels, and water quality. These features enable focused interventions, efficient allocation of resources, and enhanced decision-making, effectively tackling issues pertaining to water scarcity and inefficiency in agriculture. In addition, unmanned aerial vehicles (UAVs) offer valuable information on hydrological processes, including the analysis of river flow, forecast of floods, and management of watersheds. This contributes to the development of more efficient and sustainable strategies for water resource management. Although UAVs hold great potential, their widespread use is hindered

Qurat-ul-ain Maken · S. Hussain
Department of Environmental Sciences, COMSATS University Islamabad, Vehari Campus, Vehari, Pakistan

M. Mubeen (✉)
Department of Biotechnology, COMSATS University Islamabad, Vehari Campus, Vehari, Pakistan
e-mail: muhammadmubeen@cuivehari.edu.pk

S. Tariq
Department of Environmental Sciences, COMSATS University Islamabad, Vehari Campus, Vehari, Pakistan

Urban Unit, Lahore, Pakistan

S. A. Qaisrani (✉)
Department of Environmental Sciences, COMSATS University Islamabad, Vehari Campus, Vehari, Pakistan

Plant Breeding Institute (PBI), The University of Sydney, Sydney, NSW, Australia
e-mail: saeed.qaisrani@cuivehari.edu.pk

M. Mubeen et al. (eds.), *Innovations in Agricultural Water Management*,
https://doi.org/10.1007/978-3-031-91883-4_20

by obstacles such as regulatory barriers, restricted flying time, and the requirement for specialized data processing skills. Furthermore, limited financial resources and logistical limitations may impede small-scale farmers from effectively harnessing UAV technology. Nevertheless, the capacity of UAVs to transform agricultural practices through precision agriculture and smart farming techniques remains substantial. This paper focusses on the present uses, difficulties, and future areas of study of UAVs in agricultural water management, emphasizing their ability to improve water productivity, support sustainable methods, and tackle the increasing needs of worldwide agriculture in light of climate change.

Keywords Unmanned aerial vehicles · Soil analysis · Hydrological processes · Sustainable strategies · Climate change · Irrigation management

1 Introduction

Unmanned aerial vehicles (UAVs), also referred to as drones, are a groundbreaking technology that has the capacity to revolutionize multiple industries, including agricultural water management. UAVs have rapidly evolved from their military origins to serve diverse civilian applications. Their capacity to collect high-resolution data, their flexibility in deployment, and their decreasing costs have made them invaluable tools for precision water management. In the realm of water management, UAVs offer novel solutions that can optimize water usage, monitor irrigation systems, assess crop health, and aid in the overall decision-making process for farmers, water resource managers, and policymakers (Upadhyaya et al., 2022).

Satellite and UAV imagery share a common objective but have distinct differences in terms of spatial and temporal resolution. The selection between the two options is based on the level of precision and accuracy that the user needs to attain. Unmanned aerial vehicles (UAVs) outperform space-based sensors due to their capacity to capture high-resolution images at a lower cost. Additionally, customers have the flexibility to determine when the UAVs should revisit specific areas. Unmanned aerial vehicles (UAVs) are very suitable for rapidly assessing changes in crop growth and development, as well as monitoring the level of stress and overall health of crops in near-real-time. But the utilization of UAVs may provide financial and logistical difficulties for small-scale farmers, as they may lack the necessary means to acquire and upkeep UAVs, as well as the expertise and capability to operate, gather, analyze, and interpret data obtained remotely. However, the use of collective or communal ownership, together with the installation of extension services, offers a practical and economical option for spreading technology to small-scale farmers (Nhamo et al., 2020).

Unmanned aerial vehicles (UAVs) are used in a non-invasive way to remotely identify, analyze, and evaluate the form and composition of elements within river catchments, as well as hydrological activities across various geographical and temporal scales. They combine geographical, temporal, spectral, and continuous data to facilitate the characterization of soil and water and the study of hydrology. Unmanned aerial vehicles (UAVs) gather current water quality data directly at the location due to progress in robotics, sensor technologies, wireless networks, and image processing systems. This contributes to improving the precision of remote sensing in the monitoring of water quality (Khan & Al-Mulla, 2019).

For instance, a spectroradiometer installed on an unmanned aerial vehicle (UAV) was used to examine the potential occurrence of algal blooms in the Maumee River and the western basin of Lake Erie in almost real-time. Chabot (2018) outlined three notable advantages of modern unmanned aerial vehicles (UAVs) compared to conventional radio-controlled hobby airplanes: (a) advanced autopilot systems that integrate autonomous flight with user-friendly flight planning and tracking software, (b) prolonged flight duration achieved through advancements in rechargeable battery technology, and (c) lightweight, high-resolution, and cost-effective compact digital cameras. Therefore, unmanned aerial vehicles (UAVs) are highly regarded as a remote sensing platform because of their adaptable navigation, minimal upkeep, cost-effectiveness, and dependable outcomes when compared to conventional and invasive methods (Acharya et al., 2021).

This chapter delves into the intersection of UAV technology and agricultural water management, highlighting the multifaceted roles UAVs play in addressing the challenges posed by water scarcity and inefficiency in modern agriculture. By providing an overview of the key concepts, methodologies, and case studies, this chapter aims to underscore the significance of UAVs in transforming traditional agricultural practices into data-driven, precision-oriented systems. It will explore the various aspects of UAV deployment, data collection, analysis, and practical implementation, shedding light on the potential benefits, limitations, and future directions of this innovative approach.

2 UAVs in Smart Agriculture

UAVs have ushered in a new era of technological innovation in agriculture, transforming traditional farming practices into what is now commonly referred to as "Smart Agriculture" or "Precision Agriculture." The integration of UAVs into the agricultural landscape has given rise to a myriad of applications that enhance efficiency, sustainability, and productivity across various aspects of farming (Mohsan et al., 2023).

Unmanned aerial vehicles (UAVs) used in Smart Agriculture, commonly known as Agricultural Drones, are outfitted with a range of sensors and imaging technologies that allow them to record detailed data from an aerial perspective. This aerial perspective provides farmers, agronomists, and researchers with insights that were previously inaccessible, allowing for informed decision-making and precise resource allocation (Alexandris et al., 2021).

UAVs are deployed across various stages of agricultural operations, offering a wide array of applications:

1. **Crop Monitoring and Health Assessment**: UAVs equipped with multispectral or hyperspectral cameras can capture detailed imagery of crops. This data is then processed to assess crop health, detect diseases, nutrient deficiencies, and stress, enabling targeted interventions and optimized resource allocation.
2. **Precision Irrigation**: UAVs armed with sensors that measure soil moisture and analyze field conditions, aid in precision irrigation. Farmers can tailor their irrigation strategies to specific areas of need, conserving water and ensuring optimal crop growth.
3. **Field Mapping and Surveying**: UAVs are invaluable for creating accurate, up-to-date maps of fields. This mapping facilitates land-use planning, crop rotation, and even the design of drainage systems to minimize waterlogging.
4. **Pest and Disease Management**: Aerial surveys conducted by UAVs can identify pest infestations and disease outbreaks. This enables early intervention and targeted application of pesticides, reducing overall chemical usage.
5. **Yield Prediction and Monitoring**: The data collected by UAVs can be used to analyze vegetation indices, growth patterns, and other indicators, allowing for more accurate yield predictions and informed harvesting decisions (Zhang et al., 2021).
6. **Soil Analysis**: UAVs equipped with soil sensors can analyze soil composition and nutrient levels, aiding in the formulation of precise fertilizer application plans tailored to specific areas of a field.
7. **Data Integration and Analytics**: UAV-collected data can be integrated with Geographic Information Systems (GIS) and other data sources, enhancing the accuracy and scope of analysis. Advanced analytics and machine learning techniques can turn raw data into actionable insights, providing deeper understanding and aiding decision-making.
8. **Cost and Time Efficiency**: UAVs save time and reduce costs associated with traditional manual monitoring and data collection methods. They can cover large areas in a short time, especially when compared to ground-based surveys (Zhang et al., 2021).

3 Prospects of Improving Water Productivity Through UAV

The prospects of improving water productivity through unmanned aerial vehicles (UAVs) are promising and offer innovative solutions to address water scarcity, enhance agricultural efficiency, and promote sustainable resource management. UAVs provide valuable data and insights that can be leveraged to optimize water usage, monitor crop health, and inform decision-making in ways that ultimately lead to higher water productivity in agriculture (Nhamo et al., 2020). Some key prospects of UAVs improving water productivity are shown in Fig. 1 and described as follows.

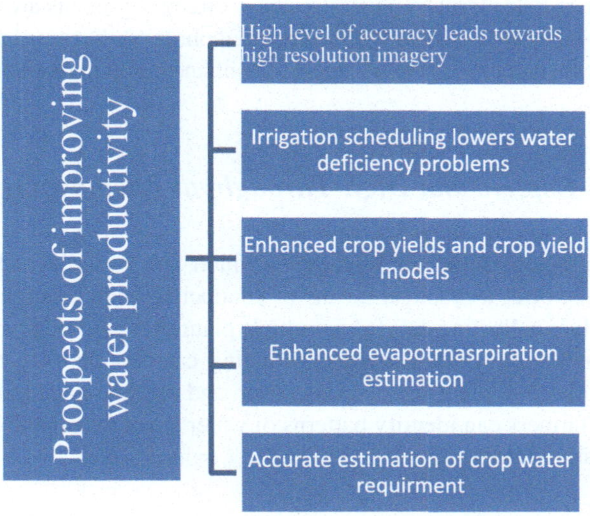

Fig. 1 Prospects of improving water productivity

3.1 Precision Irrigation Management

UAVs that have sophisticated sensors may collect up-to-date information on the moisture content in the soil, allowing for accurate management of irrigation. By identifying areas with varying moisture content, farmers can apply water only where it is needed, minimizing wastage and improving water use efficiency. This targeted approach to irrigation conserves water resources while maintaining optimal crop growth.

3.2 Thermal Imagery and Water Stress Assessment

Thermal cameras mounted on UAVs capture infrared radiation emitted by crops. This data can be processed to create thermal imagery that reveals temperature variations across a field. Elevated temperatures can indicate water stress in plants, as stressed plants often have reduced transpiration due to insufficient water uptake. By identifying these hotspots, farmers can target irrigation efforts to alleviate water stress, ensuring efficient water usage and maintaining crop health.

3.3 Chlorophyll Fluorescence for Photosynthesis and Water Use Efficiency

Chlorophyll fluorescence is an indicator of plant photosynthesis and overall health. UAVs equipped with sensors that measure chlorophyll fluorescence can provide insights into a plant's photosynthetic activity and its water use efficiency. Monitoring these parameters helps farmers understand how effectively plants are utilizing water for growth. By identifying areas with suboptimal photosynthesis or inefficient water use, farmers can adapt irrigation strategies to enhance crop performance.

3.4 Multi-Copters and High-Throughput Phenotyping

Emerging technologies like multi-copters, which are versatile and user-friendly UAV platforms, offer a practical means of conducting high-throughput phenotyping. This involves collecting data from multiple plants quickly and non-destructively. UAVs equipped with various sensors, including cameras and spectrometers, can capture data related to plant size, canopy cover, and physiological traits. By analyzing this data, farmers can identify patterns of water stress across the field, enabling targeted irrigation interventions.

3.5 Crop Trait Determination for Breeding

UAVs contribute to the rapid and non-invasive assessment of important crop traits, which is crucial in breeding programs aimed at developing drought-resistant varieties. UAVs equipped with specialized sensors can measure traits like canopy architecture, leaf size, and growth patterns. These traits are indicative of a plant's ability to tolerate water stress. By gathering such data across various crop varieties, breeders can select plants with desirable traits for water-efficient agriculture.

The ability to provide real-time, high-resolution data enables precision in water management, enhances crop health, and supports sustainable practices. By incorporating UAV technology into agricultural practices, farmers and resource managers can optimize water usage, increase yields, and contribute to a more water-efficient and resilient agricultural sector.

Fig. 1 provides a sketch of various prospects of improving water productivity.

4 UAVs in Hydrology and Water Management

UAVs are showing their ability to bring about significant changes in the realm of hydrology and water management. UAV are transforming the way water resources are monitored, appraised, and managed by capturing detailed data and offering an aerial perspective of landscapes (Acharya et al., 2021). From river flow analysis to flood prediction, UAVs offer diverse applications that contribute to more effective water management strategies. This section explores the diverse roles that UAVs play in hydrology and water management:

1. **River and Water Body Monitoring**: UAVs equipped with remote sensing instruments can capture imagery and data of rivers, lakes, and reservoirs. This information aids in tracking water levels, flow patterns, sediment transport, and water quality. By regularly monitoring these water bodies, water resource managers can make informed decisions regarding water allocation and resource planning.
2. **Flood Management and Prediction**: UAVs play a crucial role in flood management by providing real-time information during flood events. Equipped with sensors, UAVs can assess the extent of flooding, identify vulnerable areas, and provide situational awareness to emergency responders. Flood modeling and prediction benefit from UAV-captured data, helping authorities make timely decisions to mitigate damage and protect lives.
3. **Watershed Analysis**: UAVs can perform comprehensive watershed analysis by creating high-resolution topographic maps. These maps provide insights into the terrain, including elevation changes and drainage patterns. This information aids in understanding water flow dynamics, identifying erosion-prone areas, and planning effective erosion control measures.
4. **Snowpack Monitoring**: In regions where snowmelt contributes significantly to water availability, UAVs equipped with thermal sensors can monitor snowpack dynamics. By measuring temperature variations, UAVs help assess the rate of snowmelt, contributing to accurate predictions of water supply during the melt season.
5. **Wetland and Ecosystem Assessment**: UAVs are invaluable for studying wetlands and ecosystems. They can provide detailed imagery of wetland vegetation, water quality, and habitat conditions. This data aids in wetland conservation, restoration efforts, and biodiversity monitoring.
6. **Infrastructure Inspection**: UAVs can be used to inspect water infrastructure, including dams, levees, and irrigation systems. High-resolution imagery and data collected by UAVs help assess structural integrity, identify potential issues, and support maintenance planning.
7. **Water Quality Monitoring**: UAVs equipped with specialized sensors can analyze water quality parameters, such as turbidity, nutrient levels, and pollution sources. This data assists in identifying pollution hotspots and developing targeted pollution control strategies (Acharya et al., 2021).
8. **Rapid Response and Disaster Management**: In the aftermath of natural disasters, such as hurricanes or tsunamis, UAVs can quickly survey affected areas, providing critical information to assess damage, plan relief efforts, and restore water supply systems.

5 Surface Water Resources Monitoring and Assessment

Precise and timely data on the quality and amount of surface water is crucial for a range of purposes, such as irrigation in agriculture. Unmanned aerial vehicle (UAV) systems can provide valuable data on water quality and quantity in the field. This data can help fill in gaps in spatial information by generating near-real-time, high-resolution, and detailed information needed for water resources accounting. This technology offers a comprehensive and efficient way to gather crucial data for water resource management (WRM), environmental conservation, and disaster response. Here are some key aspects to consider:

- **Remote Sensing Capability**: Unmanned aerial vehicles (UAVs) equipped with advanced remote sensing technologies, such as multispectral and hyperspectral cameras, LiDAR (Light Detection and Ranging), and thermal sensors, provide accurate and detailed data on the characteristics of surface water. This includes water quality parameters (turbidity, chlorophyll levels, temperature), water levels, shoreline changes, and even underwater topography.
- **Environmental Monitoring**: UAVs enable real-time assessment of water quality and pollution sources. They can identify areas of concern, monitor the spread of contaminants, and help pinpoint potential pollution origins. This information is vital for formulating effective pollution control and management strategies (Ryu, 2022).
- **Leakage Detection**: UAVs can capture high-resolution imagery of the canal's banks and infrastructure. This data can be analyzed to identify leaks, cracks, or other structural issues that might lead to water loss.
- **Flow Velocity and Direction**: Using specialized sensors, UAVs can estimate water flow velocity and direction within the canals. This information assists in understanding how water is moving and whether adjustments are needed for efficient water distribution.
- **Sediment Accumulation Monitoring**: UAVs can capture imagery that reveals sediment buildup in canals. Excessive sedimentation can impede water flow and reduce the canal's capacity. Regular monitoring helps plan maintenance activities.
- **Precision Water Management**: The data collected by UAVs contributes to precision water management by providing insights into water distribution, identifying water wastage, and enabling better planning for irrigation and water allocation (Yao et al., 2019).

6 Real-Time Water Management and Distribution

UAVs equipped with advanced sensors and imaging capabilities gather up-to-the-minute data on factors such as water levels, flow rates, and quality indicators. The data is sent to central control systems, where it is promptly processed and analyzed. This dynamic information allows water managers, authorities, and stakeholders to make timely adjustments in water distribution strategies, ensuring that supply meets demand while minimizing wastage and maximizing resource utilization (Mishra et al., 2023).

The Green Water Drone (GWD) utilizes the latest breakthroughs in unmanned aerial vehicles (UAVs) to provide a state-of-the-art system that combines airborne observations of canopy temperatures. This method enables the timely and accurate assessment of agricultural irrigation needs (crop water stress detection) and continuous monitoring of crop growth. Groundwater monitoring technology (e.g., GWD) produces precise readings with both precise timing and detailed coverage, while also minimizing the requirement for human involvement in crop activities and decreasing system maintenance expenses. The GWD approach is especially suitable for relatively small fields with increased crop diversity that cannot be efficiently addressed by satellite remote sensing. A micrometeorological device installed on board an aircraft allows for the precise evaluation of the crop water stress index (CWSI) for different crops in the area.

The parallel utilization of thermal imagery and spatial measurements by infrared radiometers offers the advantage of allowing us to accurately compare and calibrate the techniques. Furthermore, it is crucial to emphasize that the determination of the crop water stress index (CWSI) on a practical level is achieved through the use of infrared thermometers attached to unmanned aerial vehicles (UAVs), which are also referred to as Airborne Multispectral Measuring Systems (AMMS) (Ryu, 2022). Figure shows the first pilot flight with AMMS system.

As the world grapples with the intricate challenges of water availability, the convergence of technology and agriculture presents a promising path forward. The utilization of UAVs in real-time water management signifies a paradigm shift toward proactive and adaptive approaches. This integration not only addresses the demands of the present but also safeguards the future by ensuring that agricultural practices align with both the needs of a growing population and the imperatives of environmental stewardship. In essence, the marriage of UAV technology and agriculture heralds a new era in water management that holds the promise of a more resilient and thriving agricultural industry (Alexandris et al., 2021).

A flowchart of key aspects of surface water resources monitoring and assessment is given in Fig. 2.

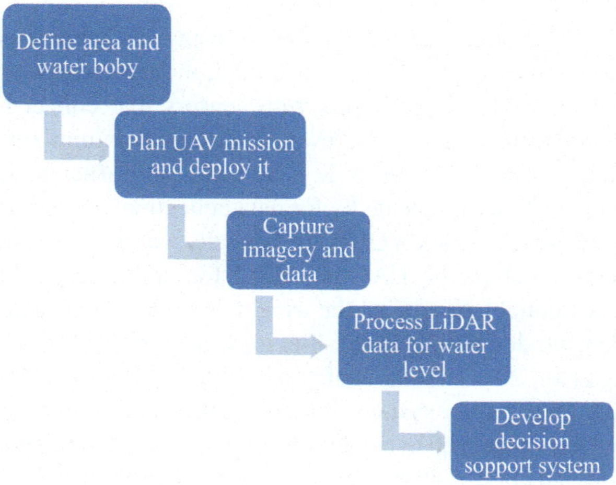

Fig. 2 Managing surface water through UAVs

7 Water Management Through Low-Altitude Remote Sensing

Low-altitude remote sensing (LARS) has emerged as a transformative approach to revolutionize water management practices, offering a versatile and efficient method to monitor, assess, and optimize water resources. By harnessing the capabilities of modern technology, particularly unmanned aerial vehicles (UAVs), this approach provides a comprehensive solution to address water scarcity, pollution, and the sustainable utilization of this precious resource. Here are key aspects of water management through low-altitude remote sensing:

- **Precision Data Collection**: Low-altitude remote sensing utilizes UAVs equipped with advanced sensors to gather precise and high-resolution data on water bodies, such as rivers, lakes, reservoirs, and aquifers. This data includes water levels, quality parameters, temperature, and other critical indicators.
- **Infrastructure Inspection**: UAVs conduct inspections of water infrastructure, such as dams and pipelines, identifying potential issues and maintenance needs. This preventive approach reduces downtime and improves safety.
- **GIS Integration**: Data collected through low-altitude remote sensing is integrated into Geographic Information Systems (GIS), enhancing spatial analysis and facilitating data-driven decision-making.
- **Community Engagement**: Visual data generated by UAVs is easily understandable by the general public, fostering community engagement and awareness about water management and conservation (Chapman et al., 2014).

7.1 UAV-LARS Platform

The UAV-LARS is an advanced aerial remote sensing system that uses a UAV as a platform to carry sensor payloads, such as digital, multispectral, and hyperspectral cameras. The UAV-LARS is equipped with sensors that collect high-resolution images, providing accurate data from low altitudes while minimizing the impact of cloud interference. The UAV-LARS is a versatile and efficient remote sensing equipment that may be used repeatedly for data collection. The system architecture of an unmanned aerial vehicle (UAV) is typically tailored to specific missions. Prioritizing the weight of the UAV is crucial for optimizing energy efficiency. Therefore, it is essential to minimize the weight of the UAV as much as feasible. To achieve an adequate duration of flight, it is imperative to have a robust propulsion system (Zhang et al., 2021). Additionally, other modules of the UAV, including the autopilot, inertial navigation system, imaging sensor and its mounting, and route planning software, are deemed to be of utmost importance. Drones come in a diverse range of designs and combinations. Figure 3 gives a schematic diagram of the UAV-LARS monitoring system.

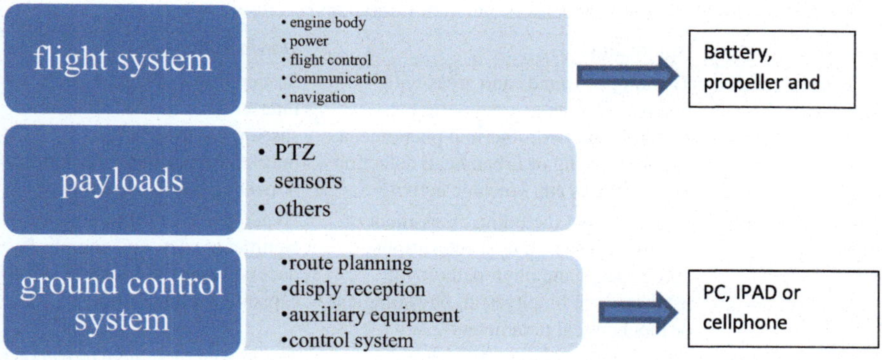

Fig. 3 The UAV-LARS monitoring system is composed of an unmanned aerial vehicle for low-altitude remote sensing

8 Challenges and Limitations of UAVs in Water Management

While unmanned aerial vehicles (UAVs) offer innovative solutions for water management, they also encounter several challenges and limitations that must be addressed to ensure their effective and sustainable implementation. Understanding these constraints is crucial for maximizing the potential of UAV technology in the field of water resource management (Gago et al., 2015).

Here are some key challenges and limitations:

1. **Regulatory Hurdles**: Unmanned aerial vehicle (UAV) operations are governed by stringent laws and limitations on airspace usage. Adhering to aviation regulations and acquiring required authorizations can be a laborious and intricate process.
2. **Limited Flight Time**: UAVs have limited battery life, resulting in relatively short flight durations. This constrains the area covered in a single flight and may require multiple trips for comprehensive monitoring.
3. **Weather Dependency**: Unfavorable weather conditions, such as high winds, precipitation, or poor visibility, might impede UAV flights and data collecting, affecting their dependability and accessibility.
4. **Payload Capacity**: The weight of sensors and equipment carried by UAVs affects their flight capabilities. Heavier payloads can reduce flight time and maneuverability.
5. **Data Processing Complexity**: The raw data collected by UAVs require processing to extract meaningful information. Data analysis and interpretation can be complex, requiring specialized expertise (Gago et al., 2015) (Table 1).

Table 1 Different UAV onboard sensors and their limitations

Sensors	Applications	Drawbacks
Thermal infrared sensors	Tracking of islands and areas with potential dangers; research into hydrothermal phenomena and monitoring of urban heat; detection of forest fires and volcanic activity	Susceptible to distance and interference; limited precision; susceptible to sunshine interference; suitable for indoor or evening use exclusively
Hyperspectral sensors	Identification, including assessment of destruction caused by a catastrophe; agroecology and phytopathology; estimation of biophysical, physiological, or biochemical parameters	Hyperspectral imaging incurs significant costs and complexity, necessitating a substantial storage capacity for huge data
Light detection and ranging (LiDAR)	Cartography of cultural and heritage sites; assimilation; forest carbon estimation; vegetation canopy study	The impact of vehicle movement is influenced by atmospheric conditions, calibration mistakes, and sensor noise
Global positioning system (GPS)	Timing; mapping; tracking; navigation; localization	The issues include flaws in the receiver clock, problems in the satellite's orbit, delays, and vulnerability to magnetic interference

Source: Mohsan et al. (2023)

9 Future Research Direction

Given the future climate change scenario and the observed increase in the severity of both heavy rainfall and drought events in the Northern Hemisphere, there is a growing need for UAV technology to assist in global water resource management decision-making. We are confident that there are significant market prospects for unmanned aerial vehicle (UAV) platforms that can deliver high-quality, real-time data with precise spatial and temporal resolution across many water resource management (WRM) domains. The previous sections have already highlighted the significant prospects that UAV science can offer (Yao et al., 2019).

In some places where water might become scarcer due to climate change, having a user-friendly and portable remote sensing system like a UAV could really help keep an eye on things like landscape water patterns and changes in water resources.

- The use of unmanned aerial vehicle (UAV) technology is already being investigated to monitor the water reserves in glaciers. Researchers have utilized photogrammetric image analysis methods to measure glaciological processes and gain insights into glacial dynamics. In arid mountainous locations, such as the Andes, the cryosphere contains concealed water behind geological formations like rock glaciers. However, our understanding of these characteristics remains limited.
- In certain regions of China, there are occurrences of reduced intensity rainfall, resulting in prolonged periods of water scarcity. Using unmanned aerial vehicles (UAVs) to improve the monitoring of water resources and control usage has the potential to greatly expand current understanding. For example, regularly con-

ducting assessments of reservoir resources and measuring soil moisture using thermal imaging, or applying precision agriculture techniques to optimize consumption, could be beneficial strategies.

- Being able to accurately identify the intricate patterns of soil moisture reserves would greatly benefit non-governmental organizations and local groups in arid regions like Sub-Saharan Africa. Understanding the rate at which aquifers are replenished is essential for comprehending how it impacts the supply of safe drinking water for local communities. Moreover, possessing a spatial understanding of soil moisture will facilitate accurate irrigation, specifically in areas where there is a scarcity of soil moisture. This will ensure that the costly procedure of irrigating agricultural land is reduced to a minimum (Kim et al., 2008).
- In areas affected by disasters, where there is a high concentration of displaced individuals, it would be beneficial to quickly map and measure the soil water resource. This would aid in humanitarian relief efforts by identifying accessible sources of clean water near the surface and establishing suitable water supplies. There exists a pre-existing group of humanitarian "UAViators" who are a worldwide network of volunteer UAV pilots. These pilots can be professionals, civilians, or responsible hobbyists. Their main purpose is to facilitate the sharing of information, coordination, and operational safety to support various humanitarian efforts. Further research in this field has the potential to offer valuable knowledge for water resource management (WRM) in different contexts.

References

Acharya, B. S., Bhandari, M., Bandini, F., Pizarro, A., Perks, M., et al. (2021). Unmanned aerial vehicles in hydrology and water management: Applications, challenges, and perspectives. *Water Resources Research, 57*, e2021WR029925. https://doi.org/10.1029/2021WR029925

Alexandris, S., Psomiadis, E., Proutsos, N., Philippopoulos, P., Charalampopoulos, I., Kakaletris, G., Papoutsi, E.-M., Vassilakis, S., & Paraskevopoulos, A. (2021). Integrating drone technology into an innovative agrometeorological methodology for the precise and real-time estimation of crop water requirements. *Hydrology, 8*, 131. https://doi.org/10.3390/hydrology8030131

Chapman, S. C., Merz, T., Chan, A., Jackway, P., Hrabar, S., Dreccer, M. F., Holland, E., Zheng, B., Ling, T. J., & Jimenez-Berni, J. (2014). Pheno-copter: A low-altitude, autonomous remote-sensing robotic helicopter for high-throughput field-based phenotyping. *Agronomy, 4*, 279–301. https://doi.org/10.3390/agronomy4020279

Gago, J., Douthe, C., Coopman, R. E., Gallego, P. P., Ribas-Carbo, M., Flexas, J., Escalona, J., & Medrano, H. (2015). UAVs challenge to assess water stress for sustainable agriculture. *Agricultural Water Management, 153*, 9–19.

Khan, A. I., & Al-Mulla, Y. (2019). Unmanned aerial vehicle in the machine learning environment. *Procedia Computer Science, 160*, 46–53.

Kim, Y., Evans, R. G., & Iversen, W. M. (2008). Remote sensing and control of an irrigation system using a distributed wireless sensor network. *IEEE Transactions on Instrumentation and Measurement, 57*(7), 1379–1387.

Mishra, V., Avtar, R., Prathiba, A. P., Mishra, P. K., Tiwari, A., Sharma, S. K., Singh, C. H., Yadav, B. C., & Jain, K. (2023). Uncrewed aerial systems in water resource management and monitoring: A review of sensors, applications, software, and issues. *Advances in Civil Engineering, 2023*, 3544724, 28 pages. https://doi.org/10.1155/2023/3544724

Mohsan, S. A. H., Othman, N. Q. H., Li, Y., et al. (2023). Unmanned aerial vehicles (UAVs): Practical aspects, applications, open challenges, security issues, and future trends. *Intelligent Service Robotics, 16*, 109–137. https://doi.org/10.1007/s11370-022-00452-4

Nhamo, L., Magidi, J., Nyamugama, A., Clulow, A. D., Sibanda, M., Chimonyo, V. G. P., & Mabhaudhi, T. (2020). Prospects of improving agricultural and water productivity through unmanned aerial vehicles. *Agriculture, 10*, 256. https://doi.org/10.3390/agriculture10070256

Ryu, J. H. (2022). UAS-based real-time water quality monitoring, sampling, and visualization platform (UASWQP). *HardwareX, 11*, e00277. https://doi.org/10.1016/j.ohx.2022.e00277. issn:2468-0672. https://www.sciencedirect.com/science/article/pii/S2468067222000220

Upadhyaya, A., Jeet, P., Sundaram, P. K., Singh, A. K., Saurabh, K., & Deo, M. (2022). Efficacy of drone technology in agriculture: A review: Drone technology in agriculture. *Journal of AgriSearch, 9*(3), 189–195.

Yao, H., Qin, R., & Chen, X. (2019). Unmanned aerial vehicle for remote sensing applications—A review. *Remote Sensing, 11*, 1443. https://doi.org/10.3390/rs11121443

Zhang, H., Wang, L., Tian, T., & Yin, J. (2021). A review of unmanned aerial vehicle low-altitude remote sensing (UAV-LARS) use in agricultural monitoring in China. *Remote Sensing, 13*, 1221. https://doi.org/10.3390/rs13061221

Harmonizing Nature's Flow: Ecohydrology for Sustainable Agricultural Water Management

Tariq Aziz

Abstract This chapter explores the applications and significance of ecohydrology in the context of sustainable agricultural water management. Ecohydrology, an interdisciplinary field bridging ecological and hydrological sciences, addresses pressing concerns related to water scarcity, environmental degradation, and ecosystem health in agriculture. By understanding the bidirectional relationship between ecosystems and hydrological processes, ecohydrology offers a holistic approach to optimizing water resources and preserving ecological integrity. The chapter discusses key ecohydrological concepts, approaches in agriculture, and their relevance to sustainable irrigation, ecosystem-based water management, and appropriate land use management. It also highlights the importance of ecohydrological tools such as remote sensing, geographic information systems (GIS), and hydrological models in decision-making processes. Emphasizing ecohydrology's potential to offer sustainable solutions, the chapter advocates integrating these principles into agricultural water management to create a more resilient and ecologically balanced future.

Keywords Ecohydrology · Bidirectional Relationships · Remote Sensing, GIS, and Hydrological Models · Sustainable Water Management

1 Introduction

Agriculture, often hailed as the cornerstone of human civilization, has been instrumental in sustaining our ever-expanding population by providing essential resources such as food, fiber, and fuel. However, the intensification of agricultural practices and the escalating demands for water resources have given rise to

T. Aziz (✉)
Ausenco, Burlington, ON, Canada

Ecohydrology Research Group, Water Institute and Department of Earth and Environmental Sciences, University of Waterloo, Waterloo, ON, Canada
e-mail: tariq.aziz@uwaterloo.ca

M. Mubeen et al. (eds.), *Innovations in Agricultural Water Management*,
https://doi.org/10.1007/978-3-031-91883-4_21

pressing concerns regarding water scarcity, environmental degradation, and the overall health of ecosystems. These challenges necessitate a paradigm shift in agricultural water management toward more sustainable and eco-friendly practices (Okasha et al., 2022).

Water pollution is a pervasive global concern that exerts adverse effects on both ecosystems and human well-being. Of note, agriculture stands as a primary contributor to water pollution, outpacing urban and industrial sources in several regions. In the United States, agriculture predominantly drives river pollution, while in China, agriculture tightly regulates groundwater nitrogen (Ramião et al., 2022). Within the European Union, a substantial portion of surface water bodies failed to attain a satisfactory ecological state in 2016, primarily due to agricultural pressures (Ramião et al., 2022). Agricultural expansion, intensified by technological advancements and increased nutrient input, has been necessary to meet the escalating food demands driven by population growth and dietary shifts. The overuse of nitrogen and phosphorus in agriculture results in water contamination via leaching and surface runoff, leading to eutrophication of both inland and coastal waters. Eutrophication, characterized by excessive plant and algal growth, has far-reaching economic consequences, impacting tourism, fisheries, and water treatment costs (Wurtsbaugh et al., 2019). Anticipated population growth in the coming decades further accentuates the urgency of addressing the agricultural sector's environmental footprint. Although various policy frameworks have sought to mitigate agricultural impacts on water quality, substantial challenges persist. Sustainable agricultural practices, including cover crops, reduced tillage, and vegetated filter strips, have been explored to mitigate water pollution, with filter strips demonstrating notable efficacy in reducing sediment and nutrient loads (Ramião et al., 2022). Nevertheless, questions linger regarding the optimal combination of measures to achieve water quality objectives, and empirical data collection is constrained by cost and lag time. Modeling approaches offer promise in predicting the long-term effectiveness of management strategies, identifying critical source areas, and informing decision-making processes in support of water quality improvement (Ramião et al., 2022).

In response to these mounting concerns, ecohydrology has emerged as a pivotal and interdisciplinary field that seeks to bridge the gap between ecological and hydrological sciences to tackle the challenges posed by agricultural water management. Ecohydrology recognizes the bidirectional relationship between ecosystems and hydrological processes, where water availability, movement, and quality profoundly influence the health and functioning of ecosystems, and vice versa (Kundzewicz, 2002). By understanding and harnessing these dynamic interactions, ecohydrology offers an innovative and integrated approach to addressing the complex challenges facing modern agriculture (Cao et al., 2022).

The primary objective of this chapter is to explore the applications and significance of ecohydrology in the context of sustainable agricultural water management. By analyzing the principles, tools, and technologies that define the field of ecohydrology, I aim to shed light on how this emerging discipline can contribute to the optimization of water resources and the preservation of ecological integrity in agricultural landscapes.

2 Understanding Ecohydrology

2.1 *Ecohydrological Concepts*

Ecohydrology is the study of the complex interactions between ecosystems and water processes. It recognizes the bidirectional relationships between ecological systems and hydrological processes, where vegetation, soil, and land use influence water flow and availability, while water availability and dynamics impact the distribution and functioning of ecosystems. The key concepts of ecohydrology include water use efficiency, water balance, and the influence of land cover changes on hydrological processes.

At the heart of ecohydrology lies the understanding of how water availability and quality influence ecological systems, and vice versa (Hannah et al., 2004). One of the key concepts in this field is the water–ecology nexus, which emphasizes the dynamic feedback between water and ecosystems (Tan et al., 2021). Water is the lifeblood of ecosystems, as it sustains plant and animal populations, drives nutrient cycles, and shapes habitat structures (Gordon et al., 2008). Conversely, ecosystems play a crucial role in regulating hydrological processes. Vegetation, for example, influences rainfall interception, evapotranspiration (ET), and soil moisture dynamics, ultimately affecting local and regional hydrology. Understanding this nexus is essential for designing sustainable water management strategies that consider both ecological and human needs.

2.1.1 Ecohydrological Processes

Ecohydrological processes encompass a wide range of interactions between water and ecosystems (Tan et al., 2021). These processes include:

(a) **Evapotranspiration**: The combined process of evaporation from the land surface and transpiration from plants (Gordon et al., 2008). It is a critical component of the hydrological cycle, regulating the movement of water vapor from the land to the atmosphere. Vegetation types, land use, and climate all influence evapotranspiration rates, making it a key variable in ecohydrological studies (Tan et al., 2021).

(b) **Infiltration**: The process by which water penetrates the soil surface and moves into the subsurface (Beven, 2021). Infiltration rates are influenced by soil properties, vegetation cover, and land use practices, and they play a vital role in groundwater recharge and surface water runoff (Gordon et al., 2008).

(c) **Streamflow generation**: The movement of water through river networks, which is influenced by precipitation patterns, land cover, and the geomorphology of the landscape. Understanding the factors that control streamflow is essential for managing water resources and maintaining ecological health in river ecosystems (Gordon et al., 2008).

(d) **Nutrient cycling**: The movement and transformation of nutrients within ecosystems, including processes such as nutrient uptake by plants, mineralization, and leaching. Ecohydrology plays a crucial role in understanding how water transport influences nutrient availability and cycling in aquatic and terrestrial ecosystems (Gordon et al., 2008).

2.2 Ecohydrological Approaches in Agriculture

In agricultural settings, ecohydrology provides a holistic approach to sustainably understanding and managing water resources. It combines ecological factors such as plant physiology, species composition, and biodiversity with hydrological processes like evapotranspiration, runoff, and groundwater recharge.

Societal alterations of water and biogeochemical cycles aim to enhance ecosystem productivity and mitigate climatic risks. However, these modifications often yield significant adverse environmental consequences, necessitating strategies for sustainable resource utilization. The use of dynamic models is imperative to tackle the complexities of modeling the bidirectional interaction between society and the environment. These models highlight the role of soil and water quality in supporting societal growth while also showcasing how excessive pressure can lead to critical transitions and sustained economic cycles. Emphasis should be placed on the coupled dynamics of soil water and solutes (nutrients or contaminants), underscoring the modeling challenges stemming from nonlinearities within the soil–plant system and unpredictable hydroclimatic forces (Porporato et al., 2015). Furthermore, applications of these models must be extended to issues such as irrigation, soil salinization, and fertilization. The potential for employing methods from stochastic control, informed by rigorous physical and mathematical descriptions of ecohydrological and biogeochemical interactions, supports optimal solutions for large-scale, long-term soil and water resource management in agroecosystems under conditions of uncertainty (Porporato et al., 2015).

3 Ecohydrology and Agricultural Water Management

Addressing water resource conservation in the face of climate change is increasingly becoming urgent, especially in arid and semi-arid regions. Ecohydrology plays a pivotal role in understanding the critical zone, optimizing water usage, and mitigating climate change impacts. The growing applications of ecohydrology suggest that environmental sustainability and socio-economic development efforts centered around water resource management hold promising prospects for the future (Castellini et al., 2022), as discussed below.

3.1 Sustainable Irrigation Practices

Ecohydrology emphasizes the importance of efficient irrigation practices to conserve water and optimize agricultural productivity. Implementing technologies like drip irrigation and precision agriculture, which deliver water directly to plant roots based on their specific needs, reduces water wastage, and enhances water use efficiency (Velasco-muñoz et al., 2019).

Sustainable irrigation practices are of paramount importance in the face of mounting challenges such as population growth, food security concerns, and the increasing strain on water resources. These practices aim to balance the needs of agriculture with environmental conservation, ensuring that irrigation does not deplete water sources or harm ecosystems. In this chapter, I present key sustainable irrigation practices, their significance, and their potential to foster resilient agricultural systems while safeguarding the environment.

3.2 Efficient Water Management

Efficient water management is a cornerstone of sustainable irrigation practices (Scott et al., 2014). It involves optimizing the use of water resources to minimize waste and improve irrigation efficiency. Modern irrigation methods, such as drip and sprinkler systems, significantly reduce water losses compared to traditional methods like flood irrigation. Implementing precision agriculture techniques, which utilize data and technology to tailor irrigation to crop needs, further enhances water use efficiency (Velasco-muñoz et al., 2019). By conserving water, efficient management not only promotes sustainable agricultural production but also mitigates the over-extraction of groundwater and protects aquatic ecosystems (Ingrao et al. 2023).

3.3 Soil Health and Conservation

Sustainable irrigation practices extend beyond water management to encompass soil health and conservation. Implementing practices such as reduced tillage, cover cropping, and mulching helps maintain soil structure, reduce erosion, and enhance water infiltration (Farmaha et al., 2022). Healthy soils with improved water-holding capacity require less irrigation water to support crop growth. Moreover, these practices contribute to carbon sequestration, mitigating climate change impacts (Farmaha et al., 2022). Sustainable soil management not only boosts agricultural productivity but also safeguards the ecological integrity of surrounding landscapes (Robertson & Swinton, 2005).

3.4 Tailored Irrigation Scheduling

Tailored irrigation scheduling involves applying water when crops need it most, based on factors like crop type, growth stage, and local weather conditions. This practice minimizes over-irrigation and reduces the risk of waterlogging and soil salinity, which can occur when excess water is applied. Modern tools and technologies, such as soil moisture sensors and weather forecasting, enable farmers to make informed decisions about irrigation timing, optimizing resource use. By aligning water application with crop demand, tailored scheduling conserves water and mitigates the negative environmental impacts associated with excessive irrigation (Levidow et al., 2014).

3.5 Integrated Pest and Disease Management

Integrated pest and disease management (IPM) is an integral component of sustainable irrigation practices (Karlsson Green et al., 2020). IPM emphasizes the use of ecologically friendly methods, such as biological control and crop rotation, to minimize the reliance on chemical pesticides (Karlsson Green et al., 2020). Over-irrigation can create favorable conditions for pest and disease development, and excessive use of pesticides can harm non-target species and contaminate water bodies (Tudi et al., 2021). By adopting IPM strategies, farmers can maintain a balance between pest control and environmental preservation while reducing the ecological footprint of agriculture (Karlsson Green et al., 2020).

3.6 Ecosystem-Based Water Management

Adopting ecosystem-based water management strategies can improve water availability and quality in agricultural landscapes. Integrating natural elements like wetlands, buffer zones, and riparian vegetation into agricultural areas can enhance water purification and retention, mitigating the impacts of agricultural runoff on downstream ecosystems (Wu et al., 2023). The realization of conventional climate adaptation shortcomings sparks interest in sustainable alternatives, fostering ecosystem health and social equity. Ecosystem-based adaptation garners attention as an integrative approach, bolstering resilience and reducing vulnerability in social–ecological systems. Yet, understanding institutional prerequisites for this shift remains insufficient. Adaptive governance, a flexible, multi-level mechanism, offers promise in addressing governance challenges by uniting actors across scales, managing value conflicts and uncertainties in ecosystem management. Akamani (2023) explored four roles of adaptive governance in water resource systems' transition toward ecosystem-based adaptation, First, policies should foster an environment conducive to integrating diverse knowledge, developing relevant climate change impact indicators, and testing mitigation and adaptation strategies through adaptive management. Second, these policies should offer a range of incentives, both economic and non-economic, to galvanize collective action at various societal levels. Third, they should support capacity-building through funding, information dissemination, and technical training. Lastly, policies geared toward advancing ecosystem-based adaptation should facilitate the meaningful engagement of diverse actors across scales in decision-making and the implementation of adaptation processes, by establishing suitable institutional frameworks (Akamani, 2023).

One example to manifest the importance of natural systems in water management is through showcasing the differences between earthen and brick water courses. Earthen water courses offer unique advantages in water ecohydrology compared to masonry or brick structures, driven by the presence of vegetation and the inherent characteristics of earthen materials that contribute to their effectiveness in supporting ecological and hydrological processes. These advantages include natural filtration facilitated by vegetation along the banks, where plant roots absorb nutrients and pollutants, enhancing water quality as it moves through the system. The presence of vegetation further promotes nutrient cycling, with plants absorbing and subsequently returning nutrients to the ecosystem, ensuring a balanced nutrient supply and overall ecosystem health.

Moreover, the sponge-like properties of earthen materials enable a higher water retention capacity compared to masonry or brick, maintaining a consistent water flow and preventing rapid runoff and soil erosion (Zhang et al., 2023). This water retention capability also contributes significantly to the recharge of groundwater aquifers, crucial for ensuring a reliable water supply during dry periods. Additionally, the natural characteristics of earthen water courses, combined with the presence of vegetation, foster diverse ecosystems, and serve as wildlife corridors, facilitating species movement between different habitats. The construction and maintenance of earthen water courses, with their lower environmental impact and potential cultural and historical significance, align with sustainable water management practices. Nevertheless, the choice between earthen and masonry structures should consider factors such as proper design, maintenance, and erosion risks, guided by local conditions and specific project requirements (Nilimaa, 2023). Overall, recognizing the ecological functions of earthen water courses underscores the importance of integrating these natural systems into comprehensive water management strategies.

3.7 Sustainable Land Use and Crop Selection

To meet global sustainability goals (the United Nations Sustainable Development Goals and Paris Agreement), attention to key sectors such as forestry and agriculture is crucial. These sectors impact emissions, biodiversity, and human well-being (Zerriffi et al., 2023). Ecohydrology encourages the adoption of appropriate land use practices and crop selection to match the local hydrological conditions. Certain crops may be more suitable for regions with limited water availability, reducing the risk of water stress and supporting food security (Ingrao et al. 2023).

Global population growth, particularly in developing nations, will likely necessitate increased agricultural production. However, sustainability hinges on preserving ecosystem functions. Farmers must navigate the trade-offs between agricultural productivity and ecosystem services. Sanou et al. (2023) reviewed 184 articles categorizing them into tools, practices, models, or other topics. While global in scope, they show that the approach to trade-off analysis varies by region, with Europe and Asia most studied. Few studies are conducted in sub-Saharan Africa. Most tools focus on specific ecosystem services (water, biodiversity, climate regulation) and different land use types, scales, and ecosystems. The study recommends collaboration among practitioners, organizations, and scientists for developing practical tools aligned with Sustainable Development Goals for sustainable strategies, productivity, and ecosystem service balance, tailored to specific contexts through innovation platforms (Sanou et al., 2023).

4 Ecohydrological Tools and Technologies

4.1 Remote Sensing and GIS

Remote sensing and geographic information systems (GIS) play a crucial role in ecohydrology by providing valuable data for water resource monitoring and management. Satellite imagery can be used to assess vegetation health, evapotranspiration rates, and land use changes, aiding in decision-making processes (Almalki et al., 2022).

This interdisciplinary approach integrates the capabilities of remote sensing technology, which gathers data from a distance using various sensors, and GIS, which facilitates the analysis and visualization of spatial data. In this chapter, I explore the critical role of Remote Sensing and GIS in ecohydrology, emphasizing their applications, advantages, and contributions to advance the understanding of ecological and hydrological processes.

4.1.1 Monitoring Land Cover and Land Use Changes

Remote sensing offers a unique vantage point for monitoring land cover and land use changes over time. Satellite imagery, aerial photography, and LiDAR (Light Detection and Ranging) data can capture high-resolution snapshots of landscapes, enabling the tracking of changes in vegetation cover, urban expansion, and deforestation (Almalki et al., 2022). By integrating these data into GIS platforms, researchers can analyze and quantify the impacts of land cover and land use changes on local and regional hydrological processes, including altered runoff patterns and changes in evapotranspiration rates (García-Santos et al., 2022; Zhao et al., 2022).

4.1.2 Estimating Evapotranspiration

Remote sensing plays a pivotal role in estimating evapotranspiration (ET), a key ecohydrological process. Sensors onboard satellites and aircraft can measure surface temperature and vegetation indices, which are critical inputs for ET models (García-Santos et al., 2022). GIS allows for the integration of remote sensing-derived data with meteorological information and ground-based measurements to produce spatially explicit ET maps. These maps help in assessing water use by different land cover types and guiding water resource management decisions (Mashala et al., 2023).

4.1.3 Watershed Analysis

GIS is instrumental in ecohydrological modeling and watershed analysis. It provides a platform for integrating diverse datasets, including topography, soil properties, land cover, and precipitation, into comprehensive hydrological models (García-Santos et al., 2022). With GIS, researchers can simulate the movement of water through watersheds, predict streamflow, and identify areas susceptible to erosion and flooding (Shuaibu et al., 2023). This information is invaluable for watershed management and conservation efforts.

4.1.4 Assessing Ecosystem Services

Remote sensing and GIS enable the assessments of ecosystem services—benefits that ecosystems provide to society. These services include water purification, carbon sequestration, and habitat provision (Aziz, 2023). Through satellite imagery and GIS-based analysis, researchers can quantify the extent and quality of ecosystems, identifying areas that are critical for providing specific services (Shuaibu et al., 2023). This information supports informed decision-making in environmental planning and policy development.

4.1.5 Climate Change Impact Assessment

Monitoring the impact of climate change at the watershed scale is a complex endeavor that requires a multifaceted approach. Remote sensing technologies and geographic information systems (GIS) have emerged as valuable tools for assessing climate change impacts on watersheds. Remote sensing methodologies present a highly effective and thorough means of monitoring climate change impacts at both local and regional levels. Furthermore, these techniques empower the generation of a range of indicators related to plant water status, thereby assuming a pivotal role in the assessment of water stress within terrestrial ecosystems (Le et al., 2023).

Utilizing satellite data, LiDAR technology, and high-resolution climate models, researchers can monitor changes in land cover, vegetation, hydrological patterns, and temperature, providing critical insights into the evolving state of watersheds (Muhadi et al., 2020; Taiwo et al., 2023). Additionally, long-term monitoring programs, such as the National Ecological Observatory Network (NEON) in the United States, offer comprehensive datasets for tracking climate-related alterations within watersheds (NEON, 2013). Furthermore, incorporating hydrological models and machine learning techniques can help predict future impacts and enhance our understanding of climate change effects on water resources at the watershed scale (Taye et al., 2011). A combination of these approaches provides a robust framework for ongoing climate change impact assessment in watersheds, facilitating informed decision-making for sustainable water resource management.

Remote sensing and GIS are essential tools for assessing the impact of climate change on ecohydrological systems. They enable the monitoring of temperature changes, vegetation shifts, and alterations in precipitation patterns. By integrating climate data with ecohydrological models, researchers can predict how climate change may affect water availability, biodiversity, and ecosystem health (Shuaibu et al., 2023).

4.2 Hydrological Models

Researchers have been employing diverse methodologies for the evaluation of water resources, spanning from rudimentary water balance models to intricate hydrological models that encompass a myriad of components pertaining to water systems (Norvanchig & Randhir, 2021). Globally, there is a push to develop and apply models to help quantify water ecosystem services. Toward this, a number of tools are emerging with ecosystem services concepts at their core, such as ARtificial Intelligence for Environment and Sustainability (ARIES), Co$ting Nature, WaterWorld, Soil and Water Assessment Tool (SWAT), Water Evaluation And Planning (WEAP) model, Envision, Land Utilisation Capability Indicator (LUCI), and Integrated Valuation of Ecosystem Services and Tradeoffs (InVEST). I used the search terms "Ecosystem Services," "Water," and "Mapping" to search in ISI Web of Science—Core Collection. I reviewed papers published between 2010 and 2022 that map one or more water ecosystem services using a model. The results indicate that the InVEST model is by far the most prominent in the scientific literature (Table 1).

Table 1 Models used for mapping of water ecosystem services in a selection of the literature

Author(s)/ year	Title	Study site	Water ecosystem services mapped	Model used
Schmidt et al. (2020)	Development of a mapping framework for ecosystem services: The case of sediment control supply at a watershed scale in Newfoundland, Canada	Harry's river watershed, Newfoundland, Canada	Sediment control service	Arc Soil and Water Assessment Tool (SWAT)
Dhar et al. (2018)	Large scale biotic damage impacts forest ecosystem services	British Columbia, Canada	Water provisioning	ArcGIS
Guerry et al. (2012)	Modelling benefits from nature: Using ecosystem services to inform coastal and marine spatial planning	Lemmens inlet, BC, Canada	Water quality	InVEST
Wu et al. (2021)	A collaborated framework to improve hydrologic ecosystem services management with sparse data in a semi-arid basin	Yixunhe River basin, China	Water yield; sediment losses	Variable infiltration capacity (VIC); SWAT; InVEST
Benra et al. (2021)	Mapping water ecosystem services: Evaluating InVEST model predictions in data scarce regions	224 watersheds in Chile	Water ecosystem services	InVEST
Balist et al. (2022)	Detecting land use and climate impacts on water yield ecosystem service in arid and semi-arid areas. A study in Sirvan River basin-Iran	Sirvan River basin, Iran	Water yield	SWAT; InVEST; ARIES
Anjinho et al. (2022)	Evaluation of InVEST's water ecosystem service models in a Brazilian subtropical basin	Jacaré-Guaçu River basin, Brazil	Water yield; nutrient and sediment delivery	InVEST
Roces-Díaz et al. (2018)	Assessing the distribution of forest ecosystem services in a highly populated Mediterranean region	Catalonia, Spain	Water provision; water regulation	ArcGIS 10.4
Timilsina et al. (2013)	A framework for identifying carbon hotspots and forest management drivers	Parque da Paz, Almada, Portugal	Erosion prevention	InVEST

(continued)

Table 1 (continued)

Author(s)/ year	Title	Study site	Water ecosystem services mapped	Model used
Shrestha et al. (2020)	Evaluation of land use change and its impact on water yield in Songkhram River basin, Thailand	Lower Songkhram River basin, Thailand	Water yield/ availability	InVEST
Sharps et al. (2017)	Comparing strengths and weaknesses of three ecosystem services modelling tools in a diverse UK river catchment	Conwy catchment, North Wales, UK	Water supply	InVEST; ARIES; LUCI
Goldenberg et al. (2017)	Distinction, quantification, and mapping of potential and realized supply-demand of flow-dependent ecosystem services	Stockholm region, Sweden	Natural stormwater regulation and associated flood protection (carried by water flow processes)	ArcGIS
Carvalho-Santos et al. (2016)	Ecosystem services and biodiversity conservation under forestation scenarios: Options to improve management in the Vez watershed, NW Portugal	Vez watershed, Portugal	Hydrological services	ArcSWAT interface
Signorello et al. (2020)	Assessing the hidden impacts of hypothetical eruption events at Mount Etna	Mt Etna, Italy	Water yield	InVEST; GIS
Minh et al. (2020)	Forest governance and economic values of forest ecosystem services in Vietnam	Northwestern uplands of Vietnam	Water yield	InVEST
Bagstad et al. (2014)	From theoretical to actual ecosystem services: Mapping beneficiaries and spatial flows in ecosystem service assessments	Puget Sound, Washington State, USA	Riverine flood regulation; sediment regulation for reservoirs	ARIES
Schirpke et al. (2019)	Integrating supply, flow, and demand to enhance the understanding of interactions among multiple ecosystem services	European Alps and surrounding foothills and lowlands, Austria, Switzerland	Filtration of surface water	InVEST

Hydrological models, combined with ecological data, offer predictive capabilities to assess the potential impacts of agricultural practices on water availability and quality. These models assist policymakers and water managers in designing sustainable agricultural water management strategies (Russo et al., 2014).

The use of advanced integrated watershed models (e.g., fully integrated surface–subsurface water models) can enhance the understanding of hydrological intricacies within a given basin, both presently and prospectively. These watershed models hold the potential to catalyze the inception of timely initiatives aimed at ensuring the sustainable provision of water resources in the future. It is noteworthy that watershed modeling stands as a conventional instrument employed to elucidate the dynamic behavior of catchments under varying environmental and climatic conditions (Norvanchig & Randhir, 2021).

Water ecosystem services range from water supply, flood protection, erosion control, to carbon storage (Sun et al., 2017). Initially used in ecological restoration and conservation, the concept has expanded to integrated water resource management, particularly aligning with UN Sustainable Development Goal 6.5 (Vollmer et al., 2022). Given the variable nature of water resources, mapping water ecosystem services is crucial for sustainable water management. Mapping of water ecosystem services constitutes an indispensable tool for the enhancement of water resource management and the preservation of ecosystem services on a regional scale. Within this context, a plethora of models and sub-models have been developed and applied in the practice of mapping water ecosystem services, each demonstrating context-specific merits and demerits. Consequently, the selection of an appropriate model poses a considerable challenge. Figure 1 is based on a comprehensive review of existing literature to facilitate the establishment of a selection criterion for use of hydrologic and ecosystem services models.

Most contemporary hydrological models exhibit a notable deficiency in comprehensively addressing the contribution of three distinct water storages, including groundwater, soil water, and surface water, in the support and sustenance of terrestrial ecosystem services (Fig. 2). Simultaneously, the distinct role played by each of these storages in sustaining evapotranspiration remains largely shrouded in ambiguity due to the intricate interconnections between them. Consequently, I advocate for

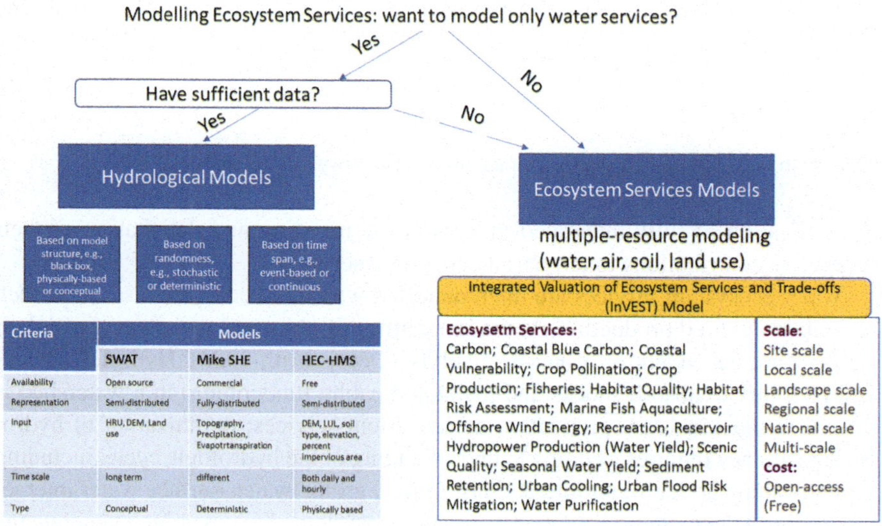

Fig. 1 Selection criteria for using most common hydrologic and ecosystem services models

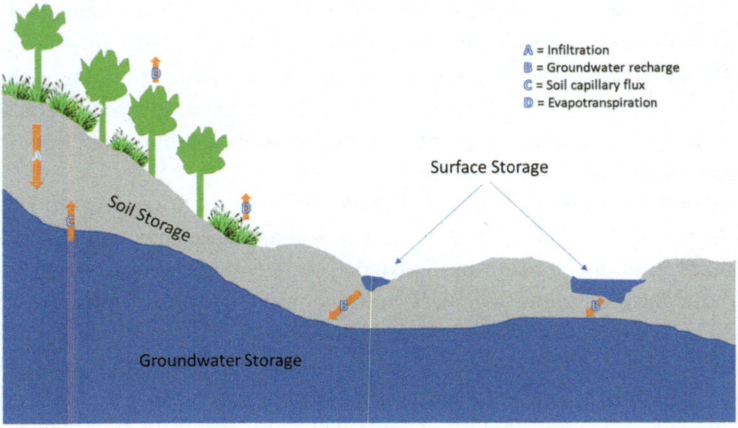

Fig. 2 Key water storages and fluxes that contribute to ecosystems and their services

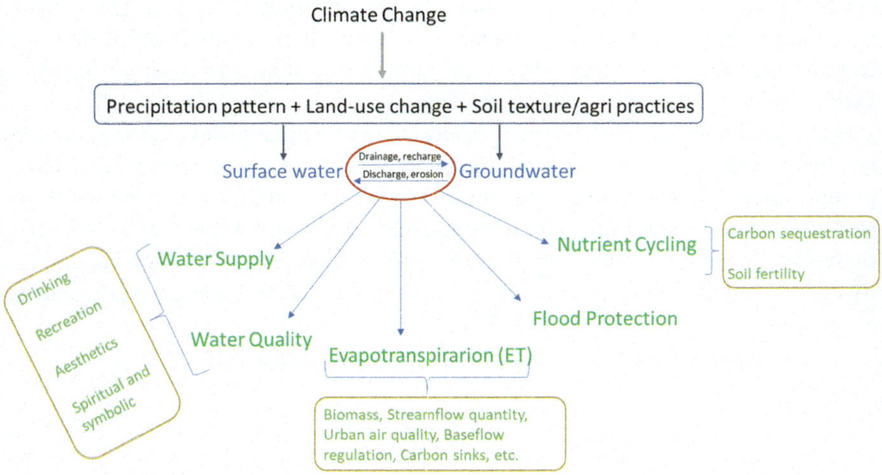

Fig. 3 Groundwater–surface water interactions control ecosystem services

the utilization of fully integrated models in the mapping and delineation of their respective contributions to terrestrial ecosystem services.

Water ecosystem services are interconnected with other ecosystem services such as timber and food production. At the same time, all terrestrial ecosystems and their services are dependent on water availability (Decsi et al., 2022). Hydrologic resiliency (ability to absorb shocks and maintain healthy functioning) in a watershed is crucial to the consistent production of ecosystem services; optimization of hydrologic resiliency requires a complete understanding of the hydrologic cycle, including groundwater–surface water interactions. In fact, groundwater–surface water interactions control the magnitude of many primary ecosystem services, which generate the final ecosystem services (Fig. 3). For example, evapotranspiration is a key process that supports a number of terrestrial ecosystem services and in many cases, is directly supported by groundwater–surface water interactions. Climate change can also affect water storages and fluxes through its influence on evapotranspiration across the landscape (Mehrnegar et al., 2021). For instance, climate warming increases evapotranspiration, which, in turn, depends on water availability (Condon et al., 2020). Because soil types and climate and land use changes alter regional hydrological cycle, the expected consequence will be changes in the groundwater–surface water interactions at a watershed scale (Scibek et al., 2007). In hydrologic evaluations involving active interactions between groundwater and surface water, it is imperative that the employed models consider the spatial and temporal variations present in the surface, subsurface, and climate domains. This is necessary to ensure a comprehensive and accurate representation of the hydrologic system under consideration.

5 Conclusions

This chapter delves into the transformative potential of ecohydrology in revolutionizing agricultural water management for a more sustainable and resilient future. As the agricultural sector grapples with escalating challenges related to water scarcity, environmental degradation, and ecosystem health, the integration of ecohydrological principles emerges as a crucial paradigm shift.

The intricate interplay between ecosystems and hydrological processes, as explored through ecohydrology, offers a holistic understanding of the dynamics shaping our agricultural landscapes. Addressing critical concerns such as water pollution originating from agricultural practices, ecohydrology stands as a beacon of hope by proposing innovative solutions grounded in scientific inquiry.

The examination of ecohydrological concepts, processes, and approaches presented in this chapter underscores the profound implications of this interdisciplinary field. From the dynamic water–ecology nexus to the ecohydrological processes governing evapotranspiration, infiltration, streamflow generation, and nutrient cycling, each facet contributes to a comprehensive framework for sustainable agricultural water management.

Ecohydrology's applications in agriculture are diverse, ranging from the implementation of sustainable irrigation practices to efficient water management, soil health conservation, tailored irrigation scheduling, and integrated pest and disease management. Moreover, the incorporation of ecosystem-based water management and sustainable land use practices highlights the multifaceted benefits of aligning agricultural activities with ecological principles.

Ecohydrological tools and technologies play pivotal roles in monitoring land cover changes, estimating evapotranspiration, conducting watershed analysis, assessing ecosystem services, and evaluating climate change impacts at the watershed scale. The integration of hydrological models, exemplified by the prevalence of the InVEST model in scientific literature, further emphasizes the relevance and applicability of these advancements in modeling water ecosystem services.

Nevertheless, considering surface–subsurface interactions, fully integrated water models are essential for mapping ecosystem services.

As we navigate the complexities of a changing climate, burgeoning global population, and evolving agricultural practices, ecohydrology emerges as a beacon of knowledge and innovation. The emphasis on sustainable irrigation practices, efficient water management, soil health, and ecosystem-based approaches provides a roadmap for agricultural stakeholders, researchers, and policymakers to collectively steer toward a harmonious coexistence between agricultural productivity and ecological integrity.

In short, by embracing the principles and applications elucidated in this chapter, the agricultural community can harness the transformative potential of ecohydrology for fostering a paradigm shift toward sustainable water management practices. Through collaborative efforts and the integration of cutting-edge tools, technologies, and methodologies, ecohydrology stands poised to redefine the future of agricultural water management and thus offers a path toward resilience, efficiency, and ecological harmony.

References

Anjinho, P. da S., Barbosa, M.A.G.A., & Mauad, F.F. (2022). Evaluation of InVEST's Water Ecosystem Service Models in a Brazilian Subtropical Basin. *Water (Switzerland) 14*. https://doi.org/10.3390/w14101559

Akamani, K. (2023). The roles of adaptive water governance in enhancing the transition towards ecosystem-based adaptation. *Water, 15*, 2341. https://doi.org/10.3390/w15132341

Almalki, R., Khaki, M., Saco, P. M., & Rodriguez, J. F. (2022). Monitoring and mapping vegetation cover changes in arid and semi-arid areas using remote sensing technology: A review. *Remote Sensing, 14*, 5143. https://doi.org/10.3390/rs14205143

Aziz, T. (2023). Accounting impacts of renewable energy expansions on ecosystem services to balance the trade-offs. *Science of the Total Environment, 879*, 162990. https://doi.org/10.1016/j.scitotenv.2023.162990

Bagstad, K.J., Villa, F., Batker, D., Harrison-Cox, J., Voigt, B., Johnson, G.W. (2014). From theoretical to actual ecosystem services. *Ecol. Soc. 19*, 64.

Balist, J., Malekmohammadi, B., Jafari, H.R., Nohegar, A., & Geneletti, D. (2022). Detecting land use and climate impacts on water yield ecosystem service in arid and semi-arid areas. A study in Sirvan River Basin-Iran. *Appl. Water Sci. 12*, 1–14. https://doi.org/10.1007/s13201-021-01545-8

Benra, F., De Frutos, A., Gaglio, M., Álvarez-Garretón, C., Felipe-Lucia, M., & Bonn, A. (2021). Mapping water ecosystem services: Evaluating InVEST model predictions in data scarce regions. *Environ. Model. Softw. 138*, 1–10. https://doi.org/10.1016/j.envsoft.2021.104982

Beven, K. (2021). The era of infiltration. *Hydrology and Earth System Sciences, 25*, 851–866. https://doi.org/10.5194/hess-25-851-2021

Cao, Z., Wang, S., Luo, P., Xie, D., & Zhu, W. (2022). Watershed ecohydrological processes in a changing environment: Opportunities and challenges. *Water (Switzerland), 14*, 1502. https://doi.org/10.3390/w14091502

Carvalho-Santos, C., Sousa-Silva, R., Gonçalves, J., & Honrado, J.P. (2016). Ecosystem services and biodiversity conservation under forestation scenarios: options to improve management in the Vez watershed, NW Portugal. *Reg. Environ. Chang. 16*, 1557–1570. https://doi.org/10.1007/s10113-015-0892-0

Castellini, M., Di Prima, S., Stewart, R., Biddoccu, M., Rahmati, M., & Alagna, V. (2022). Advances in ecohydrology for water resources optimization in arid and semi-arid areas. *Water (Switzerland), 14*, 1830. https://doi.org/10.3390/w14121830

Condon, L. E., Atchley, A. L., & Maxwell, R. M. (2020). Evapotranspiration depletes groundwater under warming over the contiguous United States. *Nature Communications, 11*, 873. https://doi.org/10.1038/s41467-020-14688-0

Decsi, B., Ács, T., Jolánkai, Z., Kardos, M. K., Koncsos, L., Vári, Á., & Kozma, Z. (2022). From simple to complex—Comparing four modelling tools for quantifying hydrologic ecosystem services. *Ecological Indicators, 141*, 109143. https://doi.org/10.1016/j.ecolind.2022.109143

Dhar, A., Parrott, L., & Heckbert, S. (2018). Large scale biotic damage impacts on forest ecosystem services. Scand. *J. For. Res. 33*, 741–755. https://doi.org/10.1080/02827581.2018.1495256

Farmaha, B. S., Sekaran, U., & Franzluebbers, A. J. (2022). Cover cropping and conservation tillage improve soil health in the southeastern United States. *Agronomy Journal, 114*, 296–316. https://doi.org/10.1002/agj2.20865

García-Santos, V., Sánchez, J. M., & Cuxart, J. (2022). Evapotranspiration acquired with remote sensing thermal-based algorithms: A state-of-the-art review. *Remote Sensing, 14*, 1–36. https://doi.org/10.3390/rs14143440

Gordon, L. J., Peterson, G. D., & Bennett, E. M. (2008). Agricultural modifications of hydrological flows create ecological surprises. *Trends in Ecology & Evolution, 23*, 211–219. https://doi.org/10.1016/j.tree.2007.11.011

Goldenberg, R., Kalantari, Z., Cvetkovic, V., Mörtberg, U., Deal, B., & Destouni, G. (2017). Distinction, quantification and mapping of potential and realized supply-demand of flow-dependent ecosystem services. Sci. Total Environ. 593–594, 599–609. https://doi.org/10.1016/j.scitotenv.2017.03.130

Guerry, A. D., Ruckelshaus, M. H., Arkema, K. K., Bernhardt, J. R., Guannel, G., Kim, C. K., Marsik, M., Papenfus, M., Toft, J. E., Verutes, G., Wood, S. A., Beck, M., Chan, F., Chan, K. M. A., Gelfenbaum, G., Gold, B. D., Halpern, B. S., Labiosa, W. B., Lester, S. E., Levin, P. S., McField, M., Pinsky, M. L., Plummer, M., Polasky, S., Ruggiero, P., Sutherland, D. A., Tallis, H., Day, A., & Spencer, J. (2012). Modeling benefits from nature: Using ecosystem services to inform coastal and marine spatial planning. *Int. J. Biodivers. Sci. Ecosyst. Serv. Manag. 8*, 107–121. https://doi.org/10.1080/21513732.2011.647835

Hannah, D. M., Wood, P. J., & Sadler, J. P. (2004). Ecohydrology and hydroecology: A "new paradigm"? *Hydrological Processes, 18*, 3439–3445. https://doi.org/10.1002/hyp.5761

Ingrao, C., Strippoli, R., Lagioia, G., & Huisingh, D. (2023). Water scarcity in agriculture: An overview of causes, impacts and approaches for reducing the risks. *Heliyon, 9*, e18507. https://doi.org/10.1016/j.heliyon.2023.e18507

Jackson, B., Masante, D., Thomas, A., Cosby, B., Emmett, B., & Jones, L. (2017). Comparing strengths and weaknesses of three ecosystem services modelling tools in a diverse UK river catchment. Sci. Total Environ. 584–585, 118–130. https://doi.org/10.1016/j.scitotenv.2016.12.160

Karlsson Green, K., Stenberg, J. A., & Lankinen, Å. (2020). Making sense of integrated Pest management (IPM) in the light of evolution. *Evolutionary Applications, 13*, 1791–1805. https://doi.org/10.1111/eva.13067

Kundzewicz, Z. W. (2002). Ecohydrology—Seeking consensus on interpretation of the notion. *Hydrological Sciences Journal, 47*, 799–804. https://doi.org/10.1080/02626660209492982

Le, T. S., Harper, R., & Dell, B. (2023). Application of remote sensing in detecting and monitoring water stress in forests. *Remote Sensing, 15*, 3360. https://doi.org/10.3390/rs15133360

Levidow, L., Zaccaria, D., Maia, R., Vivas, E., Todorovic, M., & Scardigno, A. (2014). Improving water-efficient irrigation: Prospects and difficulties of innovative practices. *Agricultural Water Management, 146*, 84–94. https://doi.org/10.1016/j.agwat.2014.07.012

Mashala, M. J., Dube, T., Mudereri, B. T., Ayisi, K. K., & Ramudzuli, M. R. (2023). A systematic review on advancements in remote sensing for assessing and monitoring land use and land cover changes impacts on surface water resources in semi-arid tropical environments. *Remote Sensing, 15*, 3926. https://doi.org/10.3390/rs15163926

Mehrnegar, N., Jones, O., Singer, M. B., Schumacher, M., Jagdhuber, T., Scanlon, B. R., Rateb, A., & Forootan, E. (2021). Exploring groundwater and soil water storage changes across the CONUS at 12.5 km resolution by a Bayesian integration of GRACE data into W3RA. *Science of the Total Environment, 758*, 143579. https://doi.org/10.1016/j.scitotenv.2020.143579

Muhadi, N. A., Abdullah, A. F., Bejo, S. K., Mahadi, M. R., & Mijic, A. (2020). The use of LiDAR-derived DEM in flood applications: A review. *Remote Sensing, 12*, 1–20. https://doi.org/10.3390/rs12142308

NEON. (2013). National Ecological Observatory Network (NEON): Revolutionizing Ecological Research.

Nguyen, M. D., Ancev, T., & Randall, A. (2020). Forest governance and economic values of forest ecosystem services in Vietnam. *Land use policy 97*, 1–10. https://doi.org/10.1016/j.landusepol.2018.03.028

Nilimaa, J. (2023). Smart materials and technologies for sustainable concrete construction. *Developments in the Built Environment, 15*, 100177. https://doi.org/10.1016/j.dibe.2023.100177

Norvanchig, J., & Randhir, T. O. (2021). Simulation of ecohydrological processes influencing water supplies in the Tuul River watershed of Mongolia. *Journal of Hydroinformatics, 23*, 1130–1145. https://doi.org/10.2166/hydro.2021.056

Okasha, A. M., Deraz, N., Elmetwalli, A. H., Elsayed, S., Falah, M. W., Farooque, A. A., & Yaseen, Z. M. (2022). Effects of irrigation method and water flow rate on irrigation performance, soil salinity, yield, and water productivity of cauliflower. *Agriculture, 12*, 1164. https://doi.org/10.3390/agriculture12081164

Porporato, A., Feng, X., Manzoni, S., Mau, Y., Parolari, A. J., & Vico, G. (2015). Ecohydrological modeling in agroecosystems: Examples and challenges. *Water Resources Research, 51*, 5081–5099. https://doi.org/10.1002/2015WR017289

Ramião, J. P., Carvalho-Santos, C., Pinto, R., & Pascoal, C. (2022). Modeling the effectiveness of sustainable agricultural practices in reducing sediments and nutrient export from a river basin. *Water, 14*, 3962.

Roces-Díaz, J. V., Vayreda, J., Banqué-Casanovas, M., Cusó, M., Anton, M., Bonet, J.A., Brotons, L., De Cáceres, M., Herrando, S., Martínez de Aragón, J., de-Miguel, S., & Martínez-Vilalta, J. (2018). Assessing the distribution of forest ecosystem services in a highly populated Mediterranean region. *Ecol. Indic. 93*, 986–997. https://doi.org/10.1016/j.ecolind.2018.05.076

Robertson, G. P., & Swinton, S. M. (2005). Reconciling agricultural productivity and environmental integrity: A grand challenge for agriculture. *Frontiers in Ecology and the Environment, 3*, 38. https://doi.org/10.1890/1540-9295(2005)003[0038:RAPAEI]2.0.CO;2

Russo, T., Alfredo, K., & Fisher, J. (2014). Sustainable water management in urban, agricultural, and natural systems. *Water (Switzerland), 6*, 3934–3956. https://doi.org/10.3390/w6123934

Sanou, J., Tengberg, A., Bazié, H. R., Mingasson, D., & Ostwald, M. (2023). Assessing trade-offs between agricultural productivity and ecosystem functions: A review of science-based tools? *Land, 12*, 1329. https://doi.org/10.3390/land12071329

Schirpke, U., Candiago, S., Egarter Vigl, L., Jäger, H., Labadini, A., Marsoner, T., Meisch, C., Tasser, E., & Tappeiner, U. (2019). Integrating supply, flow and demand to enhance the understanding of interactions among multiple ecosystem services. *Sci. Total Environ. 651*, 928–941. https://doi.org/10.1016/j.scitotenv.2018.09.235

Schmidt, A., Fournier, R. A., Luther, J. E., Trudel, M. (2020). Development of a mapping framework for ecosystem services: The case of sediment control supply at a watershed scale in Newfoundland. *Canada. Ecol. Indic. 117*, 1–11. https://doi.org/10.1016/j.ecolind.2020.106518

Scibek, J., Allen, D. M., Cannon, A. J., & Whitfield, P. H. (2007). Groundwater-surface water interaction under scenarios of climate change using a high-resolution transient groundwater model. *Journal of Hydrology, 333*, 165–181. https://doi.org/10.1016/j.jhydrol.2006.08.005

Scott, C. A., Vicuña, S., & Meza, F. (2014). Irrigation efficiency and water-policy implications for river basin resilience. *Hydrology and Earth System Sciences, 18*, 1339–1348. https://doi.org/10.5194/hess-18-1339-2014

Shrestha, M., Shrestha, S., & Shrestha, P. K. (2020). Evaluation of land use change and its impact on water yield in Songkhram River basin, Thailand. *Int. J. River Basin Manag. 18*, 23–31. https://doi.org/10.1080/15715124.2019.1566239

Shuaibu, A., Mujahid Muhammad, M., Bello, A. A. D., Sulaiman, K., & Kalin, R. M. (2023). Flood estimation and control in a micro-watershed using GIS-based integrated approach. *Water (Switzerland), 15*, 1–20. https://doi.org/10.3390/w15244201

Signorello, G., Marzo, A., Prato, C., Sturiale, G., & De Salvo, M. (2020). Assessing the hidden impacts of hypothetical eruption events at Mount Etna. *Environ. Sustain. Indic. 8*. https://doi.org/10.1016/j.indic.2020.100056

Sun, G., Hallema, D., & Asbjornsen, H. (2017). Ecohydrological processes and ecosystem services in the anthropocene: A review. *Ecological Processes, 6*, 35. https://doi.org/10.1186/s13717-017-0104-6

Taiwo, B. E., Al Kafy, A., Samuel, A. A., Rahaman, Z. A., Ayowole, O. E., Shahrier, M., Duti, B. M., Rahman, M. T., Peter, O. T., & Abosede, O. O. (2023). Monitoring and predicting the influences of land use/land cover change on cropland characteristics and drought severity using remote sensing techniques. *Environmental and Sustainability Indicators, 18*, 100248. https://doi.org/10.1016/j.indic.2023.100248

Tan, Y., Dong, Z., Guzman, S. M., Wang, X., & Yan, W. (2021). Identifying the dynamic evolution and feedback process of water resources nexus system considering socioeconomic development, ecological protection, and food security: A practical tool for sustainable water use. *Hydrology and Earth System Sciences, 25*, 6495–6522.

Taye, M. T., Ntegeka, V., Ogiramoi, N. P., & Willems, P. (2011). Assessment of climate change impact on hydrological extremes in two source regions of the Nile River basin. *Hydrology and Earth System Sciences, 15*, 209–222. https://doi.org/10.5194/hess-15-209-2011

Timilsina, N., Escobedo, F.J., Cropper, W.P., Abd-Elrahman, A., Brandeis, T.J., Delphin, S., & Lambert, S. (2013). A framework for identifying carbon hotspots and forest management drivers. *J. Environ. Manage. 114*, 293–302. https://doi.org/10.1016/j.jenvman.2012.10.020

Tudi, M., Ruan, H. D., Wang, L., Lyu, J., Sadler, R., Connell, D., & Chu, C. (2021). Agriculture development, pesticide application and its impact on the environment. *International Journal of Environmental Research and Public Health, 18*, 1112.

Velasco-muñoz, J. F., Aznar-s, A., & Batlles-delafuente, A. (2019). Sustainable irrigation in agriculture: An analysis of global research. *Water (Switzerland), 11*, 1758. https://doi.org/10.3390/w11091758

Vollmer, D., Burkhard, K., Adem Esmail, B., Guerrero, P., & Nagabhatla, N. (2022). Incorporating ecosystem services into water resources management—Tools, policies, promising pathways. *Environmental Management, 69*, 627–635. https://doi.org/10.1007/s00267-022-01640-9

Wu, Y., Xu, Y., Yin, G., Zhang, X., Li, C., Wu, L., Hao, F., Wang, X., & Hu, Q. (2021). A collaborated framework to improve hydrologic ecosystem services management with sparse data in a semi-arid basin. *Hydrol. Res. 52*, 1159–1172. https://doi.org/10.2166/nh.2021.146

Wu, S., Bashir, M. A., Raza, Q. U. A., Rehim, A., Geng, Y., & Cao, L. (2023). Application of riparian buffer zone in agricultural non-point source pollution control—A review. *Frontiers in Sustainable Food Systems, 7*, 985870. https://doi.org/10.3389/fsufs.2023.985870

Wurtsbaugh, W. A., Paerl, H. W., & Dodds, W. K. (2019). Nutrients, eutrophication and harmful algal blooms along the freshwater to marine continuum. *WIREs Water, 6*, 1–27. https://doi.org/10.1002/wat2.1373

Zerriffi, H., Reyes, R., & Maloney, A. (2023). Pathways to sustainable land use and food systems in Canada. *Sustainability Science, 18*, 389–406. https://doi.org/10.1007/s11625-022-01213-z

Zhang, P., Xiao, M., Zhang, Z., Dai, Y., Liu, G., & Senge, M. (2023). Hydrological characteristics of different organic materials mulches. *Scientific Reports, 13*, 1–9. https://doi.org/10.1038/s41598-023-28124-y

Zhao, M., Aa, G., Liu, Y., & Konings, A. (2022). Evapotranspiration frequently increases during droughts. *Nature Climate Change, 12*, 1024–1030. https://doi.org/10.1038/s41558-022-01505-3

Better Soil Moisture Management May Improve Soil Properties in the Era of Climate Change

Khurram Shahzad, Sana Ullah, Qamar Sarfaraz, Muhammad Mubeen, and Henry Y. Sintim

Abstract Soil, a crucial component in the global carbon cycle, stores approximately 1500 gigatons of carbon in the top 1 meter and 900 gigatons in the 1–2 m depth. Climate change significantly impacts soil health, which is defined as the soil's ability to sustain plants, animals, and humans through its chemical, physical, and biological properties. Key threats to soil health include erosion, organic matter decline, contamination, compaction, and salinization, all exacerbated by climate change. Elevated temperatures increase soil organic matter decomposition, leading to higher CO_2 emissions, altered soil moisture dynamics, and disrupted nutrient availability. Soil temperature influences water and nutrient dynamics, while changes in soil moisture impact plant growth, nutrient uptake, and microbial activity. Increased temperatures and altered precipitation patterns further exacerbate soil erosion, particularly in arid regions where wind erosion can become severe. To miti-

K. Shahzad (✉)
Department of Soil Science, Faculty of Agriculture, Lasbela University of Agriculture, Water, and Marine Sciences, Uthal, Balochistan, Pakistan

Department of Crop and Soil Sciences, Washington State University, Pullman, WA, USA

S. Ullah
Department of Agronomy, Faculty of Agriculture, University College Dera Murad Jamali, Lasbela University of Agriculture, Water, and Marine Sciences, Uthal, Balochistan, Pakistan

Q. Sarfaraz
Department of Soil Science, Faculty of Agriculture, Lasbela University of Agriculture, Water, and Marine Sciences, Uthal, Balochistan, Pakistan

Department of Crop, Soil and Environmental Sciences, University of Arkansas, Fayetteville, USA

M. Mubeen (✉)
Department of Biotechnology, COMSATS University Islamabad, Vehari Campus, Vehari, Pakistan
e-mail: muhammadmubeen@cuivehari.edu.pk

H. Y. Sintim
Department of Crop & Soil Sciences, University of Georgia, Tifton, GA, USA

M. Mubeen et al. (eds.), *Innovations in Agricultural Water Management*,
https://doi.org/10.1007/978-3-031-91883-4_22

gate these effects, adopting climate-resilient soil practices is essential. These include enhancing biomass production, implementing conservation tillage, applying nitrogen fertilizers judiciously, and utilizing soil mulching and stable carbon compounds like biochar. Additionally, promoting plant growth-promoting rhizobacteria (PGPR) and effective moisture management can improve soil health and sustainability. This chapter reviews the impacts of climate change on soil properties, with a particular focus on soil moisture, and proposes strategies for adaptation and mitigation to support sustainable agriculture and soil management.

Keywords Climate change · Erosion · Soil moisture · Sustainable agriculture

1 Introduction

Soils are the main sink and storage of atmospheric CO_2 and store about 1500 gigatons of carbon (GT-C) in the top 1 m soil and about 900 GT-C at 1–2 m depth (Batjes, 1996; Guo & Gifford, 1979). Therefore, understanding the effect of climate change on soil properties or soil health is very important. Soil health is generally defined as the continued ability of soil to function as a vital living ecosystem that sustains plants, animals, and humans (Bünemann et al., 2018). Soil chemical, physical, and biological properties are the main attributes of soil health, and any external factor(s) affecting these properties would eventually affect soil health (Doran & Zeiss, 2000). Soil erosion, soil organic matter decline, contamination, sealing, compaction, biodiversity loss, salinization, landslides, and floods because of climate change are the main threats to soil health. These threats could affect soil functions like element recycling, decomposition, biological population regulation, water recycling, organic matter cycling, and climate regulation (Davidson & Janssens, 2006). The climate is one of five major factors, along with parent material, relief, soil organism, and time, in the development of some specific soil characteristics (Hussain et al., 2022, 2023a, 2023b, 2024a, 2024b, 2024c; Naz et al., 2023; Muhammad et al., 2023; Khurshid et al., 2023; Huggett, 1998). About thousands of years are required in the formation of any soil and most soils are still in the developing phase followed by climate and vegetation changes (Sevink, 1991). Climate plays a direct role in the development of soil structure, water-holding capacity, erosion, and nutrient availability. Further, indirect changes may include the water use efficiency, which directly relates to the increased or decreased evaporation rate induced by climate changes. The temperature, evaporation, and precipitation could lead to a dramatic change in the dynamics of soil organic matter and CO_2. The high temperature could accelerate the decomposition of soil organic matter (SOM), which would result in high CO_2 flux (Raich & Schlesinger, 1999). Further, the continuous climate change could become severe with time scale (Fig. 1). Therefore, understanding of climate change and its effect on soil physical and chemical properties is very important for sustainable agriculture. This chapter aims to give the brief impact of climate change on soil physical, chemical, and biological properties, their adaption, and mitigation strategies.

Time scale and effect on Soil Parameters

Temperature, moisture contents, bulk density, total porosity, infiltration rate, permeability, composition of soil air and nitrate contents	Total water capacity, field capacity, hydraulic conductivity, microbiota, pH, nutrients status and composition of soil solution.	Wilting percentage, soil acidity, cation exchange capacity and exchangeable cations.	Specific surface, clay mineral association and organic matter contents	Primary mineral composition and chemical composition of mineral part	Texture, particle-size distribution and particle density
10^{-0}	$<10^{-1}$-10^{0}	10^{0}-10^{1}	10^{1}-10^{2}	10^{2}-10^{3}	$>10^{3}$
Compaction, drainage and workability			Tree roots soil biota, sialic, calcareous, sodic and vertic properties	Trees root and Colour (yellowish/reddish), iron concentrations, soil depth cracking, soft powdered lime and indurated sub soil.	Parent material, depth and abrupt textural change

Time scale and effect on soil properties and characteristics

Fig. 1 Time scale and effect of climate change on soil parameters, properties, and characteristics. Climate change impact will become serious on soil health with the passage of time. (Modified from Karmakar et al., 2016)

2 Climate Change Impact on Important Soil Properties

2.1 Soil Organic Matter

The quality of soil mostly depends on the soil's organic matter contents. The soil organic matter not only provides a habitat for most soil microorganisms but also acts as a source of carbon and energy for them. Further, soil organic matter due to having a high surface area acts as a filter for the most toxic chemicals, otherwise these chemicals would leach down into groundwater and cause carcinogenic diseases. The rising temperature due to climate change is causing high decomposition of added soil organic matter and releasing CO_2 into the atmosphere (Lal, 2004). The high decomposition of organic matter adversely affects soil properties like porosity, aggregate stability, water-holding capacity, and erosion (Abid & Lal, 2008, 2009; Blanco-Canqui & Ruis, 2018; Sabagh et al., 2020a, 2020b; El Sabagh et al., 2020; Akram et al., 2019, 2019; Fahad et al., 2019; Amanet et al., 2019; Sami ul din et al., 2019; Ali et al., 2019; Tariq et al., 2019; Zahoor et al., 2019). The decrease of soil organic matter also affects the tensile strength and water infiltration rate (Blanco-Canqui & Ruis, 2018; Six et al., 1999). According to Schmidt et al. (2015), persistence of soil organic matter depends on ecosystem properties instead of only molecular structure or composition of SOM.

2.2 Soil Moisture

Soil moisture is usually defined as the water contents in the unsaturated zone. This water is either up taken by the plant or evaporates into the atmosphere. The up taken water is used by plant and unused water in plant tissues escape into the atmosphere through transpiration. The evapotranspiration returns 60% of precipitation back to the atmosphere (Oki & Kanae, 2006; Seneviratne et al., 2010). The climate change could also affect the soil moisture contents. The increase in temperature causes high losses of soil water contents in the form of evapotranspiration. This leads to low availability of water for crop and soil-residing microorganisms. Further, the presence of moisture on surface soil controls the rise of soil temperature by observing the sun rays (Teasdale & Mohler, 1993).

2.3 *Soil Temperature*

Soil temperature is an important soil property that plays a crucial role in regulating water and nutrient dynamics in the soil (Fig. 2). The sudden change in soil temperature could lead to increased water loss through evaporation and nutrient loss through vocalization (Kirschbaum, 1995). In addition to its impact on water and nutrient regulation, soil temperature significantly influences seed germination in soil. High soil temperature can accelerate the decomposition of organic matter leading to the release of CO_2 into the atmosphere (Davidson & Janssens, 2006). Furthermore, elevated soil temperature can enhance the weathering of soil minerals (Karmakar et al., 2016). An increase in the soil temperature because of climate change has been observed in numerous studies (Qian et al., 2011). Qian et al. (2011) reported soil warming increase in Canada dating back to 1950, based on analysis of data spanning from 1958 to 2008 at various depths (5, 10, 20, 50, 100, and 150 cm) across 30 stations. Their findings indicated that atmospheric warming was a significant driver of soil temperature increase. However, it was worth noting that the rate of temperature increase was less pronounced at greater depths, particularly at 100 and 150 cm. Zhang et al. (2005) projected an increase of 0.6 °C in soil temperature at a 20 cm depth with a 1 °C increase in atmospheric temperature using the northern ecosystem soil temperature (NEST) model across a climate gradient in Canada.

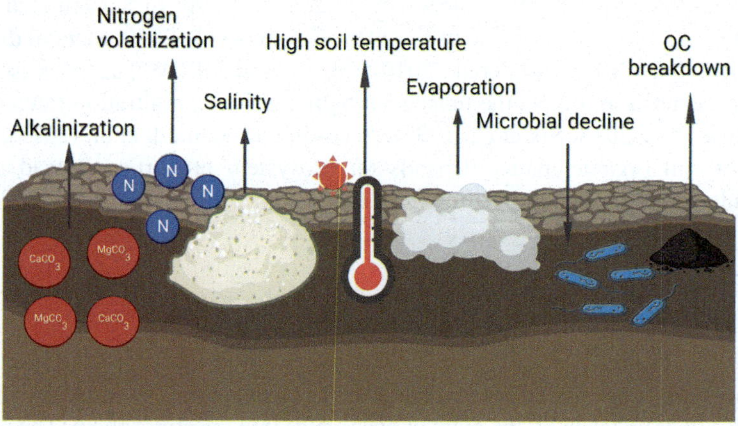

Fig. 2 The impact of climate change on the soil properties

2.4 Soil Structure

The soil structure is more important property of soil that facilitates the movement of water, air, and chemical through soil atmospheric interaction (Guo & Gifford, 1979). The stability of structure increases with the soil organic matter contents as the soil organic matter acts as binding agents between the soil particles (Naz et al., 2023; Hussain et al., 2022; Lal, 2004). The higher temperature causes decomposition of soil organic matter contents, which could result in the formation of soil microaggregates and these microaggregate could enter the macropores and become cause of sealing of soil surface (Abid & Lal, 2008). The compaction of soil surface promotes the runoff water and decreases the infiltration rate (Abid & Lal, 2009). The low infiltration and high water runoff result in the low availability of water and nutrient contents for growing roots.

2.5 Soil Microflora

Thousands of species inside the soil are playing their important role in mineralization of complex organic materials, fixing of atmospheric N, and solubilization of nutrients with excretion of hormones and many more functions (Nannipieri et al., 2004; Akram et al., 2017a, 2017b, 2018a, 2018b; Hasnain et al., 2017; Hashmi et al., 2017; McDermid et al., 2015; Ahmad et al., 2015). Climate change could affect the soil microflora through increase in soil temperature. The high soil temperature results in the high activity and growth of microflora. The increasing population of microflora require more carbon and nutrients for their energy and food needs (Kladivko, 2001). Therefore, it could result in more decomposition of organic matter in soil. This high decomposition of organic matter increases the flux of CO_2 from soil to the atmosphere (Oechel et al., 2013). Further, quick release of nutrients as a result of decomposition also causes a big loss of nutrient pool in the soil. More nutrient inputs would be required to rebuild this pool and it could increase production costs.

2.6 Salinization and Alkalinization

Salinization and alkalinization occur due to the accumulation of salts in the soil horizons. When the soil contains a high concentration of salts, it becomes unsuitable for crop production. The geochemistry, hydrology, and climate of an area play their role in the movement and salt balance of the soil (Szabolcs, 2018). Therefore, humid areas with well-drained soil typically have low levels of salts, whereas regions with arid climates often experience salinity issues. Climate change could increase the areas of the world affected by salinization and alkalinization both directly and indirectly. Directly, this could occur through the rise in atmospheric temperature and increased aridity, as well as the expansion of coastal areas due to rising sea levels. Indirectly, these effects could be exacerbated by the increased use of irrigation water (Szabolcs, 2018; Shahzad et al., 2022; Din et al., 2022; Balal et al., 2022; Abbas et al., 2020; Habib Ur Rehman et al., 2020). Irrigation water leads to secondary salinization and alkalinization. Every year, millions of acres of irrigated land become unsuitable for agricultural production due to salinization and alkalinization. With the expansion of irrigated agriculture areas resulting from climate change, the risk of salinization and alkalinization is likely to further increase.

2.7 Nutrient Acquisition

The uptake of the most mobile nutrients like nitrate, sulfate, Ca, Mg, and Si takes place through the mass flow mechanism (Matimati et al., 2014). As climate change affects the moisture regime of the soil due to high evaporation losses, the movement of these nutrients decreases due to the deficiency of moisture contents in the root areas (Baldwin, 1975). Therefore, plants would use extra energy to build the root hair to capture these nutrients and moisture present far away from the root area of the plants. Therefore, this extra consumption of energy affects the plant growth and yield parameters (Kaspar & Taylor, 1994). Further, due to surface compaction because of low stability of soil aggregates, intake and outflux of oxygen and CO_2 are badly affected, which could make a direct effect on the activities of soil-residing microorganisms especially nitrogen fixers (Zumft, 1997; Hussain et al., 2020; Sabagh et al., 2020a, 2020b; Ahmed et al., 2020). The low influx of oxygen promotes the breakdown of nitrate compound as the microbes use nitrate as an electron acceptor. This high denitrification results in the loss of soil nitrogen and could result in a deficiency of nitrogen for growing plants.

2.8 Soil Erosion

Climate change can significantly affect soil erosion rate due to changes in precipitation and temperature. However, climate change's effect on soil erosion rates is influenced by topography, ecosystem, and management practices. The uneven distribution of rainfall is a consequence of climate change. In some areas of the world, rainfall events have been increasing, while in some areas they are decreasing. High rainfall events can lead to a 1.7-fold increase in soil erosion rates due to the high runoff of rainwater, while low rainfall events can cause low biomass production and make affected areas more prone to erosion (Nearing et al., 2004). In semiarid and arid regions, wind erosion rates can increase with rising temperatures and a decrease in precipitation. Wind erosion damaged cultivated lands by 25%. The sandy soils are more subject to wind erosion due to low moisture content and high wind events. The high temperature due to climate change is increasing the incidents of drought (Ziadat

& Taimeh, 2013). The drought favors wind erosion. Wind erosion could result in the loss of fertilized soils and contamination of water bodies when the air-carried nutrients are deposited in them. The enrichment of water bodies with nutrients promotes the eutrophication or growth of unwanted plants, and ultimately the water in the rivers becomes too low quality to use for human consumption (Lal, 2003).

3 Climate-Resilient Soil Practices

The changing climate demands measures to conserve soil health for sustainable agriculture production.

3.1 Higher Biomass Production

The plants are the main absorbers of atmospheric CO_2 and 30–60% of this absorbed CO_2 is deposited in the form of organic carbon in the soil (Karmakar et al., 2016). The decay of the roots is also a great source of soil carbon and many minerals. Thus, growing more plants gives a dual benefit: decreases the level of CO_2 in the atmosphere and improves the carbon level of soils to compete with the climate change affect.

3.2 Conservation Tillage

The continuous soil manipulation with conventional tillage causes soil erosion, loss of soil moisture, and exposure of deep soil organic matter to degradation and requires high energy cost. However, thorough adaption of conservation tillage could increase soil organic matter through the retention of 30% crop residue on the surface (Höflich et al., 1999). Further, the deep soil organic matter is less exposed to sunlight and ultimately, the degradation of soil organic matter could be reduced to a maximum extent (Mloza-Banda et al., 2014).

3.3 Adequate Nitrogen Fertilizer Application

The nitrogen should be applied according to plants' need. The extra applied nitrogen could be used by soil microbes and will accelerate the decomposition of organic matter in the soil (Her & Huang, 1995). The normal C:N ratio of microbes is 30:1. It means that microorganisms consume 30 tons of soil carbon with 1 ton of applied nitrogen. Further, nitrogen should be applied in such form that could release nitrogen in a slow phase.

3.4 Soil Mulching

The mulches could improve the soil conditions by controlling the soil moisture and soil temperature, and cover the soil organic matter (Hatfield et al., 2001). The low temperature due to the presence of mulch could extend the stability of soil organic matter for the long term, otherwise it would be exposed to the atmosphere for quick degradation (Berger et al., 2013).

3.5 Stable Carbon Compounds

The use of stable carbon compounds like compost and biochar could be effective in improving the soil structure and soil environment. The biochar is produced under a limited supply of oxygen (Liang et al., 2006). The biochar could be used as a stable C source and in mitigation of greenhouse gases. The biochar could reduce the greenhouse gas emission of about 1.8 Pg CO_2-C without affecting food security (Qin et al., 2016; Wang et al., 2016). The biochar could remain in the soil for a millennium. The stability of organic amendment will not only improve the soil environment but also reduce the addition of CO_2 in the atmosphere. The low decomposition of SOM favors carbon sequestration (Kimetu & Lehmann, 2010).

3.6 Plant Growth-Promoting Rhizobacteria

The use of plant growth-promoting rhizobacteria (PGPR) could improve the mineralization of soil organic material and production of stable organic compounds like humus and glomalin (Akhtar et al., 2009). These compounds could improve the aggregate stability, porosity, and passage of gases in and out of the soil (Verzeaux et al., 2017). Further, the use of pesticides and other chemicals should be minimized to avoid killing of PGPR so that microbes could survive and work for better soil environment.

4 Soil Moisture Impact on Soil Chemical, Physical, and Biological Properties

Adequate soil moisture is essential for enhancing the chemical, physical, and biological properties of soil. We will explore each of these properties in detail below.

4.1 Soil Moisture and Chemical Properties

Soil moisture plays a critical role in influencing soil chemical properties by affecting nutrient availability and soil pH (Venterea et al., 2005). Adequate moisture levels facilitate the dissolution and mobility of nutrients in the soil, making them more accessible to plant roots. When soil moisture is optimal, nutrients like potassium, calcium, and magnesium are more readily available, leading to improved plant growth and crop yields (Cavagnaro, 2016). Conversely, both excessive and insufficient moisture can hinder nutrient uptake. Excessive moisture can lead to leaching, where nutrients are washed away from the root zone, while inadequate moisture can cause nutrient deficiencies due to reduced solubility and availability (Van Iersel & Chappell, 2009).

Additionally, soil moisture directly impacts soil pH, which in turn affects nutrient availability and microbial activity (Venterea et al., 2005). In wet conditions, the accumulation of organic matter and microbial activity can lead to increased soil acidity, while dry conditions often result in higher soil pH (Zhang & Wienhold, 2002). Changes in soil pH can influence the availability of essential nutrients such as phosphorus, which becomes less available in acidic soils and more available in alkaline conditions. Thus, managing soil moisture is crucial for maintaining optimal soil chemical properties and ensuring the efficient functioning of soil ecosystems.

4.2 Soil Moisture and Physical Properties

Soil moisture significantly affects soil physical properties such as texture, structure, and compaction (Grzesiak et al., 2002). When soil moisture levels are adequate, the soil particles are better lubricated, which helps in maintaining optimal soil structure and reduces the risk of compaction. This enhances soil aeration and root penetration, allowing plants to access water and nutrients more effectively (Nawaz et al., 2013). Conversely, too much moisture can lead to soil compaction as the soil becomes heavier and more prone to becoming waterlogged. This can reduce pore space and hinder root growth and water infiltration.

On the other hand, insufficient moisture can lead to soil hardening and increased soil erosion. When soil dries out, it can become compacted and form a hard crust on the surface, making it difficult for water to infiltrate and for plant roots to penetrate (Lepore et al., 2024). Additionally, dry conditions can exacerbate soil erosion as the lack of moisture reduces the soil's ability to hold together. Effective moisture management is therefore essential to maintaining good soil structure, preventing compaction, and promoting healthy plant growth.

4.3 Soil Moisture and Biological Properties

Soil moisture has a profound impact on soil biological properties, including microbial activity, soil organic matter decomposition, and overall soil health (Chen et al., 2007). Optimal moisture levels promote the proliferation of soil microorganisms such as bacteria and fungi, which are crucial for nutrient cycling and organic matter decomposition (Doudill et al., 1998). Adequate moisture supports microbial processes by providing a suitable environment for their activities, leading to enhanced nutrient availability for plants. This, in turn, fosters a healthier and more productive soil ecosystem.

Conversely, extreme moisture conditions—either too wet or too dry—can disrupt soil biological functions (Chen et al., 2007). Excessive moisture can create anaerobic conditions, leading to reduced microbial diversity and activity (Paradelo & Barral, 2009), and potentially causing the accumulation of harmful substances like methane and hydrogen sulfide. On the other hand, insufficient moisture can lead to microbial desiccation and decreased enzymatic activity, slowing down the decomposition of organic matter and nutrient cycling (Paradelo & Barral, 2009). Maintaining balanced soil moisture is therefore critical for sustaining soil biological health and ensuring effective soil ecosystem functioning.

5 Conclusion

The climate change is badly affecting the soil organic matter, moisture contents, and availability of essential nutrients. Both soil organic matter and moisture are the most limiting growth factors that could cause less soil aggregate stability, high soil compaction, salinization, and less nutrient acquisition. Therefore, use of conservation tillage, stable soil organic amendments, and PGPR can be a better adaptation to minimize the adverse effect of climate change on soil properties.

Soil moisture significantly affects soil chemical, physical, and biological properties. Adequate moisture enhances nutrient availability, soil structure, and microbial activity, leading to better plant growth and soil health. Conversely, both excessive and insufficient moisture can hinder nutrient uptake, cause soil compaction or hardening, and disrupt biological functions, affecting overall soil quality.

References

Abbas, G., et al. (2020). Applications of crop modeling in cotton production. In S. Ahmad, & M. Hasanuzzaman (Eds.) *Cotton production and uses*. Print isbn:978-981-15-1471-5, Online isbn:978-981-15-1472-2.

Abid, M., & Lal, R. (2008). Tillage and drainage impact on soil quality. I. Aggregate stability, carbon and nitrogen pools. *Soil and Tillage Research, 100*, 89–98. https://doi.org/10.1016/j.still.2008.04.012

Abid, M., & Lal, R. (2009). Tillage and drainage impact on soil quality: II. Tensile strength of aggregates, moisture retention and water infiltration. *Soil and Tillage Research, 103*, 364–372. https://doi.org/10.1016/j.still.2008.11.004

Ahmad, A., et al. (2015). *Impact of climate change on the rice–wheat cropping system of Pakistan* (pp. 191–221). World Scientific Publishing. isbn:978-1-78326-563-3.

Ahmed, M., et al. (2020). Abiotic stress tolerance in wheat and the role of silicon: An experimental evidence. In M. Hasanuzzaman (Ed.), *Agronomic crops*. Springer.

Akhtar, M. J., Asghar, H. N., Shahzad, K., & Arshad, M. (2009). Role of plant growth promoting rhizobacteria applied in combination with compost and mineral fertilizers to improve growth and yield of wheat (Triticum Aestivum L.). *Pakistan Journal of Botany, 41*, 381–390.

Akram, R., Nasim, W., et al. (2019). Global trends of E-waste pollution and its impact on environment. In M. Z. Hashmi, & A. Varma (Eds.) *Electronic waste pollution*. isbn:978-3-030-26614-1.

Akram, R., et al. (2017a). Role of antibiotics in climate change. In M. Hashmi, V. Strezov, & A. Varma (Eds.), *Antibiotics and antibiotics resistance genes in soils* (Soil biology) (Vol. 51). Springer.

Akram, R., et al. (2017b). Fate of antibiotics in soil. In M. Hashmi, V. Strezov, & A. Varma (Eds.), *Antibiotics and antibiotics resistance genes in soils* (Soil biology) (Vol. 51). Springer.

Akram, R., et al. (2018a). Fate of organic and inorganic pollutants in paddy soils. In M. Z. Hashmi & A. Varma (Eds.), *Environmental pollution of paddy soils* (Soil biology) (Vol. 53). Springer.

Akram, R., et al. (2018b). Paddy land pollutants and their role in climate change. In M. Z. Hashmi & A. Varma (Eds.), *Environmental pollution of paddy soils* (Soil biology) (Vol. 53). Springer.

Akram, R., et al. (2019). Plant growth and morphological changes in rice under abiotic stress. In M. Hasanuzzaman, M. Fujita, & J. K. Biswas (Eds.) *Advances in rice research for abiotic stress tolerance*. isbn:978-0-12-814332-2.

Ali, M., et al. (2019). Role of ICT in crop management. In M. Hasanuzzaman (Ed.) *Agronomic crops*. isbn:978-981-32-9150-8.

Amanet, K., Mubeen, M., et al. (2019). *Cotton production in Africa* (pp. 359–369). Wiley.

Balal, R. M., et al. (2022). Morphological, physiological, and biochemical modulations in crops under salt stress. In W. N. Jatoi, M. Mubeen, A. Ahmad, M. A. Cheema, Z. Lin, & M. Z. Hashmi (Eds.), *Building climate resilience in agriculture*. Springer.

Baldwin, J. P. (1975). A quantitative analysis of the factors affecting plant nutrient uptake from some soils. *Journal of Soil Science, 26*, 195–206. https://doi.org/10.1111/j.1365-2389.1975.tb01943.x

Batjes, N. H. (1996). Total carbon and nitrogen in the soils of the world. *European Journal of Soil Siceince, 47*, 151–163. https://doi.org/10.1111/j.1365-2389.1996.tb01386.x

Berger, S., Kim, Y., Kettering, J., & Gebauer, G. (2013). Plastic mulching in agriculture—Friend or foe of N2O emissions? *Agriculture, Ecosystems and Environment, 167*, 43–51. https://doi.org/10.1016/j.agee.2013.01.010

Blanco-Canqui, H., & Ruis, S. J. (2018). No-tillage and soil physical environment. *Geoderma, 326*, 164–200. https://doi.org/10.1016/j.geoderma.2018.03.011

Bünemann, E. K., Bongiorno, G., Bai, Z., Creamer, R. E., De Deyn, G., de Goede, R., Fleskens, L., Geissen, V., Kuyper, T. W., Mäder, P., Pulleman, M., Sukkel, W., van Groenigen, J. W., & Brussaard, L. (2018). Soil quality—A critical review. *Soil Biology and Biochemistry, 120*, 105–125. https://doi.org/10.1016/j.soilbio.2018.01.030

Cavagnaro, T. R. (2016). Soil moisture legacy effects: Impacts on soil nutrients, plants and mycorrhizal responsiveness. *Soil Biology and Biochemistry, 95*, 173–179. https://doi.org/10.1016/j.soilbio.2015.12.016

Chen, M. M., Zhu, Y. G., Su, Y. H., Chen, B. D., Fu, B. J., & Marschner, P. (2007). Effects of soil moisture and plant interactions on the soil microbial community structure. *European Journal of Soil Biology, 43*, 31–38. https://doi.org/10.1016/j.ejsobi.2006.05.001

Davidson, E. A., & Janssens, I. A. (2006). Temperature sensitivity of soil carbon decomposition and feedbacks to climate change. *Nature, 440*, 165–173. https://doi.org/10.1038/nature04514

Din, M. S. U., et al. (2022). World nations priorities on climate change and food security. In W. N. Jatoi, M. Mubeen, A. Ahmad, M. A. Cheema, Z. Lin, & M. Z. Hashmi (Eds.), *Building climate resilience in agriculture*. Springer.

Doran, J. W., & Zeiss, M. R. (2000). Soil healht and sustainibility: Managing the biotic component of soil quality. *Applied Soil Ecology, 15*, 3–11. https://doi.org/10.1016/S0929-Get

Doudill, A. J., Heathwaite, A. L., & Thomas, D. S. G. (1998). Soil water movement and nutrient cycling in semi-arid rangeland: Vegetation change and system resilience. *Hydrological Processes, 12*, 443–459.

El Sabagh, A., et al. (2020). Nitrogen fixation of legumes under the family Fabaceae: Adverse effect of abiotic stresses and mitigation strategies. In M. Hasanuzzaman, S. Araújo, & S. Gill (Eds.), *The plant family Fabaceae*. Springer.

Fahad, S., et al. (2019). Abiotic stress and rice grain quality. In M. Hasanuzzaman, M. Fujita, & J. K. Biswas (Eds.) *Advances in rice research for abiotic stress tolerance*. isbn:978-0-12-814332-2.

Grzesiak, S., Grzesiak, M. T., Filek, W., Hura, T., & Stabryta, J. (2002). The impact of different soil moisture and soil compaction on the growth of triticale root system. *Acta Physiologiae Plantarum, 24*, 331–342.

Guo, L. B., & Gifford, R. M. (1979). Soil carbon stocks and land use change: A meta analysis. *Südwestdeutsche Pilzrundschau, 15*, 1–10. https://doi.org/10.1046/j.1354-1013.2002.00486.x

Habib Ur Rehman, M., et al. (2020). Climate resilient cotton production system: A case study in Pakistan. In M. Hasanuzzaman (Ed.) *Cotton production and uses*. Print isbn:978-981-15-1471-5, Online isbn:978-981-15-1472-2.

Hashmi, M. Z., et al. (2017). Management and regulation of antibiotics and antibiotics resistance genes in soils. In M. Hashmi, V. Strezov, & A. Varma (Eds.), *Antibiotics and antibiotics resistance genes in soils* (Soil biology) (Vol. 51). Springer.

Hasnain, A., et al. (2017). Antibiotics resistance genes. In M. Hashmi, V. Strezov, & A. Varma (Eds.), *Antibiotics and antibiotics resistance genes in soils* (Soil biology) (Vol. 51). Springer.

Hatfield, J. L., Sauer, T. J., & Prueger, J. H. (2001). Managing soils to achieve greater water use efficiency: A review. *Agronomy Journal, 280*, 271–280.

Her, J.-J., & Huang, J.-S. (1995). Influence of carbon source and C/N ratio on nitrate/nitrite denitification and carbon breakthrough. *Bioresource Technology, 54*, 45–51.

Höflich, G., Tauschke, M., Kühn, G., Werner, K., Frielinghaus, M., & Höhn, W. (1999). Influence of long-term conservation tillage on soil and rhizosphere microorganisms. *Biology and Fertility of Soils, 29*, 81–86. https://doi.org/10.1007/s003740050528

Huggett, R. J. (1998). Soil chronosequences, soil development, and soil evolution: a critical review. *Catena, 32*, 155–172. https://doi.org/10.1016/S0341-8162(98)00053-8

Hussain, S., et al. (2020). Irrigation scheduling for cotton cultivation. In S. Ahmad, & M. Hasanuzzaman (Eds.) *Cotton production and uses*. Print isbn:978-981-15-1471-5, Online isbn:978-981-15-1472-2.

Hussain, S., et al. (2022). Using space–time scan statistic for studying the effects of COVID-19 in Punjab, Pakistan: A guideline for policy measures in regional agriculture. *Environmental Science and Pollution Research, 30*, 42495–42508.

Hussain, S., et al. (2023a). Assessment of land use/land cover changes and its effect on land surface temperature using remote sensing techniques in southern Punjab, Pakistan. *Environmental Science and Pollution Research., 30*, 99202–99218.

Hussain, S., et al. (2023b). Investigation of irrigation water requirement and evapotranspiration for water resource management in Southern Punjab, Pakistan. *Sustainability, 15*, 1768.

Hussain, S., et al. (2024a). Agricultural land suitability analysis of southern Punjab, Pakistan using analytical hierarchy process (AHP) and multi-criteria decision analysis (MCDA) techniques. *Cogent Food & Agriculture, 10*(1), 2294540.

Hussain, S., et al. (2024b). Assessment of future prediction of urban growth and climate change in district Multan, Pakistan using CA-Markov method. *Urban Climate, 53*, 101766.

Hussain, S., et al. (2024c). Assessing the impact of land use land cover changes on soil moisture and vegetation cover in southern Punjab, Pakistan using multi-temporal satellite data. *Geology, Ecology, and Landscapes*. https://doi.org/10.1080/24749508.2024.2338970.

Karmakar, R., Das, I., Dutta, D., & Rakshit, A. (2016). Potential effects of climate change on soil properties: A review. *Science International, 4*, 51–73. https://doi.org/10.17311/sciintl.2016.51.73

Kaspar, C., & Taylor, M. (1994). Soil compaction and root growth: A review. *Agronomy Journal, 86*, 759–766. https://doi.org/10.2134/agronj1994.00021962008600050004x

Khurshid, N., et al. (2023). Exogenously applied nicotinic acid alleviates drought stress by enhancing morpho-physiological traits and antioxidant defense mechanisms in wheat. *Ecotoxicology and Environmental Safety, 263*, 115350.

Kimetu, J. M., & Lehmann, J. (2010). Stability and stabilisation of biochar and green manure in soil with different organic carbon contents. *Australian Journal of Soil Research, 48*, 577–585. https://doi.org/10.1071/Sr10036

Kirschbaum, M. U. F. (1995). The temperature dependence of soil organic matter decomposition, and the effect of global warming on soil organic storage. *Soil Biology and Biochemistry, 27*, 753–760. https://doi.org/10.13140/RG.2.2.34459.67369

Kladivko, E. J. (2001). Tillage systems and soil ecology. *Soil and Tillage Research, 61*, 61–76. https://doi.org/10.1016/S0167-1987(01)00179-9

Lal, R. (2003). Soil erosion and the global carbon budget. *Environment International, 29*, 437–450. https://doi.org/10.1016/S0160-4120(02)00192-7

Lal, R. (2004). Soil carbon sequestration to mitigate climate change. *Geoderma, 123*, 1–22. https://doi.org/10.1016/j.geoderma.2004.01.032

Lepore, E., Schmidt, O., Fenton, O., Tracy, S., Bondi, G., & Wall, D. P. (2024). Traffic induced compaction and physical quality of grassland soil under different soil moisture deficits. *Soil and Tillage Research, 244*, 106205. https://doi.org/10.1016/j.still.2024.106205

Liang, B., Lehmann, J., Solomon, D., Kinyangi, J., Grossman, J., O'Neill, B., Skjemstad, J. O., Thies, J. F., Luizão, J., Petersen, J., & Neves, E. G. (2006). Black carbon increases cation exchange capacity in soils. *Soil Science & Plant Nutrition, 70*, 1719–1730. https://doi.org/10.2136/sssaj2005.0383

Matimati, I., Verboom, G. A., & Cramer, M. D. (2014). Nitrogen regulation of transpiration controls mass-flow acquisition of nutrients. *Journal of Experimental Botany, 65*, 159–168. https://doi.org/10.1093/jxb/ert367

McDermid, S. P., et al. (2015). *The AgMIP coordinated climate-crop modeling project [C3MP]: Methods and protocols* (pp. 191–221). World Scientific Publishing.

Mloza-Banda, M. L., Cornelis, W. M., Mloza-Banda, H. R., Makwiza, C. N., & Verbist, K. (2014). Soil properties after change to conservation agriculture from ridge tillage in sandy clay loams of mid-altitude Central Malawi. *Soil Use and Management, 30*, 569–578. https://doi.org/10.1111/sum.12139

Muhammad, H., et al. (2023). A paradigm shift towards beneficial microbes enhancing the efficiency of organic and inorganic nitrogen sources for a sustainable environment. *Land, 12*(3), 680.

Nannipieri, P., Ascher, J., Ceccherini, M. T., Landi, L., Pietramellara, G., & Renella, G. (2004). Microbial diversity and soil functions. *Eurasian Journal of Soil Science, 4*, 655–670. https://doi.org/10.1046/j.1365-2389.2003.00556.x

Nawaz, M. F., Bourrié, G., & Trolard, F. (2013). Soil compaction impact and modelling. A review. *Agronomy for Sustainable Development, 33*, 291–309. https://doi.org/10.1007/s13593-011-0071-8

Naz, T., et al. (2023). Bio-fortification of two wheat cultivars with iron and zinc through their soil and foliar application in salt-factored soil: Growth, ionic, physiological, and biochemical modifications. *Journal of Plant Growth Regulation, 42*, 5727–5745.

Nearing, M., Pruski, F. F., & O'Neal, M. R. (2004). Expected climate change impacts on soil erosion rates: A review. *Journal of Soil and Water Conservation, 59*, 43–50.

Oechel, W. C., Vourlitis, G. L., Hastings, S. J., & Bochkarev, S. A. (2013). Change in arctic CO2 flux over two decades: Effects of climate change at Barrow, Alaska. *Ecological Applications, 5*, 846–855.

Oki, T., & Kanae, S. (2006). Global hydrological cycles and world water resources. *Science, 1979*(313), 1068–1072. https://doi.org/10.1126/science.1128845

Paradelo, R., & Barral, M. T. (2009). Effect of moisture and disaggregation on the microbial activity of soil. *Soil and Tillage Research, 104*, 317–319. https://doi.org/10.1016/j.still.2009.02.010

Qian, B., Gregorich, E. G., Gameda, S., Hopkins, D. W., & Wang, X. L. (2011). Observed soil temperature trends associated with climate change in Canada. *Journal of Geophysical Research Atmospheres, 116*, D02106. https://doi.org/10.1029/2010JD015012

Qin, X., Li, Y., Wang, H., Liu, C., Li, J., Wan, Y., Gao, Q., Fan, F., & Liao, Y. (2016). Long-term effect of biochar application on yield-scaled greenhouse gas emissions in a rice paddy cropping system: A four-year case study in south China. *Science of the Total Environment, 569*, 1390–1401. https://doi.org/10.1016/j.scitotenv.2016.06.222

Raich, J. W., & Schlesinger, W. H. (1999). The global carbon dioxide flux in soil respiration and its relationship to vegetation and climate. *Tellus, 44*, 81–99.

Sabagh, A., et al. (2020a). Drought and heat stress in cotton (Gossypium hirsutum L.): Consequences and their possible mitigation strategies. In M. Hasanuzzaman (Ed.) *Agronomic crops*. Print isbn:978-981-15-0024-4, Online isbn:978-981-15-0025-1.

Sabagh, A., et al. (2020b). Prospective role of plant growth regulators for tolerance to abiotic stresses. In T. Aftab, & K. R. Hakeem (Eds.) *Plant growth regulators*

Sami ul din, M., et al. (2019). Agronomic cropping systems in relation to climatic variability. In M. Hasanuzzaman (Ed.) *Agronomic crops*. isbn:978-981-32-9150-8.

Schmidt, M. W. I., Torn, M. S., Abiven, S., Dittmar, T., Guggenberger, G., Janssens, I. A., Kleber, M., Gel-Knabner, I. K., Lehmann, J., Manning, D. A. C., Nannipieri, P., Rasse, D. P., Weiner, S., & Trumbore, S. E. (2015). Persistence of soil organic matter as an ecosystem property. *Nature, 528*, 49–56. https://doi.org/10.1038/nature15744

Seneviratne, S. I., Corti, T., Davin, E. L., Hirschi, M., Jaeger, E. B., Lehner, I., Orlowsky, B., & Teuling, A. J. (2010). Investigating soil moisture-climate interactions in a changing climate: A review. *Earth-Science Reviews, 99*, 125–161. https://doi.org/10.1016/j.earscirev.2010.02.004

Sevink, J. (1991). Soil development in the coastal dunes and its relation to climate. *Landscape Ecology, 6*, 49–56. https://doi.org/10.1007/BF00157744

Shahzad, K., et al. (2022). Importance of carbon sequestration in the context of climate change. In W. N. Jatoi, M. Mubeen, A. Ahmad, M. A. Cheema, Z. Lin, & M. Z. Hashmi (Eds.), *Building climate resilience in agriculture*. Springer.

Six, J., Elliott, E. T., & Paustian, K. (1999). Aggregate and soil organic matter dynamics under conventional tillage and no-tillage. *Soil Science Society of America Journal, 63*, 1350–1358.

Szabolcs, I. (2018). Impact of climate change on soil attributes influence on salinization and alkalinization. *Agrokémia és Talajtan, 67*, 121–141.

Tariq, M., et al. (2019). Fundamentals of crop rotation in agronomic management. In M. Hasanuzzaman (Ed.) *Agronomic crops*. isbn:978-981-32-9150-8.

Teasdale, J. R., & Mohler, C. L. (1993). Light transmittance, soil temperature, and soil moisture under residue of hairy vetch and rye. *Agronomy Journal, 85*, 673. https://doi.org/10.2134/agronj1993.00021962008500030029x

Van Iersel, M. W.,& Chappell, M. (2009). Soil moisture sensor-based irrigation reduces water use and nutrient leaching in a commercial nursery.

Venterea, R. T., Rolston, D. E., & Cardon, Z. G. (2005). Effects of soil moisture, physical, and chemical characteristics on abiotic nitric oxide production. *Nutrient Cycling in Agroecosystems, 72*, 27–40. https://doi.org/10.1007/s10705-004-7351-5

Verzeaux, J., Hirel, B., Dubois, F., Lea, P. J., & Tétu, T. (2017). Agricultural practices to improve nitrogen use efficiency through the use of arbuscular mycorrhizae: Basic and agronomic aspects. *Plant Science, 264*, 48–56. https://doi.org/10.1016/j.plantsci.2017.08.004

Wang, J., Xiong, Z., & Kuzyakov, Y. (2016). Biochar stability in soil: Meta-analysis of decomposition and priming effects. *GCB Bioenergy, 8*, 512–523. https://doi.org/10.1111/gcbb.12266

Zahoor, S. A., et al. (2019). Improving water use efficiency in agronomic crop production. In M. Hasanuzzaman (Ed.), *Agronomic crops*. Springer.

Zhang, Y., Chen, W., Smith, S. L., Riseborough, D. W., & Cihlar, J. (2005). Soil temperature in Canada during the twentieth century: Complex responses to atmospheric climate change. *Journal of Geophysical Research D: Atmospheres, 110*, 1–15. https://doi.org/10.1029/2004JD004910

Zhang, R., & Wienhold, B. J. (2002). The effect of soil moisture on mineral nitrogen, soil electrical the effect of soil moisture on mineral nitrogen, soil electrical conductivity, and pH conductivity, and pH. *Nutrient Cycling in Agroecosystem, 63*, 251–254.

Ziadat, F. M., & Taimeh, A. Y. (2013). Effect of rainfall intensity, slope, land use and antecedent soil moisture on soil erosion in an arid environment. *Land Degradation and Development, 24*, 582–590. https://doi.org/10.1002/ldr.2239

Zumft, W. G. (1997). Cell biology and molecular basis of denitrification. *Microbiology and Molecular Biology Reviews, 61*, 533–616.

Innovative Solutions for Agricultural Water Management in Future

Parveen, Muhammad Anwar-ul-Haq, Muhammad Mubeen, Irfan Iftikhar, Muhammad Nadeem, Muhammad Awais, Muhammad Usama Mustafa, and Syed Aftab Wajid

Abstract Agricultural water management is a critical aspect of global food production, as it directly impacts crop yields, food security, and environmental sustainability. In the face of increasing population and climate change, innovative solutions are required to optimize water use in agriculture, while mitigating associated risks. This manuscript explores the current state of agricultural water management, identifies key challenges and risks, and delves into innovative solutions that address these issues. By examining cutting-edge technologies, practices, and policies, we highlight the potential for sustainable and resilient agricultural water management systems in the twenty-first century.

Parveen · M. U. Mustafa
University of Agriculture, Faisalabad Sub-Campus, Depalpur, Pakistan

M. Anwar-ul-Haq (✉)
Institute of Soil and Environmental Sciences, University of Agriculture, Faisalabad, Pakistan
e-mail: haq.gondal@uaf.edu.pk

M. Mubeen (✉)
Department of Biotechnology, COMSATS University Islamabad, Vehari Campus, Vehari, Pakistan
e-mail: muhammadmubeen@cuivehari.edu.pk

I. Iftikhar
University of Agriculture, Faisalabad Sub-Campus, Depalpur, Pakistan

Ministry of Commerce, Climate Division, Islamabad, Pakistan

M. Nadeem
Institute of Agro-Industry and Environment, The Islamia University of Bahawalpur, Bahawalpur, Pakistan

M. Awais
University of Agriculture, Faisalabad Sub-Campus, Depalpur, Pakistan

Pir Mahir Ali Shah Arid Agriculture University, Rawalpindi, Pakistan

S. A. Wajid
Agro-Climatology Lab, University of Agriculture, Faisalabad, Pakistan

M. Mubeen et al. (eds.), *Innovations in Agricultural Water Management*,
https://doi.org/10.1007/978-3-031-91883-4_23

Keywords Agricultural water management · Innovation · Risk mitigation · Sustainable agriculture · Water-efficient technologies

1 Introduction

In today's rapidly changing world, the significance of accurate weather forecasts cannot be overstated. Climate change is causing shifts in weather patterns, leading to increased uncertainty and variability in rainfall and temperature hence water insecurity. This poses a major challenge to agricultural productivity, especially for smallholder farmers in developing countries (Haq et al., 2024; Nadeem et al., 2024). Traditionally, these farmers relied on local knowledge to predict rainfall and temperature patterns (Abbam et al., 2018). However, with the changing climate, this is no longer effective. As a result, farmers are increasingly exposed to climate-related risks such as droughts, floods, and timing of rainfall. These risks can have devastating impacts on crop yields and livelihoods (Atta-Aidoo et al., 2022). To address these challenges, technological innovations and interventions have been developed, including effective irrigation systems. However, these innovations often come with significant financial barriers that prevent smallholder farmers from adopting them. For countries where smallholder farming dominates the agricultural sector and rural poverty levels are high, technological innovation is essential for transforming agriculture in the face of climate change and achieving food security. One of the key impediments to the adoption of these innovative technologies is the perceived risk and risk aversion among smallholder farmers. To overcome this hurdle, it is important to identify and understand the risks associated with agricultural water management innovations and develop sustainable solutions. One approach to improving agricultural water management is through the implementation of large-scale irrigation schemes and farmer-led micro irrigation technologies (Afokpe et al., 2022). These innovations, such as the use of bucket kits, sprinklers, and tube emitters, have been proven to optimize irrigated agriculture for smallholder farmers in wetlands, along perennial rivers and streams. Another innovation is the construction of dykes to divert water into rivers in flood-prone areas. These dykes can help manage water flow and prevent crop damage during heavy rainfall. Additionally, the use of climate information and extension services is crucial for smallholder farmers to make informed decisions about crop choices, timing of planting, and water management practices. These interventions can be delivered through mobile phones and computers owned by farmers or extension officers, ensuring that vital climate information reaches those who need it most.

In summary, the adoption of innovative agricultural water management technologies is crucial for smallholder farmers in response to climate change.

1.1 Objectives

1. Identify key risks associated with agricultural water management.
2. Explore innovative solutions to mitigate these risks.
3. Provide practical recommendations for implementing these innovations on farms.

2 Current State of Agricultural Water Management

2.1 Traditional Irrigation Practices

When compared to modern approaches, traditional irrigation methods that have been used since ancient times have shown to be more efficient and cost-effective. Some of these ancient methods of irrigation are apparently still in use today (Ali, 2022).

The traditional methods of irrigation include the following:

2.1.1 Subsurface Irrigation

Vegetable and field crop production have long used seepage irrigation, commonly referred to as subterranean irrigation, especially in areas with high water tables. This kind of irrigation can happen naturally or with the use of artificial means.

2.1.2 Furrow Irrigation Method

Using this technique, water is allowed to run through the furrows made along the rows of crops. In slope landscape if these furrows are set along the land's contour, it works well. These furrows usually have a width of 3–6 m (Li et al., 2020).

2.1.3 Basin Irrigation Method

Orchards and other high-value crops are among the best choices for this strategy, especially if the irrigated area is modest. The roots of the trees are built up into a small bund that is about 15–22 cm tall and spaced around 30–60 cm apart. This aids in maintaining the dryness of the soil surrounding the tree.

2.2 Water Scarcity and Competition

The relationship between income and water needs can be explained by the fact that rich countries are able to achieve higher levels of efficiency in the distribution, transportation, and use of water. There is a water deficit in a nation when its water resources, measured at the point of origin, produce less water per capita each year. These resources are under pressure by the need for water, and when resources are few, this demand strains society (Cosgrove et al., 2015).

There is a new kind of water conflict arising that is not the same as the disputes over renewable resources that are seen in many parts of the world. Non-renewable water resources, which are found beneath the borders of many Middle Eastern and North African nations, are the subject of this conflict. In order to fulfill the growing demand for municipal water, some of these countries have depleted their renewable supplies due to extreme water stress.

Of course, Jordan is a perfect example. The nation has been planning ways to draw from the freshwater aquifer known as Desi, which is not renewable. This aquifer also reaches Saudi Arabia, where it has been extensively used for over 30 years in agricultural production. There are comparable circumstances involving Egypt, Sudan, and Chad.

2.3 Environmental Degradation

Degradation of the land affects water availability, quality, and retention by reducing water productivity at the field and landscape levels. Because land and water production are closely related, improving agricultural water management requires reducing or eliminating soil degradation. Positive estimates suggest that agriculture will probably need 30–40% more freshwater by 2050 than it does today.

Thirty-one percent of pastureland in Africa is affected by pasture degradation, which is more severe than in Asia and Latin America where the percentages are 20% and 14%, respectively. Asia has the highest rates of deforestation (27%), followed by Africa (19%) and Latin America (14%).

3 Risks in Agricultural Water Management

3.1 Depletion of Aquifers

The health of human society and aquatic and terrestrial ecosystems depends on coordinated water management, which acknowledges the interconnectedness of surface water and groundwater and treats them as a single resource. The steady migration of groundwater into streams provides a consistent supply of water for domestic, industrial, and agricultural usage by humans, as well as supporting natural populations (Zipper et al., 2022).

3.2 Water Pollution

At least 17 pesticides were found to be present in detectable amounts in drinking water wells located in agricultural areas, according to a survey conducted across 23 states. Farmers run the risk of overfertilization because of a lack of information and comprehension of nutrient quality if they do not modify fertilizer use to account for the nutrient content in municipal sewage sludge, ocean dumping, and leaching (Logan, 1993). Shortage of water leads the farmer community to use wastewater for crop irrigation in urban and peri urban areas whcih deteriorates the soil and subsurface water quality. Subsurface water quality of different urban and periurban area was found below standards (Hussain et al., 2025).

The purpose of cultural techniques such as cover crops, contour crops, and conservation tillage is to protect the soil surface and stop soil erosion.

3.3 Conflicts Over Water Resources

Water conflict may be at three different levels:

1. Global
2. Regional
3. Upstream–downstream

3.3.1 Global Level of Water Conflicts

The global-scale water conflicts result from the following three issues:

- Inaccessibility of water
- Growing concern on the natural world
- Economic importance of water

3.3.2 Regional Level of Water Conflicts

Disparities in the availability of water resources, global pollution, and disputes over the management of worldwide river basins are the three primary categories of regional-scale water conflicts. The third category includes confrontations that happen more frequently than the first two, which are comparatively rare.

3.3.3 Upstream–Downstream Level of Water Conflicts

The main underlying cause of water conflicts involving international river basin management usually stems from the relationship between states' upstream and downstream. The impact of human activities between areas is also carried by water as it flows from one to another.

3.3.4 Analysis of Causes of Conflicts in Water Management

Water management conflicts can be viewed through the lens of a management system comprising three essential spheres:

1. Water
2. Economic
3. Political

4 Innovation in Agricultural Water Management

4.1 Precision Agriculture (PA)

Overall, 40% of the world's food is produced by irrigated agriculture, which uses less than 20% of the total farmed land. This highlights the critical role that irrigation plays in ensuring global food security. Forecasts suggest that freshwater supplies will soon become scarce globally, underscoring the need to improve water management strategies and increase water use efficiency (Adeyemi et al., 2017). As a result, more stringent laws are becoming increasingly necessary to protect the environment and maintain freshwater supplies.

4.2 *Drip and Micro Irrigation*

4.2.1 Drip Irrigation

Using emitters and a system of pipelines, water is only supplied to the crop's root zone in this method. It functions as an accurate metering system that ensures the plants receive both water and vital nutrients at the same time. This method can result in significant water savings of 40–80% and an irrigation efficiency of up to 90%.

4.2.2 Micro Irrigation

With the use of a technique called micro-irrigation, it is possible to precisely distribute nutrients and water to the rhizosphere, allowing plants to flourish even in areas with a low moisture content. This relatively new method became well-known throughout the world in the latter decade of the twentieth century, especially in areas that were severely short on water.

4.3 *Rainwater Harvesting and Water Recycling*

The importance of unconventional technologies such as the use of waste energy from sources like warm wastewater has come to light more and more in recent years.

The basic method of extracting waste energy from wastewater is to install drain water heat recovery devices, frequently in conjunction with a heat pump.

Gray water can be used as an unusual source of water, especially in arid areas. The development of technology has made it possible to deploy gray water recycling systems more widely. This development has been made possible by a change in public opinion about the reuse of treated gray water and a greater understanding of the value of freshwater conservation.

4.4 *Climate-Resilient Crop Varieties*

Global food security is greatly affected by climate change, which also has a significant impact on agricultural productivity globally. Its effects on plant growth and productivity, as well as on soil fertility, carbon sequestration, microbial activity, and variety, are clear indicators of this.

However, compared to autumn-sown cereals, climate-related production changes are typically more noticeable in grain legumes like soybean and broad-leaved crops. In addition to these, several shotgun strategies of soil and crop management under water deficit conditions i.e. application of organic and inorganic amendments may increase the water use effciency of crops resulting better agricultural water mnanagement (Nadeem et al., 2022; 2025).

5 Solution and Risk Mitigation

5.1 Policy and Governance

In Pakistan, provincial administrations are in charge of water and sanitation. They separated the provision of services to local governments in 2001. Water administration and management are influenced by a number of institutional actors in the provinces of Punjab and Khyber Pakhtunkhwa (KP).

Government and official water institutions make efforts; however, not all residents are reached by their water and sanitation programs (Cooper, 2018). Therefore, self-provisioning is common in Punjab and KP. For example, in Punjab Province, of the 60 million people living in rural areas, only 32% are served by programs developed by the Provincial Public Health Engineering Department (PHED).

Water management customs in the Chitral District of KP are clarified through a case study. This mostly deals with irrigation and includes techniques for allocating water to homes and communities. It also highlights mechanisms for maintaining irrigation ditches, collective water collecting, and informal governing institutions known as "grams."

It is useful to make the case for an analytical distinction between irrigation system management and governance. Stressing the kind of management provides a new viewpoint. The obvious differences between freshly moved irrigation districts and the self-managed irrigation systems that are currently in place in Mexico and Central America make this differentiation vital.

5.2 Sustainable Farming Practices

The *Sustainable Farming Practices in Water Management Guidebook* aims to provide farmers with comprehensive guidelines and recommendations on effective and environmentally friendly approaches to managing water resources in agriculture. These guidelines promote sustainable farming practices that minimize water waste, prioritize conservation, and ensure efficient use of water in agricultural activities (Shamsudheen, 2017).

Some key sustainable farming practices in agriculture water management include implementing drip irrigation systems to minimize water loss through evaporation and runoff, practicing crop rotation and cover cropping to improve soil health and water retention, using precision farming techniques to optimize water usage based on crop needs, and implementing water harvesting and storage systems to capture water.

5.3 International Cooperation

A large number of international non-governmental organizations devoted to scientific and professional pursuits have arisen within the last several decades (Chitale, 1995). These organizations include the International Commission on Irrigation and Drainage (ICID), International Association of Hydrology (IAH), International Water Resources Association (IWRA), International Geographical Union (IGU), International Association for Water Quality (IAWQ), International Association of Hydraulic Research (IAHR), International Water Supply Association (IWSA), International Association of Hydrology (IAH), and the International Lake Environment Committee (ILEC). Certain organizations only allow participation through officially recognized National Committees; individual members are not admitted directly. The International Commission on Large Dams, the World Energy Council, and ICID are a few examples of these organizations (Palerm-Viqueira, 2009).

6 Case Study

6.1 Israel's Agricultural Water Innovation

The Ministry of Energy and Water has assumed the role of lead ministry in charge of developing and implementing policies related to the water sector since the 2000 reform. It has direct control over organizations like Mekorot and the Israel Water

Table 1 Laws enacted in Israel for managing the water sector

Law	Year	Implementations
1. Water measurement	1955	Establishes the crucial importance of metering water to enable water management and control of water flows and uses
2. Supervision of water drillings	1955	Passed to assert national control over the production of the water. Before the establishment of Israel, part of the coastal aquifer had already been depleted, and saline intrusion was occurring, threatening long-term damage of the aquifer
3. Drainage and flood prevention	1957	Enacted to help reduce and prevent floods due to the rapid urbanization
4. Water law—the cornerstone of Israel's legal water framework	1959	Setting the overall principles for managing the sector. It specifies that all water resources are public properties controlled by the state (there is no private ownership of water resources in Israel, even beneath privately owned land) and establishes the mechanisms for allocation of water rights

Authority (IWA), as well as drainage and river basin administrations. Furthermore, it uses IWA to indirectly oversee providers of water and sanitation. IWA is the national regulator for the water industry, which includes irrigation, wastewater treatment, potable water delivery, and water resource management. It is the regulating body for water and sewage. IWA essentially serves as the highest authority in Israel's water industry, acting as both a national planning organization and regulator.

The committee's two key findings served as the foundation for the creation of IWA: (a) there were too many organizations involved in managing the water sector, which resulted in a hazy allocation of duties; and (b) there was too much political influence over day-to-day operations, including the setting of tariffs that were largely motivated by political concerns. It was determined that this circumstance had a major role in the sector's vulnerability and water issue (Marin et al., 2017).

It was essential to establish IWA as an autonomous organization in order to distinguish the roles at the professional level, which manages the water sector's operational elements, and the political level, which concentrates on policy-making.

In the 10 years following the creation of the State of Israel, a set of four laws was enacted to set the guidelines for managing the water sector (Table 1).

6.2 Precision Agriculture in the United States

A conceptual framework known as precision agriculture (PA) is used to manage variability in farming operations at various scales, including fields, regions, and even global watersheds. Since the public's first access to the US Global Positioning System (GPS) in the 1980s, PA's goals, methods of use, and enabling technologies have all experienced substantial changes (Evett et al., 2020).

In the beginning, PA mostly included integrating GPS with GIS (geographic information system) technology, which were first created for satellite photography. Through the application of fertilizer and then expanding to include sowing rates, seed varieties, pesticide application, and now site-specific irrigation, this allowed farmers to strategically plan certain agricultural activities.

Over time, this passive form of precision agriculture has progressed to offer more sophisticated solutions. For example, it now allows for variable-rate application of fertilizers, based on georeferenced soil sampling that generates prescription maps indicating fertilizer requirements.

Precision agriculture solutions that are highly effective often use on-board sensor systems that transmit data to embedded computing platforms. These platforms make it easier to modify equipment operations in real time.

By applying modern technology, these active and direct precision agricultural systems can do tasks like turning on spray equipment only when weeds are present or triggering variable-rate irrigation nozzles in reaction to abiotic stress sensors that identify crop water stress.

Modern precision agriculture is further complicated by issues with data standards, data sharing, ownership rights, and striking a balance between public and private interests. However, both public and private institutions are aggressively addressing these concerns.

6.3 Sustainable Farming in India

Organic farming, natural farming, regenerative agriculture, agroecology, and other methods are included in the category of sustainable agriculture. These phrases are frequently used to describe various approaches to attaining agricultural sustainability (Gupta et al., 2021).

6.3.1 Organic Farming

Gandhi's Sewagram hosted the first-ever national meeting of proponents and practitioners of organic farming in India in 1984. The Other India Press Book *Organic Farming Source Book* provides a thorough history of the organic farming movement in India. The establishment of a broad network, which in turn facilitated the creation of the Organic Farmers Association of India (OFAI), was greatly aided by this campaign.

6.3.2 Regenerative Agriculture

The phrase "regenerative agriculture" has become quite popular during the last 10 years, especially in reaction to growing worries about climate change. This phrase mostly refers to ecological restoration, with an emphasis on soil preservation, topsoil carbon sequestration, and biodiversity enhancement, among other important features (Bossio et al., 2010).

6.3.3 Natural Farming

The concept of "nature farming" was first established by Mokichi Okada in 1935, and here is where the practice of natural farming originated. It was Okada who first pushed for farming without the use of pesticides and fertilizers, even though it was Masanobu Fukuoka who popularized the term "shizen noho" (natural farming in English).

6.3.4 Agroecology

At the beginning of the twentieth century, agronomist Basil Bensin created the term "agroecology" to refer to the ecological techniques used in agriculture. Agroecology expanded to include complete food systems by the 2000s. Its purview included social and economic aspects in addition to environment.

7 Future Perspectives and Challenges

The principal agent behind environmental changes that have a significant impact on our planet's climate is now human activity.

These modifications are changing the amount and quality of our essential freshwater resources, in addition to the changes in landscapes brought about by the growth of food and energy production as well as urbanization. These resources are essential to our physical and financial life. In addition to being essential to life, water is also the foundation of our economic success because it is necessary for the production of all that we produce. Water is a renewable resource, although it has a limited supply.

In addition to satisfying the present unmet power needs of a billion people, our daunting task for 2050 is to provide energy and food for an additional 2–2.5 billion people. In order to meet the nutritional demands of this expanding population, water usage in the production of different items, especially food and energy, needs to be carefully considered.

It is true that significant amounts of energy and water are used in the production, distribution, processing, and transportation of food goods. The *Comprehensive Assessment of Water Management in Agriculture* provides a thorough analysis of this topic.

New laws and agreements governing cross-border trade in products and services are being adopted as a result of the rapid growth of globalization. This change highlights the increasing influence of global firms that use and transfer water in an indirect manner. The effects of trade globalization are extensive and affect consumers, governments, and the environment.

Engaging in trade involving high water-consuming goods from regions with water scarcity may yield short-term economic gains, but it proves unsustainable in the long run and poses a threat to achieving other water-related objectives.

Previously, historical data on climate and hydrology, along with consumption trends, have led the development of water infrastructure and management systems. After that, this data is statistically analyzed and interpreted to determine how likely it is that particular events will occur. For example, infrastructure is frequently designed to withstand occurrences that have a given likelihood of occurring, which is ascertained by analyzing the oldest historical data series that is available. For instance, based on historical records, infrastructure constructed to resist a 100-year flood is intended to manage a flood event with a 1% chance of happening in any given year.

The ecosystems of wetlands and forests are significantly impacted by floods and droughts. These cyclical patterns are fundamental to these ecosystems; they influence them and help them adapt. Floods are essential for the renewal of natural ecosystems because they provide more water and nutrient-rich soil to support plant life, especially food crops, along with the silt they deliver. The amount and quality of water in ecosystems can be disturbed by factors like industrialization, subpar farming methods, urbanization, and changes in land use. These ecosystems may suffer negative changes as a result of this.

Compared to our current practices, we have the knowledge, technology, and financial resources necessary to greatly improve the efficacy and efficiency of our water resource management. Our understanding of the structure and behavior of the physical and natural world is being systematically expanded by ongoing scientific research, which also gives us useful instruments for this goal.

8 Conclusion

Agricultural water management is a complex challenge, but innovation offers hope for a sustainable and resilient future. By adopting the latest water-efficient technologies, implementing sound policies, and fostering international cooperation, we can mitigate the risks associated with agricultural water management. These innovations pave the way for a more water-secure world, ensuring food production remains sustainable in the face of growing global demands and environmental changes.

References

Abbam, T., Johnson, F. A., Dash, J., & Padmadas, S. S. (2018). Spatiotemporal variations in rainfall and temperature in Ghana over the twentieth century, 1900–2014. Earth and Space Science, 5(4), 120–132.

Adeyemi, O., Grove, I., Peets, S., & Norton, T. (2017). Advanced monitoring and management systems for improving sustainability in precision irrigation. Sustainability, 9(3), 353.

Afokpe, P. M., Phiri, A. T., Lamore, A. A., Toure, H. M., Traore, R., & Kipkogei, O. (2022). Progress in climate change adaptation and mitigation actions in sub-Saharan Africa farming systems. Cahiers Agricultures, 31, 4.

Ali, M. H. (2022). Practices of irrigation & on-farm water management: vol 2, Agricultural Engineering Division Bangladesh Institute of Nuclear Agriculture (BINA) Mymensingh, BAU Campus Bangladesh.

Atta-Aidoo, J., Antwi-Agyei, P., Dougill, A. J., Ogbanje, C. E., Akoto-Danso, E. K., & Eze, S. (2022, October 10). Adoption of climate-smart agricultural practices by smallholder farmers in rural Ghana: An application of the theory of planned behavior. PLoS Climate, 1(10), e0000082.

Bossio, D., Geheb, K., & Critchley, W. (2010). Managing water by managing land: Addressing land degradation to improve water productivity and rural livelihoods. Agricultural Water Management, 97(4), 536–542.

Chitale, M. A. (1995). Institutional characteristics for international cooperation in water resources. International Journal of Water Resources Development, 11(2), 113–124.

Cooper, R. (2018). Water management/governance systems in Pakistan.

Cosgrove, W. J., & Loucks, D. P. (2015). Water management: Current and future challenges and research directions. Water Resources Research, 51, 4823–4839.

Evett, S. R., O'Shaughnessy, S. A., Andrade, M. A., Kustas, W. P., Anderson, M. C., Schomberg, H. H., & Thompson, A. (2020). Precision agriculture and irrigation: Current US perspectives. Transactions of the ASABE, 63(1), 57–67.

Gupta, N., Pradhan, S., Jain, A., & Patel, N. (2021). Sustainable agriculture in India 2021. CEEW Report, 122p. Sustainable-Agriculture-in-India-2021-20Apr21.

Hussain, S. A., Nadeem, M., Ayub, M. A., Ali, M., Abbasi, G. H., ul Haq, M. A., ... & Irfan, M. (2025). Health Risk Assessment Associated with Heavy Metals Pollution in Sandy Clay Loam Soils and Vegetables Irrigated by Wastewater: a Case Study of Bahawalpur City. Eurasian Soil Science, 58(2), 16.

Li, C., Li, C., Ma, B., & Wu, W. (2020, January 10). The role of ridge-furrow with plastic film mulching system on stem lodging resistance of winter wheat in a dry semi-humid region. Agronomy Journal, 112, 885.

Logan, T. J. (1993). Agricultural best management practices for water pollution control: Current issues. Agriculture, Ecosystems & Environment, 46(1–4), 223–231.

Marin, P., Tal, S., Yeres, J., & Ringskog, K. B. (2017). Water management in Israel: Key innovations and lessons learned for water scarce countries. World Bank.

Muhammad Anwar, ul Haq Muhammad, Nadeem Munaza, Naseem Ghulam Hasan, Abbasi Muhammad, Ali Muhammad Ashar, Ayub Irfan, Iftikhar Muhammad Tayyab, Mehmood Muhammad, Awais Rizwan, Ullah Tasawar, Ali (2024) Role of exogenous silica fertilization and genotype selection in attenuating oxidative and osmotic stress in wheat (Triticum aestivum L.) under water deficit conditions Plant and Soil 505(1-2) 283-298 10.1007/s11104-024-06673-z.

Nadeem, M., Anwar-ul-Haq, M., Ayub, M. A., Abbasi, G. H., Nawaz, H., Ali, M., ... & Bhutta, W. M. (2024). Climate change induced salinization and quality deterioration of groundwater resources in Pakistan; Current scenario and future prospects. Journal of Health and Climate Change, 3(1).

Nadeem, M., ul Haq, M. A., Saqib, M., Maqsood, M., Iftikhar, I., Ali, T., ... & He, Z. (2022). Nutrients, osmotic and oxidative stress management in bread wheat (Triticum aestivum L.) by exogenously applied silicon fertilization under water deficit natural saline conditions. Silicon, 14(17), 11869-11880.

Nadeem, M., Anwar-Ul-Haq, M., Xin, X., Shohag, M. J. I., Ayub, M. A., Ahmad, W., & He, Z. (2025). Exogenous Application of Silica and Nano Silica Particles Improves Nutrient Acquisition and Growth of Maize (Zea Mays L.) Under Saline and Water Deficit Conditions. Communications in Soil Science and Plant Analysis, 1-17.

Palerm-Viqueira, J. (2009). Governance and management of irrigation systems. Water Policy, 11(3), 330–347.

Shamsudheen, M., Dayal, D., & Machiwal, D. Soil and (2017) water management strategies for drylands. ICAR—Central Arid Zone Research Institute, Regional Research Station Kukma.

Zipper, S. C., Farmer, W. H., Brookfield, A., Ajami, H., Reeves, H. W., Wardropper, C., et al. (2022). Quantifying streamflow depletion from groundwater pumping: A practical review of past and emerging approaches for water management. JAWRA Journal of the American Water Resources Association, 58(2), 289–312.

Correction to: Knowledge Transfer to Stakeholders in Use of Wastewater for Irrigation

Asif Sardar, Sidra Javed, and Aneela Afzal

Correction to: Chapter 12 in: M. Mubeen et al. (eds.), *Innovations in Agricultural Water Management*, https://doi.org/10.1007/978-3-031-91883-4_12

The original version of the chapter "Knowledge Transfer to Stakeholders in Use of Wastewater for Irrigation" was inadvertently published with incorrect chapter authors' information. The chapter authors' information has now been updated.

The updated version of this chapter can be found at https://doi.org/10.1007/978-3-031-91883-4_12

M. Mubeen et al. (eds.), *Innovations in Agricultural Water Management*,
https://doi.org/10.1007/978-3-031-91883-4_24

Index

M. Mubeen et al. (eds.), *Innovations in Agricultural Water Management*, https://doi.org/10.1007/978-3-031-91883-4

The manufacturer's authorised representative in the EU is Springer Nature Customer Service Centre GmbH, Europaplatz 3, 69115 Heidelberg, Germany. If you have any concerns regarding our products, please contact ProductSafety@springernature.com

Printed and bound by CPI Group (UK) Ltd, Croydon, CR0 4YY
01/10/2026
02225871-0018